PRENTICE HALL **Informal Geometry**

Philip L. Cox

PRENTICE HALL
Needham, Massachusetts   Englewood Cliffs, New Jersey

Author

Philip L. Cox, Mathematics Teacher, Walled Lake Central School,
Walled Lake, Michigan

Reviewers

Bettye C. Hall, Director of Mathematics (retired),
Houston Independent School District, Houston, Texas

Debbie Glaser, Mathematics Teacher, Alief
Independent School District, Alief, Texas

Terry Kent, Mathematics Teacher, Downers Grove
South High School, Downers Grove, Illinois

David Pesapane, Mathematics Teacher, St. Bernard's
High School, Uncasville, Connecticut

Credits

Publisher: Stephanie Rogalin
Editorial Director: Mimi Jigarjian
Marketing Director: Bridget A. Hadley

Project Editors: Mary A. Bright, Sylvia Gelb
Editors: Mary A. Costich, Lois J. McDonald
Teacher's Edition Editor: Joan Piper

Assistant Production Manager: Pauline P. Wright
Book Production Editor: Andrew Healy Walker

Buyer: Roger Powers
Manufacturing: Bill Wood

Art and Design Director: L. Christopher Valente

Project Design Manager: Samuel S. Wallace
Cover Design: L. Christopher Valente, Dick Hannus

Interior Design: George McLean, Linda Dana
Willis, Marie McAdam

Production Designers: Marie McAdam,
Linda Dana Willis, Mark McKertich

Photo Research Manager: Russell Lappa/ Karen
McDonough, Assistant

Photo Research: Sue C. Howard

Electronic Publishing Department: Edwin Zeitz,
Peter Brooks, Linda Johnson, Pearl Weinstein,
Darren Tong

This book was entirely produced and illustrated by the Electronic Publishing
Department at Prentice Hall, Needham, using QuarkXPress® and Aldus Freehand®
on Macintosh® computers.

The Geometric Supposer is created by Educational Development Center, Inc., and
published by Sunburst Communications, Inc. QuarkXPress is a registered trademark of
Quark, Inc. Aldus Freehand is a registered trademark of Aldus Corp. Macintosh is a
registered trademark of Apple Computer, Inc.

Contents

Contents

Contents

Contents

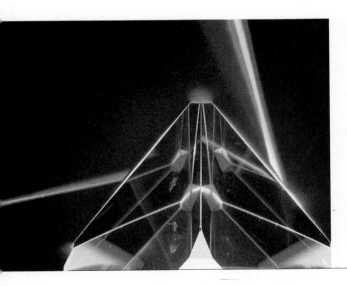

Contents

Contents

Extensions

Prentice Hall

dedicates this mathematics

program to

all mathematics educators

and

their students.

Chapter 1

Language of Plane Geometry

On this circuit board, electrical signals follow the paths that appear as *lines* connecting the *points* at which paths intersect. Miniature circuit elements are located at points along these paths.

Focus on Skills

ARITHMETIC

Add, subtract, multiply, or divide.

Example 1: $0.7 \times 10 = 7$ **Example 2:** $0.7 \div 10 = 0.07$

1. $1.4 + 0.06$ **2.** $2.5 + 1.06$ **3.** $1.6 - 0.8$ **4.** $5.8 - 4.9$

5. 3.4×1.3 **6.** 1.2×6.8 **7.** $0.98 \div 0.04$ **8.** $15.05 \div 2.15$

9. 1.7×10 **10.** 6×10 **11.** 13.6×10 **12.** 8.0×10

13. 3×100 **14.** 0.14×100 **15.** 6.3×100 **16.** 8.0×100

17. $525 \times 1,000$ **18.** $15 \times 1,000$ **19.** $7.8 \times 1,000$ **20.** $8.0 \times 1,000$

21. $3 \div 10$ **22.** $120 \div 10$ **23.** $6.3 \div 10$ **24.** $87 \div 10$

25. $4 \div 1,000$ **26.** $13 \div 100$ **27.** $160 \div 100$ **28.** $67.5 \div 100$

29. $6 \div 1,000$ **30.** $180 \div 1,000$ **31.** $8,200 \div 1,000$ **32.** $63.25 \div 1,000$

ALGEBRA

Evaluate the expression.

1. a^2 if $a = 6$ **2.** lw if $l = 6$ and $w = 7.8$ **3.** $n^2 + n + 6$ if $n = 5$

4. Is $x = 4$ a solution for $x + 4 = 8$?   **5.** Is $r = -7$ a solution for $9r = 63$?

6. Mary drove her car at 55 mi/h for 4 h. Use the formula $d = rt$ (where d is the distance traveled, r is the rate of travel, and t is the time traveled) to determine how far Mary traveled.

7. Kevin wants to put a fence around his back yard, which measures 150 ft by 75 ft. Use the formula $p = 2l + 2w$ (where p is the perimeter of the yard, l is the length of the yard, and w is the width of the yard) to determine how much fencing Kevin needs.

Write an equation for each problem, then solve.

8. Some number x increased by 10 equals 41.

9. Half the length, m, of a segment equals 18 in.

10. Fran wants to hang a picture exactly in the middle of a wall. The wall is 7 m long. How far from the corner is the middle of the wall?

1.1 Points, Lines, and Planes

Objective: To name points, lines, segments, rays, planes, collinear points, and coplanar points.

The word *geometry* comes from two Greek words, *geo* and *metric*, meaning "to measure the earth." In this lesson you will learn about some basic geometric figures and their properties.

Figure	Properties	Drawing	Symbol	Read as
Point	• has no size • indicates a definite location	• *A*		point *A*
Line	• is straight • has no thickness • extends indefinitely in two opposite directions		$\overleftrightarrow{AB}$ ($\overleftrightarrow{BA}$) l	line *AB* or *BA* line *l*
Segment	• is part of a line • has two endpoints		$\overline{AB}$ ($\overline{BA}$)	segment *AB* or *BA*
Ray	• is part of a line • has only one endpoint • extends indefinitely in one direction		$\overrightarrow{AB}$	ray *AB*
Plane	• is a flat surface • has no thickness • extends indefinitely in all directions			plane *ABC* plane *X*

Example 1: **Name the points, lines, segments, rays, and planes shown in the figure below.**

Solution: points: *X, Y, Z*
line: $\overleftrightarrow{XY}$ ($\overleftrightarrow{YX}$)
segment: $\overline{XY}$ ($\overline{YX}$)
rays: $\overrightarrow{XY}$, $\overrightarrow{YX}$
plane: *XYZ*

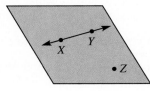

Points that are on the same line are ***collinear points.*** Points that are not on the same line are noncollinear. *If a line can be drawn through a set of points, then the points are collinear even though the line is not shown.*

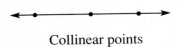

Collinear points

Noncollinear points

Points in the same plane are ***coplanar points.*** Points that are not in the same plane are noncoplanar. *If a plane can be drawn through a set of points, then the points are coplanar even though the plane is not shown.*

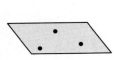

Coplanar points

Noncoplanar points

Example 2: **Name the following.**

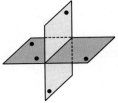

    a. three noncollinear points

    b. three coplanar points

    c. four noncoplanar points

Solution: **a.** S, U, and R are noncollinear points.

    b. R, S, and T are coplanar points.

    c. Q, R, S, and U are noncoplanar points.

EXPLORING

1. Use pencil and paper to draw a line. Erase the line. How many points do you need to again locate the line in the same position?

2. How many points do you need to draw a line?

3. Use a piece of cardboard and several pens. With a friend, try to balance the cardboard on the points of the pens. Can you balance the cardboard on four pens? on three pens?

A *postulate* is a statement that is accepted without proof. The **EXPLORING** activity leads to the following postulates.

POSTULATE 1: **Two points determine exactly one line.**

POSTULATE 2: **Three noncollinear points determine exactly one plane.**

Thinking Critically

1. Can two points be noncollinear? Can three points be noncollinear?

2. Are there points on $\overline{AB}$ other than point A and point B?

3. Can collinear points be noncoplanar? Why or why not?

4. Can coplanar points be noncollinear? Why or why not?

Class Exercises

1. Name objects that suggest a point, a line, a plane, a segment, and a ray.

State whether the points are collinear.

2. D, E, F

3. C, B, D

4. D, C, F

5. A, C, B

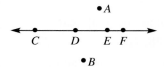

State whether the points are coplanar.

6. L, M, O

7. L, N, R

8. P, Q, M, L

9. L, M, N, O

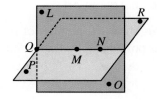

10. Name a line that is not shown in the figure.

11. Name two segments with N as an endpoint.

Exercises

1. Does $\overleftrightarrow{RS}$ pass through T?

2. Is R on $\overrightarrow{ST}$? on $\overrightarrow{UT}$?

3. Write the name for the ray through R with endpoint S.

4. Write six different correct names for the line.

5. Name all segments shown.

6. Name three different segments with S as an endpoint.

State whether the points are collinear.

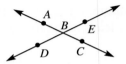

7. *A, B*   **8.** *A, E*   **9.** *A, C*

10. *D, B, E*   **11.** *D, C*   **12.** *A, D, B*

State whether the points are coplanar.

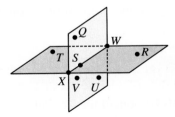

13. *Q, V, U*   **14.** *Q, S, W*

15. *R, V, U*   **16.** *X, S, W*

17. *Q, W, X, R*   **18.** *Q, S, W, U*

True or false?

19. $\overleftrightarrow{AB}$ and $\overleftrightarrow{BA}$ are the same line.

20. $\overrightarrow{AB}$ and $\overrightarrow{BA}$ are the same ray.

21. $\overline{AB}$ and $\overline{BA}$ are the same segment.

22. Any three points *A*, *B*, and *C* must lie in exactly one plane.

If possible, draw and label a figure to fit each description. Otherwise state *not possible*.

23. four points that are collinear   **24.** two points that are noncollinear

25. three points that are noncollinear   **26.** three points that are noncoplanar

27. $\overleftrightarrow{HI}$ **28.** $\overrightarrow{HI}$ **29.** $\overrightarrow{IH}$ **30.** $\overline{HI}$ **31.** $\overline{IH}$

32. three points that are coplanar and are contained in exactly one plane

33. point *C* on both $\overleftrightarrow{AB}$ and $\overrightarrow{AB}$

34. The diagrams show the number of lines determined by three noncollinear points and by four points, no three of which are collinear.

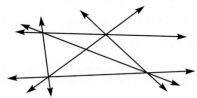

Make a table like the one shown. To complete the table, use your answers from part (a).

Points	2	3	4	5	6
Lines	1	3	6	▓	▓

a. Draw five points, no three of which are collinear. Draw all lines determined by these points. How many lines did you draw? Repeat for six points, no three of which are collinear.

b. Predict the number of lines determined by seven points, no three of which are collinear. Check your prediction.

35. Astronomy Look at the sketch of the Big Dipper. How are points (stars) and segments used to represent constellations?

36. Art Pointillism is a style of painting in which the artist uses small dots of paint to create pictures. In which of the pictures shown did the artist use pointillism?

Everyday Geometry

When you place a four-legged table on an uneven surface, it wobbles. You can use three legs to support something that must be kept steady. Why do you think this is true? Name some three-legged objects that use this principle.

1.2 Angles

Objective: To name angles, right angles, opposite rays, and straight angles.

Two rays with a common endpoint form an ***angle*** (∠). The two rays are the ***sides*** of the angle. The common endpoint is the ***vertex*** of the angle.

Example 1: Sides: $\overrightarrow{OF}$ and $\overrightarrow{OA}$

Vertex: O

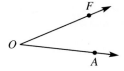

You can name an angle in several ways. If three letters are used, the middle letter always names the vertex.

Example 2: Name the angle in as many ways as possible.

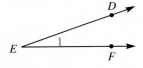

Solution: ∠E, ∠DEF, ∠FED, ∠1

EXPLORING

1. Draw an angle. Label the angle. Erase the drawing.

2. How many points would you need in order to redraw the angle in the same position? Where would these points be?

Angles that make a square corner are called ***right angles***. In drawings, the ⌐ symbol is used to indicate a right angle. Lines, segments, or rays that meet to form right angles are ***perpendicular*** (⊥).

Opposite rays are two rays that have the same endpoint and form a line. $\overrightarrow{OR}$ and $\overrightarrow{OM}$ are opposite rays. They form a ***straight angle***.

Example 3: **Name the following in the figure.**

    a. all right angles   **b.** opposite rays

    c. all straight angles   **d.** a pair of perpendicular rays

Solution: **a.** ∠SOQ and ∠SOT   **b.** $\overrightarrow{OT}$ and $\overrightarrow{OQ}$

    c. ∠TOQ (∠QOT)   **d.** $\overrightarrow{OS}$ and $\overrightarrow{OQ}$ or $\overrightarrow{OS}$ and $\overrightarrow{OT}$

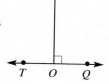

Thinking Critically

1. $\overrightarrow{CA}$ and $\overrightarrow{BC}$ form a line. Are they opposite rays? Why or why not?

2. Must two rays with the same endpoint be opposite rays? Why or why not?

3. Why is ∠Q a poor name for any of the angles shown?

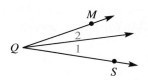

Class Exercises

1. Name some objects that suggest angles.
2. Name some objects that suggest perpendicular segments.
3. Name the angle in four different ways.
4. Name the vertex and the sides of the angle.

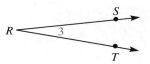

Draw a figure to fit each description.

5. a pair of opposite rays   **6.** two perpendicular lines

Exercises

Name the vertex and the sides of each angle.

1.

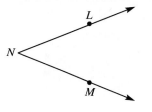

2.

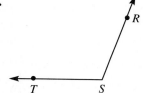

3.

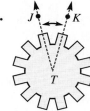

4. Name the angle in four different ways.

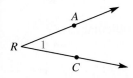

Name each of the following.

  5. all angles shown

  6. all angles with $\overrightarrow{OB}$ as one side

  7. two opposite rays

  8. two other names for $\angle 3$

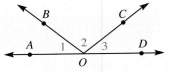

Choose all the correct names for the angle indicated. More than one answer is possible for each exercise.

9. $\angle R$
$\angle RMB$
$\angle BRM$
$\angle B$
$\angle MRB$
$\angle BMR$

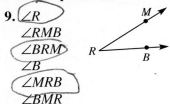

10. $\angle S$
$\angle SRT$
$\angle RST$
$\angle RTS$
$\angle TSR$
$\angle STR$

11. $\angle 1$
$\angle Q$
$\angle QXB$
$\angle B$
$\angle BQX$
$\angle QBX$

12. $\angle DRT$
$\angle DRS$
$\angle SDR$
$\angle R$
$\angle SRT$
$\angle D$

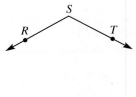

13. $\angle X$
$\angle TXR$
$\angle 2$
$\angle RXS$
$\angle TXS$
$\angle RXT$

14. $\angle M$
$\angle X$
$\angle 3$
$\angle MNX$
$\angle N$
$\angle MXN$

15. $\angle 1$
$\angle RZN$
$\angle 2$
$\angle NZR$
$\angle RZP$
$\angle Z$

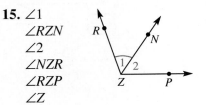

16. $\angle CFB$
$\angle CFD$
$\angle CFE$
$\angle DFB$
$\angle F$
$\angle EFD$

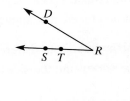

Draw and label a figure to fit each description.

17. an angle whose name is $\angle RBT$

18. an angle with sides $\overrightarrow{OR}$ and $\overrightarrow{OB}$

19. an angle whose vertex is B

20. a right angle

21. $\overrightarrow{RS} \perp \overrightarrow{RT}$

22. opposite rays $\overrightarrow{OR}$ and $\overrightarrow{OT}$

23. two rays that have the same endpoint but are not opposite rays

24. a straight angle with sides $\overrightarrow{BG}$ and $\overrightarrow{BH}$

25. two perpendicular lines $\overleftrightarrow{DB}$ and $\overleftrightarrow{SR}$

26. Two rays drawn from the same endpoint determine one angle. Three rays drawn from the same endpoint determine three angles.

   a. Assume that rays are drawn from the same endpoint. How many angles are determined by four rays? five rays? six rays? seven rays?

   b. Describe a pattern or write a formula for your results in part (a).

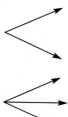

APPLICATIONS

27. **Design**   Create a class emblem using only lines and angles.

28. **Architecture**   Buildings include many examples of angles. Choose a house or other building near where you live. Give five examples of angles. Include at least two angles that are not right angles.

Seeing in Geometry

Which figure is the same as the first one in each row?

1. 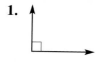   a.    b.    c.

2.   a.   b.   c.

3.   a.   b.   c.

1.3 Parallel and Intersecting Lines and Planes

Objective: To understand parallel lines and parallel planes and intersecting lines and intersecting planes.

The yardage markings on a football field never intersect. They are an example of *parallel segments*.

EXPLORING

1. Draw two lines that will not intersect. What can you say about the distance between these two lines?

2. Draw two lines that intersect. Will they ever intersect in more than one point?

3. Look at the walls on opposite sides of your classroom. What do you notice about the distance between the walls?

4. Look at the intersection of a wall and the floor. Describe the intersection.

The **EXPLORING** activity leads to the following postulates.

POSTULATE 3: **If two lines intersect, then their intersection is a point.**

POSTULATE 4: **If two planes intersect, then their intersection is a line.**

Intersecting lines are lines that meet in a point. *Parallel lines* (∥ lines) are coplanar lines that do not intersect. Segments and rays are parallel if the lines that contain them are parallel. We will use pairs of arrowheads to indicate parallel lines in figures.

Example 1: **Name all lines parallel to $\overleftrightarrow{AB}$; to $\overleftrightarrow{BD}$.**

    Solution: $\overleftrightarrow{CD} \parallel \overleftrightarrow{AB}$; $\overleftrightarrow{AC} \parallel \overleftrightarrow{BD}$

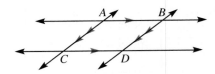

Example 2: **At what point(s) do $\overleftrightarrow{AB}$ and $\overleftrightarrow{CD}$ intersect?**

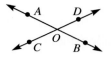

Solution: point O

If two planes do not intersect, they are parallel.

These are parallel planes.

These are intersecting planes.

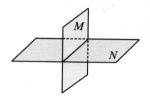

Plane $A \parallel$ plane B. Plane $C \parallel$ plane D. Plane M intersects plane N. Plane O intersects plane P in $\overleftrightarrow{XY}$.

Some lines that do not intersect are not parallel. These lines are noncoplanar. They are called *skew lines*.

Example 3: **Name the following in the box at the right.**

  a. all pairs of parallel planes
  b. all segments skew to $\overline{KN}$
  c. all segments parallel to $\overline{KL}$
  d. the intersection of plane $KLMN$ and plane $KNJG$

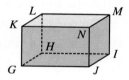

Solution: **a.** plane $HGKL \parallel$ plane $IJNM$
  plane $KGJN \parallel$ plane $LHIM$
  plane $KNML \parallel$ plane $GJIH$
  b. $\overline{LH}, \overline{MI}, \overline{GH}, \overline{JI}$
  c. $\overline{GH}, \overline{NM}, \overline{JI}$
  d. $\overleftrightarrow{KN}$

Thinking Critically

1. Do planes A and B have points in common other than points C and D?

2. Are there other planes that intersect planes A and B in $\overleftrightarrow{CD}$?

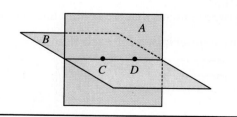

Class Exercises

Name the following.

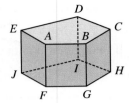

1. all lines parallel to $\overleftrightarrow{AB}$
2. two segments skew to $\overline{EJ}$
3. two segments that intersect $\overline{AF}$
4. all lines parallel to $\overleftrightarrow{AF}$
5. the intersection of planes $ABGF$ and $EAFJ$

Exercises

Name the following.

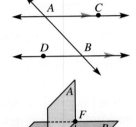

1. a pair of parallel lines
2. two points of intersection
3. two pairs of intersecting lines
4. a line that intersects two lines

Plane A intersects plane B. Name the following.

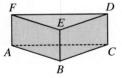

5. the intersection of planes A and B
6. four points in plane A
7. three noncollinear points
8. four noncoplanar points

Name the following.

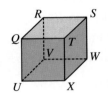

9. three segments parallel to $\overline{QR}$
10. three pairs of intersecting segments
11. four segments skew to $\overline{QR}$
12. a plane parallel to plane $QTXU$
13. two planes that intersect in $\overleftrightarrow{TX}$
14. the intersection of planes $QRST$ and $RVWS$

Describe each pair of segments as parallel, intersecting, or skew.

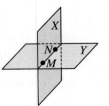

15. $\overline{BC}, \overline{ED}$
16. $\overline{BC}, \overline{AF}$
17. $\overline{BC}, \overline{AB}$
18. $\overline{AF}, \overline{EF}$

Draw and label a figure to fit each description.

19. $\overleftrightarrow{XY}$ intersecting $\overleftrightarrow{XZ}$
20. $\overline{AB} \parallel \overline{CD}$
21. plane $ABCD$ intersecting plane $DEFG$
22. two segments that are neither intersecting nor skew

Write a description for each figure.

23.
24.
25.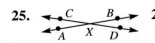
26.

27. Two lines may have either one point or zero points of intersection. Three lines may have zero to three points of intersection.

0 points   1 point   0 points   1 point   2 points   3 points

Find all possibilities for the number of points of intersection for four lines; for five lines.

APPLICATIONS

28. Design   Describe the parts of the chair that are marked in red as parallel, intersecting, or skew lines. Are they coplanar or noncoplanar?

29. Drafting   Name one pair of each of the following: parallel lines, parallel planes, intersecting lines, intersecting planes, and skew lines.

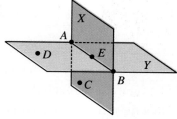

Test Yourself

Plane *X* intersects plane *Y*. Name the following.

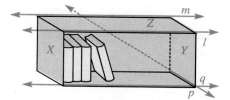

1. four noncoplanar points
2. three collinear points
3. the intersection of planes *X* and *Y*
4. four points in plane *X*

True or false?

5. Two points are always collinear.
6. Skew lines are always in the same plane.
7. Three points are always coplanar.
8. Four points are always coplanar.
9. A line contains exactly two points.
10. Two rays that intersect form opposite rays.

Draw and label a figure to fit each description.

11. $\overleftrightarrow{XY}$ **12.** $l \parallel m$ **13.** $\angle BXR$ **14.** opposite rays $\overrightarrow{AB}$ and $\overrightarrow{AC}$

1.4 Measuring Length

Objective: To measure length using the metric system.

The metric system is widely used around the world. The basic unit of length in the metric system is the **meter** (m). Other common units of length are the **centimeter** (cm) and **millimeter** (mm).

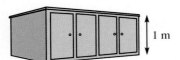

A kitchen counter is about 1 m high.

A paperclip is about 1 cm wide

A dime is about 1 mm thick.

You measure longer distances in **kilometers** (km). The distance from Chicago to Denver is about 1,680 km.

The table below shows the prefixes, symbols, and relationships for some metric units of length.

Prefix	Meaning	Unit	Symbol	Distance
kilo-	1,000	kilometer	km	1 km = 1,000 m
hecto-	100	hectometer	hm	1 hm = 100 m
deka-	10	dekameter	dam	1 dam = 10 m
	1	meter	m	1 m
deci-	0.1	decimeter	dm	1 dm = 0.1 m
centi-	0.01	centimeter	cm	1 cm = 0.01 m
milli-	0.001	millimeter	mm	1 mm = 0.001 m

The **length** of $\overline{AB}$, written as AB, is the distance between A and B.

Example 1: To the nearest centimeter, $AB = 4$ cm. To the nearest millimeter, $AB = 43$ mm. To the nearest tenth of a centimeter, $AB = 4.3$ cm.

EXPLORING

1. Draw segments you think are 3 cm, 8 cm, and 20 cm long. Use a metric ruler to check the length of your segments.

2. Estimate, to the nearest centimeter, the length of four classroom objects. Measure to check your estimates.

3. Estimate, to the nearest meter, three lengths greater than 1 m. Measure to check your estimates.

You multiply or divide by powers of ten to change between metric units.

Example 2: Multiply to change from a larger unit to a smaller unit.

$$\overset{\times\ 10}{3\text{ cm} = 30\text{ mm}} \qquad \overset{\times\ 100}{6\text{ m} = 600\text{ cm}} \qquad \overset{\times\ 1{,}000}{8\text{ km} = 8{,}000\text{ m}}$$

Example 3: Divide to change from a smaller unit to a larger unit.

$$\overset{\div\ 10}{80\text{ mm} = 8\text{ cm}} \qquad \overset{\div\ 100}{450\text{ cm} = 4.5\text{ m}} \qquad \overset{\div\ 1{,}000}{3500\text{ m} = 3.5\text{ km}}$$

Thinking Critically

1. Suppose you had a broken ruler that started at 68 mm. Could you use it to measure? Why or why not?

2. A pencil is 132 mm long. Can you tell, without measuring, what the length is to the nearest centimeter? Explain.

Class Exercises

Estimate the length of each segment to the nearest centimeter.

1. ⟋⟍

2. ⟋⟍

3.-4. Measure the length of each segment in Exercises 1 and 2 to the nearest centimeter, millimeter, and tenth of a centimeter.

Use a metric ruler to draw a segment with the given length.

5. 2 cm   **6.** 34 mm   **7.** 74 mm   **8.** 3.6 cm

Complete each statement.

9. 50 mm = ▓ cm   **10.** 10 cm = ▓ mm   **11.** 410 mm = ▓ cm   **12.** 8.9 cm = ▓ mm

Exercises

Estimate the length of each segment to the nearest centimeter.

1. ⎯⎯⎯⎯⎯⎯⎯⎯⎯   **2.** ⎯⎯⎯⎯⎯⎯

3. ⎯⎯⎯⎯⎯   **4.** ⎯⎯⎯⎯⎯⎯⎯⎯⎯

5.-8. Measure the length of each segment in Exercises 1-4 to the nearest centimeter, millimeter, and tenth of a centimeter.

Complete each statement.

9. 1.7 cm = ▓ mm   **10.** 6 cm = ▓ mm   **11.** 82 mm = ▓ cm

12. 38 mm = ▓ cm   **13.** 13.6 cm = ▓ mm   **14.** 98 mm = ▓ cm

15. 320 mm = ▓ cm   **16.** 11.2 cm = ▓ mm   **17.** 4.0 cm = ▓ mm

Use a metric ruler to draw a segment with the given length.

18. 4 cm   **19.** 7 mm   **20.** 58 mm   **21.** 6.7 cm   **22.** 4.8 cm

23. Arrange in order from shortest to longest:
     1 m   99 cm   0.3 km   2,000 mm

Choose the best unit for measuring each distance. Write km, m, cm, or mm.

24. the distance traveled on a trip   **25.** your height

26. the height of a mountain   **27.** the width of a classroom

28. the length of an automobile   **29.** the thickness of a quarter

30. the width of a street   **31.** the thickness of a piece of thread

32. the distance from New York to Dallas   **33.** the length of your arm

34. the thickness of a human hair   **35.** the height of a fence

Is the statement reasonable? Write *yes* or *no*.

36. My foot is 1.5 mm long.

37. The door is 2 km high.

38. I live 10 cm from school.

39. My thumb is 2.5 m wide.

40. My textbook is 20 cm wide.

41. My teacher is 5.7 km tall.

Choose the most likely measure. Write a, b, or c.

42. the length of a pen

    a. 14 mm **b.** 14 cm **c.** 14 m

43. the width of a goldfish

    a. 1 cm   **b.** 1 m   **c.** 1 km

44. the thickness of a 200-page book

    a. 1.75 mm **b.** 1.75 cm **c.** 1.75 m

45. the depth of a pond

    a. 30 mm   **b.** 30 m   **c.** 30 km

APPLICATIONS

Consumer   Electricity is measured in watt-hours (W•h) or kilowatt-hours (kW•h). A kilowatt-hour is 1,000 watt-hours. (1 kW•h = 1,000 W•h)

An appliance rated at 40 watts will use 40 watts of electricity per hour. Therefore a 100-watt bulb will use 1 kW•h of electricity in 10 hours. (100 watts × 10 hours = 1,000 watt-hours = 1 kW•h)

46. How many hours will it take for two 100-watt bulbs to use 1 kW•h of electricity?

47. An electric fan is rated at 40 watts. How many hours will it take the fan to use 2 kW•h of electricity?

Everyday Geometry

Pacing is a quick way to estimate distances when an exact measurement is not needed.

1. Count the number of strides you take to walk the length of a fairly long room.

2. Measure the length of the room in meters.

3. Divide the length of the room by the number of strides to find the distance per stride.

4. To estimate a distance, pace it off and use the length of your stride to find the distance.

Give an example of a situation in which measurement by pacing would be useful. Give an example of a situation in which it would not.

1.5 Congruent Segments

Objective: To construct a segment congruent to a given segment and to bisect a segment.

The spokes on a bike wheel all are the same length. They are examples of ***congruent segments***. The symbol $\cong$ means *is congruent to*. In a figure, we use hatch marks to show that segments are congruent. Two segments are congruent *if and only if they are equal in length*.

Example 1:

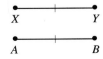

$XY = AB$, so $\overline{XY} \cong \overline{AB}$. (Read as "$\overline{XY}$ is congruent to $\overline{AB}$.")

You can copy a segment by measuring or tracing. You can also ***construct*** a segment congruent to a given segment using only a compass and a straightedge.

CONSTRUCTION 1

Construct a segment congruent to $\overline{AB}$.

1. With a straightedge, draw a segment longer than $\overline{AB}$. Label the endpoints as C and D.

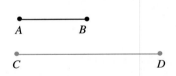

2. Open your compass to the length of $\overline{AB}$. With the compass point on C, draw an arc intersecting $\overline{CD}$.

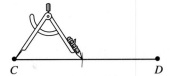

3. Label the point of intersection as E. Are the lengths of $\overline{AB}$ and $\overline{CE}$ the same? How can you verify this?

The *midpoint* of a segment is the point that divides the segment into two congruent segments.

Example 2: $\overline{AM} \cong \overline{MB}$, so M is the midpoint of $\overline{AB}$.

A *bisector* of a segment is a line, segment, or ray that passes through the midpoint of the segment.

Example 3: $\overleftrightarrow{NM}$, $\overrightarrow{MR}$, and $\overleftrightarrow{RM}$ are bisectors of $\overline{AB}$.

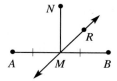

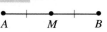

EXPLORING

1. Use the *Geometric Supposer: Triangles* disk to create any triangle *ABC*. Each side of the triangle is a segment. Measure $\overline{AB}$.

2. Label the midpoint of $\overline{AB}$.

3. Measure $\overline{AD}$ and $\overline{DB}$. Compare *AB* to *AD* + *DB*.

4. Now measure $\overline{BC}$.

5. If you label midpoint *E* on $\overline{BC}$, what are the lengths of $\overline{BE}$ and $\overline{EC}$?

Thinking Critically

M is the midpoint of $\overline{AB}$

1. Does $\overleftrightarrow{AB}$ have a midpoint?

2. How many lines can you draw through *M*?

3. How many bisectors does $\overline{AB}$ have?

4. Is every line through *M* a bisector of $\overline{AB}$?

Class Exercises

Draw and label a figure to fit each description.

1. $\overleftrightarrow{MN}$ bisecting $\overline{PQ}$

2. $\overline{AB}$ with midpoint *P*

3. $\overrightarrow{CD}$ bisecting $\overline{EF}$

Exercises

Draw segments like the ones shown.

1. Construct a segment congruent to $\overline{AB}$.
2. Construct a segment congruent to $\overline{CD}$.

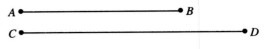

M is the midpoint of $\overline{CD}$ and $\overline{PQ}$.

3. Name two pairs of congruent segments.
4. Name three different bisectors of $\overline{CD}$.
5. Does $\overleftrightarrow{PQ}$ have a midpoint?

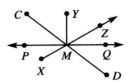

True or false?

6. $\overleftrightarrow{RM}$ is a bisector of $\overline{AB}$.
7. $\overleftrightarrow{AM}$ is a bisector of $\overline{RS}$.
8. $\overleftrightarrow{AB}$ is a bisector of $\overline{RS}$.
9. M is the midpoint of $\overline{CD}$.
10. $\overline{AM} \cong \overline{MB}$ 11. $\overline{CM} \cong \overline{AC}$ 12. $\overline{RM} \cong \overline{MS}$ 13. $\overline{CM} \cong \overline{DB}$

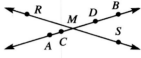

$RM = 4 \quad AC = 1 \quad MD = 2$
$MS = 4 \quad CM = 1 \quad DB = 2$

Draw segments like the ones shown.

14. Construct a segment whose length is $c + d$.
15. Construct a segment whose length is $3\,d$.
16. Construct a segment whose length is $c - d$.

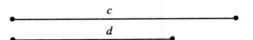

APPLICATION

17. **Computer**   Draw $\overline{AB}$ with midpoint D as described in the Exploring activity. Label the midpoint of $\overline{AD}$ as E. Label the midpoint of $\overline{DB}$ as F. What is the relationship between AB and EF?

Computer

1. Use the *Geometric Supposer: Triangles* disk to create any triangle ABC.
2. Label the midpoints of $\overline{AB}$ and $\overline{AC}$. Draw $\overline{DE}$. Does $\overline{DE}$ appear to be parallel to $\overline{BC}$?
3. Try to label the intersection of $\overline{DE}$ and $\overline{BC}$. Is $\overline{DE}$ parallel to $\overline{BC}$?
4. Repeat with other triangles. Will $\overline{DE}$ always be parallel to $\overline{BC}$?

What can you conclude about the relationship between the segment connecting the midpoints of two sides of a triangle and the third side of the triangle?

Perpendiculars and Perpendicular Bisectors

Objective: To construct the perpendicular bisector of a segment and perpendiculars to a line through a point on or not on the line.

A line that is perpendicular to a segment at its midpoint is the *perpendicular bisector* of the segment.

You can find the perpendicular bisector of a segment by folding a segment onto itself. You can also measure to find the midpoint of the segment and then use a square corner to draw the perpendicular bisector. The constructions that follow show how to construct perpendiculars using a compass and straightedge.

CONSTRUCTION 2

Construct the perpendicular bisector of $\overline{AB}$.

A •————————• B

1. Draw a segment like the one shown. Open your compass to more than half the length of $\overline{AB}$. With the compass point on A, draw an arc intersecting $\overline{AB}$.

2. Using the same compass setting, repeat Step 1 at point B. Label the points of intersection of the arcs as C and D.

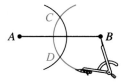

3. Draw $\overleftrightarrow{CD}$. Label the point where $\overleftrightarrow{CD}$ intersects $\overline{AB}$ as M. $\overleftrightarrow{CD}$ is the perpendicular bisector of $\overline{AB}$.

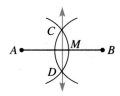

4. What do you know about M? What is true about the lengths of $\overline{AM}$ and $\overline{MB}$?

**Construct the perpendicular to a line
at a point on the line.**

1. Draw a line like the one shown. With
   the compass point on A, draw an arc
   intersecting l in two points. Label these
   points as B and C.

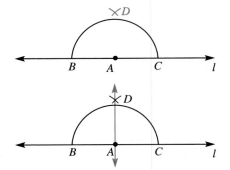

2. Open your compass to more than half the
   length of $\overline{BC}$. Draw arcs from B and C.
   Label the point of intersection as D.

3. Draw $\overleftrightarrow{DA}$. $\overleftrightarrow{DA}$ is perpendicular to line l. $\overleftrightarrow{DA}$
   is the perpendicular bisector of $\overline{BC}$.

**Construct the perpendicular to a line
from a point not on the line.**

1. Draw a figure like the one shown. With the
   compass point on P, draw an arc intersecting
   l in two points. Label these points as Q and R.

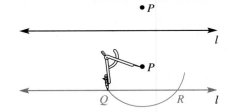

2. Using the same compass setting, draw arcs from
   Q and R on the side of the line opposite from P.
   Label the point of intersection of the arcs as S.

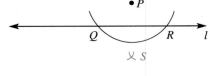

3. Draw $\overleftrightarrow{PS}$. Is $\overleftrightarrow{PS}$ the perpendicular bisector
   of l? of $\overline{QR}$? Why or why not?

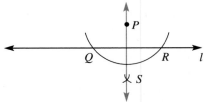

1. Look at line *l*. How many lines can you draw through *P*? How many lines through *P* will be perpendicular to *l*? How many lines in a plane can you draw that pass through *P* and are perpendicular to *l*?

2. Look at line *m*. How many lines can you draw through *R*? How many lines passing through *R* will be perpendicular to *m*?

3. Would Construction 2 work if the compass setting was not more than half the length of $\overline{AB}$? Why or why not?

4. In what ways are Constructions 2, 3, and 4 similar?

Class Exercises

1. Draw a segment. Use Construction 2 to construct its perpendicular bisector.

2. Draw a line and choose any point on it. Use Construction 3 to construct a perpendicular to the line at that point.

3. Draw a line and choose any point not on the line. Use Construction 4 to construct a perpendicular from the point to the line.

Exercises

1. How many lines can you draw through *M*?

2. Is every line through *M* a bisector of $\overline{AB}$?

3. How many bisectors does $\overline{AB}$ have?

4. How many perpendicular bisectors does $\overline{AB}$ have?

5. Draw a segment at least 8 cm long. Construct its perpendicular bisector.

6. Draw a segment at least 8 cm long. Divide it into four congruent parts using only a compass and straightedge.

7. Draw a line *l* and choose any point *P* on the line. Construct the line through *P* that is perpendicular to *l*.

8. Draw a line *l* and choose any point *P* not on the line. Construct the line through *P* that is perpendicular to *l*.

Draw segments like the ones shown.

9. Construct a segment whose length is $\frac{3}{4} \cdot MN$. (*Hint*: Use Construction 2 twice.)

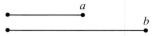

10. Construct a square with sides of length *MN*.

11. Construct a rectangle with two opposite sides of length *a* and two opposite sides of length *b*.

Draw a figure similar to, but larger than, the one at the right for Exercises 12 and 13.

12. Construct the perpendicular from *C* to $\overleftrightarrow{AB}$. (*Hint*: Extend $\overline{AB}$.)

13. Construct the perpendicular from *A* to $\overleftrightarrow{BC}$.

14. Follow the directions to construct a symbol like the one shown.

    a. Draw a circle and mark its center.

    b. Draw a horizontal diameter. The diameter consists of two radii. Locate the midpoints of the two radii by constructing the perpendicular bisector of each radius.

    c. Using the two midpoints for centers, construct a semicircle above the left radius and another below the right radius. Shade the figure as shown.

APPLICATION

15. **Home Improvement**   In most buildings, the distance between the ceiling and the floor varies by some small amount. For this reason, paperhangers use a plumb line to be sure that the wallpaper they hang is straight. Find out how to use a plumb line. Why is it important, when hanging wallpaper, to use a plumb line?

Historical Note

The Greek philosopher Plato (430–349 B.C.) is credited with introducing the idea that only a compass and straightedge may be used in geometric constructions. It is said that above the entrance to The Academy, Plato's school, was carved the inscription, "Let no one ignorant of geometry enter here."

1.7 Reasoning in Geometry

Objective: To use deductive reasoning and inductive reasoning.

In science class, you conduct experiments to help discover facts. From your observations and experimental results you draw conclusions. When you reach a general conclusion from specific examples you are using *inductive reasoning*.

EXPLORING

1. Consider the expression $n^2 + n + 11$. If you substitute 1 for n, the result is $1 \cdot 1 + 1 + 11 = 13$, a prime number. Evaluate this expression for the next eight counting numbers. Is your result always a prime number?

2. Evaluate the expression for $n = 10$. Is your result a prime number?

This **EXPLORING** activity demonstrates that inductive reasoning may result in an incorrect conclusion. You might be tempted to conclude from Step 1 of the **EXPLORING** activity that $n^2 + n + 11$ is prime for every counting number. Step 2 provides a *counterexample* to disprove the conclusion reached using inductive reasoning.

Example 1: **State a reasonable conclusion based on inductive reasoning.**

We have had a test every Friday for four months.
Today is Friday.

Solution: We will have a test today.

When you use inductive reasoning to reach a conclusion, be sure you try a sufficient number and variety of examples. If you do, your conclusion will probably (but not necessarily) be true.

Deductive reasoning is the process of reasoning from known or accepted facts to a new conclusion. We often use deductive reasoning to prove that conclusions made by inductive reasoning are true.

Example 2: **State a reasonable conclusion based on deductive reasoning.**

A prime number has exactly two factors, itself and 1.
Thirteen is a prime number.

Solution: Thirteen has exactly two factors, itself and 1.

Deductive reasoning does not require looking at several cases or examples. The conclusion reached in Example 2 is true because it is based on two true statements.

Properties and major conclusions in this textbook are labeled as postulates and theorems. Recall that a postulate is a statement that is accepted to be true without proof. A ***theorem*** is a statement that can be proved. In general, in this textbook, we will not provide proofs of theorems.

Thinking Critically

1. Jane knows that the number 7 is prime, and that all prime numbers have exactly two factors. She concludes that the number 7 has exactly two factors.

  a. Which method of reasoning did Jane use to reach her conclusion?

  b. Is her conclusion valid? Why or why not?

2. Tim notices that 3, 5, and 7 are prime numbers. He concludes that all odd numbers are prime.

  a. Which method of reasoning did Tim use to reach his conclusion?

  b. Is his conclusion valid? Why or why not?

Class Exercises

True or false?

1. One counterexample is enough to disprove a statement.

2. Inductive reasoning involves looking at several examples.

3. Inductive reasoning involves reasoning from known or accepted facts.

4. Using the results of a customer survey to plan a new product would be an example of deductive reasoning.

What is wrong with the following conclusions?

5. Pete is a cat.
   Pete has spots.
   All cats have spots.

6. Melanie is over 21.
   Melanie is a registered voter.
   Registered voters must be over 21.

Exercises

State a reasonable conclusion. Tell whether you used inductive reasoning or deductive reasoning.

1. All tests for the first six chapters had 20–30 questions. The Chapter 7 test will be given tomorrow.

2. 33 is divisible by 3; 3 + 3 = 6   69 is divisible by 3; 6 + 9 = 15
21 is divisible by 3; 2 + 1 = 3   15 is divisible by 3; 1 + 5 = 6

3. 1 + 3 = 4   3 + 5 = 8   9 + 21 = 30   113 + 69 = 182

State or draw a counterexample that shows each statement to be false.

4. All rectangles are squares.

5. Every fraction is less than one.

6. The product of any two fractions is less than one.

Is the conclusion reasonable?

7. To play basketball, students must be passing all their courses.
Tom plays basketball.
Tom is passing all of his courses.

8. Bob plays the piano very well.
Bob never practices.
Playing the piano well does not require practice.

State a reasonable conclusion based on deductive reasoning. Accept the given statements as true.

9. All rectangles are parallelograms. *ABCD* is a rectangle.

10. An acute angle has a measure between 0° and 90°.
$\angle ABC$ is an acute angle.

11. In a scalene triangle, no two sides are the same length.
Triangle *DEF* is a scalene triangle.

12. If equal quantities are added to equal quantities, the resulting sums are equal. Suppose that x and y are equal, and 7 is added to both x and y.

13. Parallel lines do not intersect.
Lines a and b are parallel.

14. What kind of number do you get if you multiply a positive integer and a negative integer? Give three examples that would help you use inductive reasoning to draw a conclusion. Write the conclusion.

15. Complete the deductive proof following the statement below. "The sum of any two even integers is an even integer."

Any two even integers can be represented by $2m$ and $2n$, where m and n are integers. Thus their sum is $2m + 2n = 2(m + n)$.

   a. Is $m + n$ an integer? Why or why not?

   b. Is $2(m + n)$ an integer? Why or why not?

   c. Is the above argument true regardless of the values of m and n? Does it prove the statement for any two even integers?

APPLICATIONS

16. Travel   You have a flight leaving Boston at 1:05 P.M., and you know it normally takes 25 min to drive to the airport. You must be at the airport $\frac{1}{2}$h before flight time. At what time should you leave your house? During rush hour, it takes 55 min to drive to the airport. At what time should you leave your house if your flight is at 5:35 P.M.?

17. Patterns   Find the next three items in each pattern.

   a. 3, 7, 11, 15, 19, . . .   **b.** 5, 5.3, 5.6, 5.9, 6.2, . . .

   c. 1, 1, 2, 3, 5, 8, . . .   **d.**

 , , , . . .

Seeing in Geometry

Fold a piece of paper horizontally and label the fold as $\overleftrightarrow{AB}$. Mark a point C on the paper, not on $\overleftrightarrow{AB}$. Then fold the paper vertically, so that the fold passes through the point C. Label this fold as $\overleftrightarrow{CD}$. What is the relationship between $\overleftrightarrow{AB}$ and $\overleftrightarrow{CD}$?

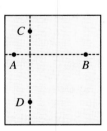

Repeat this activity, marking another point, E, on the paper (not on $\overleftrightarrow{AB}$ or $\overleftrightarrow{CD}$). Fold the paper vertically again so that the fold passes through point E. Label this fold as $\overleftrightarrow{EF}$. What is the relationship between $\overleftrightarrow{EF}$ and $\overleftrightarrow{AB}$? between $\overleftrightarrow{EF}$ and $\overleftrightarrow{CD}$?

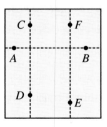

If you marked more points and made more vertical folds, do you think the relationships would hold true for these additional folds?

Thinking About Proof

What is Proof in Geometry?

In a court of law, attorneys provide evidence (or *reasons*) to prove their cases. In a geometry proof, you provide definitions, theorems, and postulates to make a convincing argument to prove that a conclusion is true.

In this lesson, you will provide convincing arguments in this deductive manner.

1. Start with accepted facts labeled the *Given*.
2. Apply a definition, postulate, or theorem to the *Given* information to prove that a *Conclusion* is true.

Example: **Supply a *reason* that proves that the conclusion is true.**

> **Given:** M is the midpoint of $\overline{DE}$.
>
> **Conclusion:** $\overline{DM} \cong \overline{ME}$

Solution: definition of midpoint

Exercises

Supply a *reason* that proves that the conclusion is true.

1. **Given:** $\overrightarrow{AR}$ and $\overrightarrow{AB}$ are opposite rays.

   Conclusion: $\angle RAB$ is a straight angle.

2. **Given:** $\angle GQY$ is a right angle.

   Conclusion: $\overleftrightarrow{XY} \perp \overleftrightarrow{GF}$

Write a *conclusion*. Give a reason to prove that your conclusion is true.

3. **Given:** CK = KD

4. **Given:** Line n bisects $\overline{CD}$.

5. **Given:** Coplanar segments $\overline{CD}$ and $\overline{EF}$ do not intersect.

6. **Given:** K is the midpoint of $\overline{CD}$.

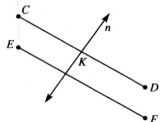

Objective: To solve a problem by drawing a diagram or making a model.

You can solve many problems by drawing a diagram or making a model.

Example: Newspapers, books, and magazines often are printed in groups of 8, 16, or 32 pages, called *signatures*. The pages are positioned to print on both sides of the paper that is fed through a printing press. When the paper is folded, the pages are in order.

As editor of an 8-page school newspaper, it is your job to determine a layout that will work.

Solution: Making a model of the finished layout will help. Use a piece of paper to represent the signature.

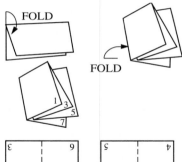

1. Fold the paper in half horizontally, then vertically.

2. With the longer folded edge on the left, begin numbering both sides of the pages in order.

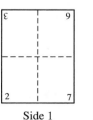
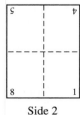

3. Unfold the paper. You have a diagram to show the order of the pages. Notice that some of your pages will appear upside down.

4. Try another folding pattern. Can another page order work?

Side 1   Side 2

Class Exercise

Draw a diagram or make a model to solve.

1. Many books are printed in 16-page signatures. Make a diagram to show one way to position the pages in a 16-page book.

Exercises

Draw a diagram or make a model to solve.

1. Trace the figures shown. Copy the figures onto cardboard and cut them out. Can you use either figure repeatedly to cover the surface of a sheet of paper without leaving any spaces between figures?

2. A singles tennis tournament has 32 players. When a player loses, that player is eliminated. How many matches must be played before there is a winner?

3. A 234-ft pole is driven 26 ft into the bottom of a lake that is 89 ft deep. On the pole is a rust mark 41 ft below the surface of the water. A hook hangs 5 ft from the top of the pole. How far is the hook from the rust mark?

4. A ball is dropped from a height of 12 m. Each time it hits the ground it bounces to half the previous height. The ball is caught when the height of its bounce is 3 m. What is the total distance the ball has traveled?

5. Josie has four rods that measure 9 cm, 7 cm, 2 cm, and 5 cm. How can she use them to measure a length of 1 cm?

6. Suppose you have three pairs of pants and two shirts. How many different outfits are possible?

Vocabulary and Symbols

You should be able to write a brief description, draw a picture, or give an example to illustrate the meaning of each of the following.

Vocabulary

angle (p. 8)
bisector of a segment (p. 21)
collinear points (p. 4)
congruent segments
  (p. 20)
construct (p. 20)
coplanar points (p. 4)
counterexample (p. 27)
deductive reasoning (p. 27)
inductive reasoning (p. 27)
intersecting lines (planes)
  (pp. 12, 13)
line (p. 3)
midpoint of a segment
  (p. 21)
noncollinear points (p. 4)
noncoplanar points (p. 4)
opposite rays (p. 8)

parallel lines (planes)
  (pp. 12, 13)
perpendicular bisector of a
  segment (p. 23)
perpendicular lines (p. 8)
plane (p. 3)
point (p. 3)
postulate (p. 5)
ray (p. 3)
right angle (p. 8)
segment (p. 3)
sides of an angle (p. 8)
skew lines (p. 13)
straight angle (p. 8)
theorem (p. 28)
vertex of an angle (p. 8)

Units of length
centimeter (p. 16)
kilometer (p. 16)

meter (p. 16)
millimeter (p. 16)

Symbols

$\angle ABC$	(angle ABC) (p. 8)
cm	(centimeter) (p. 16)
$\cong$	(is congruent to)
	(p. 20)
km	(kilometer) (p. 16)
$\overleftrightarrow{AB}$	(line AB) (p. 3)
m	(meter) (p. 16)
mm	(millimeter) (p. 16)
$\parallel$	(is parallel to) (p. 12)
$\perp$	(is perpendicular to)
	(p. 8)
$\overrightarrow{AB}$	(ray AB) (p. 3)
$\overline{AB}$	(segment AB) (p. 3)
⌐	(right angle) (p. 8)
➤➤	(parallel) (p. 12)

Summary

The following list indicates the major skills, facts, and results you should have mastered in this chapter.

1.1 Name points, lines, segments, rays, planes, collinear points, and coplanar points. (pp. 3–7)

1.2 Name angles, right angles, opposite rays, and straight angles. (pp. 8–11)

1.3 Understand parallel lines and parallel planes and intersecting lines and intersecting planes. (pp. 12–15)

1.4 Measure length using the metric system. (pp. 16–19)

1.5 Construct a segment congruent to a given segment and bisect a segment. (pp. 20–22)

1.6 Construct the perpendicular bisector of a segment and perpendiculars to a line through a point either on or not on the line. (pp. 23–26)

1.7 Use deductive reasoning and inductive reasoning. (pp. 27–30)

1.8 Solve a problem by drawing a diagram or making a model. (pp. 32–33)

Exercises

Name each of the following.

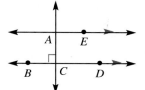

1. opposite rays **2.** all segments shown

3. three noncollinear points **4.** two parallel lines

5. a right angle **6.** a straight angle

Name each of the following.

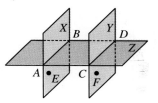

7. two parallel planes **8.** two intersecting planes

9. four noncoplanar points **10.** the intersection of two planes

Measure each segment to the nearest centimeter, millimeter, and tenth of a centimeter.

11. **12.**

Complete each statement.

13. 30 mm = ▉ cm **14.** 3.2 m = ▉ cm **15.** 52 cm = ▉ mm

Draw a figure like the one shown.

16. Construct a segment congruent to $\overline{AB}$.

17. Construct the perpendicular bisector of $\overline{AB}$.

18. Construct a perpendicular to $\overleftrightarrow{AB}$ through C.

State a reasonable conclusion. Tell whether you used inductive or deductive reasoning.

19. All rectangles have four right angles. A square is a rectangle.

PROBLEM SOLVING

20. A farmer is using fenceposts at the corners and at 4-ft intervals around a field. How many posts are needed for a 48 ft by 32 ft field?

Chapter 1 Test

Write the best word or phrase to describe the geometric figure(s).

1. coplanar lines that do not intersect
2. the intersection of two nonparallel planes
3. a line that intersects a segment at its midpoint

Name each of the following.

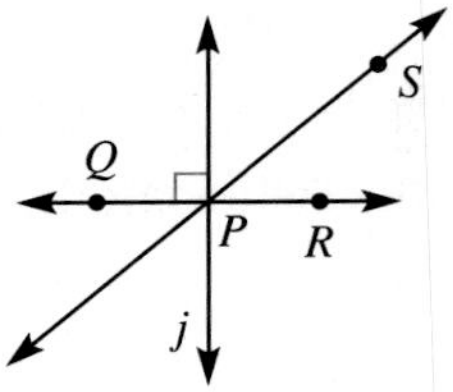

4. three collinear points
5. perpendicular lines
6. a segment
7. an angle with vertex P

Name each of the following.

8. four coplanar points
9. two planes
10. the intersection of two planes

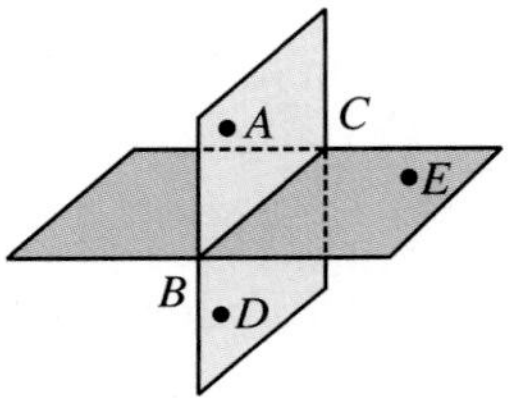

Measure the lengths to the nearest centimeter and millimeter.

11. 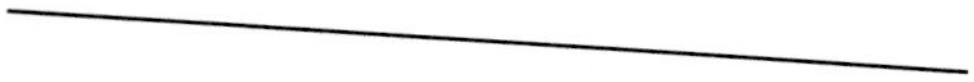

12. 

Complete each statement.

13. 4.27 m = ■ cm
14. 112 mm = ■ cm
15. 0.8 cm = ■ mm

16. Draw $\overline{PQ}$ and construct its perpendicular bisector, j.
17. Draw $\overleftrightarrow{AB}$ and construct $\overleftrightarrow{BC}$ perpendicular to $\overleftrightarrow{AB}$.
18. Draw a line m and a point X not on m. Construct a line, j, through X perpendicular to m.
19. Find 3×5, 13×7, 9×5, and 123×7. Find the product of any two odd numbers. State a reasonable conclusion based on inductive reasoning.

True or false?

20. Any two points are collinear.
21. A segment has exactly one bisector.

PROBLEM SOLVING

22. An artificial lake is 1.5 m deep at the end of a dock. The lake gets 1.75 m deeper every 10 m from the dock. How deep is the lake 30 m from the dock?

Angles

The sharp angles formed by the steeply pitched roofs in the foreground stand in stark contrast to the complex curves in the building behind them.

Focus on Skills

ARITHMETIC

Write three fractions equivalent to the given fraction.

Example: $\dfrac{4}{6} = \dfrac{4 \cdot 2}{6 \cdot 2} = \dfrac{8}{12}$ $\qquad$ $\dfrac{4}{6} = \dfrac{4 \div 2}{6 \div 2} = \dfrac{2}{3}$ $\qquad$ $\dfrac{4}{6} = \dfrac{4 \cdot 3}{6 \cdot 3} = \dfrac{12}{18}$

1. $\dfrac{3}{5}$ $\qquad$ **2.** $\dfrac{1}{4}$ $\qquad$ **3.** $\dfrac{3}{8}$ $\qquad$ **4.** $\dfrac{15}{20}$ $\qquad$ **5.** $\dfrac{3}{12}$

6. $\dfrac{18}{20}$ $\qquad$ **7.** $\dfrac{20}{25}$ $\qquad$ **8.** $\dfrac{12}{16}$ $\qquad$ **9.** $\dfrac{3}{7}$ $\qquad$ **10.** $\dfrac{10}{15}$

Express each fraction in simplest form.

Example: $\dfrac{6}{10} = \dfrac{6 \div 2}{10 \div 2} = \dfrac{3}{5}$

11. $\dfrac{6}{9}$ $\qquad$ **12.** $\dfrac{18}{30}$ $\qquad$ **13.** $\dfrac{12}{16}$ $\qquad$ **14.** $\dfrac{12}{60}$ $\qquad$ **15.** $\dfrac{32}{60}$

16. $\dfrac{45}{60}$ $\qquad$ **17.** $\dfrac{60}{360}$ $\qquad$ **18.** $\dfrac{30}{360}$ $\qquad$ **19.** $\dfrac{50}{360}$ $\qquad$ **20.** $\dfrac{135}{360}$

Multiply. Write each answer in simplest form.

21. $\dfrac{1}{2} \cdot \dfrac{1}{5}$ $\qquad$ **22.** $\dfrac{3}{4} \cdot \dfrac{2}{3}$ $\qquad$ **23.** $\dfrac{3}{5} \cdot \dfrac{3}{4}$ $\qquad$ **24.** $\dfrac{12}{18} \cdot \dfrac{6}{15}$

25. $\dfrac{5}{8} \cdot 8$ $\qquad$ **26.** $\dfrac{2}{3} \cdot 360$ $\qquad$ **27.** $\dfrac{6}{16} \cdot 360$ $\qquad$ **28.** $\dfrac{1}{6} \cdot 360$

ALGEBRA

Solve.

29. $x + (x + 8) = 18$ $\qquad$ **30.** $y + 17 = 90$ $\qquad$ **31.** $2x + x = 102$

GEOMETRY

32. Name the angle at the right in four different ways.

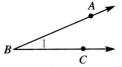

Draw and label a figure to fit each description.

33. an angle named $\angle DEF$ $\qquad$ **34.** an angle with sides $\overrightarrow{AR}$ and $\overrightarrow{AG}$

35. perpendicular lines j and k $\qquad$ **36.** a right angle with vertex C

37. opposite rays $\overrightarrow{BT}$ and $\overrightarrow{BR}$ $\qquad$ **38.** a straight angle

2.1 Circles and Amount of Turn

Objective: To use amount of turn to measure angles and use the relationships among parts of circles.

A *circle* (⊙) is the set of all points in a plane that are the same distance from some point called the *center.* A circle is named for its center.

The table defines some segments and their relationship to a circle.

Term	Definition
Radius	A segment whose endpoints are the center of the circle and any point on the circle; the length of such a segment is called *the* radius of the circle.
Chord	A segment whose endpoints are any two points on a circle.
Diameter	A chord that passes through the center of a circle; the length of such a segment is called *the* diameter of the circle.

Example 1: **a.** Name all radii, chords, and diameters shown for ⊙C.

b. Find the radius and the diameter of the circle.

Solution: **a.** radii: $\overline{AC}$, $\overline{BC}$, and $\overline{DC}$; chords: $\overline{AB}$ and $\overline{EF}$; diameter: $\overline{AB}$

b. radius: 6; diameter: 12

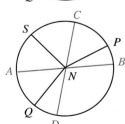

EXPLORING

1. Use a compass to draw a circle with center N. Label points on the circle as *P, S,* and *Q.* Draw and measure $\overline{PN}$, $\overline{SN}$, and $\overline{QN}$.

2. Draw two different diameters. Label them as $\overline{AB}$ and $\overline{CD}$. What kinds of figures are $\overline{CN}$, $\overline{DN}$, $\overline{AN}$, and $\overline{BN}$?

3. Measure $\overline{CN}$, $\overline{DN}$, $\overline{AN}$, and $\overline{BN}$. How do the measures compare to *PN, QN,* and *SN*?

4. Measure $\overline{AB}$ and $\overline{CD}$. What do you discover?

5. What relationship do you see between the radius and diameter of a circle?

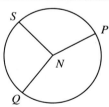

In this lesson we will look at an angle as the figure produced by turning a ray about its endpoint. For example, ∠*BAC* is produced by turning $\overrightarrow{AC}$ about point *A*.

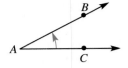

We will use *amount of turn* to measure an angle. In a figure, a single-headed arrow will show the direction and amount of turn.

$\frac{1}{4}$ turn

1 full turn

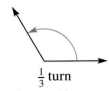

$\frac{1}{3}$ turn

An angle with its vertex at the center of a circle is a ***central angle.*** In this lesson we look at central angles to make it easier to measure amount of turn. Though it is possible to have angles greater than one full turn, in this textbook we will confine our work to angles of $\frac{1}{2}$ turn or less.

    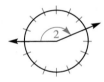

These are central angles.

Example 2: **a.** What is the measure of ∠1? **b.** What is the measure of ∠2?

  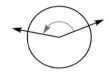

Solution: **a.** The circle is divided into 8 equal-size sections. Each section is $\frac{1}{8}$ of the circle. The measure of ∠1 is $\frac{3}{8}$ turn.

b. The circle is divided into 16 equal-size sections. Each section is $\frac{1}{16}$ of the circle. The measure of ∠2 is $\frac{7}{16}$ turn.

Thinking Critically

1. Does changing the lengths of the sides of an angle change the measure of the angle? Explain.

2. How many radii, chords, and diameters can be drawn through each point described? Draw a figure to support your answer.
   a. point *A* on ⊙*D* **b.** point *B* inside ⊙*D*

Class Exercises

1. Name all radii, chords, and diameters shown for $\odot X$.

2. If $BX = 3$ cm, then $AC = $ ▦.

3. If $AC = 16$ cm, then $XC = $ ▦.

Find the amount of turn for each angle.

4.   **5.**   **6.**   **7.**

Exercises

Name all radii, chords, and diameters shown.

1.   **2.**   **3.**

4. The radius of a circle is 4 cm. What is the diameter?

5. The diameter of a circle is 10 cm. What is the radius?

Find the amount of turn for each angle.

6.   **7.**   **8.**   **9.**   **10.**

For each angle tell whether its measure is (a) less than $\frac{1}{4}$ turn, (b) $\frac{1}{4}$ turn, (c) between $\frac{1}{4}$ and $\frac{1}{2}$ turn, or (d) $\frac{1}{2}$ turn.

11.   **12.**   **13.**   **14.**   **15.**

True or false?

16. Every diameter of a circle is a chord.

17. The longest chord of a circle is a diameter.

18. Every radius of a circle is a chord.

19. A circle has an unlimited number of chords.

20. Every chord of a circle is a diameter.

21. All radii of the same circle are congruent.

22. The radius of ⊙D is 5 cm. Indicate whether the point described is inside, outside, or on ⊙D.

a. point X: 6 cm from D **b.** point Y: 4 cm from D **c.** point Z: 5 cm from D

Draw a central angle having the given measure.

23. $\frac{3}{8}$ turn   **24.** $\frac{3}{12}$ turn   **25.** $\frac{1}{5}$ turn   **26.** $\frac{1}{6}$ turn   **27.** $\frac{1}{4}$ turn

28. Consider all possible positions of points A and B inside ⊙M.

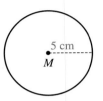

5 cm
M

a. Can $\overline{AB}$ intersect ⊙M?

b. Must $\overline{AB}$ be located entirely inside ⊙M?

c. What can be said about the length of a segment whose endpoints are any two points inside ⊙M?

29. Consider all possible positions for points C and D outside ⊙M.

a. Can $\overline{CD}$ intersect ⊙M?

b. Must $\overline{CD}$ be located entirely outside ⊙M?

c. What can be said about the length of a segment whose endpoints are any two points outside ⊙M?

APPLICATIONS

Time   You can think of the hour and minute hands on a circular clock face as forming central angles. The hour markings divide the circle into 12 equal parts. The minute markings divide the circle into 60 equal parts.

30. Through what size angle does the hour hand move in the time period?

a. 1 h   **b.** 2 h   **c.** 4 h   **d.** 6 h

31. Through what size angle does the minute hand move in the time period?

a. 1 min   **b.** 10 min   **c.** 25 min   **d.** 30 min

Seeing in Geometry

Suppose you had a wheel of cheese. Make a sketch to show how you could cut the wheel into eight equal-size pieces with just three straight cuts.

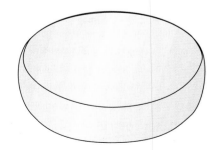

2.2 Degree Measure

Objective: To measure angles in degrees and to classify acute, right, obtuse, and straight angles.

A **degree** (°) is the most commonly used unit for measuring angles. One full turn equals 360°. We classify angles by their measures.

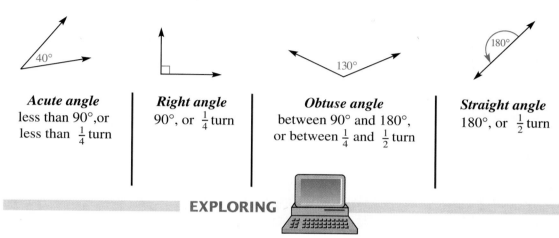

Acute angle
less than 90°, or
less than $\frac{1}{4}$ turn

Right angle
90°, or $\frac{1}{4}$ turn

Obtuse angle
between 90° and 180°,
or between $\frac{1}{4}$ and $\frac{1}{2}$ turn

Straight angle
180°, or $\frac{1}{2}$ turn

EXPLORING

1. Use LOGO to enter and run the ANGLE procedure at the right. Repeat the procedure until you have seen an acute angle, a right angle, and an obtuse angle. The random command outputs a number greater than or equal to 0 and less than the output number.

```
To angle
cs
fd 60 bk 60
rt ((random 180) + 1)
fd 60 bk 60 ht
end
```

2. Change line 4 so the turtle draws only acute angles.

3. Change the procedure so the turtle first draws an obtuse angle and then draws an acute angle.

The notation $m\angle ABC$ means *the measure of $\angle ABC$*.

You can change between amount of turn and degree measure.

Example: a. $1° = \dfrac{1}{360}$ turn

$60° = \dfrac{60}{360}$ turn

$= \dfrac{1}{6}$ turn

b. 1 turn $= 360°$

$\dfrac{3}{8}$ turn $= \dfrac{3}{8} \cdot \dfrac{360°}{1}$

$= \dfrac{1080°}{8} = 135°$

1. $\angle AOB$ is a straight angle and C is a point such that $m\angle AOC > m\angle COB$. Classify each of $\angle AOC$ and $\angle COB$ as acute, obtuse, right, or straight angles. (*Hint*: Draw a figure.)

2. Look up *horizontal* and *vertical*. Are horizontal and vertical lines perpendicular? Can horizontal and vertical lines form an acute angle? an obtuse angle? a right angle?

Class Exercises

Complete each statement.

1. 1 turn = ■

2. $1° = $ ■ turn

Change each measure from degrees to amount of turn or from amount of turn to degrees.

3. 40°   4. 90°   5. $\frac{1}{5}$ turn   6. $\frac{2}{5}$ turn   7. $\frac{1}{3}$ turn

Name each of the following.

8. three acute angles

9. three obtuse angles

10. two straight angles

11. two right angles

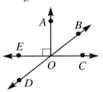

Exercises

Change each measure from degrees to amount of turn or from amount of turn to degrees.

1. $\frac{1}{12}$ turn   2. 10°   3. 24°   4. $\frac{1}{3}$ turn   5. 20°

6. 75°   7. $\frac{1}{6}$ turn   8. $\frac{2}{9}$ turn   9. 45°   10. 30°

Complete each statement.

11. An angle with measure $\frac{1}{4}$ turn is called a(n) ■ angle.

12. An angle with measure $\frac{1}{2}$ turn is called a(n) ■ angle.

13. The sides of a right angle are ■ to each other.

14. The sides of a straight angle are ■.

Classify each angle as acute, right, obtuse, or straight.

15.   16.    17.    18.

19. $m\angle 1 = 138°$ **20.** $m\angle X = 90°$ **21.** $m\angle 2 = 68°$ **22.** $m\angle ABX = 27°$

23. has a measure between 0° and 90° **24.** has a measure between 90° and 180°

Draw and label a figure to fit each description.

25. an obtuse angle named $\angle RST$ **26.** an acute angle named $\angle RDX$

27. a straight angle named $\angle BHT$ **28.** a right angle named $\angle XYZ$

Select the better estimate of the measure for each angle .

29. 60°, 80° **30.** 30°, 45° **31.** 45°, 60° **32.** 120°, 150°

Find the measure of each angle in both amount of turn and in degrees. Then classify the angle.

33. **34.** **35.** **36.**

37. A circle is divided into 24 equal-size pieces. A central angle of this circle cuts off 6 of the 24 pieces. What is the measure of the angle in amount of turn? in degrees?

APPLICATIONS

38. Nature An octopus has eight legs. If the measure of the angle formed by each consecutive pair of legs is the same, what is the angle measure?

39. Time Through how many degrees does the minute hand of a circular clock pass from 12:10 P.M. to 12:39 P.M.?

Computer

Many watches have no numbers printed on the face. However, the time can be told by examining the angles formed by the hands. Write three LOGO procedures that show the position of watch hands at three different times. Use all the facts you have learned about converting amount of turn to degrees.

2.3 Using A Protractor

Objective: To use a protractor to measure and draw angles.

You use a *protractor* to measure or draw angles in degrees. Many protractors have more than one scale of numbers. Estimating whether an angle is acute, obtuse, right, or straight before measuring will help you choose the correct scale.

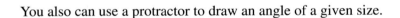 **EXPLORING**

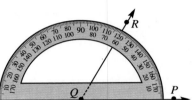

1. Draw an angle. Label it as ∠PQR.
2. Place the center point of the protractor on Q.
3. Position the protractor so that one side of the angle passes through zero on the protractor scale.
4. Estimate. Is ∠PQR acute, obtuse, right, or straight? What is the measure of the angle?
5. Measure the angle three more times, each time moving the protractor so that $\overrightarrow{QP}$ passes through a point on the scale other than zero. Is the angle measure always the same?

You also can use a protractor to draw an angle of a given size.

Example: Use a protractor to draw an angle with measure 67°.

Solution:
1. Draw a ray like the one shown.
2. Place the center point of the protractor on the endpoint (X) of the ray. Align the ray with the base line of the protractor.
3. Locate 67° on the protractor scale. Make a dot at that point and label it as Z.
4. Remove the protractor and draw $\overrightarrow{XZ}$. $m\angle YXZ = 67°$.

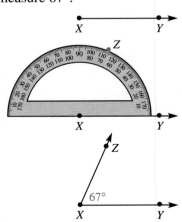

Class Exercises

Find the measure of each angle.

1. ∠AOB   2. ∠AOC

3. ∠AOD   4. ∠AOE

5. ∠BOC   6. ∠BOD

7. ∠BOE   8. ∠BOF

9. ∠COD   10. ∠COE

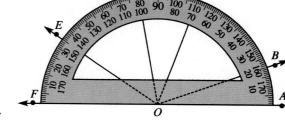

Use a protractor and straightedge to draw angles with the given measures.

11. 45°   12. 75°   13. 100°   14. 125°   15. 131°

Exercises

For each angle shown, (a) estimate the measure, (b) measure to the nearest degree, and (c) compute the difference between your estimate and measurement. Record your results in a table like the one shown.

Angle	Estimate	Measurement	Difference
1	■	■	■
2	■	■	■

1. 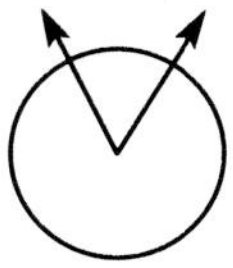   2. 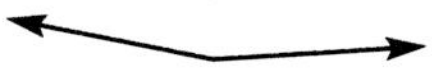   3. 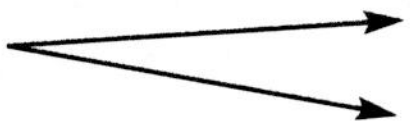    4.

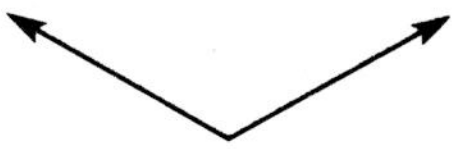

5. 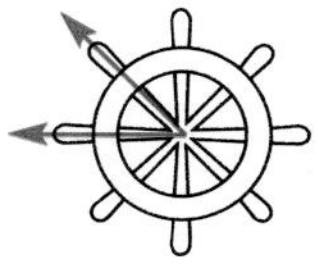   6.    7. 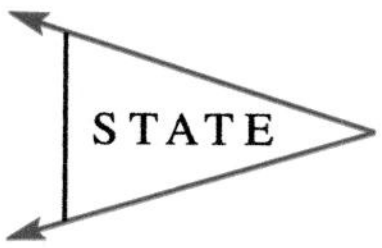   8.

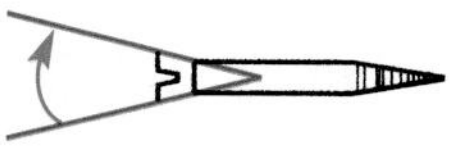

STATE

Find the measure of each angle.

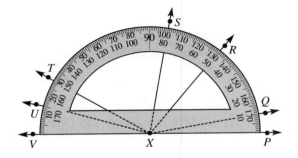

9. ∠PXQ   10. ∠PXR   11. ∠PXS

12. ∠PXT   13. ∠PXU   14. ∠PXV

15. ∠QXR   16. ∠QXS   17. ∠QXT

18. ∠QXV   19. ∠RXT   20. ∠RXU

Use a protractor and straightedge to draw angles with the given measures.

21. 78°   22. 154°   23. 23°   24. 35°   25. 127°   26. 155°

27. **a.** Draw an angle that you think has each indicated measure.

50°   115°   140°   70°

b. Check your estimates by measuring each angle. If you came within 10°, you did a very good job!

APPLICATIONS

Measurement The United States Army uses a unit of angle measure called a *mil*. A ***mil*** is defined as $\frac{1}{6400}$ of a circle. The protractor shown is marked in mils.

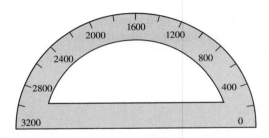

Find the measure in mils of each angle.

28. 90° angle   29. 135° angle

30. 45° angle   31. 180° angle

Seeing in Geometry

Estimate the angle measure of each section. Use your estimate to decide how many sections of the size shown can be cut from each circle.

a.

b.

c.

d.

2.4 Adding and Subtracting Angle Measures

Objective: To add and subtract angle measures.

Adjacent angles are two angles in the same plane that have the same vertex and a common side but do not have any interior points in common.

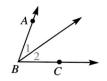

Example 1: $\angle 1$ and $\angle 2$ are adjacent angles.

$\angle 1$ and $\angle ABC$ are *not* adjacent angles.

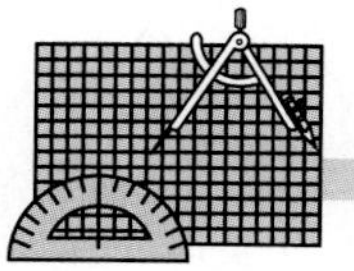

EXPLORING

1. Use a protractor to measure $\angle 3$, $\angle 4$, and $\angle PQR$.

2. What relationship do you notice among the measures of the three angles?

3. Draw three more pairs of adjacent angles. Measure the angles in each pair. Is the relationship the same as that in Step 2?

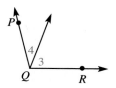

This **EXPLORING** activity leads to the following postulate.

POSTULATE 5 (Angle Addition Postulate): If point W lies in the interior of $\angle XYZ$, then $m\angle XYW + m\angle WYZ = m\angle XYZ$.

Example 2

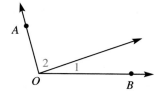   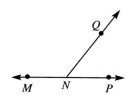

a. $m\angle 1 = 20°$
$m\angle 2 = 85°$
$m\angle AOB = \blacksquare$

b. $m\angle MNP = 180°$
$m\angle MNQ = 127°$
$m\angle QNP = \blacksquare$

Solution   a. $m\angle 1 + m\angle 2 = m\angle AOB$
$20° + 85° = m\angle AOB$
$105° = m\angle AOB$

b. $180° - m\angle MNQ = m\angle QNP$
$180° - 127° = m\angle QNP$
$53° = m\angle QNP$

Thinking Critically

1. $m\angle 1 + m\angle 2 = 180°$. Must $\angle 1$ be a right angle? Must $\angle 2$ be a right angle? Explain.

2. Let x, y, and z represent the measures of the angles in the figure. Which equation is true?

   a. $x + y = z$ **b.** $x + y + z = 180°$

   c. $x + y + z = m\angle JKL$ **d.** $x + y = y + z$

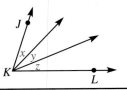

Class Exercises

Find the measure of each angle.

1. $m\angle BAC = 40°$
   $m\angle CAF = 55°$
   $m\angle BAF = \blacksquare$

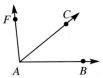

2. $m\angle YWZ = 120°$
   $m\angle YWX = 58°$
   $m\angle ZWX = \blacksquare$

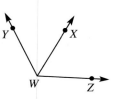

3. $m\angle MEO = 35°$
   $m\angle GEO = \blacksquare$

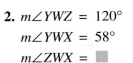

4. $m\angle RPS = 100°$
   $m\angle MPS = 60°$
   $m\angle RPM = \blacksquare$

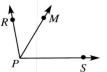

5. **a.** $m\angle BOC = \blacksquare$ **b.** $m\angle AOC = \blacksquare$
   c. $m\angle EOC = \blacksquare$ **d.** $m\angle DOC = \blacksquare$
   e. $m\angle DOE = \blacksquare$ **f.** $m\angle AOD = \blacksquare$

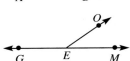

Exercises

Find the measure of each angle.

1. $m\angle RXS = 70°$
   $m\angle SXT = 50°$
   $m\angle RXT = \blacksquare$

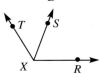

2. $m\angle 1 = 30°$
   $m\angle 2 = 45°$
   $m\angle NOT = \blacksquare$

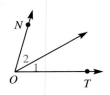

3. $m\angle XYZ = 155°$
   $m\angle XYW = 70°$
   $m\angle WYZ = \blacksquare$

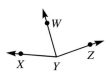

4. $m\angle 4 = 46°$
   $m\angle ABD = 98°$
   $m\angle 3 = \blacksquare$

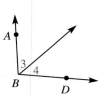

5. a. $m\angle YZX + m\angle YZW =$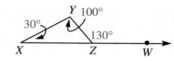

   b. $m\angle YZX =$ ▨

   c. $m\angle YXZ + m\angle YZX + m\angle XYZ =$ ▨

6. $m\angle AOB + m\angle BOC = m$ ▨

7. $m\angle AOB + m\angle BOD = m$ ▨

8. $m\angle BOE - m\angle DOE = m$ ▨

9. $m\angle AOD - m\angle BOD = m$ ▨

10. Point X is an interior point of $\angle ABC$. If $m\angle ABX = 30°$ and $m\angle ABC = 110°$, what is $m\angle XBC$?

APPLICATIONS

Find the measure of each angle.

11. Algebra

$m\angle ABC = 165°$

$m\angle ABE =$ ▨

$m\angle EBC =$ ▨

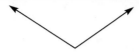

12. Algebra

$m\angle RBW + m\angle BWR + m\angle WRB = 180°$

$m\angle WRB = 135°$

$m\angle RBW =$ ▨

$m\angle RWB =$ ▨

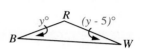

Test Yourself

Name the following for ⊙ C.

1. all radii shown   **2.** all chords shown

3. all diameters shown   **4.** If $XY = 10$ cm, then $CZ =$ ▨

Complete the following.

5. $1° =$ ▨ turn   **6.** $60° =$ ▨ turn   **7.** $150° =$ ▨ turn

8. 1 turn = ▨   **9.** $\dfrac{1}{10}$ turn = ▨   **10.** $\dfrac{3}{8}$ turn = ▨

Estimate and then measure each angle to the nearest degree.

11. 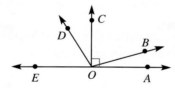   **12.**

Draw a figure to fit each description.

13. a right angle   **14.** a pair of adjacent angles   **15.** an angle with measure $68°$

16. an obtuse angle   **17.** an acute angle   **18.** an angle with measure $135°$

2.5 Congruent Angles and Angle Bisectors

Objective: To construct congruent angles and angle bisectors.

Angles with equal measures are called ***congruent angles***. Congruent angles are sometimes marked alike as shown in Example 1.

Example 1: **a.** 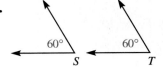 **b.**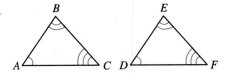

    a. $\angle S \cong \angle T$, since $m\angle S = m\angle T$.   **b.** $\angle A \cong \angle D$, $\angle B \cong \angle E$, $\angle C \cong \angle F$

You can trace over an angle to make a congruent copy. You also can construct a congruent angle with a compass and straightedge.

CONSTRUCTION 5

Construct an angle congruent to $\angle B$.

1. Draw an angle like the one shown.

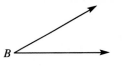

2. Draw a ray with endpoint P.

3. Using the same compass setting, draw arcs from B and P. Label the points of intersection as shown.

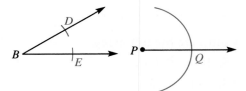

4. Adjust your compass to DE. Use this setting to draw an arc with center at Q that intersects the arc drawn in Step 3. Label the point of intersection as R. Draw $\overrightarrow{PR}$.

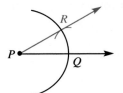

5. $m\angle P = m\angle B$ so $\angle P \cong \angle B$.

The **bisector** of an angle is the ray that divides the angle into two congruent angles.

Example 2:

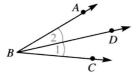

Since $\angle 1 \cong \angle 2$, $\overrightarrow{BD}$ is the bisector of $\angle ABC$.

═══════════════ **CONSTRUCTION 6** ═══════════════

Construct the bisector of an angle.

  1. Draw an angle like the one shown.

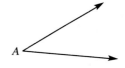

  2. With the compass point on A, draw an arc that intersects the sides of $\angle A$. Label the points of intersection as B and C.

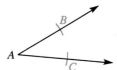

  3. Draw arcs of equal radius from B and C. Label the point of intersection of the arcs as D.

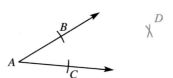

  4. Draw $\overrightarrow{AD}$. $\angle DAB \cong \angle DAC$. $\overrightarrow{AD}$ is the bisector of $\angle BAC$.

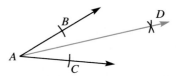

Fold your angle from Construction 6 so that $\overrightarrow{AB}$ coincides with $\overrightarrow{AC}$. If you did the construction carefully, the fold should coincide with $\overrightarrow{AD}$.

1. Draw an angle like the one shown. Construct $\overrightarrow{NP}$, the bisector of $\angle MNO$.

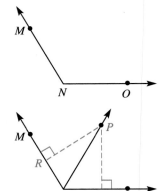

2. Construct the perpendicular from P to $\overrightarrow{NO}$. Label the intersection as Q. Construct the perpendicular from P to $\overrightarrow{NM}$. Label the point of intersection as R.

3. Measure $\overline{PQ}$ and $\overline{PR}$. What do you discover?

4. Label other points on $\overrightarrow{NP}$ and repeat Steps 2 and 3. Is your conclusion the same?

A point that is the same perpendicular distance from two lines or rays is **equidistant** from the lines or rays.

The **EXPLORING** activity leads to the following theorem.

THEOREM 2.1 (The Angle Bisector Theorem): Each point on the bisector of an angle is equidistant from the sides of the angle.

Thinking Critically

1. A segment has an unlimited number of bisectors. How many bisectors does an acute angle have? an obtuse angle? a right angle?

2. Can an acute angle have a perpendicular bisector? Can an obtuse angle have a perpendicular bisector? Explain.

Class Exercises

Draw an angle like the one shown.

1. Use Construction 5 to construct an angle congruent to $\angle B$.

2. Use Construction 6 to construct the bisector of $\angle B$.

3. Construct an angle whose measure is $2 \cdot m\angle B$.

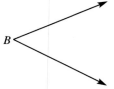

Exercises

True or false?

1. The bisector of an angle always divides the angle into two congruent angles.

2. The bisector of a right angle forms two acute angles.

3. Two acute angles are always congruent.

4. Two adjacent angles are always congruent.

5. The bisector of an angle forms a pair of adjacent congruent angles.

6. The bisector of an angle is a ray that is a common side of the angles formed.

Draw figures similar to, but larger than, the ones shown. Construct an angle congruent to each angle and construct the bisector of each angle.

7. 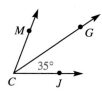　　8. 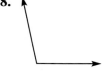　　9.

$\overrightarrow{AC}$ **is the bisector of** $\angle YAX$.

10. If $CY = 5$, then $CX = $ ▪.

11. If $BW = 4$, then $BZ = $ ▪.

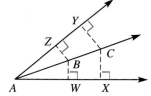

12. Construct an angle whose measure is $3 \cdot m\angle 3$.

13. Construct an angle whose measure is $m\angle 2 + m\angle 3$.

14. Construct an angle whose measure is $m\angle 2 - m\angle 3$.

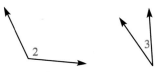

15. $\overrightarrow{CG}$ bisects $\angle MCJ$. Find $m\angle MCG$.

16. $\overrightarrow{BD}$ bisects $\angle ABC$. $m\angle DBE = 19°$. Find $m\angle EBC$.

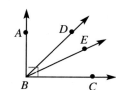

17. $m\angle CLM = 120°$. $\overrightarrow{LK}$ bisects $\angle CLG$. Find $m\angle CLK$ and $m\angle KLG$.

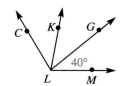

Draw a figure similar to, but larger than, the one shown. Construct the bisector of ∠ABD and the bisector of ∠DBC.

18.
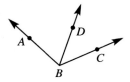

19.
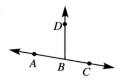

20.
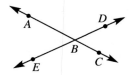

21. Draw a triangle similar to, but larger than, the one shown. Construct triangles by constructing copies of the segments and angles given.

a. $\overline{DF}$, $\overline{EF}$, and $\overline{DE}$

b. $\angle D$, $\overline{DF}$, and $\overline{DE}$

c. $\overline{DE}$, $\angle D$, and $\angle E$

d. What appears to be true about the triangles you constructed?

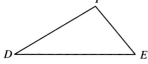

APPLICATIONS

22. Algebra $\overrightarrow{YW}$ is the bisector of $\angle XYZ$. $m\angle XYZ = (3x - 5)°$ and $m\angle XYW = x°$. Find x, $m\angle XYZ$, and $m\angle XYW$.

23. Reasoning A man walks along a path that is the bisector of an angle whose sides are Brook Street and Apple Street. When the man is 0.5 mi from Apple Street, how far is he from Brook Street?

Everyday Geometry

A helicopter pilot is flying north in search of a car somewhere below. When the road forks, the pilot must decide on a course that will allow her to search both branches of the road equally well. What course should she follow north of the fork? Explain.

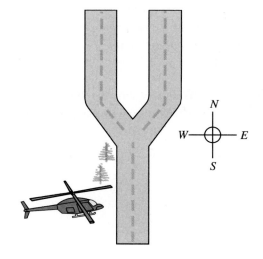

2.6 Pairs of Angles

Objective: To recognize the relationships among vertical angles, complementary angles, and supplementary angles.

Vertical angles are two angles whose sides form two pairs of opposite rays. When two lines intersect, they form two pairs of vertical angles.

Example 1: Name two pairs of vertical angles.

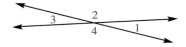

Solution: ∠1 and ∠3 are vertical angles.

∠2 and ∠4 are vertical angles.

Two angles are *complementary angles* if and only if the sum of their measures is 90°. Each angle is a *complement* of the other.

Example 2: Name the following.

    a. all pairs of complementary angles shown

    b. a complement of ∠A

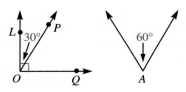

Solution: **a.** ∠*LOP* and ∠*POQ* are complementary angles.

          ∠*LOP* and ∠*A* are complementary angles.

    b. ∠*LOP*

Two angles are *supplementary angles* if and only if the sum of their measures is 180°. Each angle is a *supplement* of the other.

Example 3: Name the following.

    a. all pairs of supplementary angles shown

    b. a supplement of ∠*B*

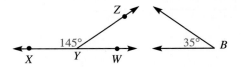

Solution: **a.** ∠*XYZ* and ∠*ZYW* are supplementary angles.

          ∠*XYZ* and ∠*B* are supplementary angles.

    b. ∠*XYZ*

EXPLORING

1. Use the *Geometric Supposer: Quadrilaterals* disk to create any trapezoid.

2. Draw $\overline{AC}$ and $\overline{BD}$. Label their intersection.

3. Use measurement as necessary to check whether the following statements are true or false.

   a. $m\angle BEC + m\angle CED = 180°$

   b. $m\angle BEC = m\angle AED$

   c. $m\angle CED + m\angle BEA = 180°$

   d. $m\angle CED = m\angle BEA$

   e. $m\angle CED + m\angle BEA = m\angle AED + m\angle BEC$

4. Angles *CED* and *BEA* and angles *BEC* and *AED* are pairs of vertical angles. How are the measures of vertical angles related?

Since angles that have the same measure are congruent, the **EXPLORING** activity demonstrates the following theorem.

THEOREM 2.2 (Vertical Angles Theorem): Vertical angles are congruent.

Thinking Critically

1. Suppose $\angle D$ and $\angle E$ are complementary angles. What kind of angles must they be?

2. Can two acute angles be complementary? supplementary? Explain.

Class Exercises

Find the measure of a complement of $\angle 1$.

  1. $m\angle 1 = 45°$ **2.** $m\angle 1 = 75°$ **3.** $m\angle 1 = 89°$ **4.** $m\angle 1 = 23°$ **5.** $m\angle 1 = 9°$

Find the measure of a supplement of $\angle 2$.

  6. $m\angle 2 = 90°$ **7.** $m\angle 2 = 125°$ **8.** $m\angle 2 = 59°$ **9.** $m\angle 2 = 117°$ **10.** $m\angle 2 = 179°$

11. Which two angles are supplements of ∠*DOC*?

12. Which two angles are complementary?

13. Name two pairs of vertical angles.

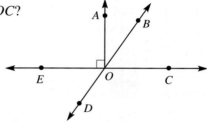

Exercises

Find the measure of a complement and a supplement of each angle, if possible.

1. $m\angle ABC = 35°$ **2.** $m\angle DEF = 55°$ **3.** $m\angle QRS = 49°$ **4.** $m\angle MNO = 63°$

5. $m\angle BHT = 119°$ **6.** $m\angle PBD = 157°$ **7.** $m\angle XYZ = 72°$ **8.** $m\angle 2 = 81°$

True or false?

9. Complementary angles must be adjacent.

10. Supplementary angles must be adjacent.

11. Vertical angles must have the same measure.

12. Any two right angles are supplementary.

13. Two obtuse angles can be supplementary.

14. If two angles are complementary, both angles are acute.

15. Two acute angles are always complementary.

16. An acute angle and an obtuse angle are always supplementary.

Match each term with its definition.

17. acute angle   **a.** an angle with measure 90°

18. obtuse angle   **b.** two angles whose measures have a sum of 90°

19. right angle   **c.** an angle with measure between 90° and 180°

20. straight angle   **d.** an angle with its vertex at the center of a circle

21. complementary angles   **e.** two angles whose measures have a sum of 180°

22. supplementary angles   **f.** an angle with measure 180°

23. central angle   **g.** two angles whose sides form two pairs of opposite rays

24. vertical angles   **h.** an angle with measure less than 90°

Are the angles complementary, supplementary, or neither?

25. 75°, 20°   **26.** 111°, 89°   **27.** 130°, 50°   **28.** 13°, 77°

29. 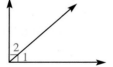   **30.**   **31.**    **32.**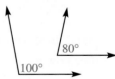

33. Name each of the following.

  a. one pair of supplementary angles

  b. two angles adjacent to ∠2

  c. one pair of vertical angles

  d. two angles complementary to ∠5

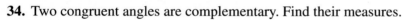

34. Two congruent angles are complementary. Find their measures.

35. Two congruent angles are supplementary. Find their measures.

Find the indicated measures.

36. a. $m\angle 1 = $ ■
     37. a. $m\angle 4 = $ ■
     38. a. $m\angle AXB = $ ■

  b. $m\angle 2 = $ ■
          b. $m\angle 5 = $ ■
          b. $m\angle CXN = $ ■

  c. $m\angle 3 = $ ■
          c. $m\angle 6 = $ ■
          c. $m\angle AXC = $ ■

  d. $m\angle 1 + m\angle 2 = $ ■
               d. $m\angle BXN = $ ■

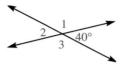
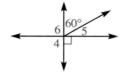
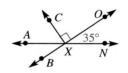

39. Twice the measure of ∠A is equal to the measure of a complement of ∠A. Find $m\angle A$.

40. A supplement of ∠D is three times as great as a complement of ∠D. Find $m\angle D$.

41. Tangrams Use a protractor as necessary to measure angles. Name the following.

  a. two pairs of vertical angles

  b. two pairs of complementary angles

  c. two pairs of supplementary angles

  d. two angles congruent to ∠15

  e. Are ∠1 and ∠2 adjacent angles? Why or why not?

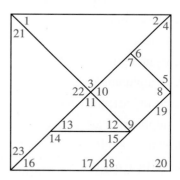

Computer

Write directions for use with the *Geometric Supposer: Quadrilaterals* disk that will tell a user how to create vertical angles in any quadrilateral.

Thinking About Proof

Analyzing a Figure

When solving geometry problems, you often can draw conclusions by examining figures. You get some information from symbols that indicate congruent segments, congruent angles, right angles, and perpendicular and parallel lines. At other times, you must rely on your knowledge of definitions, postulates, and theorems to analyze a figure.

Assume by Appearance	Do *Not* Assume by Appearance
All parts of a figure are in the same plane.	Angles are right angles.
Points are collinear.	Lines are perpendicular.
Angles are straight angles.	Segments or angles are congruent.

Example: What are some assumptions you can make from the figure?

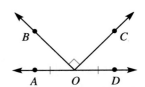

Solution:
1. Points *A*, *E*, and *B* are collinear.
2. Points *A*, *E*, and *C* are noncollinear.
3. ∠*AEB* and ∠*CED* are straight angles.
4. ∠*AED* and ∠*CEB* are vertical angles.
5. ∠*AED* ≅ ∠*CEB* and ∠*AEC* ≅ ∠*DEB*.
6. Four pairs of supplementary angles are shown.

Exercises

1. List some assumptions you can make from the figure. Support each conclusion with a reason.

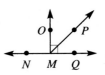

2. List some assumptions you *cannot* make from the figure. What additional information would you need to make each of these assumptions?

3. $\overrightarrow{MP}$ bisects ∠*OMQ*. Find the measure of each angle shown. Justify your answers.

4. $AB = 5x - 8$. $BC = 2x + 13$. Find *AB*, *BC*, and *AC*.

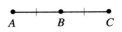

Problem Solving Strategy: Working Backwards

Objective: To solve problems by working backwards.

Sometimes you can work backwards to solve a problem.

Example: Mary designed a pattern for a stained-glass window. She drew an angle for the corner of one piece. The angle seemed too small, so Mary doubled it. Now it was too large, so she trimmed 25° from the angle. Mary was satisfied with the 55° angle that remained. Find the measure of the angle with which she started.

Solution: Your plan should be to work backwards since you know the end but want to find the beginning.

- The measure of the final angle was 55°.

- Mary *subtracted* 25° from the measure of her second angle, leaving a 55° angle. The inverse of subtraction is addition, so the measure of the second must be:

$$55° + 25° = 80°$$

- Mary *doubled* the measure of her original angle, resulting in an 80° angle. The inverse of multiplication is division, so the measure of the original angle must be:

$$80° \div 2 = 40°$$

$$\text{Check:} \quad 2 \cdot 40° = 80°$$
$$80° - 25° = 55°$$

Class Exercise

1. Judith drew a segment, $\overline{AB}$. Below it she drew a segment with length $\frac{1}{3} \cdot AB$. She labeled the second segment as $\overline{CD}$. Below $\overline{CD}$, Judith drew a 17 cm segment that was 8 cm longer than $\overline{CD}$. She labeled the third segment as $\overline{EF}$. Follow these steps to find the length of $\overline{AB}$.

a. How long was $\overline{EF}$?

b. $\overline{EF}$ was 8 cm longer than $\overline{CD}$. How long was $\overline{CD}$?

c. The length of $\overline{CD}$ was $\frac{1}{3} \cdot AB$. How long was $\overline{AB}$?

d. Check your answer. Does it satisfy the conditions of the problem? Does it seem reasonable?

Exercises

1. $\angle A$, $\angle B$, $\angle C$, and $\angle D$ are related in the following way.

   $m\angle B = \frac{1}{2} \cdot m\angle A$.

   $m\angle C = m\angle B + 40°$; $m\angle D = \frac{1}{3} \cdot m\angle C$.

   If $m\angle D = 35°$, what is the measure of $\angle A$?

2. Larry, Lionel, and Lester painted a fence. Larry painted the first $\frac{1}{3}$ of the fence. Lionel painted $\frac{3}{5}$ of what remained. Lester painted the final 40 ft. How long is the fence?

3. A snail is crawling along a meter stick. At a certain point the snail stops. The snail needs to crawl three times as far as it has already crawled in order to be 4 cm from the end. How far has the snail already crawled?

4. An antique dealer visited three shops. She spent $25 at the first shop. At the second shop, she spent half of her remaining money. At the third shop, she spent one third of her remaining money and had $60 left. How much money did the dealer have originally?

5. Find the pattern and draw the first two figures.

_____ , _____ , , , , ,

Vocabulary and Symbols

You should be able to write a brief statement, draw a picture, or give an example to illustrate the meaning of each term or symbol.

Vocabulary

acute angle (p. 43)
adjacent angles (p. 49)
amount of turn (p. 40)
bisector of an angle (p. 53)
center of a circle (p. 39)
central angle (p. 40)
chord (p. 39)
circle (p. 39)
complement of an angle (p. 57)
complementary angles (p. 57)
congruent angles (p. 52)
degree (p. 43)
diameter of a circle (p. 39)
equidistant (p. 54)

measure of an angle (pp. 40,43)
obtuse angle (p. 43)
protractor (p. 46)
radius of a circle (p. 39)
right angle (p. 43)
straight angle (p. 43)
supplement of an angle (p. 57)
supplementary angles (p. 57)
vertical angles (p. 57)

Symbols

$\odot$ (circle) (p. 39)
$^{\circ}$ (degree) (p. 43)
$m\angle ABC$ (measure of $\angle ABC$) (p. 43)

Summary

The following list indicates the major skills, facts, and results you should have mastered in this chapter.

2.1 Use amount of turn to measure angles and use the relationships among parts of circles. (pp. 39–42)

2.2 Measure angles in degrees and classify acute, right, obtuse, and straight angles. (pp. 43–45)

2.3 Use a protractor to measure and draw angles. (pp. 46–48)

2.4 Add and subtract angle measures. (pp. 49–51)

2.5 Construct congruent angles and angle bisectors. (pp. 52–56)

2.6 Recognize the relationships among vertical angles, complementary angles, and supplementary angles. (pp. 57–60)

2.7 Solve problems by working backwards. (pp. 62–63)

Exercises

Name the following that are shown for ⊙C.

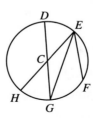

1. all radii shown
2. all diameters shown
3. two angles supplementary to ∠GCH
4. two pairs of vertical angles
5. the diameter of ⊙C if GC = 5 cm

Find the measure of each angle in amount of turn and in degrees.

6.   7.   8.

Use a protractor and straightedge to draw each angle. Classify the angle as acute, right, obtuse, or straight.

9. an angle with measure 116°   10. an angle with measure 44°

Find the measure of a complement and a supplement of each angle, if possible.

11. $m\angle A = 61°$ 12. $m\angle B = 114°$ 13. $m\angle C = 32°$ 14. $m\angle D = 90°$

Find the measure of each angle.

15. $m\angle WZY = 84°$
$\overrightarrow{ZX}$ bisects $\angle WZY$
$m\angle XZY = $ ▨

16. $m\angle 1 = 31°$
$m\angle 2 = 42°$
$m\angle ABC = $ ▨

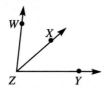

17. Draw an acute angle with vertex A. Construct an angle with measure $2 \cdot m\angle A$ and an angle with measure $\frac{1}{2} \cdot m\angle A$.

PROBLEM SOLVING

18. June has a 9:10 A.M. meeting. It takes her 8 min to walk to the bus stop. June's bus ride is 17 min long. June wants to arrive at her office 10 min before her meeting. If a bus leaves as soon as June arrives at the bus stop, what is the latest that she can leave her house?

Chapter 2 Test

True or false? Refer to ⊙P.

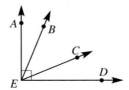

1. $\overline{PS} \cong \overline{PR}$

2. $TR = 4$ cm

3. $\overline{SR}$ is a chord of ⊙P.

4. $\angle TRS$ is a central angle of ⊙P.

5. $\angle RPQ$ and $\angle SPQ$ are supplementary angles.

6. Name all radii, chords, and diameters of ⊙P shown.

Change each measure from degrees to amount of turn or from amount of turn to degrees.

7. $\frac{1}{12}$ turn   8. $144°$ 9. $108°$ 10. $\frac{9}{20}$ turn

Measure each angle to the nearest degree. Tell whether the angle is acute, right, obtuse, or straight.

11.   12.   13.

14. Draw an angle like the one in Exercise 12. Construct $\angle A$ congruent to the one you have drawn. Then construct the bisector of $\angle A$.

15. If $m\angle AEB = 22°$ and $m\angle BEC = 45°$, find $m\angle AEC$.

16. If $m\angle DEB = 63°$ and $m\angle BEC = 45°$, find $m\angle CED$.

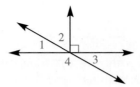

Complete.

17. $\angle 3$ and ▦ are supplementary angles.

18. If $m\angle 1 = 35°$, then $m\angle 3 = $ ▦.

19. If $m\angle 1 = 35°$, then $m\angle 2 = $ ▦.

20. $\angle 1$ and ▦ are vertical angles.

PROBLEM SOLVING

21. You are planning a 30-min radio program. You must include 3 min for public-service announcements and 9.5 min for commercials. The playing time for each music recording is about 2.5 min. About how many recordings should you plan to include?

Chapter 3

Polygons and Polyhedrons

This building rises 18 stories yet requires no internal support from pillars or beams. It is a geodesic dome — a single, continuous, self-supporting wall built up out of triangles arranged in a special pattern.

Focus on Skills

GEOMETRY

1. Give two other names for ∠1.
2. Name all segments shown.
3. Name two pairs of supplementary angles.
4. Name two pairs of vertical angles.
5. Name two angles adjacent to ∠CBE.

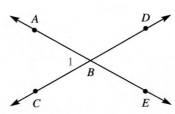

Tell whether each angle appears to be acute, right, or obtuse.

6.   7.   8.    9.

Complete. Use one of the following words: acute, right, obtuse.

10. A supplement of a right angle is a(n) ▇ angle.
11. A supplement of an acute angle is a(n) ▇ angle.
12. A supplement of an obtuse angle is a(n) ▇ angle.

What relationships are indicated by the markings in each figure?

13. 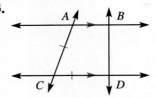   14. 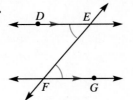   15.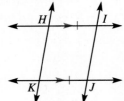

ALGEBRA

Write a mathematical expression to represent each statement. Use the variables indicated.

16. One number (c) is five more than another number (d).
17. The sum of two numbers (x and y) is 12.
18. One number (w) is three more than twice another number (z).

3.1 Polygons

Objective: To recognize convex and concave polygons.

The word *polygon* comes from the Greek words meaning "many angled."

EXPLORING

1. These are polygons.   These are *not* polygons.

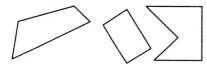   

   How are the polygons alike? How are they different from the figures that are not polygons?

2. These are *convex* polygons.   These are *concave* polygons.

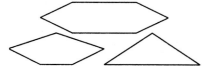   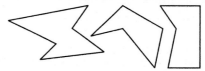

   How are the convex polygons alike? How are they different from the concave polygons?

3. If *A* and *B* are any two points inside a convex polygon, can $\overline{AB}$ intersect the polygon? If *A* and *B* are any two points in the interior of a concave polygon, can $\overline{AB}$ intersect the polygon?

4. Use a geoboard or dot paper to show a 5-sided concave polygon and a 10-sided convex polygon.

The **EXPLORING** activity leads to the following definitions.

A ***polygon*** is formed by three or more coplanar segments called ***sides*** that intersect only at their endpoints. The intersection of two sides is called a ***vertex***. No intersecting sides are collinear.

A polygon is ***convex*** when no segment connecting two vertices contains points outside the polygon. A polygon that is not convex is ***concave***. In this textbook, polygon means "convex polygon" unless otherwise stated.

Polygons are named for the number of sides.

Polygon	Number of Sides	Polygon	Number of Sides
Triangle	3	Heptagon	7
Quadrilateral	4	Octagon	8
Pentagon	5	Nonagon	9
Hexagon	6	Decagon	10

When a polygon has more than ten sides, it may be referred to as an 11-gon, 12-gon, and so on. When the number of sides of a polygon is unknown, the polygon is called an *n*-gon.

We use the following terms to describe parts of polygons.
Consecutive angles are two angles with one side in common.
Adjacent (consecutive) sides are sides that intersect.
Adjacent (consecutive) vertices are vertices of consecutive angles.
A ***diagonal*** is a segment joining a pair of nonconsecutive vertices.

Example 1: **Name the following.**

    a. five vertices and five sides
    b. all angles consecutive to $\angle A$
    c. all sides adjacent to $\overline{ED}$
    d. all vertices adjacent to D

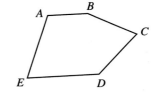

Solution: **a.** $A, B, C, D, E;\ \overline{AB}, \overline{BC}, \overline{CD}, \overline{DE}, \overline{EA}$
    b. $\angle E$ and $\angle B$
    c. $\overline{EA}$ and $\overline{DC}$
    d. E and C

When you name a polygon, you list its vertices in any *consecutive* order.

Example 2: **Look at the polygon at the right.**

    a. Name the hexagon in two ways, each starting with C.
    b. Name all diagonals with endpoint C.

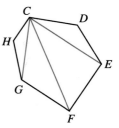

Solution: **a.** *CDEFGH*; *CHGFED*
    b. $\overline{CG}, \overline{CF},$ and $\overline{CE}$

1. Is it possible to draw a concave 3-sided polygon? Explain.

2. Does a triangle have any diagonals? Explain.

Class Exercises

Classify each figure as a convex polygon, a concave polygon, or not a polygon.

1.
2.
3.
4.

Name the following for the polygon.

5. all vertices

6. all sides

7. a pair of consecutive sides

8. a pair of nonconsecutive sides

9. a pair of consecutive vertices

10. a pair of consecutive angles

11. Name the polygon by the number of sides.

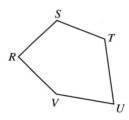

Exercises

Classify each figure as a convex polygon, a concave polygon, or not a polygon.

1.
2.
3.
4.

Name each polygon by the number of sides.

5.
6.
7.
8.

Name the following for the polygon shown.

9. eight vertices

10. eight sides

11. all sides adjacent to $\overline{PQ}$

12. all vertices adjacent to P

13. all angles consecutive to $\angle P$

14. all sides nonconsecutive to $\overline{PQ}$

15. all diagonals with endpoint Q

16. two different names, each starting with P

Draw and label a figure to fit each description.

17. polygon *FGHI*

18. polygon *ABCDE*

19. a polygon with three vertices

20. a quadrilateral

21. a pentagon

22. a convex quadrilateral

23. an octagon

24. a hexagon

25. a concave quadrilateral

26. a polygon with the least possible number of sides

27. a quadrilateral with no sides equal in length

28. a polygon with all sides equal in length

29. a polygon with $\overline{RS}$ as a side and with two sides adjacent to $\overline{RS}$

30. a polygon with $\overline{AB}$ as a side and exactly three sides nonconsecutive to $\overline{AB}$

APPLICATION

31. **Traffic**   Name the polygon represented by each traffic sign.

a.

b.

c.

Seeing in Geometry

Geometric shapes are found in abundance in nature. The dragonfly's wing is constructed of hundreds of polygons. How many quadrilaterals can you find in the photo? How many pentagons? Do you see any other kinds of polygons? If so, how many?

3.2 Classifying Triangles

Objective: To classify triangles.

You can classify a triangle (△) by the number of its congruent sides or by its angles.

Equilateral
all sides congruent

Isosceles
at least two sides congruent

Scalene
no sides congruent

Equiangular
all angles congruent

Acute
all angles acute

Right
contains one right angle

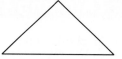

Obtuse
contains one obtuse angle

Example 1: **Judging only by appearance, classify the triangle.**

Solution: isosceles right triangle

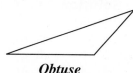

EXPLORING

1. Use the *Geometric Supposer: Triangles* disk to create an acute triangle, *ABC*. Label the midpoint of side $\overline{BC}$ and draw $\overline{AD}$.

2. What kind of triangle is *ABD*? What kind of triangle is *ADC*? Use measurement to determine your answers.

3. Repeat Step 1, creating a right triangle. How are the new triangles *ABD* and *ADC* the same? How are they different?

4. Choose *Your Own* and the *Side, Angle, Side* option to create a triangle with the following measurements: $AB = 5$, $m\angle BAC = 90°$, $AC = 5$. Label the midpoint of side $\overline{BC}$ and draw $\overline{AD}$. What kind of triangle is *ABD*? What kind is *ADC*? How do they compare?

The parts of an isosceles triangle have special names.
The *legs* are the two congruent sides of an isosceles triangle.
The *vertex angle* is formed by the two congruent sides.
The *base angles* are the angles other than the vertex angle.
The *base* is the side *opposite* the vertex angle.

Example 2: **Name the following in the isosceles triangle.**

    a. the legs
    b. the base
    c. the base angles
    d. the vertex angle
    e. the side opposite ∠B

Solution: **a.** $\overline{AC}$ and $\overline{BC}$
          b. $\overline{AB}$
          c. ∠A and ∠B
          d. ∠C
          e. $\overline{AC}$

Thinking Critically

1. Is an equilateral triangle also an isosceles triangle?

2. Is an isosceles triangle also an equilateral triangle?

Class Exercises

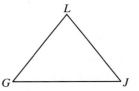

Name each of the following in the isosceles triangle.

1. the legs   **2.** the base

3. the vertex angle   **4.** the base angles

Judging only by appearance, name all triangles shown that fit each description.

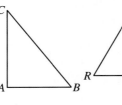
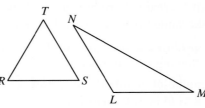
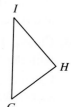

5. right triangle

6. obtuse triangle

7. acute triangle

8. equiangular triangle

9. scalene triangle

10. isosceles triangle

Exercises

Name the legs, base, vertex angle, and base angles in the isosceles triangles.

1.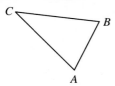

2.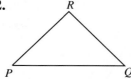

3.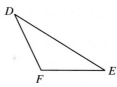

Judging by appearance, name all triangles shown that fit each description.

4. isosceles triangle

5. equilateral triangle

6. scalene triangle

7. equiangular triangle

8. acute triangle

9. right triangle

10. obtuse triangle

11. acute isosceles triangle

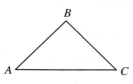

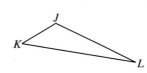

Classify each triangle as equilateral, isosceles, or scalene.

12. $\triangle XYZ$ with $XY = 8$, $YZ = 10$, and $ZX = 8$

13. $\triangle JKL$ with $JK = 10$, $KL = 8$, and $LJ = 5$

14. $\triangle RST$ with $RS = 7$, $ST = 7$, and $TR = 7$

Classify each triangle as acute, right, or obtuse.

15. $\triangle MNO$ with $m\angle M = 132°$, $m\angle N = 35°$, and $m\angle O = 13°$

16. $\triangle GHI$ with $m\angle G = 55°$, $m\angle H = 90°$, and $m\angle I = 35°$

17. $\triangle PQR$ with $m\angle P = 65°$, $m\angle Q = 58°$, and $m\angle R = 57°$

Draw and label a figure to fit each description.

18. an isosceles triangle with legs $\overline{FG}$ and $\overline{FH}$

19. an isosceles triangle with a vertex angle, $\angle A$, and base angles, $\angle R$ and $\angle S$

20. an isosceles triangle with base $\overline{XY}$ and a vertex angle, $\angle B$

21. a triangle with an obtuse angle, $\angle S$, and side $\overline{AM}$ opposite $\angle S$

Draw and label a figure to fit each description.

22. an acute triangle

23. an obtuse triangle

24. an obtuse isosceles triangle

25. an isosceles right triangle

26. an acute isosceles triangle

27. a scalene right triangle

$\overline{AB} \cong \overline{CB}$. **Name the following.**

28. eight triangles

29. three obtuse triangles

30. two isosceles triangles

31. four right triangles

Classify each triangle by its sides and by its angles.

32.

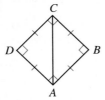

33.

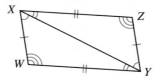

34.

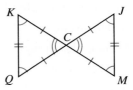

APPLICATIONS

35. Computer   Write a procedure for the *Geometric Supposer* that will tell another student how to construct an equilateral triangle starting with a 5 unit segment.

36. Logic   Can an isosceles triangle be scalene? Why or why not?

Computer

Use the *Geometric Supposer: Triangles* disk to create an obtuse triangle *ABC*. Label the midpoint of side $\overline{BC}$ and draw $\overline{AD}$. Answer the following questions by using measurement.

1. Can triangles *CAD* and *DAB* both be acute? Explain.

2. Can triangles *CAD* and *DAB* both be obtuse? Explain.

3. Can one triangle be right and the other acute? Explain.

4. Can one triangle be right and the other obtuse? Explain.

3.3 Quadrilaterals and Other Polygons

Objective: To recognize types of quadrilaterals and other polygons.

A baseball diamond is an example of a quadrilateral. Which of the quadrilaterals listed do you think best describes a baseball diamond?

Quadrilateral	Definition	Drawing
Trapezoid	a quadrilateral with exactly one pair of parallel sides	
Parallelogram	a quadrilateral with two pairs of opposite parallel sides	
Rhombus	a parallelogram with four sides congruent	
Rectangle	a parallelogram with four right angles	
Square	parallelogram that is both a rectangle and a rhombus	

Example 1: **Judging by appearance, describe the polygon in three different ways.**

Solution: It is a quadrilateral because it is a four-sided polygon.
It is a parallelogram because two pairs of opposite sides are parallel.
It is a rectangle because it is a parallelogram with four right angles.

We usually describe a figure by the name that gives the most information about the figure. Although the figure in Example 1 has three correct names, we usually call it a rectangle.

A *regular polygon* is a convex polygon that is both equilateral and equiangular.

Example 2: **Judging by appearance, state whether each figure is a regular polygon.**

a. 　　b. 　　c. 　　d.

Solution:　**a.** yes　　**b.** no　　**c.** yes　　**d.** no.

EXPLORING

1. Use the *Geometric Supposer: Quadrilaterals* disk to create a quadrilateral. In the *Your Own* menu, use the *Sides and Angles* option. The angles should measure 90°. The sides should measure 6.

2. Which of the following names apply to this figure: trapezoid, parallelogram, rhombus, rectangle, regular polygon?

3. What is the name most commonly given to this quadrilateral?

4. Repeat Step 1 using different values for the lengths of the sides and the angles. Record the measurements you choose. Use those measurements to determine the name of the quadrilateral, if possible.

Thinking Critically

1. What is the most common name for a regular quadrilateral?
2. Draw an equiangular quadrilateral that is not equilateral.
3. Draw an equilateral quadrilateral that is not equiangular.

Class Exercises

Judging by appearance, state all the names that apply to each polygon.

1. 　2. 　3. 　4. 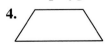　5.

6.   **7.**  **8.**   **9.**   **10.**

Exercises

Judging by appearance, state whether each figure is a regular polygon.

1. 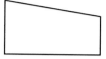 **2.**   **3.**   **4.**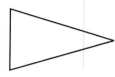

5.   **6.**  **7.**  **8.**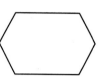

Choose all names that appear to be true for each figure. (Each exercise has more than one answer.)

9. hexagon
polygon
octagon
parallelogram
regular polygon
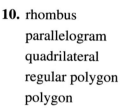

10. rhombus
parallelogram
quadrilateral
regular polygon
polygon

11. pentagon
polygon
hexagon
regular polygon
octagon

12. quadrilateral
parallelogram
square
polygon
rectangle

True or false? Explain your answer.

13. Every parallelogram is a quadrilateral.

14. Every quadrilateral is a parallelogram.

15. Every parallelogram is a rectangle.

16. Every rectangle is a parallelogram.

17. Every rhombus is a square.

18. Every square is a rhombus.

19. Every rhombus is a rectangle.

20. Every rectangle is a rhombus.

21. Every square is a rectangle.

22. Every rectangle is a square.

Draw a figure to fit each description.

23. a parallelogram that is not a rectangle

24. a quadrilateral that is not a parallelogram

25. a hexagon that is not a regular polygon

26. a quadrilateral that is not a rectangle

27. a trapezoid with two sides equal

28. a rectangle that is also a square

29. a quadrilateral that is not a square but has all sides equal

APPLICATION

30. **Computer** Use the *Geometric Supposer: Triangles* disk and the *adjustable element* option to join two congruent equilateral triangles in the manner shown. Describe the procedure. What kind of quadrilateral is formed?

Test Yourself

Name each of the following.

1. all sides adjacent to $\overline{AE}$

2. all vertices adjacent to E

3. all diagonals with endpoint B

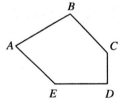

$\triangle RST$ **is isosceles. Name the following.**

4. the legs   **5.** the base angles

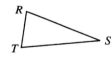

Draw a figure to fit each description.

6. a polygon with four sides

7. a scalene triangle

8. a five-sided polygon

9. a quadrilateral and all its diagonals

10. an obtuse triangle

11. an isosceles right triangle

12. an acute triangle

13. a concave hexagon

True or false?

14. Every rectangle is a square.   **15.** Every rectangle is a parallelogram.

16. State four different names for the figure.

3.4 Polyhedrons and Prisms

Objective: To name polyhedrons and prisms.

Three-dimensional figures are figures that enclose a space. A *polyhedron* is a three-dimensional figure in which each surface is shaped like a polygon.

The parts of a polyhedron have special names.
Face: a flat surface that is shaped like a polygon
Edge: the segment formed by the intersection of two faces
Vertex: a point at which three or more edges intersect

Example 1: **Name the faces, edges, and vertices.**

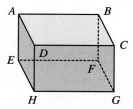

Solution: Faces: quadrilaterals *ABCD, EFGH, ABFE, BCGF, DCGH,* and *ADHE*

Edges: $\overline{AB}$, $\overline{BC}$, $\overline{CD}$, $\overline{DA}$, $\overline{EF}$, $\overline{FG}$, $\overline{GH}$, $\overline{HE}$, $\overline{AE}$, $\overline{BF}$, $\overline{CG}$, and $\overline{DH}$

Vertices: *A, B, C, D, E, F, G,* and *H*

EXPLORING

1. How are the *prisms* different from those figures that are not prisms? How are they alike?

These are prisms.

These are *not* prisms.

2. Write a definition of a prism based on your observations.

The **EXPLORING** activity leads to the following definition.

A *prism* is a polyhedron with two faces (called the *bases*) that have the same size and shape and lie in parallel planes. The *lateral faces* are the faces that are not the bases. Each lateral face is shaped like a parallelogram.

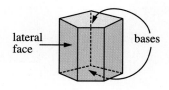

Unless otherwise indicated you may assume that any prism in this book is a *right prism*. All lateral faces of a right prism are rectangles that lie in planes perpendicular to the planes of the bases.

Right prisms are classified by the shapes of their bases. For example, the bases of a rectangular prism are rectangles; the bases of a triangular prism are triangles. What do you think could be another name for a cube?

Example 2: **Classify each prism.**

a. b. c. d.

Solution: **a.** rectangular prism **b.** triangular prism **c.** pentagonal prism **d.** rectangular prism (cube)

Thinking Critically

1. The figure at the right is a *right cylinder.* A cylinder is not a prism. Why not? Draw a cylinder that is not a right cylinder.

2. The figures at the right are *regular pyramids.* A pyramid is not a prism. Why not? Describe the shapes of the bases and lateral faces for each pyramid. Classify the pyramids.

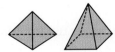

3. The figures at the right are *oblique prisms*. How does an oblique prism differ from a right prism?

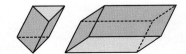

Class Exercises
For each prism (a) state the number of faces, edges, and vertices, (b) describe the shape of the bases, and (c) classify the prism.

1.   2.   3.   4.

Exercises

State whether each figure is a prism.

1. **2.** **3.** **4.** **5.**

For each prism (a) state the number of faces, edges, and vertices, (b) describe the shape of the bases, and (c) classify the prism.

6. **7.** **8.** **9.**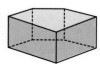

10. **11.** **12.** **13.**

True or false?

14. A cube is a rectangular prism.

15. Every right prism has a rectangular base.

16. A regular pyramid is a prism.

17. A triangular prism has three lateral faces.

18. The lateral faces of all prisms are rectangular regions.

APPLICATION

19. Linguistics A dictionary defines *lateral* as follows: "of, at, from, or toward the side; sideways." How is this definition consistent with the mathematical definition for lateral faces of a prism?

Seeing in Geometry

Leonard Euler, a Swiss mathematician, discovered a relationship among the number of faces, vertices, and edges of any polyhedron. Can you find it? (*Hint:* Count the number of faces, vertices, and edges for each polyhedron. Compare the numbers for the three parts. Write a formula for this relationship.)

A *B*

C *D*

3.5 Drawing Prisms

Objective: To make an accurate drawing of a prism.

Drawing prisms may be difficult because you are trying to show a three-dimensional object with a two-dimensional drawing. To make the drawing more realistic, use dashed lines to represent edges of the prisms that cannot be seen. For example, when you look at a block from the front, you cannot see the back edges.

 EXPLORING

Part A

1. Draw a square. Draw the same-size square over the first square as shown. Connect the vertices as shown.

2. To draw a cube as if you are seeing it from a point *above* and to the right of center, draw the dashed lines as indicated.

3. If you were looking at the cube from a point *below* and to the left of center, you would use different dashed lines.

4. Repeat Step 1, but draw the second square over the lower right section of the first square. Where would you use dashed lines so that your point of view is above and to the left of center?

Part B

1. Draw a triangle. Draw the same size triangle above and to the right of the first triangle. Connect the vertices as shown.

2. Use dashed lines to show that your point of view is above and to the right of center.

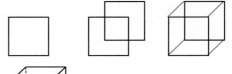

Remember, the segments closest to you are always solid lines. The dashed lines represent segments that are not visible unless the three-dimensional figure is transparent.

You can change the appearance of a triangular prism by changing the shape and position of the triangles.

Example 1: a. **b.** 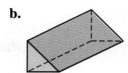 **c.**

You can draw other prisms by using the steps in the EXPLORING activity.

Example 2: **Draw a hexagonal prism viewed from a point above and directly in front of center.**

Solution: **1.** Draw two hexagons, one directly above the other. Connect the vertices.

2. Use dashed lines to indicate that the point of view is above and in front of center.

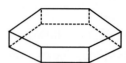

Thinking Critically

1. Look at the cube at the right.

   a. Why is it difficult to determine the point of view?

   b. Copy the cube and give it a specific point of view.

2. Consider the three drawings of a cube.

   a. Describe the point of view for each drawing.

   b. Why are there no dashed lines in cube *A*?

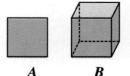

A   *B*   *C*

3. Consider the four drawings of a cube. Describe the different appearances created by using dashed lines for different segments.

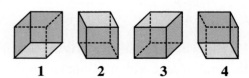

1   2   3   4

Class Exercises

Draw a triangular prism to fit each description.

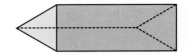

1. the bases are smaller than the bases of the triangular prism shown

2. the bases are larger than the bases of the triangular prism shown

3. standing on one of its bases

4. with right triangular bases

Exercises

Draw a polyhedron to fit each description.

1. triangular prism   2. rectangular prism

Draw a rectangular prism to fit each description.

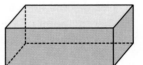

3. like the one shown

4. deeper than the one shown

5. taller (higher) than the one shown

6. like the one shown, but viewed from a point above and to the left of center

Draw a polyhedron like each one shown and classify it.

7.   8.   9.   10.   11.

12.   13.   14.   15.    16.

17. Draw three rectangular prisms, each from a different point of view.

18. Draw a cube with the front face open.

19. Spatial Perception

    a. Draw a cube. Draw lines on the three visible faces to show how the cube could be divided into 27 smaller cubes.

    b. If the cube in part (a) were painted red and then cut into 27 smaller cubes, how many of the 27 smaller cubes would contain no color?

20. Visualization   A *net* is a two-dimensional version of a polyhedron. Which nets can be folded to make a triangular prism?

a.   **b.**   **c.** 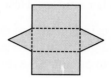   **d.**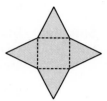

Seeing in Geometry

A *pentomino* is a figure formed by five same-size squares arranged edge to edge. All of the squares must have a complete side in common with any adjacent squares. Two pentominoes are considered the same if one pattern can be flipped or turned to form the other. There are twelve different pentominoes.

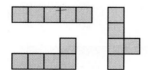   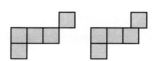   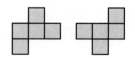

These are pentominoes.   These are not pentominoes.   These are the same.

1. Draw all twelve pentominoes on graph paper.

2. Identify the pentominoes that can be folded into a cube with an open top. (There are eight possibilities.)

3. Look at your answers for Exercise 2. For each pattern, label the square that would be the bottom of the cube as "*B*."

3.6 Problem Solving Strategy: Using a Diagram or Model

Objective: To solve a problem using a diagram or model.

You can solve some problems by making a diagram and using a table to organize information.

Example: A polygon has 267 sides. Diagonals are drawn from one vertex to all the other vertices. How many diagonals are drawn?

Solution: Because it is impractical to draw a polygon with 267 sides, your plan might be to draw several simpler polygons. You can use a table to organize the results and then look for a pattern.

Draw several simple polygons. For each polygon, choose one vertex and draw diagonals to the other vertices. Make a table like the one below.

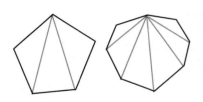

Figure	Number of sides	Number of diagonals from one vertex
Quadrilateral	4	1
Pentagon	5	2
Hexagon	▧	▧
Octagon	8	5
Nonagon	▧	▧

Do you see the pattern? You should have discovered that the number of diagonals (d) from one vertex is always 3 less than the number of sides (n). This can be expressed as $d = n - 3$.

The drawings and the table made this conclusion easier to reach. Applying the conclusion to the 267-sided polygon gives the result you are seeking.

$$d = n - 3$$
$$= 267 - 3$$
$$= 264$$

Class Exercises

Solve the problem using a model or a diagram.

1. A polygon has 113 sides. One vertex is chosen, and all possible diagonals are drawn from the vertex, dividing the polygon into triangles. How many triangles are formed?

2. If eight people are in a room and each person shakes hands with each of the others, how many handshakes take place?

Exercises

Solve each problem using a model or a diagram.

1. Gear 1 rotates clockwise as shown. In what direction does gear 5 rotate? In what direction would gear 7 rotate? In what direction will all odd-numbered gears rotate?

2. Two pennies rest side by side with both heads in a vertical position. The penny on the left is now rolled along the edge of the penny on the right until it reaches the far side as shown. In what position is the head now?

3. A dotted line is drawn in the center of a narrow length of paper. The paper is twisted once, and the ends are joined to form a ring. If the ring is now cut apart along the dotted line, how many rings are produced?

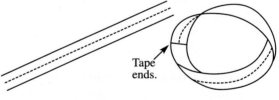

Tape ends.

4. At the end of a year, a tree has split into two branches. At the end of each succeeding year, each branch has split into two more branches. How many branches are there at the end of nine years?

5. Toothpicks were used to make an array of nine small squares, as shown. How can you remove just four toothpicks and leave five small squares?

6. How many squares, of any size, are there in the checkerboard shown?

Thinking About Proof

Conditional Statements in *if-then* Form

Statements that are to be proven in geometry are often presented as conditionals. A ***conditional*** is a statement that can be written in *if-then* form. The *if* clause contains information that is accepted as true. It is called the *hypothesis*. The *then* clause contains information to be proven true. It is called the *conclusion*.

Example: **Rewrite each conditional in if-then form. Then underline the hypothesis once and the conclusion twice.**

   a. An equilateral triangle must be isosceles.

   b. The diagonals of a rectangle are congruent.

Solution: **a.** If <u>a triangle is equilateral</u>, then <u>the triangle must be isosceles</u>.

   b. If <u>a quadrilateral is a rectangle</u>, then <u>its diagonals are congruent</u>.

In a geometry proof the hypothesis is labeled *Given* and the conclusion is labeled *Prove*. Your task is to apply known definitions, postulates, and theorems to the *Given* information in order to *Prove* the conclusion.

Exercises

Rewrite each conditional in *if-then* form. Then underline the hypothesis once and the conclusion twice.

1. Each point on the bisector of an angle is equidistant from the sides of the angle.

2. The diagonals of a parallelogram bisect each other.

3. The diagonals of a rhombus are perpendicular.

4. Opposite sides of a parallelogram are congruent.

5. The base angles of an isosceles triangle are congruent.

6. The bisector of the vertex angle of an isosceles triangle is the perpendicular bisector of the base of the triangle.

7. The sum of the angle measures of any quadrilateral is 360°.

3.7 Perspective Drawing

Objective: To draw figures using one-point and two-point perspective.

If you stand on a straight road and look into the distance, the sides of the road will appear to come together in one point. The sides appear to meet in some *vanishing point* on the *horizon line*. The photo at the right is an example of *one-point perspective*.

In *one-point perspective*, parallel lines that intersect the horizon line seem to meet in a point on the horizon line, called the ***vanishing point***.

EXPLORING

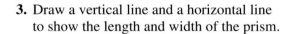

1. Draw a rectangle and a horizon line. Choose a vanishing point on the horizon line.

2. Use dashed lines to connect each corner of the rectangle to the vanishing point.

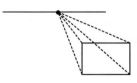

3. Draw a vertical line and a horizontal line to show the length and width of the prism.

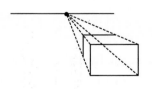

4. Draw over the dashed lines to show the visible faces of the prism. Erase the remaining dashed lines.

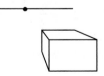

5. Repeat Steps 1–4 with the horizon line below the prism and a different vanishing point. How does the point of view change?

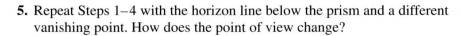

You can use *two-point perspective* to draw objects such as the corner of a building.

1. Draw a horizon line with two vanishing points. Draw a vertical segment below the horizon line.

2. Use dashed lines to connect the endpoints of the segment to each vanishing point.

3. To show length and width, draw vertical segments parallel to the first segment and between the dashed lines on each side. Use dashed lines to connect each segment to the opposite vanishing point.

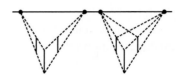

4. Draw over the dashed lines to show the visible faces of the prism. Erase the remaining dashed lines.

Thinking Critically

Look at the first Exploring activity.

1. How many faces of the prism would be visible if the vanishing point were centered above the prism? directly in back of the prism? to the right of the prism?

2. If the vanishing point were directly in back of the prism, from what point of view would you be seeing the prism?

Look at the second Exploring activity.

3. Using two-point perspective, draw a prism that lies directly on the horizon line. How many faces of the prism are visible?

Class Exercises

Draw a rectangular prism with the horizon line below the prism, using each of the following.

1. one-point perspective

2. two-point perspective

Copy the figures and locate their vanishing point(s).

3.

4.

Exercises

State whether the figure was drawn using one- or two-point perspective.

1.

2.

3.

4.

Copy each figure and locate its vanishing point(s).

5.

6.

7. Draw a rectangular prism with the horizon line above the prism, using one-point perspective.

8. Draw a triangular prism with the horizon line above the prism, using one-point perspective.

APPLICATION

9. Art   Find three pictures in magazines that show perspective. Was one- or two-point perspective used? Find the vanishing point(s).

Seeing in Geometry

Special effects can play tricks on visual perception. Wearing an outfit with vertical stripes makes a person *appear* thinner. Does the drawing show a square? How do the intercepting lines distort your perception?

Chapter 3 *Review*

Vocabulary and Symbols

You should be able to write a brief description, draw a picture, or give an example to illustrate the meaning of each term or symbol.

Vocabulary

adjacent (consecutive) sides (p.70)
adjacent (consecutive) vertices (p.70)
concave (p.69)
conclusion (p.90)
conditional (p.90)
consecutive angles (p.70)
convex (p.69)
diagonal (p.70)
horizon line (p.91)
hypothesis (p.90)
one-point perspective (p.91)
polygon (p.69)
  decagon (p.70)
  heptagon (p.70)
  hexagon (p.70)
  octagon (p.70)
  n-gon (p.70)
  nonagon (p.70)

pentagon (p.70)
regular (p.78)
side (p.69)
polyhedron (p.81)
  edge (p.81)
  face (p.81)
  vertex (p.81)
prism (p.81)
  bases (p.81)
  lateral faces (p.81)
  oblique (p.82)
  right (p.82)
pyramid (p.82)
quadrilateral (p.70)
  parallelogram (p.77)
  rectangle (p.77)
  rhombus (p.77)
  square (p.77)
  trapezoid (p.77)
  right cylinder (p.82)

three-dimensional figure (p.81)
triangle (p.70)
  acute (p.73)
  equiangular (p.73)
  equilateral (p.73)
  isosceles (p.73)
    base (p.74)
    base angles (p.74)
    legs (p.74)
    vertex angle (p.74)
  obtuse (p.73)
  right (p.73)
  scalene (p.73)
two-point perspective (p.92)
vanishing point (p.91)
vertex (p.69)

Symbol
$\triangle$ (triangle) (p.73)

Summary

The following list indicates the major skills, facts, and results you should have mastered in this chapter.

3.1 Recognize convex and concave polygons. (pp. 69–72)

3.2 Classify triangles. (pp. 73–76)

3.3 Recognize types of quadrilaterals and other polygons. (pp. 77–80)

3.4 Name polyhedrons and prisms. (pp. 81–83)

3.5 Make an accurate drawing of a prism. (pp. 84–87)

3.6 Solve a problem using a diagram or model. (pp. 88–89)

3.7 Draw figures using one-point and two-point perspective. (pp. 91–93)

Exercises

Classify each figure as a convex polygon, a concave polygon, or not a polygon. If the figure is a polygon, name it by the number of sides.

1. 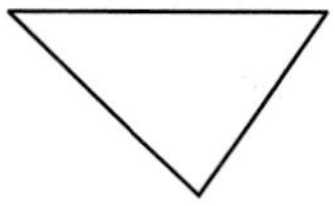

2. 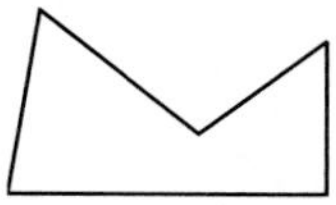

3. 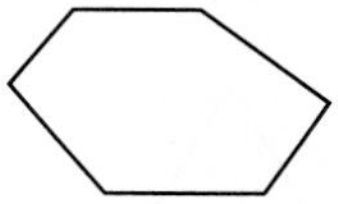

4. 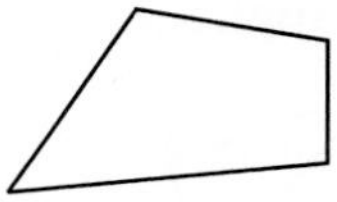

Judging by appearance, classify each triangle by its sides and by its angles.

5. 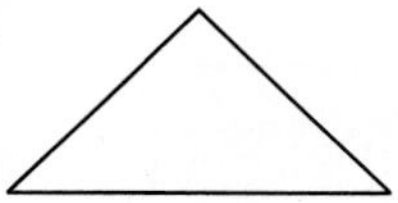

6. 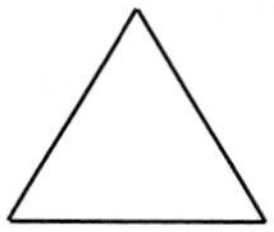

7. 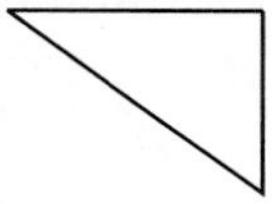

8. 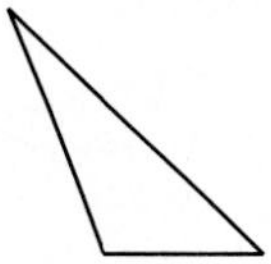

Choose the best word or words to complete each statement.

9. A ■ is a parallelogram with four right angles.

10. A ■ is a quadrilateral with exactly one pair of parallel sides.

11. A square is a parallelogram that is both a ■ and a ■.

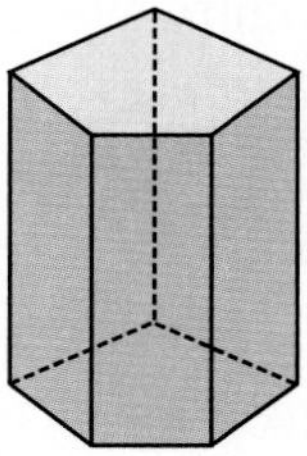

12. Refer to the figure at the right and (a) state the number of faces, edges, and vertices, (b) describe the shape of the bases, and (c) classify the prism.

13. Draw a rectangular prism. Then draw it again from another point of view.

14. Draw a rectangular prism with the horizon line above the prism.

PROBLEM SOLVING

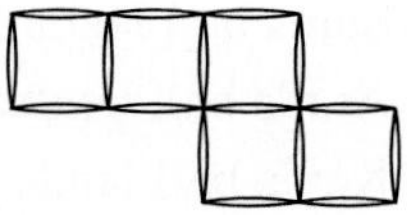

15. The figure shows five small squares made from toothpicks. How can you *remove* two toothpicks and leave four small squares?

Choose all the names that appear to be true for each figure from among the following: convex polygon, concave polygon, trapezoid, rhombus, rectangle, parallelogram, square.

1.   **2.**   **3.** 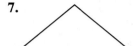   **4.**

Judging by appearance, classify each triangle by its sides and by its angles.

5. 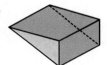   **6.**   **7.**   **8.**

For each prism (a) state the number of faces, edges, and vertices, (b) describe the shape of the bases, and (c) classify the prism.

9.   **10.**   **11.**

12. Draw a prism like the one shown in Exercise 10, but viewed from a point above and to the right of center.

Copy each figure and locate its vanishing point(s).

13.    **14.**

PROBLEM SOLVING

15. Lee, Ann, Sue, Kay, Bea, and Joy competed in the softball throw. Use these clues to determine how far they each threw the ball.

Ann threw the ball 12 m farther than Bea did.

Sue's ball landed l9 m behind Lee's ball.

Ann's ball landed 2 m in front of Joy's ball and 5 m behind Lee's ball.

Kay's ball landed halfway between Bea's ball and Ann's ball.

Kay threw the ball 50 m.

Chapter 4
Introduction to Transformations

Beyond color and proportion, the beauty we respond to in nature may arise from the harmony of a balanced arrangement of features such as we see in the wings of this butterfly and the leaves on which it has settled.

Focus on Skills

GEOMETRY

Draw and label a figure to fit each description.

1. a segment with endpoints *C* and *D*
2. a ray with endpoint *A*
3. the line through points *X* and *Y*
4. congruent segments $\overline{AB}$ and $\overline{CD}$
5. $\overleftrightarrow{AB}$ bisecting $\overline{CD}$
6. $\overrightarrow{OR}$ bisecting $\angle NOP$
7. a polygon with vertices *A*, *B*, *C*, and *D*
8. a hexagon
9. an angle with measure 60°
10. an angle with measure 135°

Measure each segment to the nearest millimeter.

11. ————————————————
12. ——————————————————————
13. ——————————
14. ————————————

Measure each angle to the nearest degree.

15.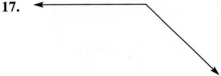
16.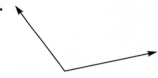

17.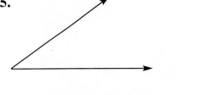
18.

19. Draw a segment, $\overline{AB}$, about 6 cm long. Construct the perpendicular bisector of $\overline{AB}$ to locate its midpoint, *M*.

20. Draw a line *j* and a point *P* not on *j*. Construct a perpendicular from *P* to *j*.

21. Draw an angle like the one shown. Construct the bisector of the angle.

4.1 Congruent Polygons and Corresponding Parts

Objective: To name corresponding parts of congruent polygons and write congruence statements for congruent figures.

Manufacturers often need to produce many parts with exactly the same size and shape. Figures that have the same size and shape are ***congruent.***

When two polygons are congruent, one fits exactly over the other. If you place *LMNO* over congruent polygon *PQRS*, the following vertices match or *correspond to* (↔) each other: *L* and *P*, *M* and *Q*, *N* and *R*, and *O* and *S*. The matching vertices determine the *corresponding angles* and *corresponding sides*.

Example 1: **Name the corresponding sides and angles.**

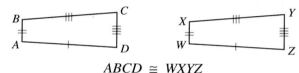

$$ABCD \cong WXYZ$$

Solution: corresponding sides

$\overline{AB} \leftrightarrow \overline{WX}$ $\overline{BC} \leftrightarrow \overline{XY}$

$\overline{CD} \leftrightarrow \overline{YZ}$ $\overline{DA} \leftrightarrow \overline{ZW}$

corresponding angles

$\angle A \leftrightarrow \angle W$ $\angle B \leftrightarrow \angle X$

$\angle C \leftrightarrow \angle Y$ $\angle D \leftrightarrow \angle Z$

EXPLORING

1. Fold a piece of paper in half. Without opening the paper, draw a polygon on one side.

2. With the paper folded, cut out your polygon. Label the vertices of both polygons with different letters.

3. Place one polygon over the other so that the sides and angles match. Make a list of the corresponding parts of your figures.

4. Measure the sides and angles of each polygon. What do you notice about the measures of the corresponding parts?

5. Repeat Steps 1–4 using different polygons. What conclusions can you make about corresponding parts of congruent polygons?

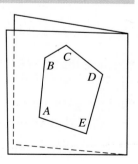

The **EXPLORING** activity leads to the following definition.

Congruent polygons are polygons whose corresponding parts are congruent.

Sometimes the positions of congruent polygons make it difficult to determine the corresponding parts. Hatch marks on the figures marking the congruent parts can help you see which parts correspond. In a congruence statement the polygons are named so that their corresponding vertices are in the same order.

Example 2: $\triangle ABC \cong \triangle DEF \cong \triangle XYZ$. **Write congruence statements for corresponding parts.**

Solution: $\overline{AB} \cong \overline{DE} \cong \overline{XY}$ $\angle A \cong \angle D \cong \angle X$
$\overline{AC} \cong \overline{DF} \cong \overline{XZ}$ $\angle B \cong \angle E \cong \angle Y$
$\overline{BC} \cong \overline{EF} \cong \overline{YZ}$ $\angle C \cong \angle F \cong \angle Z$

Thinking Critically

Suppose that $DEFG \cong QRST$. How many other correct congruence statements can you write to show that the first polygon is congruent to the second polygon? Write at least three.

Class Exercises

Suppose that $\triangle ABC \cong \triangle DEF$.

1. $\overline{AC} \cong$ ▧   2. $\overline{AB} \cong$ ▧   3. $\overline{BC} \cong$ ▧
4. $\angle A \cong$ ▧   5. $\angle B \cong$ ▧   6. $\angle C \cong$ ▧
7. $\triangle CBA \cong$ ▧   8. $\triangle ACB \cong$ ▧   9. $\triangle BAC \cong$ ▧

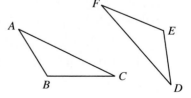

Suppose that $\triangle JKL \cong \triangle XYZ$.

10. Which side of $\triangle JKL$ is congruent to $\overline{XY}$? to $\overline{YZ}$? to $\overline{XZ}$?
11. Which angle of $\triangle JKL$ is congruent to $\angle X$? to $\angle Y$? to $\angle Z$?

Suppose that $\triangle RST \cong \triangle GHK$.

12. Write congruence statements for three pairs of congruent angles.

13. Write congruence statements for three pairs of congruent sides.

14. Are the corresponding vertices correctly named by the congruence statement $\triangle SRT \cong \triangle HGK$? by the statement $\triangle TRS \cong \triangle KGH$?

Exercises

1. Write a congruence statement for each pair of figures that appear to be congruent.

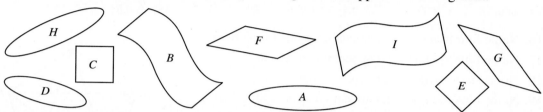

Name the corresponding sides and corresponding angles for each pair of congruent figures.

2. $\triangle WXY \cong \triangle DEF$

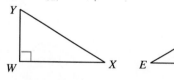

3. $\triangle KMN \cong \triangle JLP$

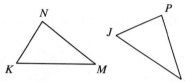

4. $ABCD \cong WXYZ$

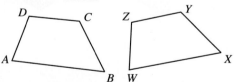

5. $RSTUV \cong MNOPQ$

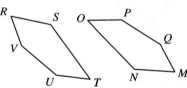

Suppose that $PQRS \cong ABCD$.

6. Write congruence statements for four pairs of congruent angles.

7. Write congruence statements for four pairs of congruent sides.

8. Are the corresponding vertices correctly named by the congruence statement $RQPS \cong CBAD$? by the statement $PSRQ \cong ADBC$?

9. $\triangle ADC \cong \triangle BCD$. Complete each congruence statement.

   a. $\angle DAC \cong$

   b. $\overline{DC} \cong$ ▨

   c. $\angle ADC \cong$ ▨

   d. ▨ $\cong \overline{BD}$

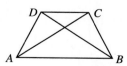

Write a congruence statement for each of the following.

10.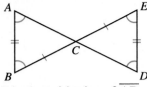
C is the midpoint of $\overline{AD}$.

11.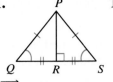
$\overrightarrow{PR}$ bisects $\angle QPS$.

12.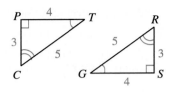

For each congruence statement, draw and label a figure or figures and list all pairs of corresponding sides.

13. $MNOP \cong ABCD$ **14.** $TVRS \cong XYRS$ **15.** $\triangle ABC \cong \triangle ABD$

16. Name four triangles that appear congruent to $\triangle BGF$.

17. Name four triangles that appear congruent to $\triangle AFB$.

18. Name four triangles that appear congruent to $\triangle ABC$.

19. Name nine triangles that appear congruent to $\triangle AGB$.

20. There are 35 different triangles in the figure. In Exercises 16–19 you named 21. Name the other 14.

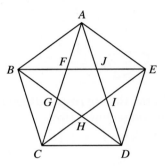

APPLICATION

21. Visualization Which cube is formed by folding the pattern?

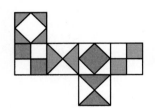

a. **b.** **c.**

Seeing in Geometry

Duplicate keys must be congruent to open the same lock. Five keys are shown. Study the shapes of the keys and tell which are the duplicates.

Thinking About Proof

Using Definitions

Congruent triangles are triangles whose corresponding parts are congruent. This definition is equivalent to the following two *if-then* sentences.

If two triangles are congruent, then the corresponding parts are congruent.

Given: $\triangle ABC \cong \triangle FED$
Conclusion: $\angle A \cong \angle F$,
$\angle B \cong \angle E$, $\angle C \cong \angle D$,
$\overline{AB} \cong \overline{FE}$, $\overline{BC} \cong \overline{ED}$,
$\overline{AC} \cong \overline{FD}$

If there is a correspondence between two triangles so that all corresponding parts are congruent, then the triangles are congruent.

Given: $\angle S \cong \angle Z$, $\angle R \cong \angle Y$,
$\angle T \cong \angle X$, $\overline{RS} \cong \overline{YZ}$,
$\overline{ST} \cong \overline{ZX}$, $\overline{RT} \cong \overline{YX}$
Conclusion:
$\triangle RST \cong \triangle YZX$

You can also use other definitions as reasons to justify conclusions.

Example: **Supply a reason for each conclusion.**

   a. Given: *RSTW* is a rectangle.
   Conclusion: $\angle T$ is a right angle.

   b. Given: *RSTW* is a parallelogram such that $\overline{RS} \cong \overline{ST} \cong \overline{TW} \cong \overline{WR}$.
   Conclusion: *RSTW* is a rhombus.

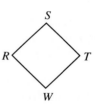

Solution: **a.** definition of a rectangle   **b.** definition of a rhombus

Exercises

Supply a reason for each conclusion.

1. **Given:** *ABCD* is a parallelogram.
   Conclusion: $\overline{AB} \parallel \overline{DC}$ and $\overline{BC} \parallel \overline{AD}$

2. **Given:** *KLMN* is a square.
   Conclusion: *KLMN* is a rectangle.

3. **Given:** *ABCD* is a quadrilateral such that $\overline{AB} \parallel \overline{DC}$ and $\overline{BC} \parallel \overline{AD}$.
   Conclusion: *ABCD* is a parallelogram.

4. **Given:** *PQRS* is a parallelogram with four right angles.
   Conclusion: *PQRS* is a rectangle.

5. Complete in two different ways:
   $\triangle RST \cong$ ▨

6. Complete in as many ways as possible: $ABCD \cong$ ▨

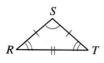   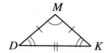   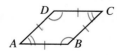   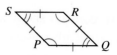

4.2 Reflections

Objective: To name and draw reflection images.

The likeness of an object in a mirror or a pond is an example of a *reflection image*.

EXPLORING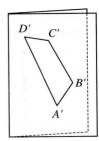

1. Fold a sheet of tracing paper in half vertically as shown. Without opening the paper, draw a quadrilateral that is not regular. Label it as *ABCD*.

2. Turn the folded paper over and trace the image of *ABCD* onto the opposite side of the paper. Label the vertex corresponding to *A* as *A'* (read as *A prime*). Label the other corresponding vertices as *B'*, *C'*, and *D'*.

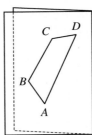

3. Open the paper and draw line *l* along the fold between the two quadrilaterals. Line *l* is the *line of reflection*.

4. Record the measures of the sides and angles of each quadrilateral. How do the measures of corresponding parts compare? What is the relationship between the two quadrilaterals?

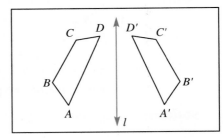

5. Draw $\overline{AA'}$, $\overline{BB'}$, $\overline{CC'}$, and $\overline{DD'}$. Use a protractor to measure the angle formed by the intersection of $\overline{AA'}$ and line *l*. Repeat for the angles formed by $\overline{BB'}$, $\overline{CC'}$, and $\overline{DD'}$. What does the angle measure tell you about the relationship between each segment and line *l*?

6. Construct the perpendicular bisector of $\overline{AA'}$. Repeat for $\overline{BB'}$, $\overline{CC'}$, and $\overline{DD'}$. What do you discover?

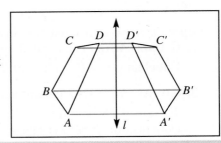

The **EXPLORING** activity suggests the following definition.

Two points *A* and *A'* are **reflections** of each other over line *l* if line *l* is the perpendicular bisector of $\overline{AA'}$.

△ *D'E'F'* is the reflection image of △ *DEF.* Line *l* is the ***line of reflection*** because it is the perpendicular bisector of the segments formed by connecting each vertex to its image.

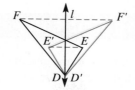

Example 1: **Draw and label the reflection image of △ *PQR* over line *l*.**

Solution: 1. Locate a point *P'* on the opposite side of line *l* such that *l* is the perpendicular bisector of $\overline{PP'}$.

2. In the same way, locate *Q'*, the image of *Q*. *R'*, the image of *R*, coincides with *R* since *R* lies on line *l*.

3. △ *P'Q'R'* is the reflection of △ *PQR* over line *l*.

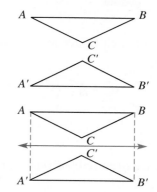

Example 2: **△ *A'B'C'* is the reflection image of △ *ABC*. Find the line of reflection.**

Solution: 1. Draw a segment joining one vertex to its reflection image.

2. Construct the perpendicular bisector of the segment.

3. Repeat Steps 1 and 2 with another vertex to check your result.

A triangle and its reflection image have reverse *orientation*. This means that the corresponding vertices are ordered in opposite directions. In Example 1, the vertices *P, Q,* and *R* have a *clockwise* orientation, and the corresponding vertices *P', Q',* and *R'* have a *counterclockwise* orientation.

Example 3: **Determine whether the figures in each pair have the same or reverse orientation.**

a.

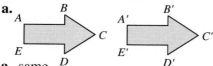

b.

Solution: a. same

b. reverse

1. How does the length of a segment compare to that of its reflection image? How does the measure of an angle compare to that of its reflection image?

2. If four points are collinear, will their reflection images be collinear?

3. Does a reflection *always*, *sometimes*, or *never* change the orientation of a figure?

Class Exercises

Do the figures in each pair have the same or reverse orientation?

1.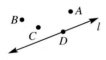

2.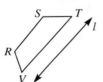

3.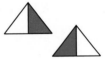

Trace each figure and draw its reflection image over line *l*.

4.

5.

6.

Trace each pair of figures and find the line of reflection.

7.

8.

9.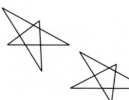

Exercises

Do the figures in each pair have the same or reverse orientation?

1.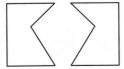

2.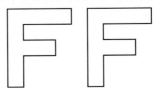

3.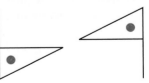

Name each reflection (if one exists) of points A, B, C, D, and E over line l.

4.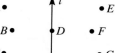

5.

6.

Trace each figure and draw its reflection image over line l.

7.

8.

9.

10.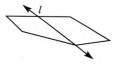

11.

12.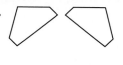

13.

14.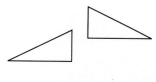

Trace each pair of figures and find the line of reflection.

15.

16.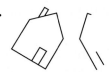

17.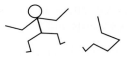

Trace each pair of figures and determine whether the figures are reflection images of each other. If so, draw the line of reflection.

18.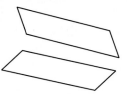

19.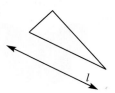

20.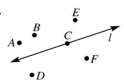

Trace each pair of figures. Draw the line of reflection and complete each reflection image.

21.

22.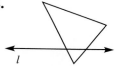

23.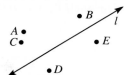

If the image of a figure is the same as the original, then the line of reflection is also a *line of symmetry*. For each figure, tell whether the dashed line is a line of symmetry.

24.

25.

26.

27.

28.

29.

30.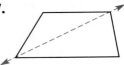

31.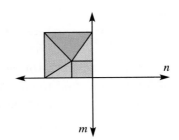

APPLICATIONS

32. **Quilting**   Jane started to design a quilt pattern. She has finished one fourth of it. Using the two lines of reflection shown, help Jane complete the quilt pattern.

33. **Codes**   The distress message *SOS* is represented in *dots and dashes* by the code ••• − − − •••. Would both the letters and their code version remain the same if reflected over a vertical line? Explain.

Seeing in Geometry

You may have noticed that the word *AMBULANCE* often appears backwards on the front of an ambulance. This allows a driver viewing the ambulance in a rear-view mirror to see the word correctly.

Some words would not change when viewed in a mirror.

1. Which letters in the words OTTO and BOX appear the same backwards and forwards?

2. For a word to read the same both backwards and forwards, what properties must it have aside from using only letters such as O and X?

4.3 Translations

Objective: To name and draw translation images.

A *translation* slides each point of the plane the same distance in the same direction. The sliding of a checker on a checkerboard is an example of a translation.

EXPLORING

1. Draw △*ABC* on graph paper. Label *A′*, the point that is 6 units to the right of *A*.

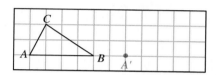

2. Label *B′* and *C′*, the points that are 6 units to the right of *B* and *C* respectively. Draw △*A′B′C′*.

3. What is the relationship between △*ABC* and △*A′B′C′* ?

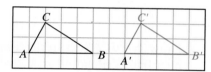

4. Repeat Steps 1–3, letting each new image be the point 6 units to the right and 4 units up from the original point.

5. How does the position of *B′* with respect to *B* compare to that of *A′* with respect to *A*? *C′* with respect to *C*?

Example: Draw the image of △*DEF* under the translation that takes *D* to *D′*.

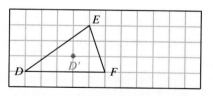

Solution: *D′* is 3 units to the right and 1 unit up from *D*. Find *E′* and *F′* by making the same moves in the same direction from *E* and *F* respectively. Draw △*D′E′F′*, the image of △*DEF*.

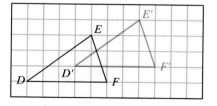

1. Fold a piece of tracing paper into thirds. Open the paper and draw lines *m* and *n* on the folds so that *m* ∥ *n*. Draw △*ABC* in the first section.

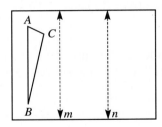

2. Draw △*A'B'C'*, the reflection of △*ABC* over line *m*.

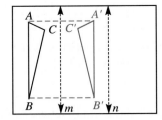

3. Draw △*A"B"C"*, the reflection of △*A'B'C'* over line *n*. (Read *A"* as *A double prime*.)

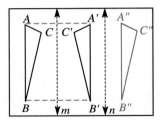

4. What is the relationship between △*A"B"C"* and △*ABC*? Does △*A"B"C"* have the same orientation as △*A'B'C'* ? as △*ABC*?

This **EXPLORING** activity demonstrates the following theorem.

THEOREM 4.1:   **Two consecutive reflections over parallel lines result in an image that is equivalent to a translation image.**

Thinking Critically

1. How does the length of a segment compare to that of its translation image? How does the measure of an angle compare to that of its translation image?

2. Does a translation *always, sometimes,* or *never* change the orientation of a figure?

Class Exercises

Choose the figure that could be a translation image of the given figure.

1.
   a.
   b.
   c.

Trace each figure on graph paper and draw its image under the translation that takes B to B'.

2.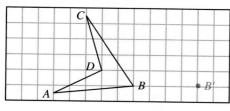
   3.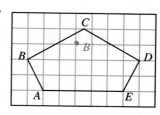

Exercises

Choose the figure that could be a translation image of the given figure.

1.
   a.
   b.
   c.

2.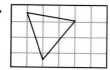
   a.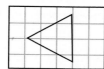
   b.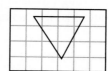
   c.

Trace each figure on graph paper and draw its image under the translation that takes B to B'.

3.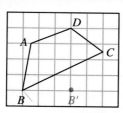
   4.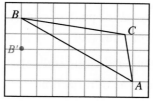
   5.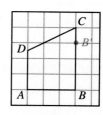

6.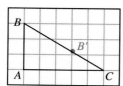
   7.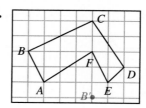
   8.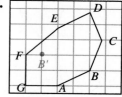

Trace each figure and draw the translation image obtained by first reflecting the polygon over line *m* and then over line *n*.

9.

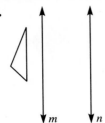

10.

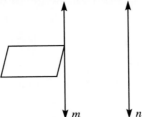

11.

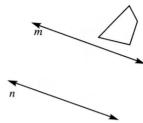

Name the translation image of each triangle obtained by first reflecting over line *m* and then over line *n*.

12. △GHI

13. △DEF

14. △ABC

15. △JKL

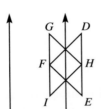

APPLICATION

16. Interior Design   Many wallpaper patterns consist of simple designs repeated continuously through translations. Use translations to design a wallpaper that you would like to see hanging in your bedroom.

Test Yourself

Suppose that △ ABC ≅ △ XYZ.

1. Write congruence statements for three pairs of congruent angles.

2. Write congruence statements for three pairs of congruent sides.

Trace each figure and draw the image or line indicated.

3. reflection image over line *l*

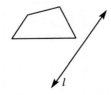

4. line of reflection

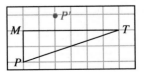

5. image under a translation that takes *P* to *P'*

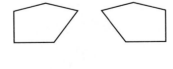

Rotations

Objective: To name and draw rotation images.

The turning motion of an object about a point is called a *rotation.* Each rotor blade on a helicopter is a *rotation image* of the other blades. Rotations can be clockwise (CW) or counterclockwise (CCW). When no direction of rotation is given, assume that the rotation is counterclockwise.

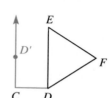

EXPLORING

1. Draw △*DEF* and point *C*.

2. To rotate △*DEF* 90° CCW about point *C*, begin by drawing $\overline{CD}$. Moving counterclockwise, draw a ray that makes a 90° angle with $\overline{CD}$. On that ray, mark *D'* so that *CD'* = *CD*.

3. Draw $\overline{CE}$. Continuing in a counterclockwise direction, draw a ray that makes a 90° angle with $\overline{CE}$. On that ray mark *E'* so that *CE'* = *CE*.

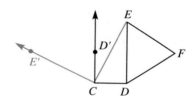

4. In a similar way, find *F'* so that *m∠FCF'* = 90° and *CF'* = *CF*. Draw △*D'E'F'*.

5. What is the relationship between △*DEF* and △*D'E'F''*?

6. Repeat Steps 1–5, rotating △*DEF* 120° CW about point *C*.

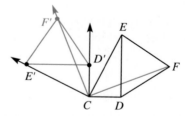

7. Repeat Steps 1–5, rotating △*DEF* 180° CCW about point *C*. Would you get the same image if you rotated △*DEF* 180° CW about point *C*?

The **EXPLORING** activity suggests the following definition.

Point *P'* is the image of point *P* under an *x*° **rotation** about point *C* if *m∠PCP'* = *x*° and *CP'* = *CP*.

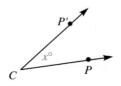

You have seen that a 180° CW rotation is *equivalent* to a 180° CCW rotation. Clockwise rotations are sometimes described in terms of equivalent counterclockwise rotations greater than 180°. That is, a 120° CW rotation is equivalent to a 240° CCW rotation.

Example: △R'S'T' is the image of △RST. Describe the rotation about point C that takes R to R'.

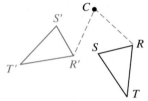

Solution: Draw the segments connecting C to R and R'. Measure ∠RCR'. The rotation is 70° CW (or 290° CCW).

EXPLORING

1. Trace △ABC, intersecting lines m and n, and △A'B'C', the reflection image of △ABC over line m.

2. Draw the reflection image of △A'B'C' over line n and label it as △A"B"C".

3. What is the relationship between △ABC and △A"B"C"? Do the triangles have the same orientation?

The **EXPLORING** activity demonstrates the following theorem.

THEOREM 4.2: Two consecutive reflections over intersecting lines result in an image equivalent to a rotation image.

1. How does the length of a segment compare to that of its rotation image? How does the measure of an angle compare with that of its rotation image?

2. Explain why a polygon and its rotation image have the same orientation.

3. Explain why the sum of the measures of equivalent clockwise and counterclockwise rotations is 360°.

Class Exercises

Choose the figure(s) that could be a rotation image of the given figure.

1.    **a.** 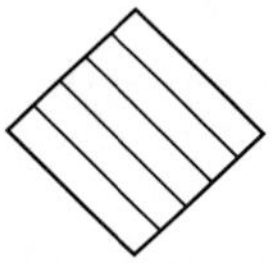   **b.**    **c.**

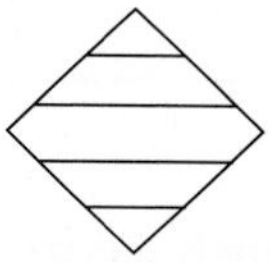

2.    **a.**    **b.** 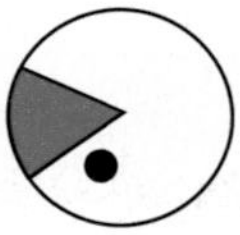   **c.**

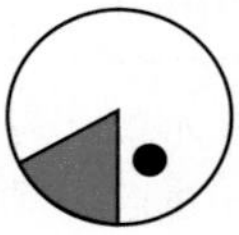

Describe the counterclockwise rotation about point *C* that takes *P* to *P'*.

3. 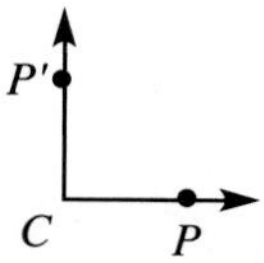   **4.** 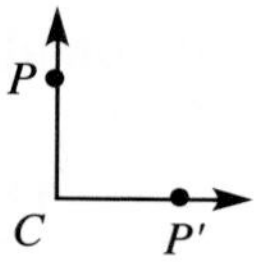   **5.** 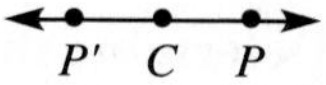   **6.**

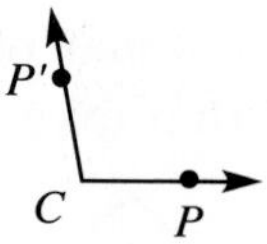

Trace each figure and draw its image under the given rotation about point *R*.

7. 180° CCW   **8.** 90° CW 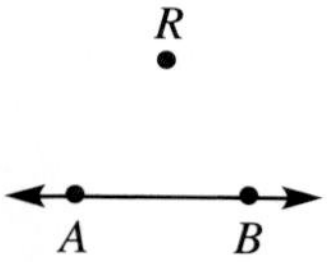   **9.** 90° CCW   **10.** 100° CCW

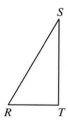   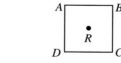   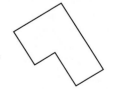

Exercises

Choose the figure(s) that could be a rotation image of the given figure.

1.   **a.**   **b.**   **c.**

2.   **a.**   **b.**   **c.**

Describe the counterclockwise rotation about point C that takes P to P'.

3.   **4.**   **5.**   **6.**

Trace each figure and draw its image under the given rotation about point R.

7. 90° CCW   **8.** 270° CW   **9.** 180° CCW   **10.** 120° CW

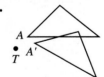   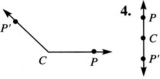

Trace each pair of figures. Determine the clockwise rotation about point T that takes A to A'.

11. 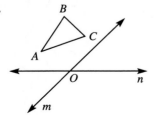   **12.**   **13.**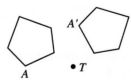

Trace each figure. Draw the rotation image of each figure about point O by first reflecting over line m and then over line n.

14. 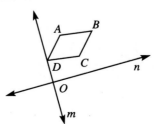   **15.**

A figure has *point symmetry* if it coincides with its image under a rotation of 180°. Trace each figure and draw its image under a 180° rotation about point *C*.

16.

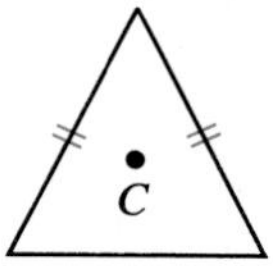

17.

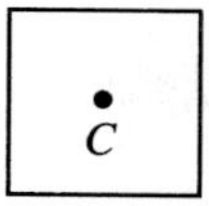

18.

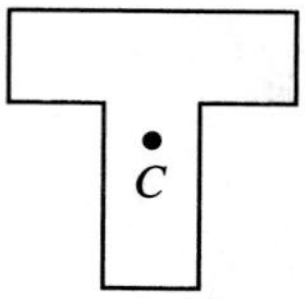

19.

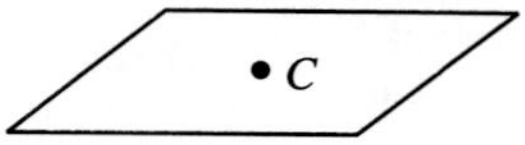

20.

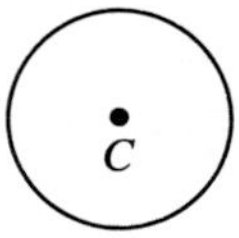

21.

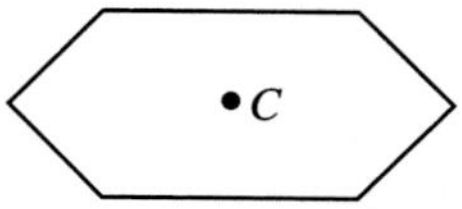

22.

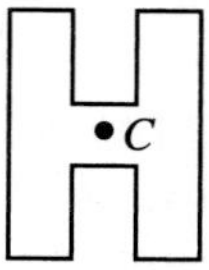

23.

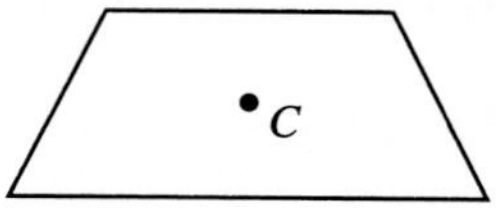

24. Which figures in Exercises 16–23 have point symmetry?

APPLICATION

25. Computer Use LOGO to enter the following procedures.

```
To triangle
fd 60 rt 90 fd 80 home
end
```

```
To rotate :anglemeasure
triangle
rt :anglemeasure
triangle
end
```

a. Run ROTATE, using 0 as the input for ANGLEMEASURE. Then clear the screen and run ROTATE again, using 30 as the input. Describe the two results.

b. Run ROTATE for several different inputs for ANGLEMEASURE.

c. Change the third line of the ROTATE procedure so that the triangle will be rotated counterclockwise. Run the procedure several times for different inputs for ANGLEMEASURE.

d. Enter the procedure shown at the right. Replace TRIANGLE in the ROTATE procedure above with SQUARE. Run the new procedure for several different inputs for ANGLEMEASURE. Describe the results.

```
To square
pu fd 30 pd
repeat 4 [fd 40 rt 90]
pu bk 30 pd
end
```

Computer

Write LOGO procedures that draw a square and its rotation image under a rotation about a point at the center of the square.

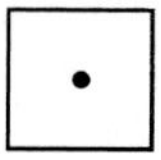

4.5 Problem Solving Application: Reflections in Miniature Golf

Objective: To solve problems involving reflections.

You can apply what you know about reflections to playing miniature golf. You can *bank* or reflect the ball off one side of the course to ensure that the ball will go in the right direction.

Example 1: You want to hit the golf ball, *B*, into the hole, *H*. Describe one way to use reflections to get a hole-in-one.

Solution: There is a corner between the golf ball and the hole. Thus, you will have to bank the golf ball off one side of the playing area, in this case, $\overline{PN}$.

1. Copy the figure. Draw the reflection of *H* over $\overline{PN}$ and label it as *H′*.

2. Draw $\overline{BH′}$. Label the intersection of $\overline{BH′}$ and $\overline{PN}$ as *R*. By aiming at *R*, you can get a hole-in-one.

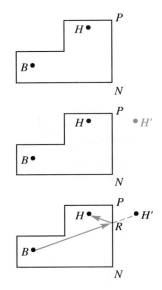

Example 2: You want to hit the golf ball, *B*, into the hole, *H*. Describe how you can get a hole-in-one.

Solution: The corner prevents you from aiming at *H′*, so you will need a double bank shot, in this case, off side $\overline{MN}$ and then off side $\overline{PN}$.

1. Copy the figure. Extend $\overline{MN}$ so that you can draw the reflection of *H′* over $\overleftrightarrow{MN}$. Draw $\overline{BH″}$. Label the intersection of $\overline{BH″}$ and $\overline{MN}$ as *S*.

2. Connect *B*, *S*, *R*, and *H* to show the path of the ball when you aim at *S*.

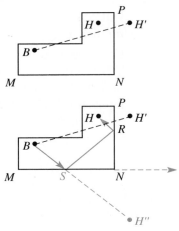

Class Exercises

Copy the figure and find the points P the golf ball must hit in order for the player to get a hole-in-one. Draw segments to show the path of the golf ball.

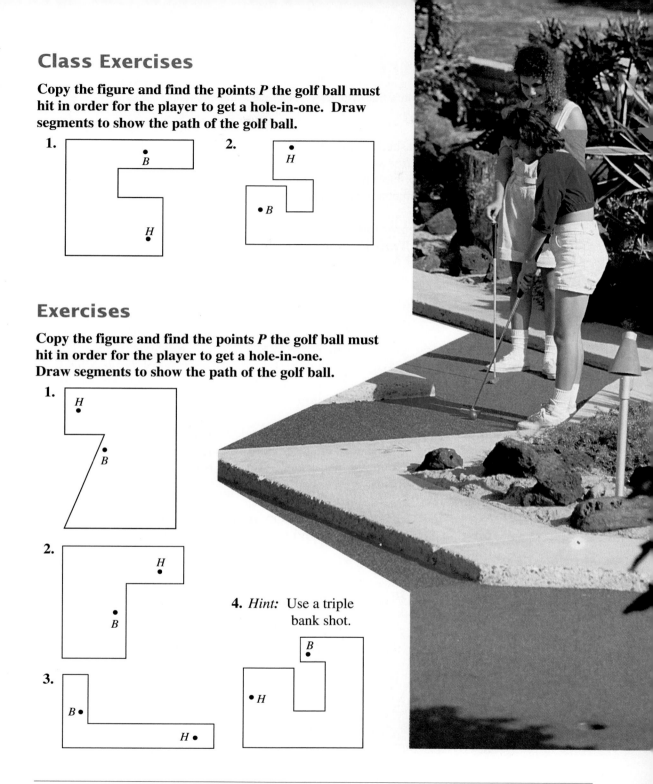

1.

B

H

2.

H

• B

Exercises

Copy the figure and find the points P the golf ball must hit in order for the player to get a hole-in-one. Draw segments to show the path of the golf ball.

1.

H

• B

2.

H

• B

3.

B •

H •

4. *Hint:* Use a triple bank shot.

B

• H

4.6 Symmetric Figures

Objective: To identify line symmetry, rotational symmetry, and point symmetry.

If you fold the square on the dashed line, the two halves of the figure are congruent. The square has *line symmetry,* and the dashed line is a *line of symmetry.* A figure may have more than one line of symmetry. A square has four lines of symmetry. Can you find them?

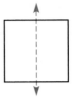

Example 1: **Tell which figures have line symmetry. Draw all the lines of symmetry. Trace the figures and fold if necessary.**

a.   b.    c.

Solution: **a.** none   b.   c.

A figure has *rotational symmetry* if there is some rotation less than 360° about a point *C* for which the figure and its image coincide. A square has 90°, 180°, and 270° rotational symmetry. The center of rotation is the intersection of the square's diagonals.

Example 2: **Which figures have rotational symmetry and for what rotations? Trace the figures and try various rotations if necessary.**

a. 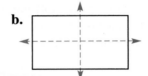   b. 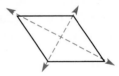   c.

d.   e.   f.

Solution: **a.** none   **b.** 180°   **c.** 180°
              d. 120° and 240°   **e.** 180°   **f.** none

A figure that has 180° rotational symmetry is said to have **point symmetry.** Every rectangle has point symmetry.

Example 3: Which figures in Example 2 have point symmetry?

   Solution: (b), (c), and (e)

EXPLORING

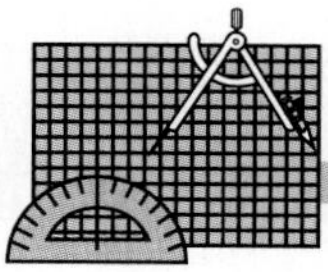

Part A

1. Draw an angle with measure less than 180°. Label it as ∠ABC.

2. Construct the angle bisector and label it as $\overrightarrow{BD}$. Fold ∠ABC along $\overrightarrow{BD}$.

3. Is the line containing the angle bisector a line of symmetry? Why or why not?

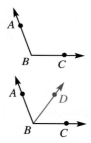

Part B

1. Draw any segment and label it as $\overline{AB}$. Construct the perpendicular bisector of $\overline{AB}$. Draw three other lines that also bisect $\overline{AB}$.

2. Are any of the lines you drew lines of symmetry for $\overline{AB}$?

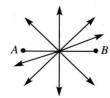

Thinking Critically

1. Must a figure with rotational symmetry have point symmetry? Explain.

2. Must a figure with point symmetry have rotational symmetry? Explain.

3. In the figure at the right, is the dashed line a line of symmetry for $\overline{AB}$?

Class Exercises

Trace each figure and draw all the lines of symmetry.

1.

2.

3.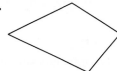

4.

5.

6.

7. Which figures in Exercises 1–6 have rotational symmetry and for what rotations? Which have point symmetry? Trace the figures and try various rotations if necessary.

Exercises

Trace each figure and draw all the lines of symmetry.

1.

2.

3.

4.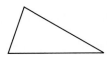

5.

6.

7.

8.

9. Which figures in Exercises 1–8 have rotational symmetry and for what rotations? Which have point symmetry? Trace the figures and try various rotations if necessary.

Which figures have rotational symmetry and for what rotations? Trace each figure and try various rotations if necessary. Does the figure have point symmetry?

10.

11.

12.

13.

14. Draw two intersecting lines.

   a. Draw all lines of symmetry for the angles.

   b. Describe the relationship of the lines of symmetry. Why is this relationship true?

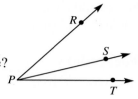

15. Draw a figure like the one shown.

   a. Draw the lines of symmetry for ∠*RPS* and ∠*SPT*.

   b. How does the measure of the angle formed by the lines of symmetry compare to the measure of ∠*RPT* ? Why is this true?

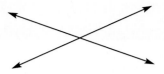

Draw a figure to fit each description.

16. a triangle that has exactly one line of symmetry

17. a triangle that has more than one line of symmetry

18. a hexagon that has no lines of symmetry

19. a hexagon that has at least one line of symmetry

20. a quadrilateral that has exactly two lines of symmetry

21. a quadrilateral with one line of symmetry but no point symmetry

22. a quadrilateral with point symmetry but no line of symmetry

APPLICATION

23. Design   For each flag, describe any line symmetry, rotational symmetry, and point symmetry.

a.

Quebec

b.

Tennessee

c.

North Atlantic Treaty Organization (NATO)

Seeing in Geometry

1. The word *DOCK* has a horizontal line of symmetry. Print three other words that have a horizontal line of symmetry.

2. The word *HUT*, when printed vertically, has a vertical line of symmetry. Print three other words vertically that have a vertical line of symmetry.

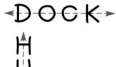

Vocabulary and Symbols

You should be able to write a brief description, draw a picture, or give an example to illustrate the meaning of each term or symbol.

Vocabulary

clockwise (p. 105)
congruent (p. 99)
congruent polygons
   (p. 100)
corresponding angles
   (p. 99)
corresponding sides (p. 99)
counterclockwise (p. 105)
line of reflection (p. 105)

line of symmetry (p. 120)
line symmetry (p. 120)
orientation (p. 105)
point symmetry (p. 121)
reflection (p. 104)
reflection image (p. 104)
rotation (p. 113)
rotation image (p. 113)
rotational symmetry
   (p. 120)

translation (p. 109)
translation image (p. 110)

Symbols

CW	clockwise (p. 113)
↔	corresponds to (p. 99)
CCW	counterclockwise (p. 113)

Summary

The following list indicates the major skills, facts, and results you should have mastered in this chapter.

4.1 Name corresponding parts of congruent polygons and write congruence statements for congruent figures. (pp. 99–102)

4.2 Name and draw reflection images. (pp. 104–108)

4.3 Name and draw translation images. (pp. 109–112)

4.4 Name and draw rotation images. (pp. 113–117)

4.5 Solve problems involving reflections. (pp. 118–119)

4.6 Identify line symmetry, rotational symmetry, and point symmetry. (pp. 120–123)

Exercises

Suppose that △ *TRM* ≅ △ *ANG*.

1. Name three pairs of corresponding angles.

2. Name three pairs of corresponding sides.

3. Complete the congruence statements:
   △ *TMR* ≅ ▣ and △ *RTM* ≅ ▣

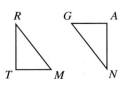

Name the image of each figure.

4. reflection image of $\overline{AB}$ over line l

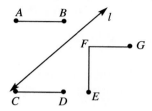

5. rotation image of $\triangle HIJ$ under 90° CCW rotation about H

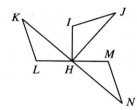

Trace each figure and find the line or image indicated.

6. reflection image over line k

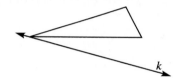

7. image under a translation that takes P to P'

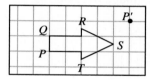

8. line of reflection

9. rotation image about O by reflecting over line m and then over line n

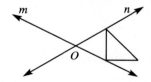

10. Trace the figure at the right and draw all the lines of symmetry. Does the figure have point symmetry? rotational symmetry? for what rotations?

PROBLEM SOLVING

11. Copy the figure and find the point the golf ball must hit in order for the player to get a hole-in-one. Draw segments to show the path of the golf ball.

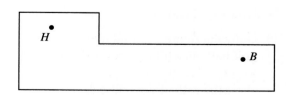

Suppose $QUAD \cong POLY$. **Complete each of the following.**

1. $\overline{QU} \cong$ ■   2. $\angle A \cong$ ■   3. $\overline{PY} \cong$ ■   4. $\angle O \cong$ ■

5. $\overline{DA} \cong$ ■   6. $\overline{OL} \cong$ ■   7. $QDAU \cong$ 8. $YLOP \cong$ ■

For each pair of figures tell whether the figure on the right could be the image of the figure on the left.

9. reflection image   10. translation image   11. rotation image

Trace each figure and find the line or image indicated.

12. reflection image over line m

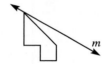

13. image under a translation that takes P to P'.

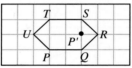

14. rotation image under a 180°
    counterclockwise rotation about P

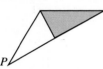

15. line of reflection

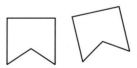

Complete each statement.

16. Two consecutive reflections over
    intersecting lines result in an image
    equivalent to a ■ image.

17. Two consecutive reflections over
    parallel lines result in an image
    equivalent to a ■ image.

18. Trace the figure at the right and draw all the lines of symmetry.
    Does the figure have point symmetry? rotational symmetry?
    for what rotations?

PROBLEM SOLVING

19. Copy the figure and find the point the
    golf ball must hit in order for the player
    to get a hole-in-one. Draw segments to
    show the path of the golf ball.

Chapter 5
Triangles and Inequalities

The simplest of all polygons, the triangle is a favorite of artists, architects, and engineers. This basic, stable shape has both visual appeal and the rigidity needed to provide support for bridges, buildings, and other structures.

Focus on Skills

ALGEBRA

Complete. Use <, >, or =.

1. $11 \cdot 3$ ▓ $27 + 5$ **2.** $15 - 8$ ▓ $32 \div 4$ **3.** $4(12 - 7)$ ▓ 50 **4.** 11 ▓ $5(7 - 4)$

Find the value of each expression if $x = 5$.

5. $4x$ **6.** $x + 16$ **7.** $2x + 7$ **8.** $5x - 11$

Solve each equation.

9. $3y - 5 = 82$ **10.** $6x + 11 = 53$ **11.** $14t + 12 = 180$

12. $12z - 8 = 5z + 13$

GEOMETRY

13. $\triangle BND$ is an isosceles triangle. Name the vertex angle, the base, and the base angles.

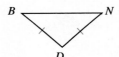

14. Draw any obtuse angle and construct its angle bisector.

15. Draw a segment and construct its perpendicular bisector.

16. Draw a figure like the one shown. Construct a perpendicular from point P to line l.

17. Draw a segment, $\overline{AB}$, and use construction to locate its midpoint, M.

18. In the figure at the right, $\overleftrightarrow{AC}$ is a line of symmetry. Write a congruence statement relating two of the triangles shown.

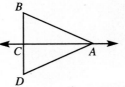

19. In $\triangle XYZ$, $XY = (2a - 5)$ cm, $YZ = (a + 8)$ cm, and $XZ = (a + 3)$ cm. If the perimeter of $\triangle XYZ$ is 38 cm, find the length of each side.

20. In $\triangle PQR$, $m\angle P = x°$, $m\angle Q = (2x + 6)°$, and $m\angle R = 3x°$. If the sum of the measures of the angles is 180°, find the measure of each angle.

5.1 Isosceles and Equilateral Triangles

Objective: To use the Isosceles Triangle Theorem and related theorems.

Semaphore flags, used to send messages from ship to ship or from ship to shore, are made up of isosceles triangles. By holding the flags in different positions, a messenger can represent numbers and letters.

EXPLORING

Part A

1. Draw a large isosceles triangle, △ABC, with *exactly* two congruent sides, $\overline{AB}$ and $\overline{AC}$. Fold △ABC to find all the lines of symmetry. How many are there?

2. Draw the line of symmetry through A intersecting $\overline{BC}$. Label the point of intersection as D.

3. Write a congruence statement involving each angle or segment.
   a. ∠B   **b.** ∠BAD   **c.** $\overline{BD}$ **d.** ∠ADB

4. What is the relationship between $\overleftrightarrow{AD}$ and $\overline{BC}$?

Part B

1. Draw a large △XYZ with *exactly* two congruent angles, ∠Y and ∠Z. Find all the lines of symmetry.

2. What, if anything, can you conclude about the sides of △XYZ?

The **EXPLORING** activity demonstrates the following theorems.

THEOREM 5.1 (Isosceles Triangle Theorem): The base angles of an isosceles triangle are congruent.

THEOREM 5.2: The line of symmetry for an isosceles triangle bisects the vertex angle and is the perpendicular bisector of the base.

THEOREM 5.3: If two angles of a triangle are congruent, then the sides opposite those angles are congruent.

Example: $\overleftrightarrow{EC}$ **is a line of symmetry for isosceles** $\triangle MCJ$.
$m\angle MCJ = 72°$. **Find** $m\angle 1$, $m\angle CEM$, **and** EJ.

Solution: $m\angle 1 = 36°$ $m\angle CEM = 90°$ $EJ = 3$

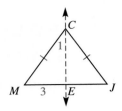

EXPLORING

1. Draw a large equilateral triangle, $\triangle EFG$. Fold $\triangle EFG$ to find all the lines of symmetry. How many are there?
2. Draw one line of symmetry. Which angles of $\triangle EFG$ must be congruent? Draw a different line of symmetry. Which angles of $\triangle EFG$ must be congruent?
3. What can you conclude about the angles of $\triangle EFG$?

Because an equilateral triangle is also isosceles, the next three theorems follow from the theorems about isosceles triangles.

THEOREM 5.4: **An equilateral triangle is also equiangular.**

THEOREM 5.5: **Each of the three lines of symmetry for an equilateral triangle bisects an angle of the triangle and is the perpendicular bisector of the side opposite that angle.**

THEOREM 5.6: **An equiangular triangle is also equilateral.**

Thinking Critically

1. Can an isosceles triangle be equilateral? Must it be equilateral?
2. Can an isosceles triangle have three lines of symmetry? Must it have three lines of symmetry?

Class Exercises

Find each angle measure or segment length.

1. a. $m\angle A$
   b. $m\angle C$
   c. BC

2. a. LK
   b. $m\angle K$

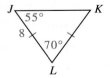

3. EF

4. $m\angle XYZ = 40°$, $XZ = 10$, and $\overleftrightarrow{YW}$ is a line of symmetry.

   a. $m\angle Z$

   b. $m\angle XYW$

   c. XW

   d. ZW

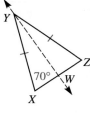

5. $\triangle RST$ is equilateral. $\overleftrightarrow{TX}$ is a line of symmetry.

   a. $m\angle 1$ **b.** $m\angle 2$

   c. $m\angle 3$ **d.** RS

   e. RX **f.** SX

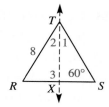

Exercises

True or false?

  1. In $\triangle ABC$, if $\overline{AB} \cong \overline{AC}$, then $\angle A \cong \angle B$.

  2. If $m\angle R = m\angle S = m\angle T$, then $\triangle RST$ is equilateral.

  3. Every isosceles triangle has exactly three lines of symmetry

  4. Every equilateral triangle has exactly three lines of symmetry.

  5. In $\triangle XYZ$, shown at the right, $YZ = 41$.

  6. Every equilateral triangle has point symmetry.

For each figure, name the angles that must be congruent.

7.

8.

9.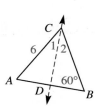

Find each angle measure or segment length.

10. $\overleftrightarrow{KN}$ is a line of symmetry for isosceles $\triangle KET$, and $m\angle TKE = 48°$.

   a. $m\angle T$ **b.** $m\angle TKN$

   c. $m\angle KNE$ **d.** $m\angle EKN$

   e. NT **f.** KT

11. $\overleftrightarrow{CD}$ is a line of symmetry for equilateral $\triangle ABC$.

   a. $m\angle BDC$ **b.** $m\angle CAD$

   c. $m\angle ACB$ **d.** $m\angle 1$

   e. $m\angle 2$ **f.** AB

   g. BC **h.** AD

12. Name three pairs of congruent angles in the figure shown. $\overleftrightarrow{CD}$ is a line of symmetry.

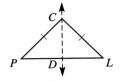

13. $\overline{ME}$ bisects $\angle GEO$. Complete the following.

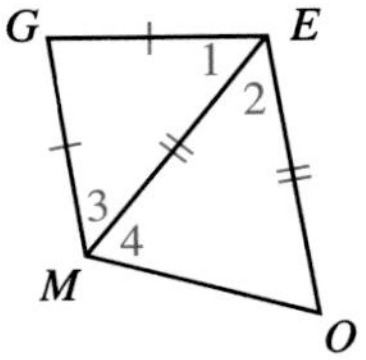

   a. If $m\angle 1 = 50°$, then $m\angle 2 = \blacksquare$ and $m\angle GEO = \blacksquare$.

   b. If $m\angle O = 65°$, then $m\angle 4 = \blacksquare$.

   c. If $GM = 6$ and $ME = 8$, then $GE = \blacksquare$ and $OE = \blacksquare$.

   d. Name two isosceles triangles.

   e. Name four pairs of congruent angles.

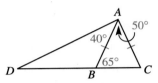

14. Name each of the following.

   a. three triangles   **b.** an obtuse triangle

   c. a right triangle   **d.** an isosceles triangle

   e. an acute triangle   **f.** two scalene triangles

15. In $\triangle JKL$, $\overline{KL} \cong \overline{JL}$. Which angles must be congruent?

16. In $\triangle RST$, $m\angle R = 67°$, $m\angle S = 46°$, and $m\angle T = 67°$. Which sides are congruent?

17. In $\triangle ABC$, $\overline{AB} \cong \overline{AC}$. If $m\angle A = (6x - 10)°$, $m\angle B = (3x + 5)°$, and $m\angle C = (x + 35)°$, find the measure of each angle.

18. In $\triangle DEF$, $\angle D \cong \angle E$. The length of $\overline{EF}$ is 2 cm less than the length of $\overline{DE}$. The length of $\overline{DF}$ is 9 cm less than twice the length of $\overline{DE}$. Find the lengths of the three sides.

APPLICATIONS

19. Algebra Find x.

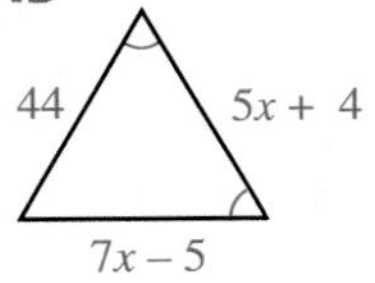

20. Computer Use LOGO to run the following procedure. Describe the result.

```
To triangle
fd 10 bk 70 fd 10 rt 90 fd 40 home
bk 50 lt 90 fd 40 home
end
```

Seeing in Geometry

How many lines of symmetry does each figure have? Does the figure have point symmetry?

1.   **2.** 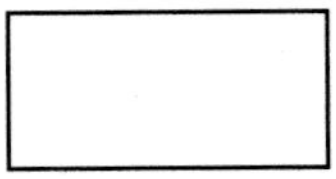   **3.** 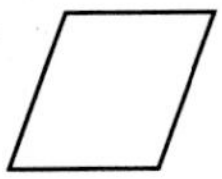   **4.**

5.2 Altitudes and Medians of Triangles

Objective: To construct or name altitudes and medians of triangles.

An *altitude* of a triangle is a perpendicular segment from a vertex to the line containing the opposite side. The altitudes of △ RST are $\overline{RX}$, $\overline{SY}$, and $\overline{TZ}$.

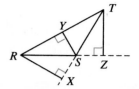

EXPLORING

1. Use the *Geometric Supposer: Triangles* disk to create an acute scalene triangle. Draw the three altitudes. Do the altitudes meet in a point? Repeat using an equilateral triangle.

2. Repeat using an obtuse scalene triangle. Where is the intersection of the altitudes? What must you do to the altitudes so that they intersect at this point?

3. Where is the intersection of the altitudes of a right triangle?

The lines containing the three altitudes of a triangle will always meet in one point. This point of intersection is called the *orthocenter.*

A *median* of a triangle is a segment from a vertex to the midpoint of the opposite side. Every triangle has three medians. In △ABC, $\overline{AM}$ is the median from A to M, the midpoint of $\overline{BC}$.

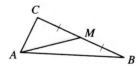

EXPLORING

1. Use the *Geometric Supposer: Triangles* disk to create an acute scalene triangle. Draw the three medians. Do the medians meet in one point?

2. Repeat, using a scalene right triangle and an obtuse scalene triangle. Describe each point of intersection.

The point of intersection of the three medians of a triangle is called the *centroid* of the triangle.

If three or more lines intersect in a point, then the lines are *concurrent*.

The EXPLORING activities demonstrate the following theorems.

THEOREM 5.7: **The three altitudes of a triangle are concurrent.**

THEOREM 5.8: **The three medians of a triangle are concurrent.**

Thinking Critically

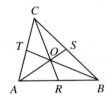

1. Draw acute, obtuse, and right scalene triangles. Construct the medians to the three sides and label the triangles as shown. Cut out the triangles. Use paper folding to investigate the relationship between the length of each median and the distance from the corresponding vertex to the centroid.

   a. What seems to be the relationship between the length of each median and the location of the centroid?

   b. $\overline{AS}$, $\overline{BT}$, and $\overline{CR}$ are medians of $\triangle ABC$, and point O is the centroid. If $AS = 24$, what is AO?

2. The *distance from a point to a line* is the length of the perpendicular segment from the point to the line.

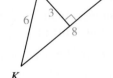

   a. The area of a triangle is half the product of the *base length* and the *height*. Any side of the triangle can be considered its base. The corresponding height is the length of what segment?

   b. In $\triangle JKL$, $JK = 6$ and $KL = 8$. The length of the altitude from vertex J is 3. What is the length of the altitude from vertex L?

3. $\triangle RST$ is isosceles. $\overleftrightarrow{TZ}$ is its line of symmetry.

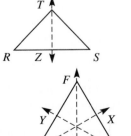

   a. Is $\overleftrightarrow{TZ}$ an altitude? a median? Why or why not?

   b. Is $\overleftrightarrow{TZ}$ the perpendicular bisector of a side? Is it an angle bisector? Why or why not?

4. $\triangle DEF$ is equilateral. $\overleftrightarrow{DX}$, $\overleftrightarrow{EY}$, and $\overleftrightarrow{FZ}$ are its lines of symmetry.

   a. Are $\overline{DX}$, $\overline{EY}$, and $\overline{FZ}$ altitudes? medians? Why or why not?

   b. Are $\overline{DX}$, $\overline{EY}$, and $\overline{FZ}$ perpendicular bisectors of sides? Are they angle bisectors? Why or why not?

Class Exercises

Name the segment that is the altitude to each indicated side.

1. side $\overline{AB}$ of $\triangle ABC$

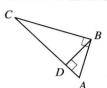

2. side $\overline{XY}$ of $\triangle XYZ$

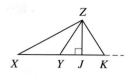

3. side $\overline{ST}$ of $\triangle RST$

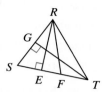

Draw a triangle similar to, but larger than, each one shown. Construct the altitude to $\overline{AB}$.

4.

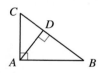

5.

6.

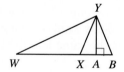

7–9. Draw triangles similar to those in Exercises 4–6. For each triangle, construct the midpoint of $\overline{AB}$ and draw the median from C to $\overline{AB}$.

Exercises

Name the segment that is the altitude to each indicated side.

1. side $\overline{AB}$ of $\triangle ABC$

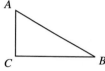

2. side $\overline{RT}$ of $\triangle RST$

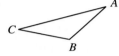

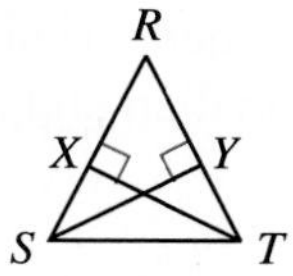

3. side $\overline{WX}$ of $\triangle WXY$

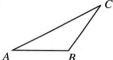

Draw a triangle similar to, but larger than, each one shown. Use construction to find the midpoints of the sides, draw the medians, and locate the centroid.

4.

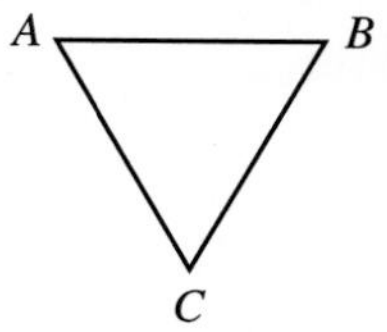

5.

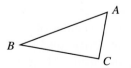

6.

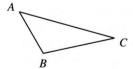

7–9. Draw triangles similar to those in Exercises 4–6. Locate the orthocenter of each triangle.

Using graph paper, draw a triangle to fit each description.

10. acute triangle: base length = 6, height = 3

11. acute triangle: base length = 4, height = 3

12. right triangle: base length = 4, height = 3

13. obtuse triangle: base length = 5, height = 4

14. three noncongruent triangles: base length = 6, height = 4

15. three noncongruent triangles: base length = 5, height = 3

Draw a triangle that fits each description.

16. Exactly one of the three medians is also an altitude.

17. All three medians are altitudes.

18. No median is an altitude.

19. Exactly one median determines a line of symmetry.

20. The triangle is a right triangle in which one altitude is the perpendicular bisector of a side.

21. The triangle is an obtuse triangle in which one altitude is also an angle bisector.

22. All three medians, altitudes, and angle bisectors coincide.

APPLICATION

23. **Computer** Use the *Geometric Supposer: Triangles* disk and the *Your Own* option to create a scalene triangle. Draw and measure the altitude and median from the same vertex. Which is longer? Repeat with a different scalene triangle. Is there any kind of triangle in which the median and altitude will be the same length? Explain.

Everyday Geometry

The *center of gravity*, or balance point, of an object is important to such different activities as using a seesaw or balancing a mobile. The centroid is the center of gravity of a triangular region. You can verify this by cutting a large triangle from a piece of cardboard. You should be able to balance the triangle on a pencil point at the centroid. What do you think is the center of gravity of a parallelogram region?

Thinking About Proof

Properties of Congruence

Congruence is *reflexive*, *symmetric*, and *transitive*. Examples are given below for segments and angles, but the properties of congruence also apply to other congruent figures, such as triangles and circles. If you think about the meaning of congruence, these properties will probably not surprise you.

Reflexive Property: $\overline{AB} \cong \overline{AB}$ $\angle A \cong \angle A$

Symmetric Property: If $\overline{AB} \cong \overline{CD}$, then $\overline{CD} \cong \overline{AB}$. If $\angle A \cong \angle B$, then $\angle B \cong \angle A$.

Transitive Property: If $\overline{AB} \cong \overline{CD}$ and $\overline{CD} \cong \overline{EF}$, If $\angle A \cong \angle B$ and $\angle B \cong \angle C$, then $\overline{AB} \cong \overline{EF}$. then $\angle A \cong \angle C$.

The property we will use the most is the reflexive property, which we will sometimes state in one of the ways shown on the right below.

Statement	Reason
$\overline{AB} \cong \overline{AB}$	A segment is congruent to itself.
$\angle A \cong \angle A$	An angle is congruent to itself.

Exercises

Give a reason for each statement.

1. If $\triangle DEF \cong \triangle RST$, then $\triangle RST \cong \triangle DEF$.

2. $\angle X \cong \angle X$

3. If $\odot A \cong \odot B$ and $\odot B \cong \odot C$, then $\odot A \cong \odot C$.

4. $\overline{EF} \cong \overline{EF}$

For each pair of triangles write congruence statements involving congruent parts.

5.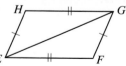

6.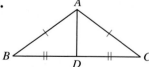

7.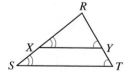

8. **Given:** $\angle 2 \cong \angle 3$
   Prove: $\angle 1 \cong \angle 3$
   a. Why is $\angle 1 \cong \angle 2$?
   b. Why is $\angle 1 \cong \angle 3$?

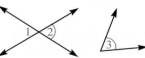

9. Tell whether each statement must be true for all real numbers a, b, and c.

   a. $a < a$ **b.** If $a < b$, then $b < a$.   **c.** If $a < b$ and $b < c$, then $a < c$.

5.3 Perpendicular Bisectors and Angle Bisectors

Objective: To construct the circumcenter and incenter of a triangle and use facts about the perpendicular bisector of a segment.

In this lesson you will examine the relationships between a point equidistant from two points and the segment that is determined by the two points.

EXPLORING

Part A

1. Use the *Geometric Supposer: Triangles* disk to create any triangle. Erase side $\overline{AC}$, side $\overline{BC}$, and point C. You now have $\overline{AB}$.

2. Draw the perpendicular bisector of $\overline{AB}$. Make $\overline{DE}$ 9 units long.

3. Put a random point on $\overline{DE}$. Draw segments connecting the point to the endpoints of $\overline{AB}$. Measure $\overline{GA}$ and $\overline{GB}$. What is true about the measures?

4. Use the *Repeat* option to repeat Steps 1–3 for a new segment. Are the results the same as for the first segment? Will they always be the same?

Part B

1. Use paper and pencil. Draw any segment and label it as $\overline{CD}$. With the compass point on C and then on D use the same compass opening to draw arcs that intersect at a point *equidistant* from C and D.

2. Locate three or four more points that are equidistant from C and D.

3. Draw a line through all of these points. Describe the line.

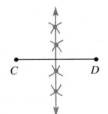

The **EXPLORING** activities demonstrate the following theorems.

THEOREM 5.9: **If a point lies on the perpendicular bisector of a segment, then the point is equidistant from the endpoints of the segment.**

THEOREM 5.10: **If a point is equidistant from the endpoints of a segment, then the point lies on the perpendicular bisector of the segment.**

1. Draw large acute, right, and obtuse scalene triangles. Construct the perpendicular bisectors of all three sides of each triangle.

2. For each triangle, extend the perpendicular bisectors if necessary until they intersect. This point of intersection is called the ***circumcenter*** of the triangle. Describe the location of each circumcenter.

3. For each triangle, measure the distance from the circumcenter to each vertex. How do these distances compare?

4. Draw large acute, right, and obtuse scalene triangles. Construct the three angle bisectors for each triangle.

5. The point of intersection of the angle bisectors of a triangle is called the ***incenter*** of the triangle. For each triangle, measure the distance from the incenter to each side. How do these distances compare?

This **EXPLORING** activity demonstrates the following theorems.

THEOREM 5.11:   **The perpendicular bisectors of the three sides of a triangle intersect in a point that is equidistant from the vertices of the triangle.**

THEOREM 5.12:   **The three angle bisectors of a triangle intersect in a point that is equidistant from the sides of the triangle.**

Thinking Critically

1. Line *l* is the perpendicular bisector of $\overline{XY}$, and line *m* is the perpendicular bisector of $\overline{YZ}$.

    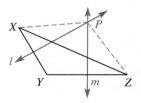

    a. Why is *PX* equal to *PZ*?

    b. To construct the circumcenter of △*XYZ*, is it necessary to also construct the perpendicular bisector of $\overline{XZ}$? Explain.

2. To construct the incenter of △*XYZ*, do you need to construct all three angle bisectors? Why or why not?

Class Exercises

Line *l* is the perpendicular bisector of $\overline{CD}$.

1. If $RC = 5$, then $RD = $ ▨.
2. If $SD = 6$, then $SC = $ ▨.
3. If $CM = 4$, then $DM = $ ▨ and $CD = $ ▨.
4. What kind of triangles are $\triangle RCD$ and $\triangle SCD$?

Draw a triangle similar to, but larger than, each one shown. Construct the perpendicular bisectors of $\overline{AB}$ and $\overline{BC}$.

5.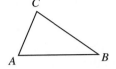

6.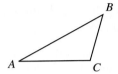

7–8. Draw triangles similar to those in Exercises 5 and 6. For each triangle, construct the bisectors of $\angle A$ and $\angle C$.

Exercises

$\overleftrightarrow{PQ}$ is the perpendicular bisector of $\overline{AB}$.

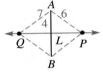

1. Find PB.
2. Find BL.
3. Find QB.
4. Suppose V is a point such that $VA = 9$ and $VB = 9$. Where is V located?
5. What kind of triangles are $\triangle APB$ and $\triangle AQB$?

$\overleftrightarrow{MO}$ is the perpendicular bisector of both $\overline{CD}$ and $\overline{AB}$. Find each length.

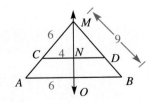

6. MD
7. DB
8. CA
9. MA
10. ND
11. OB
12. CD
13. AB

14. What kind of triangles are $\triangle CMD$ and $\triangle AMB$?

Draw a large triangle of each type indicated. Construct the perpendicular bisectors of any two sides.

15. acute scalene
16. obtuse scalene
17. isosceles right
18. acute isosceles

19–22. Draw triangles of each type indicated in Exercises 15–18. Construct the angle bisectors of any two angles.

23. a. Which of the following are always in the interior of a triangle?

orthocenter   centroid   circumcenter   incenter

b. Describe the type of triangle for which some of the points listed in part (a) are in the exterior of the triangle.

c. Describe the type of triangle for which some of the points listed in part (a) are on the triangle.

24. Are the orthocenter, the centroid, the circumcenter, and the incenter of a triangle ever the same point? Explain.

Draw a triangle similar to, but larger than, $\triangle ABC$.

25. a. Construct the point equidistant from A, B, and C.

b. Use your compass to construct the circle passing through A, B, and C.

26. a. Construct the point P equidistant from $\overline{AB}$, $\overline{BC}$, and $\overline{AC}$.

b. Construct $\overline{PD}$, the perpendicular segment from P to $\overline{AC}$.

c. Use your compass to construct the circle with center P and radius PD.

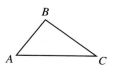

APPLICATIONS

Copy the figure at the right.

27. Community Development   The towns of Adams and Bakersville want to build picnic grounds on the shore of the lake and equidistant from the towns. Locate the appropriate place for the picnic grounds.

28. Community Development   The towns of Adams, Bakersville, and Columbus want to build a school that is equidistant from all three towns. Locate the appropriate place for the school.

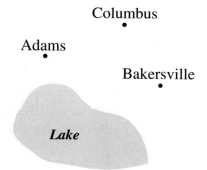

Test Yourself

In the figure, $m\angle BAC = 46°$, $BC = 6$, and $\overleftrightarrow{AD}$ is a line of symmetry. Find each angle measure or segment length.

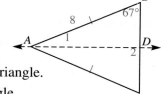

1. $m\angle 1$ **2.** $m\angle C$ **3.** $m\angle 2$ **4.** AC **5.** DC

6. Draw a large obtuse triangle. Locate the orthocenter of the triangle.

7. Draw a large acute triangle. Locate the centroid of the triangle.

5.4 Problem Solving Strategy: Guess and Check

Objective: To solve a problem by using guess and check.

When you are solving a problem, it is sometimes faster to make several guesses than it is to work out the solution. Using *guess and check* allows you to eliminate incorrect answers. You then can adjust your guesses accordingly. You can repeat the process until you solve the problem.

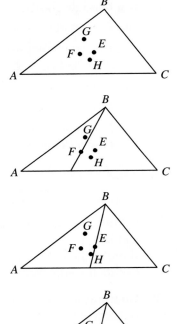

Example 1: You are taking a standardized test and have only your pencil and paper with which to work. You need to decide which point is the centroid of $\triangle ABC$.

Solution: *Guess 1:* The centroid is F.
Sketch $\overrightarrow{BF}$. You can see that $\overline{AC}$ has not been divided into congruent segments. F cannot be the centroid.

Guess 2: The centroid is E.
Sketch $\overrightarrow{BE}$. By folding along $\overrightarrow{BE}$, you can see that $\overline{AC}$ has been divided into two congruent segments. Sketch $\overrightarrow{AE}$ and $\overrightarrow{CE}$ and fold along those rays. $\overrightarrow{AE}$ and $\overrightarrow{CE}$ also divide their opposite sides into congruent segments.

Your second guess is correct.

Sometimes you can use guess and check instead of an equation.

Example 2: Two angles are supplementary. The measure of the second angle is twice that of the first. Find the measures of the angles.

Solution: *Guess 1:* $50°$ and $100°$
$50° + 100° = 150°$ Too low, so guess higher.

Guess 2: $60°$ and $120°$
$60° + 120° = 180°$ Your second guess is correct.

Class Exercises

Solve using guess and check.

1. Which point is the midpoint of $\overline{MN}$?

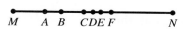

2. Two angles are complementary. The second angle has a measure twice that of the first. Find the measures of the angles.

Exercises

Solve using guess and check.

1. Which ray bisects ∠JKL?

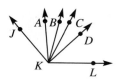

2. Which point lies on the median from B to $\overline{AC}$?

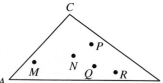

3. Which point lies on the altitude from R to the opposite side?

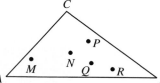

4. Two angles are supplementary. The second angle has a measure five times that of the first. Find the measures of the angles.

5. AB is 12 in. Point C lies on $\overline{AB}$ and is twice as far from A as it is from B. Find AC.

6. MN is 21 in. Point P lies on $\overline{MN}$ and is six times as far from N as it is from M. Find MP.

5.5 Inequalities for One Triangle

Objective: To use inequalities involving side lengths and angle measures in a triangle.

You can write inequalities relating the lengths of the sides of triangles and the measures of the angles opposite those sides.

=========== **EXPLORING** ===========

1. Draw a large acute scalene triangle. Measure each side to the nearest millimeter and each angle to the nearest degree.

2. List the sides in order from shortest to longest and the angles from smallest to largest. Compare the order of each list. Repeat Steps 1 and 2 for an obtuse scalene triangle.

The **EXPLORING** activity demonstrates the following theorems.

THEOREM 5.13: **If two sides of a triangle are unequal, then the measures of the angles opposite those sides are unequal, and the larger angle is opposite the longer side.**

THEOREM 5.14: **If the measures of two angles of a triangle are unequal, then the sides opposite those angles are unequal, and the longer side is opposite the larger angle.**

Example: a. If $AC > BC$, then $m\angle B > m\angle A$.
b. If $m\angle B < m\angle C$, then $AC < AB$.

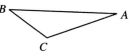

=========== **EXPLORING** ===========

1. Cut straws into four pieces with lengths of 2 cm, 4 cm, 5 cm, and 8 cm. Which pieces can you use to make a triangle?

2. Measure the three sides of each triangle shown. For each triangle, compare the length of each side to the sum of the lengths of the other two sides.

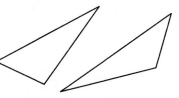

The **EXPLORING** activity demonstrates the following theorem.

THEOREM 5.15 (The Triangle Inequality): **The sum of the lengths of any two sides of a triangle is greater than the length of the third side.**

The Triangle Inequality can be thought of as another way of saying, "The shortest path between two points is a straight line."

Thinking Critically

1. The lengths of two sides of a triangle are 8 cm and 9 cm. Which of the following lengths can the third side of the triangle have?

   a. 1 cm   **b.** 1.1 cm

   c. 10 cm   **d.** 16.9 cm

   e. 17 cm   **f.** 18 cm

2. The lengths of the sides of a triangle are 4 cm, 10 cm, and x cm. Use the Triangle Inequality to complete:

   a. $x + \blacksquare > 4$, $x + \blacksquare > 10$, and $\blacksquare + \blacksquare > x$

   b. $x > \blacksquare$ and $\blacksquare > x$

   c. $\blacksquare > x > \blacksquare$

Class Exercises

1. List the angles from smallest to largest.

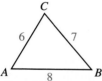

2. List the sides from shortest to longest.

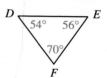

Can the lengths given be the lengths of the sides of a triangle?

3. 3 cm, 5 cm, 7 cm   **4.** 8 cm, 3 cm, 4 cm   **5.** 4 cm, 6 cm, 2 cm

Exercises

List the angles of each triangle from smallest to largest.

1.

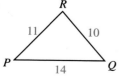

2.

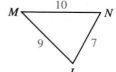

3.

List the sides of each triangle from shortest to longest.

4.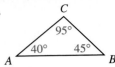

5.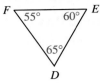

6.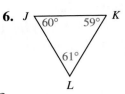

Can the lengths given be the lengths of the sides of a triangle?

7. 3 cm, 5 cm, 9 cm   **8.** 6 cm, 6 cm, 6 cm   **9.** 8 cm, 2 cm, 7 cm   **10.** 3 cm, 10 cm, 7 cm

The lengths of the sides of a triangle are given. List the angles of the triangle from smallest to largest.

11. $AB = 5$ cm, $BC = 7$ cm, $AC = 10$ cm   **12.** $DE = 12$ cm, $EF = 10$ cm, $DF = 15$ cm

The measures of the angles of a triangle are given. List the sides from shortest to longest.

13. $m\angle R = 35°$, $m\angle S = 120°$, $m\angle T = 25°$ **14.** $m\angle X = 70°$, $m\angle Y = 60°$, $m\angle Z = 50°$

The lengths of two sides of a triangle are given. What can you conclude about the length of the third side?

15. $DE = 5$ cm, $DF = 12$ cm

16. $JK = 12$ cm, $KL = 15$ cm

17. P is *any* point in the interior of $\triangle ABC$. Complete each part with the correct number.

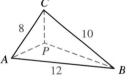

   a. $PA + PB >$ ▨   **b.** $PA + PC >$ ▨   **c.** $PB + PC >$ ▨

   d. $PA + PB + PA + PC + PB + PC >$ ▨   **e.** $PA + PB + PC >$ ▨

APPLICATIONS

18. Navigation   One ship is 25 mi from a lighthouse. Another ship is 10 mi from the lighthouse. What is the least possible distance between the ships? the greatest possible distance?

19. Logic
List the angles from smallest to largest.

Thinking in Geometry

The figure at the right is not drawn to scale. Which segment should be the longest? Why?

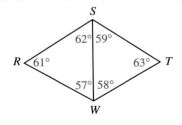

5.6 Inequalities for Two Triangles

Objective: To compare side lengths and angle measures for two triangles that have two pairs of congruent sides.

The following terms are used to describe the relationship between certain sides and angles of a triangle.

An *angle* of a triangle is ***included*** by the two sides of the triangle that determine the angle.

A *side* of a triangle is ***included*** by the two angles of the triangle whose vertices are the endpoints of the side.

Example 1: ∠*C* is included by $\overline{AC}$ and $\overline{BC}$.
$\overline{AC}$ is included by ∠*A* and ∠*C*.

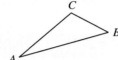

EXPLORING

1. Use a ruler and protractor to draw noncongruent angles *JMW* and *FDE* so that $\overline{MW} \cong \overline{DE}$ and $\overline{MJ} \cong \overline{DF}$.

2. What is true of *WJ* and *EF*?

The **EXPLORING** activity demonstrates the following theorem.

THEOREM 5.16 (The Hinge Theorem): If two sides of one triangle are congruent to two sides of another triangle and the measures of the included angles are unequal, then the sides opposite those angles are unequal and the longer side is opposite the larger angle.

Example 2: In △*PMN* and △*RST*, $\overline{PN} \cong \overline{RT}$, $\overline{PM} \cong \overline{RS}$, and *m*∠*P* < *m*∠*R*. What can you conclude about *MN* and *ST*?

Solution: *MN* < *ST*

The following theorem applies to two triangles with exactly two pairs of congruent sides.

THEOREM 5.17: **If two sides of one triangle are congruent to two sides of another triangle and the third sides are unequal, then the measures of the angles opposite the third sides are unequal and the larger angle is opposite the longer side.**

Example 3: What can you conclude about angles E and X of △GEF and △XYZ?

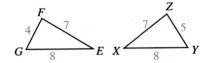

Solution: $\overline{GE} \cong \overline{XY}$, $\overline{EF} \cong \overline{XZ}$, and $ZY > FG$. Therefore $m\angle X > m\angle E$.

Thinking Critically

1. a. If $m\angle M = 40°$ and $m\angle Z = 41°$, what is true of NP and WX?

b. If $m\angle M = 41°$ and $m\angle Z = 40°$, what is true of NP and WX?

c. If $m\angle M = m\angle Z = 40°$, what do you think is true of NP and WX?

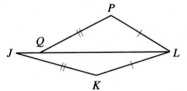

2. What can you conclude about $\angle P$ and $\angle K$? Explain.

3. $\overline{XM}$ is a median of scalene △XYZ. Can $\overline{XM}$ also be an altitude? Why or why not?

Class Exercises

1. $\angle X$ is included by sides ▦ and ▦.

2. $\overline{XZ}$ is included by ▦ and ▦.

3. The included angle for $\overline{XY}$ and $\overline{ZY}$ is ▦.

4. The included side for $\angle Y$ and $\angle Z$ is ▦.

Complete each statement with > or <.

5. AC ▦ DF

6. $m\angle J$ ▦ $m\angle P$

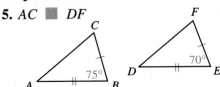

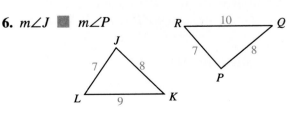

Exercises

Complete each statement with > or <.

1. RT ▨ YZ

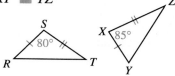

2. $m\angle F$ ▨ $m\angle O$

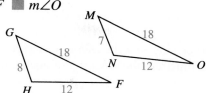

What can you conclude?

3.

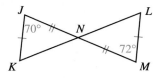

4.

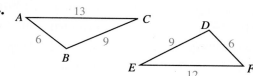

5. In $\triangle RST$, $RS = 10$ cm, $ST = 13$ cm, and $m\angle S = 40°$.
In $\triangle XYZ$, $YZ = 10$ cm, $XZ = 13$ cm, and $m\angle Z = 35°$.

6. In $\triangle ABC$, $AB = 5$ cm, $BC = 8$ cm, and $AC = 10$ cm.
In $\triangle DEF$, $DE = 5$ cm, $EF = 11$ cm, and $DF = 10$ cm.

7.

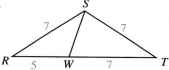

8.

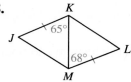

APPLICATION

9. Algebra　The perimeter of $\triangle ABC$ is 39 cm, with $AB = x$ cm,
$BC = (x + 4)$ cm, and $AC = (x + 8)$ cm. In $\triangle DEF$, $DE = x$ cm,
$EF = (x + 4)$ cm, and $DF = (2x - 3)$ cm. Which is greater, $m\angle B$
or $m\angle E$? Explain.

Seeing in Geometry

What, if anything, can you conclude about each of the following pairs?

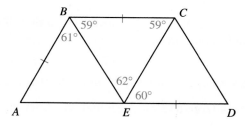

1. BE and CE　　**2.** AE and CE

3. BE and CD　　**4.** AE and CD

5. CE and CD　　**6.** $m\angle D$ and $60°$

Chapter 5 *Review*

Vocabulary

You should be able to write a brief description, draw a picture, or give an example to illustrate the meaning of each term.

Vocabulary

distance from a point to
  a line (p. 134)

concurrent lines (p. 134)

Triangles

altitude (p. 133)

centroid (p. 134)

circumcenter (p. 139)

incenter (p. 139)

included angle (p. 147)

included side (p. 147)

median (p. 133)

orthocenter (p. 133)

Summary

The following list indicates the major skills, facts, and results you should have mastered in this chapter.

5.1 Use the Isosceles Triangle Theorem and related theorems. (pp. 129–132)

5.2 Construct or name altitudes and medians of triangles. (pp. 133–136)

5.3 Construct the circumcenter and incenter of a triangle and use facts about the perpendicular bisector of a segment. (pp. 138–141)

5.4 Solve a problem by using guess and check. (pp. 142–143)

5.5 Use inequalities involving side lengths and angle measures in a triangle. (pp. 144–146)

5.6 Compare side lengths and angle measures for two triangles that have two pairs of congruent sides. (pp. 147–149)

Exercises

In △RST, m∠RTS = 48°, and $\overleftrightarrow{TY}$ is a line of symmetry. Find each angle measure or segment length.

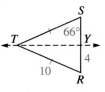

1. m∠RTY   **2.** m∠RYT   **3.** m∠R   **4.** YS

5. Draw a large triangle *ABC* and construct the altitude to $\overline{AB}$.

6. Draw a large triangle *DEF* and construct the median from *E*.

7. Draw a large triangle *GHI* and construct its circumcenter.

8. Draw a large triangle *JKL* and construct its incenter.

Complete each statement.

9. If $\overline{PQ} \cong \overline{PR}$, then ▆ $\cong$ ▆.

10. If $\angle P \cong \angle R$, then ▆ $\cong$ ▆.

11. If $\triangle PQR$ is equilateral and $m\angle P = 60°$, then $m\angle Q =$ ▆ and $m\angle R =$ ▆.

12. If $m\angle P = m\angle Q = m\angle R$ and $QR = 15$ cm, then the sum of the lengths of the sides of $\triangle PQR$ is ▆.

Refer to $\triangle MNO$. $\overleftrightarrow{DA}$, $\overleftrightarrow{DB}$, and $\overrightarrow{DC}$ are the perpendicular bisectors of the sides. Complete each statement.

13. If $PM = 6$ cm and $PN = 6$ cm, then P is on ▆.

14. If $DM = 7$ cm, then $DO =$ ▆.

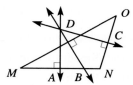

15. List the angles from smallest to largest.

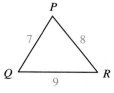

16. List the sides from shortest to longest.

Complete each statement with < or >.

17. $m\angle K$ ▆ $m\angle N$

18. TP ▆ LR

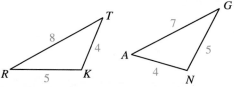

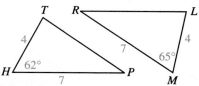

19. Can a triangle have sides of length 4 cm, 5 cm, and 8.9 cm?

PROBLEM SOLVING

Solve using guess and check.

20. Two angles are supplementary. The first angle has a measure one third that of the second. Find the measures of the angles.

Chapter 5 Test

True or false? Exercises 1–5 refer to the figure.

1. If $\overrightarrow{BA} \cong \overrightarrow{BC}$, and $\overleftrightarrow{BY}$ is a line of symmetry, then $\overline{AY} \cong \overline{YC}$.
2. If $\angle A \cong \angle ABC$, then $\overline{AC} \cong \overline{BC}$.
3. It is possible that $AB = 8$ cm, $BY = 5$ cm, and $AY = 3$ cm.
4. If $\triangle ABC$ is equilateral, then $\triangle ABC$ is equiangular.
5. If $\overleftrightarrow{BY}$ is the perpendicular bisector of $\overline{AC}$, then $BA = BC$.

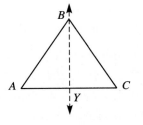

6. For every isosceles triangle there is a segment that is an angle bisector, an altitude, and a median and that determines a line of symmetry for the triangle.
7. The altitudes of any triangle intersect in a point inside the triangle.

For each exercise, draw a large obtuse scalene $\triangle DVY$.

8. Construct the altitude to $\overline{VY}$.
9. Construct the median from D.
10. Construct the incenter.
11. Construct the circumcenter.
12. List the angles from smallest to largest.
13. List the sides from shortest to longest.

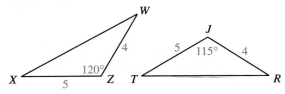

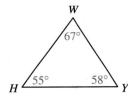

Complete each statement with < or >.

14. $XW \ \blacksquare \ TR$
15. $m\angle P \ \blacksquare \ m\angle D$

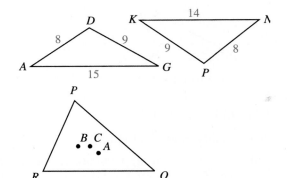

PROBLEM SOLVING

Solve using guess and check.

16. Which point is the circumcenter of $\triangle PQR$?

Chapter 6

Congruent Triangles

The polygons that make up the alternating light-and-dark pattern of this floor are all triangles of the same size and shape. The apparent shrinking of the triangles with increasing distance is a perspective effect.

Focus on Skills

ALGEBRA

Solve each equation.

1. $5x - 13 = 2x + 38$ **2.** $12t - 4 = 3t + 23$ **3.** $x + 14 = 3x - 12$

4. $16 - 5x = 1 - 2x$ **5.** $43 - 4a = 8 + 3a$ **6.** $11y + 9 = 7y + 25$

7. Find values of x and y such that $4x - 2 = x + 16$ and $x - y = 1$.

8. Find values of x and y such that $2x + 5 = 23$, $3y - 11 = y + 9$, and $x + y = 19$.

GEOMETRY

Can the lengths be the lengths of the sides of a triangle?

9. 4 cm, 4 cm, 3 cm   **10.** 1 m, 2 m, 3 m   **11.** 2 cm, 3 cm, 8 cm

Draw a sketch to illustrate the given information.

12. $\triangle ABC$ is a right triangle with $m\angle B = 90°$ and $\overline{AB} \cong \overline{BC}$.

13. $\triangle DEF$ is an acute scalene triangle.

14. $\overleftrightarrow{PQ}$ and $\overleftrightarrow{RS}$ are perpendicular lines, and $\overleftrightarrow{PQ}$ intersects $\overline{RS}$ at its midpoint, M.

15. $\triangle TRI$ is an isosceles triangle, and $\overrightarrow{RA}$ is the bisector of the vertex angle.

Draw a figure like the one shown and construct each of the following.

16. a segment congruent to $\overline{AB}$.   **17.** an angle congruent to $\angle B$

18. a perpendicular to $\overrightarrow{BA}$ from C **19.** the bisector of $\angle A$

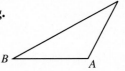

Refer to the figure, in which $\triangle GHK \cong \triangle WXY$.

20. Write three congruence statements for three pairs of congruent angles.

21. Write three congruence statements for three pairs of congruent sides.

22. Complete: $\triangle HGK \cong$ ▧ and $\triangle YWX \cong$ ▧

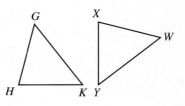

6.1 SSS

Objective: To use the SSS (Side-Side-Side) Postulate to show that two triangles are congruent.

Recall that two triangles are congruent when their six pairs of corresponding parts are congruent. In this chapter, you will learn other ways to show that two triangles are congruent.

CONSTRUCTION 7

Construct a triangle congruent to $\triangle ABC$ by constructing segments congruent to $\overline{AB}, \overline{AC},$ and $\overline{BC}$.

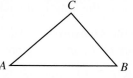

1. With a straightedge, draw a segment longer than $\overline{AB}$. Construct $\overline{PQ} \cong \overline{AB}$.

2. Set your compass for length AC. With the compass point on P, draw an arc above $\overline{PQ}$. (This arc is part of a circle with center P and radius AC.)

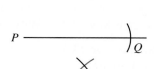

3. Set your compass for length BC. With the compass point on Q, draw an arc above $\overline{PQ}$ that intersects the arc drawn in Step 2.

4. Point R, where the two arcs intersect, is the third vertex of the triangle. With a straightedge, draw $\overline{PR}$ and $\overline{QR}$. Then $\triangle PQR \cong \triangle ABC$.

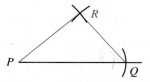

Construction 7 demonstrates that the lengths of the three sides of a triangle completely determine the size and shape of the triangle.

POSTULATE 6 (SSS Postulate): If three sides of one triangle are congruent to the corresponding sides of another triangle, then the triangles are congruent.

Example: Is △ABC congruent to △DBC?

Solution: $\overline{AC} \cong \overline{DC}$ and $\overline{AB} \cong \overline{DB}$

$\overline{BC} \cong \overline{BC}$ (A segment is congruent to itself.)

Therefore △ABC ≅ △DBC by the SSS Postulate.

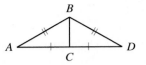

EXPLORING

1. Use the *Geometric Supposer: Triangles* disk to create any triangle. Measure the length of each side.

2. Use the *Measure* and *Adjustable Elements* options to create and move points D, E, and F so that DE = AB, DF = AC, and EF = BC.

3. Draw $\overline{DE}$, $\overline{EF}$, and $\overline{DF}$. Is △DEF ≅ △ABC? Explain.

Thinking Critically

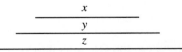

1. Could you construct a triangle whose sides have lengths x, y, and z? Why or why not?

2. Suppose you know that $\overline{RS} \cong \overline{GH}$. In order to use the SSS Postulate to show that △RST ≅ △GHI, what else must you show to be true?

Class Exercises

1. Draw segments like those shown. Construct a triangle with sides of length x, y, and z.

State whether the SSS Postulate could be used to show that each pair of triangles are congruent.

2.   3.   4.

5. To use the SSS Postulate to prove that △XYZ ≅ △RST, you must show that $\overline{XY} \cong$ ■, $\overline{YZ} \cong$ ■, and $\overline{ZX} \cong$ ■.

Exercises

For Exercises 1–5 use four segments with the following lengths.

    a = 3 cm   *b* = 4 cm   *c* = 5 cm   *d* = 8 cm

Construct a triangle to fit each description, *if possible*.

1. sides with lengths *a*, *b*, and *c*
           2. sides with lengths *b*, *c*, and *d*

3. sides with lengths *a*, *b*, and *d*
           4. all sides with length *c*

5. two sides with length *c* and
one side with length *a*

State which triangles constructed in Exercises 1–5 are of each type.
For Exercises 9–11 judge by appearance.

6. scalene   **7.** isosceles   **8.** equilateral

9. obtuse   **10.** acute   **11.** right

12. Draw a triangle similar to, but larger than, the one shown. Construct $\triangle DEF \cong \triangle XYZ$.

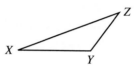

State whether the SSS Postulate could be used to show that each pair of triangles are congruent.

13.   **14.**   **15.**

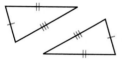      

16. To use the SSS Postulate to prove that $\triangle PQR \cong \triangle GHK$, you must show that $\overline{PQ} \cong \blacksquare$, $\overline{QR} \cong \blacksquare$, and $\overline{RP} \cong \blacksquare$.

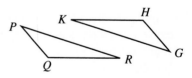

17. $\overline{AB}$ and $\blacksquare$ are corresponding sides.

18. $\angle Z$ and $\blacksquare$ are corresponding angles.

19. Why is $\triangle ABZ \cong \triangle WLC$?

20. If $m\angle A = 65°$, then the other angle that has measure 65° is $\blacksquare$.

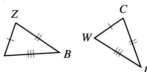

21. To use the SSS Postulate to prove that $\triangle DEF \cong \triangle MNO$, you must show that $\overline{DE} \cong \blacksquare$, $\overline{EF} \cong \blacksquare$, and $\overline{DF} \cong \blacksquare$.

22. Draw one segment 4 cm long and another 5 cm long. Then construct an isosceles triangle with a base 5 cm long and legs 4 cm long.

A triangle is called a *rigid* figure because the length of at least one side of a triangle must be changed before its shape changes. State whether each of the following is a rigid figure.

23.

24.

25.

26.

27. If $\overline{CA} \cong \overline{DB}$ and $\overline{CB} \cong \overline{DA}$, write a congruence statement involving two triangles.

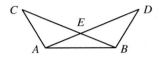

APPLICATIONS

28. **Algebra**   In $\triangle ABC$, $AB = x + 3$, $BC = 3y$, and $AC = x + y$. In $\triangle DEF$, $DE = 2x - 5$, $EF = y + 6$, and $DF = 11$.

    a. Find the values of x and y for which $\triangle ABC \cong \triangle DEF$.

    b. Find the perimeter of $\triangle ABC$.

29. **Computer**   Use the *Geometric Supposer: Triangles* disk to create a triangle. Then describe how you would use the *SSS* option to create a new triangle congruent to the first triangle.

Everyday Geometry

Triangular forms are used in construction because a triangle is a rigid figure. A triangle will always hold its shape, while other figures will collapse unless support is added.

Many roofs have a triangular cross section. Triangles are also used in constructing geodesic domes like the one shown here. A *geodesic dome* is a dome-shaped building that needs no interior support. It was one of the many ideas of the inventor R. Buckminster Fuller. Geodesic domes have been built as large as 200 ft high and 250 ft across.

Research different ways in which geodesic domes are used in modern architecture.

6.2   SAS

Objective: To use the SAS (Side-Angle-Side) Postulate to show that two triangles are congruent.

In this lesson you will learn another way to show that two triangles are congruent.

━━━━━━━━━━━━━━━━ **CONSTRUCTION 8** ━━━━━━━━━━━━━━━━

Construct a triangle congruent to △ABC by constructing an angle and segments congruent to ∠A, $\overline{AB}$, and $\overline{AC}$.

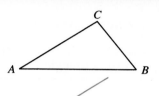

1. Draw a segment longer than $\overline{AB}$ and label one endpoint as P. Construct $\angle P \cong \angle A$.

2. On the sides of $\angle P$, construct $\overline{PQ} \cong \overline{AB}$ and $\overline{PR} \cong \overline{AC}$. Draw $\overline{RQ}$. Then $\triangle PQR \cong \triangle ABC$.

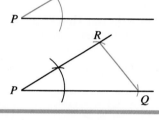

Construction 8 demonstrates the following postulate.

POSTULATE 7 (SAS Postulate):   **If two sides and the included angle of one triangle are congruent to the corresponding sides and angle of another triangle, then the triangles are congruent.**

Example 1: Which pairs of triangles are congruent by the SAS Postulate?

a. 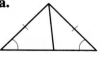   **b.**   **c.**   **d.**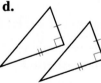

Solution: (b) and (d)

Notice that the triangles in part (d) of Example 1 are right triangles. The following terms are used to describe the sides of a right triangle.

Legs of a right triangle — sides that determine the right angle

Hypotenuse — side opposite the right angle of a right triangle

---

Example 2: The hypotenuse of △*DEF* is $\overline{DF}$.
The legs of △*DEF* are $\overline{DE}$ and $\overline{EF}$.

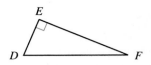

Thinking Critically

△*ABC* and △*DEF* are right triangles whose corresponding legs are congruent.

1. Why is ∠*B* ≅ ∠*E*?

2. Why is △*ABC* ≅ △*DEF*?

3. State an LL (Leg–Leg) Theorem for congruence of right triangles.

Class Exercises

State whether the SAS Postulate could be used to show that each pair of triangles are congruent.

1.

2.

3.

Draw a triangle similar to, but larger than, the one shown. Use the indicated sides and angle to construct a triangle congruent to △*RST*.

4. $\overline{RT}, \overline{RS}, ∠R$ 5. $\overline{RS}, \overline{ST}, ∠S$

Exercises

Draw a triangle similar to, but larger than, the one shown. Use the given method and indicated sides and angles to construct a triangle congruent to △*XYZ*.

1. SAS: $\overline{XY}$, ∠*X*, and $\overline{XZ}$ 2. SAS: $\overline{YZ}$, ∠*Y*, and $\overline{XY}$ 3. SSS: $\overline{XY}, \overline{XZ},$ and $\overline{YZ}$

State whether the SAS Postulate could be used to show that each pair of triangles are congruent.

4.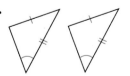

5.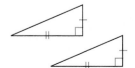

6.

Suppose you know that the indicated parts of the triangles are congruent. Which postulate or theorem, if any, could you use to prove that $\triangle ABC \cong \triangle HGK$?

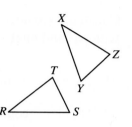

7. $\overline{AB} \cong \overline{HG}, \overline{AC} \cong \overline{HK}, \angle A \cong \angle H$

8. $\overline{AB} \cong \overline{HG}, \overline{AC} \cong \overline{HK}, \overline{BC} \cong \overline{GK}$

9. $\overline{AB} \cong \overline{HG}, \overline{BC} \cong \overline{GK}, \angle B \cong \angle G$

10. $\overline{AB} \cong \overline{HG}, \overline{AC} \cong \overline{HK}, \angle C \cong \angle K$

11. $\angle B \cong \angle G, \angle A \cong \angle H, \angle C \cong \angle K$

12. $\angle C$ and $\angle K$ are right angles; $\overline{AC} \cong \overline{HK}, \overline{BC} \cong \overline{GK}$

Suppose you wish to prove that $\triangle RST \cong \triangle XYZ$. Complete each statement.

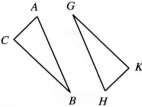

13. To use SSS, you must show that $\overline{RS} \cong$ ▪, $\overline{ST} \cong$ ▪, and $\overline{TR} \cong$ ▪.

14. To use SAS, you could show that $\overline{RT} \cong$ ▪, $\overline{ST} \cong$ ▪, and $\angle T \cong$ ▪.

15. To use SAS, you could show that $\angle Y \cong$ ▪, $\overline{YX} \cong$ ▪, and $\overline{YZ} \cong$ ▪.

Using the given information and method, what additional information would you need to prove that $\triangle PQR \cong \triangle JKL$?

16. $\overline{PR} \cong \overline{JL}$; SSS Postulate

17. $\angle P \cong \angle J$; SAS Postulate

18. $\overline{QR} \cong \overline{KL}$; SAS Postulate (2 solutions)

19. $\angle Q$ and $\angle K$ are right angles; LL Theorem

20. **a.** Is $\triangle DEF \cong \triangle MNO$?

   b. Is there an AAA (Angle-Angle-Angle) Postulate that you could use to show that triangles are congruent? Why or why not?

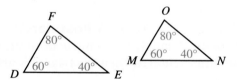

21. **Algebra and Logic**   Are there values of x and y for which the triangles shown are congruent? If so, find those values and write a congruence statement for the triangles.

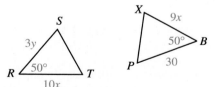

Seeing in Geometry

Write a congruence statement for the triangles and explain why they are congruent.

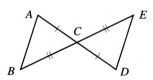

6.3 ASA and AAS

Objective: To use the ASA (Angle-Side-Angle) Postulate and the AAS (Angle-Angle-Side) Theorem to show that two triangles are congruent.

In this lesson you will learn two more ways to show that two triangles are congruent.

CONSTRUCTION 9

Construct a triangle congruent to △ABC by constructing a segment and angles congruent to $\overline{AB}$, ∠A, and ∠B.

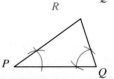

1. Draw a segment longer than $\overline{AB}$. Construct $\overline{PQ} \cong \overline{AB}$.

2. At point P, construct an angle congruent to ∠A. At Q, construct an angle congruent to ∠B. The intersection of the sides of these angles is the third vertex, R, of the triangle. Then $\triangle PQR \cong \triangle ABC$.

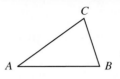

Construction 9 demonstrates the following postulate.

POSTULATE 8 (ASA Postulate): If two angles and the included side of one triangle are congruent to the corresponding angles and side of another triangle, then the triangles are congruent.

EXPLORING

1. Use the *Geometric Supposer: Triangles* disk and the *Your Own* option to create △ABC such that m∠BAC = 75°, AB = 6, and m∠CBA = 25°. Sketch the triangle and record the measurements of all the sides and angles.

2. Repeat Step 1, creating △ABC such that m∠BAC = 75°, AB = 9, and m∠CBA = 25°. What is true about the third angle in both triangles?

3. Repeat Steps 1 and 2 using other angle and side measures in the same manner. Are the results the same?

The **EXPLORING** activity demonstrates the following theorem.

THEOREM 6.1:   **If two angles of one triangle are congruent to two angles of another triangle, then the third pair of angles are congruent.**

EXPLORING

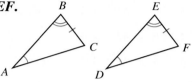

In △ABC and △DEF, ∠A ≅ ∠D, ∠B ≅ ∠E, and $\overline{BC}$ ≅ $\overline{EF}$.

  1. Why is ∠C ≅ ∠F?

  2. Why is △ABC ≅ △DEF?

The **EXPLORING** activity demonstrates the following theorem.

THEOREM 6.2 (AAS Theorem):   **If two angles and a non-included side of one triangle are congruent to the corresponding angles and side of another triangle, then the triangles are congruent.**

Example: **Which postulate or theorem, if any, could you use to show that each pair of triangles are congruent?**

a.   **b.**   **c.**

Solution: **a. AAS Theorem**   **b. SAS Postulate**   **c. ASA Postulate**

Thinking Critically

  1. Given the information for the right triangles, which postulate or theorem could you use to prove that △ABC ≅ △DEF?

  a. $\overline{AB}$ ≅ $\overline{DE}$, ∠A ≅ ∠D   **b.** $\overline{AB}$ ≅ $\overline{DE}$, ∠C ≅ ∠F

  c. $\overline{BC}$ ≅ $\overline{EF}$, ∠A ≅ ∠D   **d.** $\overline{BC}$ ≅ $\overline{EF}$, ∠C ≅ ∠F

  e. $\overline{AC}$ ≅ $\overline{DF}$, ∠A ≅ ∠D   **f.** $\overline{AC}$ ≅ $\overline{DF}$, ∠C ≅ ∠F

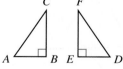

  2. Explain an LA (Leg-Angle) Theorem and an HA (Hypotenuse-Angle) Theorem that you could use to show that right triangles are congruent.

Class Exercises

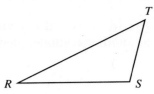

Draw a triangle similar to, but larger than, the one shown. Use the indicated side and angles to construct a triangle congruent to △RST.

1. $\overline{RS}$, ∠R, ∠S   **2.** $\overline{ST}$, ∠T, ∠S

Given the information, which postulate or theorem, if any, could you use to prove that △ABC ≅ △MNO?

3. $\overline{AB}$ ≅ $\overline{MN}$, ∠A ≅ ∠M, ∠B ≅ ∠N

4. $\overline{AB}$ ≅ $\overline{MN}$, ∠A ≅ ∠M, ∠C ≅ ∠O

5. $\overline{AB}$ ≅ $\overline{MN}$, $\overline{AC}$ ≅ $\overline{MO}$, ∠C ≅ ∠O

6. $\overline{AC}$ ≅ $\overline{MO}$, ∠A ≅ ∠M, ∠C ≅ ∠O

7. $\overline{AB}$ ≅ $\overline{MN}$, $\overline{BC}$ ≅ $\overline{NO}$, $\overline{AC}$ ≅ $\overline{MO}$

8. ∠A ≅ ∠M, ∠B ≅ ∠N, ∠C ≅ ∠O

9. ∠B and ∠N are right angles; ∠A ≅ ∠M, $\overline{AC}$ ≅ $\overline{MO}$

10. ∠B and ∠N are right angles; ∠A ≅ ∠M, $\overline{BC}$ ≅ $\overline{NO}$

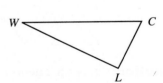

Exercises

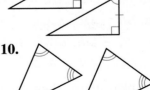

Draw a triangle similar to, but larger than, the one shown. Use the given method and indicated sides and angles to construct a triangle congruent to △WLC.

1. ASA: $\overline{WC}$, ∠W, ∠C   **2.** ASA: $\overline{LC}$, ∠L, ∠C
3. SAS: $\overline{WC}$, $\overline{LC}$, ∠C   **4.** SSS: $\overline{WC}$, $\overline{LC}$, $\overline{WL}$

Which postulate or theorem, if any, could you use to show that each pair of triangles are congruent?

5. 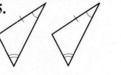   **6.** 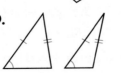   **7.**

8. 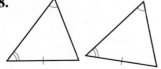   **9.**   **10.**

Some information is given about △ABC and △DEF. What additional information would enable you to use the indicated method to prove that △ABC ≅ △DEF?

11. $\overline{AC} \cong \overline{DF}$; ASA Postulate

12. ∠A ≅ ∠D, ∠B ≅ ∠E; ASA Postulate

13. ∠A ≅ ∠D; SAS Postulate

14. $\overline{AC} \cong \overline{DF}$; SSS Postulate

15. ∠B ≅ ∠E, $\overline{AB} \cong \overline{DE}$; AAS Theorem

16. ∠C and ∠F are right angles; LL Theorem

17. $\overline{BC} \cong \overline{EF}$; AAS Theorem (2 solutions)

18. $\overline{BC} \cong \overline{EF}$; SAS Theorem (2 solutions)

19. ∠C and ∠F are right angles; HA Theorem (2 solutions)

20. ∠A ≅ ∠D, ∠B ≅ ∠E; AAS Theorem (2 solutions)

21. ∠C and ∠F are right angles; LA Theorem (4 solutions)

22. ∠A ≅ ∠D; AAS Theorem (4 solutions)

23. a. Could you use the SAS Postulate to show that △ABC ≅ △RST? Why or why not?

b. Is △ABC ≅ △RST?

c. Is there an SSA (Side-Side-Angle) Postulate that you could use to show that triangles are congruent? Why or why not?

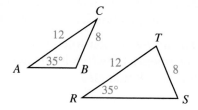

APPLICATION

24. Computer   Use LOGO to run the following. Describe the result.

fd 60 rt 117 fd 100 pu home pd rt 47 fd 100 ht

Test Yourself

Draw a triangle similar to, but larger than, △DEF. Use the indicated sides and angles to construct △ABC ≅ △DEF.

1. $\overline{DE}$, $\overline{EF}$, $\overline{DF}$ **2.** $\overline{DF}$, ∠D, $\overline{DE}$ **3.** ∠D, $\overline{DE}$, ∠E

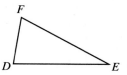

Which postulate or theorem, if any, could you use to show that each pair of triangles are congruent?

4.   **5.**   **6.**

6.4 HL and CPCTC

Objective: To use the HL (Hypotenuse-Leg) Theorem and other postulates and theorems to show that two triangles are congruent and that their corresponding parts are congruent.

Three sets of conditions for determining congruence of two right triangles were derived as special cases of SAS, ASA, and AAS. Two right triangles are congruent if one of the following is true.

1. The legs of one triangle are congruent to the legs of the other. (LL)

2. The hypotenuse and an acute angle of one triangle are congruent to the hypotenuse and an acute angle of the other. (HA)

3. A leg and an acute angle of one triangle are congruent to the corresponding leg and acute angle of the other. (LA)

The following construction demonstrates the only method for determining congruence of two right triangles that is not derived from a method which is true for all triangles.

━━━━━━━━━━━━━━━ **CONSTRUCTION 10** ━━━━━━━━━━━━━━━

Construct a triangle congruent to △ABC by constructing segments congruent to $\overline{AB}$ and $\overline{AC}$ and an angle congruent to right ∠B.

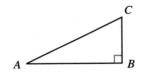

1. Use Construction 2, 3, or 4 to construct a right angle. (Construction 4 is shown.)

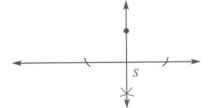

2. On one side of the right angle, construct $\overline{SR} \cong \overline{BA}$.

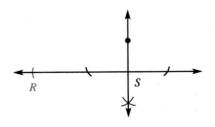

---

3. Set your compass for length *AC*. With the compass point on *R*, draw an arc intersecting the other side of the right angle at *T*. Draw $\overline{RT}$. $\overline{RT} \cong \overline{AC}$ and $\triangle RST \cong \triangle ABC$.

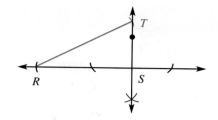

Construction 10 demonstrates the following theorem.

THEOREM 6.3 (HL Theorem): If the hypotenuse and one leg of a right triangle are congruent to the hypotenuse and one leg of another right triangle, then the right triangles are congruent.

You have learned a number of ways to show that two triangles are congruent by showing that certain pairs of corresponding parts are congruent. Recall that when two triangles are congruent, all six pairs of corresponding parts are congruent. One way to prove that two angles or segments are congruent is to show that they are corresponding parts of congruent triangles.

Using Congruent Triangles to Prove Angles or Segments Congruent

1. Find two triangles in which the angles or sides are corresponding parts.
2. Prove that the triangles are congruent.
3. State that angles or sides are congruent because corresponding parts of congruent triangles are congruent (CPCTC).

Example: **Given:** $\overline{RQ} \cong \overline{RT}$ and $\overline{RP} \cong \overline{RS}$

Prove: $\angle Q \cong \angle T$, $\angle P \cong \angle S$, and $\overline{QP} \cong \overline{TS}$

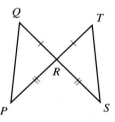

Solution: **a.** $\overline{RQ} \cong \overline{RT}$ and $\overline{RP} \cong \overline{RS}$ (Given)

b. $\angle QRP \cong \angle TRS$ (Vertical Angles Theorem)

c. $\triangle QRP \cong \triangle TRS$ (SAS)

d. $\angle Q \cong \angle T$, $\angle P \cong \angle S$, and $\overline{QP} \cong \overline{TS}$ (CPCTC)

Given: $\overline{DA} \cong \overline{BA}$ and $\overline{CD} \cong \overline{CB}$ **Prove:** $\angle 1 \cong \angle 2$

1. $\angle 1 \cong \angle 2$ if ▧ $\cong$ ▧.

2. Give a reason for each statement.

   a. $\overline{DA} \cong \overline{BA}$ and $\overline{CD} \cong \overline{CB}$
   b. $\overline{AC} \cong \overline{AC}$
   c. $\triangle ADC \cong \triangle ABC$
   d. $\angle 1 \cong \angle 2$

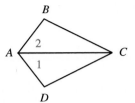

Class Exercises

Which postulate or theorem could you use to show that each pair of triangles are congruent?

1. 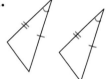   **2.**   **3.**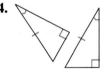

4. 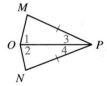   **5.**    **6.**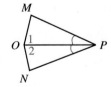

What additional congruence would you need to know in order to prove that the triangles are congruent by each indicated method?

7. a. SSS
   b. SAS

8. a. SAS
   b. ASA
   c. AAS

Name two triangles that you would prove congruent in order to prove that each congruence statement is true.

9. $\angle 1 \cong \angle 2$ **10.** $\angle C \cong \angle D$

11. $\angle CAB \cong \angle DBA$ **12.** $\overline{CE} \cong \overline{DE}$

13. $\overline{CB} \cong \overline{DA}$ **14.** $\overline{CA} \cong \overline{DB}$

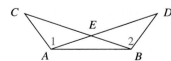

Exercises

What postulate or theorem could you use to show that each pair of triangles are congruent?

1.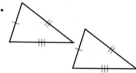

2.

3.

4.

5.

6.

7.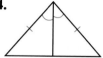

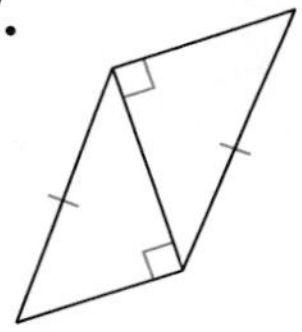

8.

9.

What additional congruence would you need to know in order to prove that the triangles are congruent by each indicated method?

10. a. SAS
   b. ASA
   c. AAS

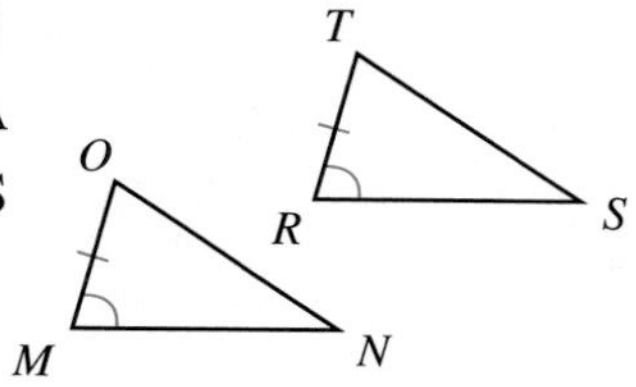

11. a. SSS
   b. SAS

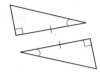

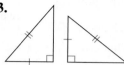

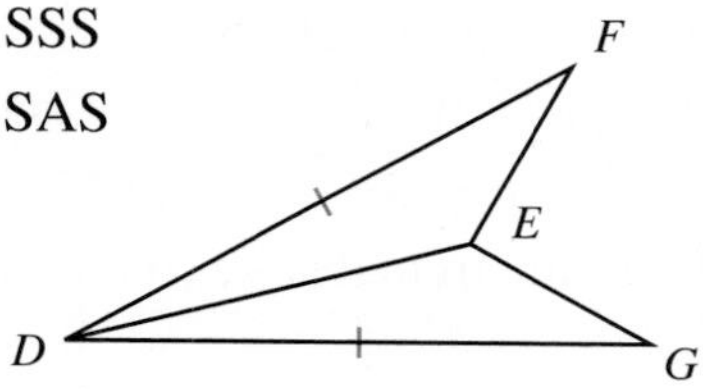

What additional congruences would you need to know in order to prove that the right triangles are congruent by each indicated method?

12. SAS   **13.** LL

14. HA (2 solutions)

15. HL (2 solutions)

16. ASA (2 solutions)

17. AAS (4 solutions)

18. LA (4 solutions)

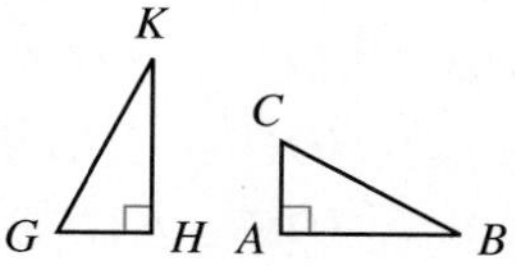

Name two triangles you would prove congruent in order to prove that each congruence statement is true.

19. $\angle 1 \cong \angle 2$

20. $\angle 3 \cong \angle 4$

21. $\angle ZWX \cong \angle YXW$

22. $\overline{WZ} \cong \overline{XY}$

23. $\overline{ZX} \cong \overline{YW}$

24. $\overline{WU} \cong \overline{XU}$

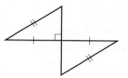

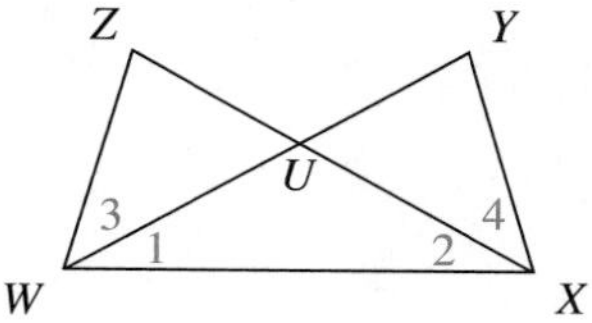

25. a. Write a congruence statement for the two triangles.

  b. Which postulate or theorem did you use?

  c. List three additional congruence statements for corresponding parts of the triangles.

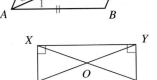

26. $\angle WXY$ and $\angle ZYX$ are right angles; $\overline{WY} \cong \overline{ZX}$.

  a. What could you prove? Explain.

  b. List any additional congruence statements that follow from your answer to part (a).

  c. Are there any other congruent triangles shown? Explain.

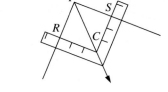

APPLICATIONS

27. Carpentry   A carpenter's square can be used to bisect an angle at the corner of a board. Mark equal lengths PS and PR along the edges. Put the carpenter's square on the board so that $CS = CR$. Mark point C and draw $\overrightarrow{PC}$.

  a. Which postulate or theorem can you use to show that $\triangle PRC \cong \triangle PSC$?

  b. Why is $\angle RPC \cong \angle SPC$?

  c. Why is $\overrightarrow{PC}$ the bisector of $\angle RPS$?

28. Algebra   In triangles RJV and WXP, $\angle J$ and $\angle X$ are right angles. $\overline{RV} \cong \overline{WP}$, $VJ = 3y - 3$, and $PX = y + 7$. For what value of y will the triangles be congruent?

Thinking in Geometry

1. Why is $\triangle ABC \cong \triangle HDC$?

2. Why is $\triangle HDC \cong \triangle HDE$?

3. Why is $\triangle HDE \cong \triangle GFE$?

4. Why is $\angle B \cong \angle F$?

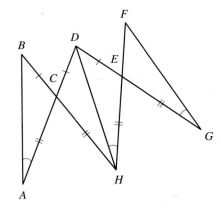

6.5 Proving Theorems

Objective: To use congruent triangles to prove theorems and to justify constructions.

A *proof* is a convincing argument. One common form of a geometric proof is a logical sequence of statements that begins with given information and ends with the statement that is to be proven. Each statement is justified with a reason such as a postulate, a definition, or a theorem.

Example 1: **Given:** $\triangle RSZ \cong \triangle TSZ$

          Prove: $\overline{SZ} \perp \overline{RT}$

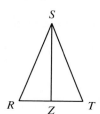

Solution: **a.** $\triangle RSZ \cong \triangle TSZ$ (Given)

             b. $\angle$RZS $\cong \angle$TZS (CPCTC)

             c. $\angle$RZS and $\angle$TZS are right angles. (Angles that are congruent and supplementary are right angles.)

             d. $\overline{SZ} \perp \overline{RT}$ (Definition of perpendicular lines)

In the reason for part (c) of Example 1, the fact that $\angle RZS$ and $\angle TZS$ are supplementary was assumed from the figure. For any proofs you write in the rest of this course, you may assume that this statement is a theorem: *Angles that are congruent and supplementary are right angles.* Another statement you will find useful as a reason is: *A segment is congruent to itself.*

Before you can prove a theorem, you need a diagram that represents the information in the theorem. The given information and what is to be proven are described in terms of the diagram.

Example 2: **For the following statement, draw an appropriate diagram. State, in terms of the diagram, what is given and what is to be proven.**

If a line bisects the vertex angle of an isosceles triangle, then the line is the perpendicular bisector of the base.

Solution: **Given:** $\triangle ABC$ is isosceles, with $\overline{AC} \cong \overline{BC}$; $\overleftrightarrow{CD}$ bisects $\angle ACB$.

          Prove: $\overleftrightarrow{CD}$ is the perpendicular bisector of $\overline{AB}$.

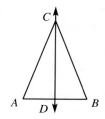

Given: $\triangle ABC$ is isosceles, with $\overline{AC} \cong \overline{BC}$;
$\overleftrightarrow{CD}$ bisects $\angle ACB$.

Prove: $\overleftrightarrow{CD}$ is the perpendicular bisector of $\overline{AB}$.

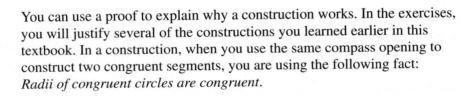

Give a reason for each statement.

1. $\overline{AC} \cong \overline{BC}$; $\overleftrightarrow{CD}$ bisects $\angle ACB$.

2. $\angle ACD \cong \angle BCD$

3. $\overline{CD} \cong \overline{CD}$

4. $\triangle ACD \cong \triangle BCD$

5. $\angle ADC \cong \angle BDC$

6. $\angle ADC$ and $\angle BDC$ are right angles.

7. $\overleftrightarrow{CD} \perp \overline{AB}$

8. $\overline{AD} \cong \overline{BD}$

9. $\overleftrightarrow{CD}$ is the perpendicular bisector of $\overline{AB}$.

You can use a proof to explain why a construction works. In the exercises, you will justify several of the constructions you learned earlier in this textbook. In a construction, when you use the same compass opening to construct two congruent segments, you are using the following fact: *Radii of congruent circles are congruent.*

Thinking Critically

For each of the following, draw an appropriate diagram. Explain in terms of the diagram what is given and what you are to prove.

1. If a point lies on the perpendicular bisector of a segment, then it is equidistant from the endpoints of the segment.

2. Each diagonal of a rhombus bisects a pair of opposite angles.

Class Exercise

1. $\angle B$ was given, and Construction 5 was used to construct $\angle P$. To prove that $\angle B \cong \angle P$, give a reason for each statement.

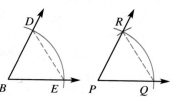

  a. $\overline{BD} \cong \overline{PR}, \overline{BE} \cong \overline{PQ}, \overline{ED} \cong \overline{QR}$

  b. $\triangle BDE \cong \triangle PRQ$

  c. $\angle B \cong \angle P$

Exercises

1. $\angle XAY$ was given, and Construction 6 was used to construct $\overrightarrow{AD}$.

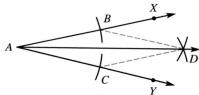

To prove that $\overrightarrow{AD}$ is the bisector of $\angle XAY$, give a reason for each statement.

  a. $\overline{AB} \cong \overline{AC}$ and $\overline{BD} \cong \overline{CD}$

  b. $\overline{AD} \cong \overline{AD}$

  c. $\triangle ABD \cong \triangle ACD$

  d. $\angle BAD \cong \angle CAD$

  e. $\overrightarrow{AD}$ is the bisector of $\angle XAY$.

2. Point A on line l was given, and Construction 3 was used to construct $\overleftrightarrow{AD}$.

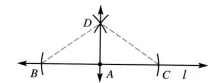

To prove that $\overleftrightarrow{AD} \perp l$, give a reason for each statement.

  a. $\overline{AB} \cong \overline{AC}$ and $\overline{BD} \cong \overline{CD}$

  b. $\overline{AD} \cong \overline{AD}$

  c. $\triangle ABD \cong \triangle ACD$

  d. $\angle BAD \cong \angle CAD$

  e. $\angle BAD$ and $\angle CAD$ are right angles.

  f. $\overleftrightarrow{AD} \perp l$

3. **To be Proven:** Isosceles Triangle Theorem: The base angles of an isosceles triangle are congruent. (*Note:* To help prove the theorem, you can construct $\overrightarrow{AD}$, the bisector of $\angle BAC$.)

  Given: $\triangle ABC$ is isosceles, with $\overline{AB} \cong \overline{AC}$; $\overrightarrow{AD}$ is the bisector of $\angle BAC$.
  Prove: $\angle B \cong \angle C$
  Give a reason for each statement.

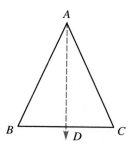

  a. $\overline{AB} \cong \overline{AC}$

  b. $\overrightarrow{AD}$ is the bisector of $\angle BAC$.

  c. $\angle BAD \cong \angle CAD$

  d. $\overline{AD} \cong \overline{AD}$

  e. $\triangle BAD \cong \triangle CAD$

  f. $\angle B \cong \angle C$

Give a reason for each statement.

4. To be Proven: If a point lies on the perpendicular bisector of a segment, then it is equidistant from the endpoints of the segment.

Given: $\overleftrightarrow{PE}$ is the perpendicular bisector of $\overline{AB}$.

Prove: $\overline{AP} \cong \overline{BP}$

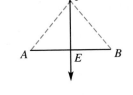

a. $\overleftrightarrow{PE}$ is the perpendicular bisector of $\overline{AB}$.

b. $\overline{AE} \cong \overline{BE}$; $\angle AEP$ and $\angle BEP$ are right angles.

c. $\triangle AEP$ and $\triangle BEP$ are right triangles.

d. $\overline{PE} \cong \overline{PE}$

e. $\triangle APE \cong \triangle BPE$

f. $\overline{AP} \cong \overline{BP}$

5. To be Proven: Each point on the bisector of an angle is equidistant from the sides of the angle.

Given: $\overrightarrow{NP}$ bisects $\angle MNO$; $\overline{PQ} \perp \overrightarrow{NM}$; $\overline{PR} \perp \overrightarrow{NO}$

Prove: $\overline{PQ} \cong \overline{PR}$

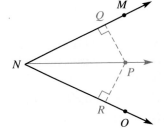

a. $\overrightarrow{NP}$ bisects $\angle MNO$.

b. $\angle QNP \cong \angle RNP$

c. $\overline{PQ} \perp \overrightarrow{NM}$ and $\overline{PR} \perp \overrightarrow{NO}$

d. $\angle PQN$ and $\angle PRN$ are right angles.

e. $\triangle PQN$ and $\triangle PRN$ are right triangles.

f. $\overline{NP} \cong \overline{NP}$

g. $\triangle QNP \cong \triangle RNP$

h. $\overline{PQ} \cong \overline{PR}$

6. $\overline{AB}$ was given, and Construction 2 was used to construct $\overleftrightarrow{XY}$. To prove that $\overleftrightarrow{XY}$ is the perpendicular bisector of $\overline{AB}$, give a reason for each statement.

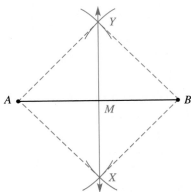

a. $\overline{XA} \cong \overline{XB}$ and $\overline{YA} \cong \overline{YB}$

b. $\overline{XY} \cong \overline{XY}$

c. $\triangle XAY \cong \triangle XBY$

d. $\angle AXY \cong \angle BXY$

e. $\overline{XM} \cong \overline{XM}$

f. $\triangle XAM \cong \triangle XBM$

g. $\angle XMA \cong \angle XMB$ and $\overline{MA} \cong \overline{MB}$

h. $\angle XMA$ and $\angle XMB$ are right angles.

i. $\overleftrightarrow{XY}$ is the perpendicular bisector of $\overline{AB}$.

7. To be Proven: If a segment is the altitude from the vertex angle of an isosceles triangle, then it is a median of the triangle.

Given: △ABC is isosceles, with $\overline{AB} \cong \overline{AC}$; $\overline{AD}$ is an altitude of △ABC.

Prove: $\overline{AD}$ is a median of △ABC.

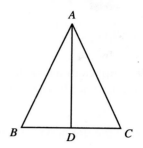

a. $\overline{AB} \cong \overline{AC}$

b. $\overline{AD} \cong \overline{AD}$

c. $\overline{AD}$ is an altitude of △ABC.

d. $\overline{AD} \perp \overline{BC}$

e. ∠ADB and ∠ADC are right angles.

f. △ADB and △ADC are right triangles.

g. △ADB ≅ △ADC

h. $\overline{BD} \cong \overline{CD}$

i. D is the midpoint of $\overline{BC}$.

j. $\overline{AD}$ is a median of △ABC.

APPLICATION

8. Visualization It is possible to prove the Isosceles Triangle Theorem without drawing the bisector of the vertex angle. (See Exercise 3.)

Given: △ABC is isosceles, with $\overline{AB} \cong \overline{AC}$.

Prove: ∠B ≅ ∠C

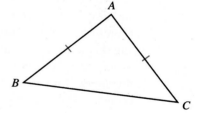

a. For the correspondence △ABC ↔ △ACB, which pairs of corresponding parts must be congruent?

b. Why is △ABC ≅ △ACB?

c. Why is ∠B ≅ ∠C?

Thinking in Geometry

To be Proven: If two angles of a triangle are congruent, then the sides opposite those angles are congruent.

Given: ∠B ≅ ∠C

Prove: $\overline{AB} \cong \overline{AC}$

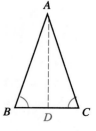

In order to prove that $\overline{AB} \cong \overline{AC}$, you would like to use some point D on $\overline{BC}$ that enables you to show that △ABD ≅ △ACD. If $\overline{AD}$ is the segment described below, could you prove that the triangles are congruent? If so, what congruence postulate or theorem would you use?

1. the bisector of ∠BAC   **2.** a median of △ABC   **3.** an altitude of △ABC

6.6 Problem Solving Application: Estimating Distances

Objective: To estimate distances using congruent triangles.

You can use congruent triangles to indirectly estimate distances that otherwise could not be measured easily. Follow these steps.

1. Locate a triangle with the unknown distance as the length of one of its sides.

2. Locate a second triangle congruent to the first one. Make sure that the side of the second triangle that corresponds to the unknown side in the original triangle can be measured. You should be able to explain why the triangles are congruent.

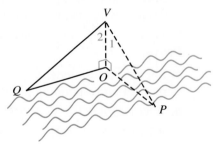

Example: Legend has it that Napoleon asked an officer to find the width of a river. The officer, standing at point O, mentally marked a point P on the opposite shore. He then pulled his cap down until his line of sight past his visor V was directly at point P. Without changing the angle of his visor, he turned and noted the corresponding point Q on his side of the river.

The officer paced off the distance, OQ, along the shore. He estimated the distance to be 75 yd. Estimate the width of the river.

Solution: **a.** $\angle VOP \cong \angle VOQ$ (Both are right angles.)

b. $\angle 1 \cong \angle 2$ (The officer turned without changing the angle of his visor.)

c. $\overline{VO} \cong \overline{VO}$ (A segment is congruent to itself.)

d. $\triangle VOP \cong \triangle VOQ$ (ASA)

e. $\overline{OP} \cong \overline{OQ}$ (CPCTC)

f. Therefore the width, OP, of the river is about 75 yd.

Class Exercise

1. A swimmer wants to measure the distance from A to B across a pond. She walks along a line that forms a right angle with $\overline{AB}$ for 90 yd. She places a post in the ground and then measures another 90 yd to D. Next, she turns 90° to the right and walks until she reaches E, where she can sight B across the post at C—a distance of 115 yd.

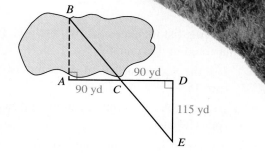

   a. Why is ∠BCA ≅ ∠ECD?

   b. Why is ∠BAC ≅ ∠EDC?

   c. Why is △BAC ≅ △EDC?

   d. How far is it from A to B?

   e. What if the swimmer wants to know the distance from C to B? Explain how she can measure the distance indirectly.

Exercises

1. Engineers have constructed a tower at the edge of a canyon. To estimate the length of wire they need for guy wire, $\overline{CT}$, they measure ∠TCV. Then they find a point D such that ∠TDV ≅ ∠TCV.

   a. Why is △TVD ≅ △TVC?

   b. How can the engineers estimate the length of guy wire $\overline{CT}$?

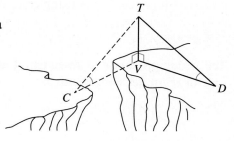

2. Suppose you want to measure the distance from A to B. You pace off AC and CD, both 80 yd. Then you pace off BC and CE, both 60 yd. Explain how you could locate congruent triangles that would enable you to estimate the unknown distance. Be sure to tell why the triangles are congruent.

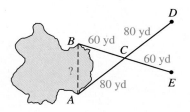

Vocabulary

You should be able to write a brief description, draw a picture, or give an example to illustrate the meaning of each of the following terms:

hypotenuse of a right triangle (p. 159)   legs of a right triangle (p. 159)

Summary

The following list indicates the major skills, facts, and results you should have mastered in this chapter.

6.1 Use the SSS Postulate to show that two triangles are congruent. (pp. 155-158)

6.2 Use the SAS Postulate to show that two triangles are congruent. (pp. 159-161)

6.3 Use the ASA Postulate and the AAS Theorem to show that two triangles are congruent. (pp. 162-165)

6.4 Use the HL Theorem and other postulates and theorems to show that two triangles are congruent and that their corresponding parts are congruent. (pp. 166-170)

6.5 Use congruent triangles to prove theorems and to justify constructions. (pp. 171-175)

6.6 Estimate distances using congruent triangles. (pp. 176-177)

Exercises

Tell which postulate or theorem, if any, you could use to prove each pair of triangles congruent.

1.
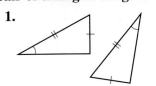

2.
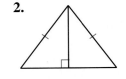

3.
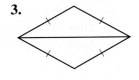

4.

5.
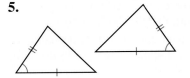

6.

Name two triangles you would prove congruent in order to prove each congruence statement.

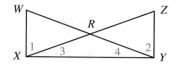

7. $\angle 1 \cong \angle 2$ **8.** $\angle 3 \cong \angle 4$

9. $\overline{XR} \cong \overline{YR}$ **10.** $\overline{WY} \cong \overline{ZX}$

Draw a triangle similar to, but larger than, the one shown. Use the given method and the indicated sides and angles to construct a triangle congruent to $\triangle HKJ$.

11. SSS: $\overline{HK}, \overline{KJ}, \overline{HJ}$

12. ASA: $\angle H, \overline{HK}, \angle K$

13. SAS: $\overline{HJ}, \angle J, \overline{KJ}$

Suppose you know that the indicated parts of the triangles are congruent. What additional information would enable you to use the indicated method to prove that $\triangle UVW \cong \triangle XYZ$?

14. $\angle V \cong \angle Y$; SAS **15.** $\overline{UV} \cong \overline{XY}$; ASA

16. $\overline{UW} \cong \overline{XZ}$; SSS **17.** $\angle U \cong \angle X, \angle W \cong \angle Z$; AAS (2 solutions)

18. $\angle U$ and $\angle X$ are right angles; HL (2 solutions)

19. Given: $ABCD$ is a quadrilateral with $\overline{AB} \cong \overline{CB}$ and $\overline{AD} \cong \overline{CD}$

  Prove: $\triangle AED \cong \triangle CED$

  Give a reason for each statement.

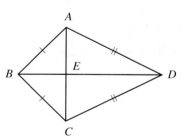

  a. $\overline{AB} \cong \overline{CB}$ and $\overline{AD} \cong \overline{CD}$

  b. $\overline{BD} \cong \overline{BD}$

  c. $\triangle ABD \cong \triangle CBD$

  d. $\angle ADE \cong \angle CDE$

  e. $\overline{ED} \cong \overline{ED}$

  f. $\triangle AED \cong \triangle CED$

PROBLEM SOLVING

20. To estimate the width of a stream, a hiker walks 40 yd along a line perpendicular to $\overline{VW}$. He marks point X and continues along $\overrightarrow{WX}$ for another 40 yd. He turns 90° to the right and walks until he reaches point Z, 30 yd from Y, where he can see V across X. Find VW. Explain your solution.

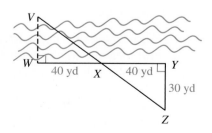

Suppose you know that the indicated parts of the triangles are congruent. Which postulate or theorem, if any, could you use to prove that $\triangle GHN \cong \triangle JKL$?

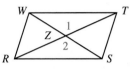

1. $\angle G \cong \angle J, \overline{HN} \cong \overline{KL}, \angle N \cong \angle L$
2. $\overline{GH} \cong \overline{JK}, \overline{GN} \cong \overline{JL}, \overline{HN} \cong \overline{KL}$
3. $\angle H$ and $\angle K$ are right angles; $\angle G \cong \angle J, \overline{GH} \cong \overline{JK}$
4. $\angle H$ and $\angle K$ are right angles; $\overline{GH} \cong \overline{JK}$
5. $\overline{GH} \cong \overline{JK}, \overline{HN} \cong \overline{KL}, \angle N \cong \angle L$

Name two triangles you would prove congruent in order to prove each congruence statement.

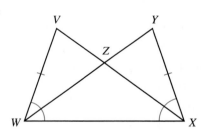

6. $\angle 1 \cong \angle 2$ 7. $\angle WRS \cong \angle STW$
8. $\overline{RZ} \cong \overline{TZ}$ 9. $\overline{RS} \cong \overline{WT}$

What additional congruences would you need to know in order to prove that $\triangle ABC \cong \triangle DEF$ by each indicated method?

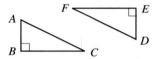

10. SAS   11. ASA (2 solutions)   12. HL (2 solutions)

13. **Given:** $\angle VWX \cong \angle YXW$ and $\overline{VW} \cong \overline{YX}$
    Prove: $\triangle VZW \cong \triangle YZX$
    Give a reason for each statement.
    a. $\angle VWX \cong \angle YXW$ and $\overline{VW} \cong \overline{YX}$
    b. $\overline{WX} \cong \overline{WX}$
    c. $\triangle VWX \cong \triangle YXW$
    d. $\angle V \cong \angle Y$
    e. $\angle VZW \cong \angle YZX$
    f. $\triangle VZW \cong \triangle YZX$

PROBLEM SOLVING

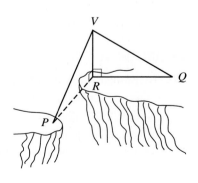

14. To estimate the distance across a canyon, a forest ranger at point R mentally marks point P on the other side of the canyon. She pulls her hat down until her line of sight past the visor is directly to point P. Keeping the angle of her visor set, she turns and notes point Q on her side. She paces off RQ, about 50 yd. Estimate RP and explain your solution.

True or false?

1. Supplementary angles must be adjacent.

2. A right triangle can be isosceles.

3. Given points A and C, $\overrightarrow{AC}$ and $\overrightarrow{CA}$ are opposite rays.

4. An isosceles triangle has at least two congruent angles.

5. Every isosceles triangle has exactly two congruent angles.

Name each of the following for the figure at the right.

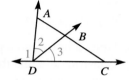

6. three segments with endpoint D

7. all angles adjacent to $\angle 2$

8. two other names for $\angle 3$

9. three noncollinear points

10. an angle bisector

Select the best choice for each question.

11. Choose the best unit for measuring the length of an eyelash.
   a. millimeter   b. centimeter
   c. meter   d. kilometer

12. $PQRS$ is an equilateral parallelogram with four right angles. Choose the name that best describes $PQRS$.
   a. parallelogram   b. rectangle
   c. rhombus   d. square

13. Given that $\triangle ABC \cong \triangle FDE$, which segment must be congruent to $\overline{AC}$?
   a. $\overline{DE}$ b. $\overline{FD}$ c. $\overline{EF}$ d. $\overline{DF}$

14. Find $m\angle DHF$.

   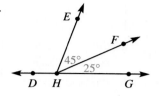

   a. 135°
   b. 110°
   c. 70°
   d. 155°

15. Which of the following results in the same image as two consecutive reflections over parallel lines?
   a. a reflection
   b. a translation
   c. a clockwise rotation
   d. a counterclockwise rotation

16. Exactly one altitude of $\triangle ABC$ is a line of symmetry. Which of the following best describes $\triangle ABC$?
   a. scalene
   b. isosceles
   c. non-equilateral isosceles
   d. equilateral

17. Choose the lengths that could be the lengths of the sides of a triangle.
   a. 3 cm, 4 cm, 7 cm
   b. 40 m, 40 m, 2 m
   c. 1 cm, 1 cm, 2 cm
   d. 5 m, 8 m, 15 m

18. List the sides of $\triangle KLM$ from shortest to longest.

   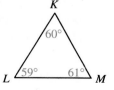

   a. $\overline{KL}, \overline{LM}, \overline{KM}$
   b. $\overline{KM}, \overline{LM}, \overline{KL}$
   c. $\overline{LM}, \overline{KM}, \overline{KL}$
   d. $\overline{KL}, \overline{KM}, \overline{LM}$

Cumulative Review

19. Given three noncollinear points J, K, and L, which of the following could you construct to locate a point equidistant from J, K, and L?

  a. the perpendicular bisectors of two sides of $\triangle JKL$

  b. the bisectors of two angles of $\triangle JKL$

  c. two altitudes of $\triangle JKL$

  d. two medians of $\triangle JKL$

20. What additional information would *not* enable you to prove that $\triangle RST \cong \triangle JKL$?

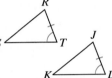

  a. $\overline{ST} \cong \overline{KL}$

  b. $\angle R \cong \angle J$

  c. $\angle S \cong \angle K$

  d. $\overline{RS} \cong \overline{JK}$

21. In triangles TRA and NGL, $\angle T \cong \angle N$, $\overline{TR} \cong \overline{NL}$, and $\overline{TA} \cong \overline{NG}$. $\triangle TRA \cong \triangle \blacksquare$.

  a. $\triangle NGL$ **b.** $\triangle LGN$

  c. $\triangle NLG$ **d.** $\triangle LNG$

22. In triangles ABC and XYZ, $\overline{AB} \cong \overline{XY}$ and $\overline{AC} \cong \overline{XZ}$. What additional information would allow you to prove that $\triangle ABC \cong \triangle XYZ$?

  a. $\overline{AB} \perp \overline{BC}$ and $\overline{XY} \perp \overline{YZ}$

  b. $\angle C \cong \angle Z$

  c. $\angle B \cong \angle Y$

  d. $\overline{AB} \cong \overline{AC}$

23. In triangles DEF and RST, $\overline{DE} \cong \overline{RS}$, $\angle E \cong \angle S$, and $\angle F \cong \angle T$. Which postulate or theorem would you use to show that the triangles are congruent?

  a. SSS   **b.** SAS

  c. ASA   **d.** AAS

24. Two angles are complementary. The measure of one angle is $6°$ less than twice that of the other. Find the measure of the larger angle.

  a. $32°$ **b.** $58°$ **c.** $28°$ **d.** $50°$

25. In $\triangle ABC$, $AB = 3$, $BC = 4$, and $AC = 5$. In $\triangle DEF$, $DE = 3$, $EF = 4$, and $DF = 6$. Which statement is *not* true?

  a. $m\angle B > m\angle A$

  b. $m\angle D > m\angle F$

  c. $m\angle B > m\angle E$

  d. $m\angle E > m\angle A$

26. Draw a line l and a point R not on l. Construct the line through R that is perpendicular to l.

27. Draw a segment, $\overline{AB}$. Construct the perpendicular bisector of $\overline{AB}$.

28. Draw an obtuse angle, $\angle DEF$. Construct the bisector of $\angle DEF$.

29. Draw a segment, $\overline{RS}$. Construct a segment whose length is $2 \cdot RS$.

30. Use a protractor to draw an angle with measure $114°$.

31. Draw a rectangular prism.

Trace the figure and draw the indicated image of the rectangle.

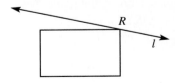

32. under reflection over line l

33. under $90°$ CCW rotation about R

Parallel Lines

Although not very common in the natural world, the idea of parallel lines has great importance for the world that humankind has constructed. For example, when a building is being put up, the beams that support the floors must be parallel.

Focus on Skills

ALGEBRA

Complete.

1. If a, b, and c are all greater than zero and $a = b + c$, then ▪ is the greatest of the three numbers.

2. If $x + a = y$ and $x + b = y$, then ▪ = ▪.

3. If $c = d$ and $d = e$, then ▪ = ▪.

Solve each equation.

4. $5x - 1 = 2x + 5$

5. $(2x + 5) + (7x - 14) = 180$

6. $6t - 24 = 4t + 14$

7. $135 = 71 + z$

8. $b + 43 + 82 = 180$

9. $(p - 2)180 = 3{,}780$

GEOMETRY

Sketch a figure to match each description. Mark all parallel sides, all congruent sides, and any right angles.

10. a rectangle that is not a square

11. a rhombus that is not a rectangle

12. a trapezoid with no congruent sides

13. a parallelogram that is not a rectangle

Draw a figure similar to, but larger than, the one shown and construct the indicated figure.

14. $\angle 2$ with vertex Q such that $\angle 2 \cong \angle 1$

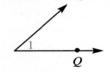

15. a triangle congruent to $\triangle ABC$

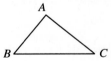

16. a perpendicular to k through P

17. a perpendicular to j through M

PROBLEM SOLVING

18. In $\triangle RST$, $m\angle T = 3 \cdot m\angle R$, $m\angle S = 2 \cdot m\angle R$, and $m\angle R + m\angle S + m\angle T = 180°$. Find the measure of each angle.

EXPLORING

1. Draw two intersecting lines *t* and *c*. Label ∠1 and point *P* on line *t* as shown.

2. Using a protractor and a straightedge, draw line *d* through point *P* so that ∠1 and ∠2 are congruent alternate interior angles.

3. What is *m*∠2 + *m*∠3? What is *m*∠1 + *m*∠3? Why?

4. Describe the relationship between ∠1 and ∠3.

5. What seems to be the relationship between lines *c* and *d*?

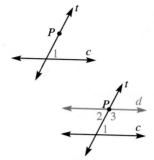

POSTULATE 10:   **If two lines are cut by a transversal so that one pair of alternate interior angles are congruent, then the lines are parallel.**

THEOREM 7.2:   **If two lines are cut by a transversal so that one pair of same-side interior angles are supplementary, then the lines are parallel.**

Example 2: **Tell whether the lines shown are parallel.**

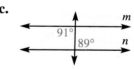

    a.   **b.**   **c.**

Solution: a. *l* ∥ *m*   **b.** *a* ∥ *b*   **c.** *m* is not parallel to *n*

Thinking Critically

1. If *P* is a point not on line *c*, how many lines parallel to *c* can be drawn through *P* in a plane?

2. Tell how the measures of each pair of angles are related.

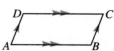

   a. ∠*A* and ∠*B*   **b.** ∠*B* and ∠*C*   **c.** ∠*C* and ∠*D*

   d. ∠*A* and ∠*D*   **e.** ∠*A* and ∠*C*   **f.** ∠*B* and ∠*D*

3. *RSTW* is a parallelogram. ∠*R* is a right angle.

   a. What is *m*∠*S*? Why?   **b.** What are *m*∠*T* and *m*∠*W*?

Class Exercises

Name each of the following.

1. an angle congruent to ∠1
2. an angle congruent to ∠4
3. two angles supplementary to ∠2
4. two angles supplementary to ∠3

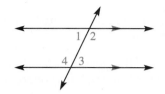

Find the measure of each angle.

  5. ∠1   **6.** ∠2
  7. ∠3   **8.** ∠4
  9. ∠5   **10.** ∠6
11. ∠7   **12.** ∠8

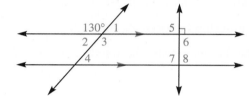

Exercises

Name each of the following.

1. an angle congruent to ∠1
2. an angle congruent to ∠4
3. two angles supplementary to ∠2
4. two angles supplementary to ∠3

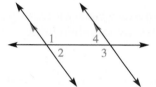

Find the measure of each angle.

  5. ∠1   **6.** ∠2
  7. ∠3   **8.** ∠4
  9. ∠5   **10.** ∠6
11. ∠7   **12.** ∠8

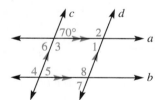

13. **a.** Name two pairs of alternate interior angles.
    b. Name two pairs of same-side interior angles.
    c. If ∠1 ≅ ▇, then *l* ∥ *m*.
    d. If $m\angle 1 + m$ ▇ = 180°, then *l* ∥ *m*.
    e. If ∠2 ≅ ▇, then *l* ∥ *m*.

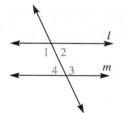

Find the measure of each angle.

14. *PMRQ* is a parallelogram.

  a. ∠M
  b. ∠Q
  c. ∠R

15. *EHGF* is a parallelogram.

  a. ∠HEF
  b. ∠HEG
  c. ∠HGE

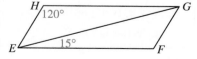

16. *DABC* is a parallelogram.

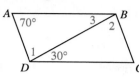

   a. ∠CDA

   b. ∠1

   c. ∠2

   d. ∠3

17. a. ∠1

   b. ∠2

   c. ∠3

   d. ∠4

18. *RECT* is a rectangle.

   a. ∠1

   b. ∠2

   c. ∠3

19. a. ∠1

   b. ∠2

   c. ∠3

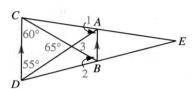

Name the segments or lines that must be parallel.

20.

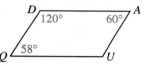

21.

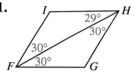

22.

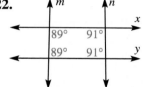

Name the lines, if any, that must be parallel if each statement is true.

23. ∠1 ≅ ∠2

24. $m\angle 1 + m\angle 3 = 180°$

25. $m\angle 3 + m\angle 5 = 180°$

26. ∠2 ≅ ∠4

27. $m\angle 2 + m\angle 3 = 180°$

28. ∠4 ≅ ∠5

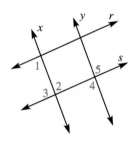

29. If ∠4 ≅ ∠6 and ∠8 ≅ ∠10, name all lines that must be parallel.

30. If *a* ∥ *b* and *b* ∥ *c*, must line *a* be parallel to line *c*? Explain.

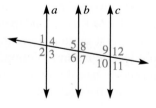

APPLICATIONS

Find the measure of each indicated angle.

31. Algebra ∠ABC and ∠BCD

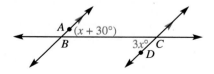

32. Algebra ∠EFG and ∠FGH

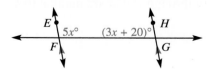

Find the measure of each indicated angle.

33. Algebra ∠RST, ∠PTS, and ∠STR

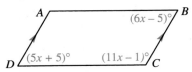

34. Algebra ∠J, ∠K, and ∠L

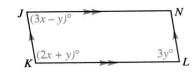

Are the indicated segments parallel? Explain.

35. Logic $\overline{AB}$ and $\overline{DC}$

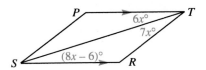

36. Logic $\overline{RW}$ and $\overline{ST}$

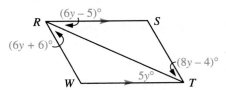

Historical Note

The following statement is equivalent to our Postulates 9 and 10.

> Given a line l and a point P not on l, there is exactly one line through P that is parallel to l.

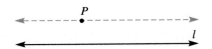

Although this statement, known as *The Parallel Postulate*, may seem obvious, mathematicians tried for more than 2,000 years to prove it. In the early 1800s, two mathematicians, Janos Bolyai of Hungary and Nicolai Lobachevsky of Russia, independently made the same discovery: The assumption that there is more than one line through P that is parallel to l does not lead to a contradiction. In fact, the mathematicians developed a new geometry based on the assumption that an infinite number of lines through P are parallel to l.

During the same period, the German mathematician Georg Riemann developed another geometry based on the assumption that there are no lines passing through P that are parallel to l.

7.3 Parallel Lines and Corresponding Angles

Objective: To use facts about parallel lines cut by a transversal and pairs of corresponding angles.

In this lesson, you will learn more about how the fact that two lines are parallel relates to the congruence between certain angles formed by a transversal.

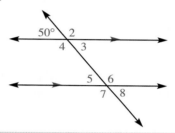

EXPLORING

The measure of ∠1 = 50°.

1. Find the measures of ∠2, ∠3, and ∠4.

2. What is the measure of ∠6? Why?

3. Find the measures of ∠5, ∠7, and ∠8.

4. Name all pairs of corresponding angles. Compare the measures of corresponding angles.

The **EXPLORING** activity demonstrates the following theorem.

THEOREM 7.3 (Corresponding Angles Theorem): If two parallel lines are cut by a transversal, then each pair of corresponding angles are congruent.

EXPLORING

1. Enter and run the following LOGO program.
   ```
   To transversal
   fd 50 bk 100 fd 50
   rt 40
   fd 50 bk 100 fd 50
   fd 20 lt 25
   fd 50 bk 100 fd 50
   end
   ```

2. Are the first and third lines created by the turtle parallel? If not, what must you change to make the turtle draw the third line parallel to the first line? Make the change(s). Run the new program.

The **EXPLORING** activity demonstrates the following theorem.

THEOREM 7.4: **If two lines are cut by a transversal so that one pair of corresponding angles are congruent, then the lines are parallel.**

Example: **a. Name the lines, if any, that must be parallel.**

b. Find the measure of ∠1.

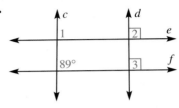

Solution: **a.** Since corresponding angles 2 and 3 are both right angles, *e* ∥ *f*. (Notice that since *m*∠3 ≠ 89°, lines *c* and *d* are not parallel.)

b. Since *e* ∥ *f*, *m*∠1 = 89°.

A statement that follows directly from a theorem is sometimes called a *corollary*. Part (a) of the Example demonstrates the following corollary of Theorem 7.4.

COROLLARY: **If two coplanar lines are perpendicular to a third line, then the two lines are parallel to each other.**

Thinking Critically

1. Describe four methods of showing that two lines are parallel.

2. Explain in your own words the difference between the Corresponding Angles Theorem and Theorem 7.4.

Class Exercises

Name each of the following.

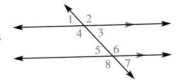

  1. four pairs of corresponding angles

  2. three angles congruent to ∠2

  3. three angles congruent to ∠5

Find the measure of each numbered angle.

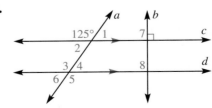

4. ∠1	**5.** ∠2
6. ∠3	**7.** ∠4
8. ∠5	**9.** ∠6
10. ∠7	**11.** ∠8

Exercises

1. Name each of the following.

    a. four pairs of corresponding angles

    b. three angles congruent to ∠1

    c. four angles supplementary to ∠1

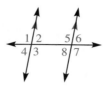

2. Complete.

    a. If ∠1 ≅ ▨, then $l \parallel n$.

    b. If $m∠3 + m$ ▨ $= 180°$, then $l \parallel n$.

    c. If ∠2 ≅ ▨, then $l \parallel n$.

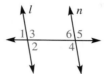

Find the measure of each numbered angle.

3. ∠1	**4.** ∠2
5. ∠3	**6.** ∠4
7. ∠5	**8.** ∠6

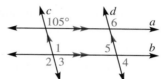

Find the measure of each numbered angle.

9.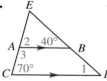

10.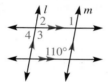

11.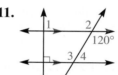

12.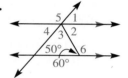

13. What must be the measure of each angle in order that $a \parallel b$?

    a. ∠1

    b. ∠2

    c. ∠3

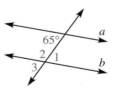

Name the segments or lines that must be parallel.

14.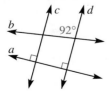

15.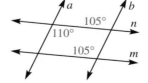

16.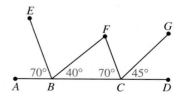

Name the lines, if any, that must be parallel if each statement is true.

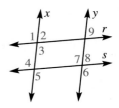

17. $\angle 1 \cong \angle 4$ **18.** $\angle 4 \cong \angle 6$ **19.** $m\angle 2 + m\angle 1 = 180°$

20. $\angle 8 \cong \angle 9$ **21.** $\angle 4 \cong \angle 5$ **22.** $m\angle 4 + m\angle 8 = 180°$

23. Draw a figure like the one shown.

   a. Construct the line through P that is perpendicular to l.

   b. Construct the line through P that is perpendicular to the line constructed in part (a).

   c. How is the line constructed in part (b) related to line l?

APPLICATIONS

24. Architecture In City Hall, Corridor One and Corridor Two are both perpendicular to Corridor Three. Why can you say that Corridor One and Corridor Two are parallel to each other?

25. Algebra Two parallel lines are cut by a transversal so that two corresponding angles have measures $(3x - 8)°$ and $(2x + 10)°$. Find the value of x and the measures of the angles.

26. Computer The following LOGO program is for a turtle that likes to travel backwards. Rewrite the program so it allows the user to decide what the angles of the parallelogram will be.

```
To parallelogram
repeat 2 [bk 50 rt 40 rt 180 bk 20 rt 140 rt 180]
end
```

Test Yourself

Name each of the following.

   1. two pairs of alternate interior angles

   2. two pairs of same-side interior angles

   3. four pairs of corresponding angles

   4. all angles supplementary to $\angle 1$

   5. all angles congruent to $\angle 1$

   6. If $m\angle 2 = 60°$, find the measures of $\angle 5$, $\angle 6$, and $\angle 7$.

What must be the measure of each angle so that $a \parallel b$?

   7. $\angle 2$ **8.** $\angle 3$ **9.** $\angle 4$

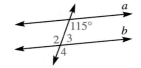

7.4 Constructing Parallel Lines

Objective: To construct the line parallel to a given line through a point not on the line.

You can use the properties of angles formed by a transversal to construct a line parallel to a given line.

CONSTRUCTION 11

Construct the line through *P* that is parallel to line *l*.

1. Draw a line *l* and a point *P* not on *l*. Draw a line through *P* intersecting *l*.

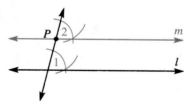

2. At vertex *P*, construct ∠2 congruent to ∠1. Draw line *m*. Then *m* ‖ *l*.

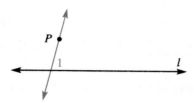

Thinking Critically

1. In Construction 11, ∠2 was constructed congruent to ∠1. Why does this guarantee that *m* ‖ *l* ?

2. The figures show an alternative construction of line *m* parallel to the given line *l* through the given point *P*.
   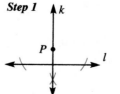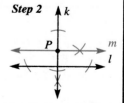
   a. Describe Step 1.
   b. Describe Step 2.
   c. Why is line *m* parallel to line *l*?
   d. Why can this construction be considered a special case of Construction 11?

Class Exercise

1. Draw a line *m* and a point *A* not on *m*. Construct the line through *A* that is parallel to *m*.

Exercises

Using only the given information, can you conclude that *a* ‖ *b*?

1.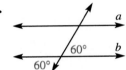

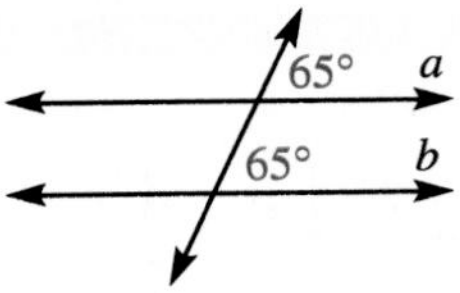

2.

3.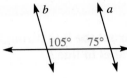

4.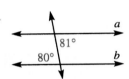

5.

6.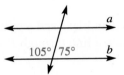

Draw a figure like the one shown. Construct the line through *A* that is parallel to $\overleftrightarrow{BC}$.

7.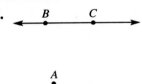

8.

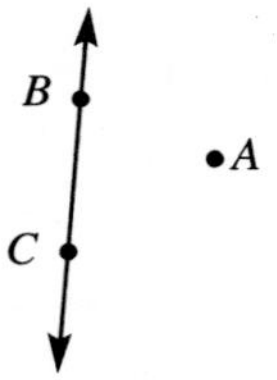

9.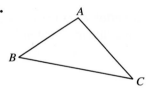

10. Draw a line *l*. Construct two lines, one on each side of *l*, that are parallel to *l*.

Draw a figure like the one shown. Construct the figure described.

11. rhombus *JKLM* with $\overline{KL}$ on line *a*

12. parallelogram *ABCD*

13. square *EFGH* with $\overline{FG}$ on line *l*

14. trapezoid *ABCD* such that $\overline{AD} ‖ \overline{BC}$ and $AD = 2 \cdot BC$

15. Draw a figure like the one shown and use it to construct square *WXYZ*.

16. Use the construction suggested by Postulate 10 on page 189 to construct the line through *P* that is parallel to *l*.

APPLICATIONS

17. Space Geometry   Given a point *P* not on line *l*, how many lines through *P* do not intersect *l*?

18. Paper Folding   Draw a line *l* and a point *P* not on *l*. Use paper folding to determine the line through *P* that is parallel to *l*.

Thinking in Geometry

The following steps show an alternative construction of a line through *P* parallel to *l*.

Step 1: From *P*, draw an arc intersecting *l* at any point *A*.

Step 2: Construct $\overline{AB} \cong \overline{PA}$ on *l*. With the same compass setting and with the compass point on *P* and then *B*, draw arcs intersecting at some point *C*.

Step 3: Draw $\overleftrightarrow{PC}$.

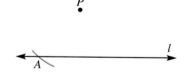

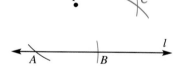

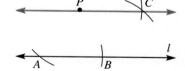

1. Why is $\triangle PAC \cong \triangle BCA$?

2. Why is $\angle 1 \cong \angle 2$?

3. Why is $\overleftrightarrow{PC} \parallel l$?

4. What kind of figure is *PABC*?

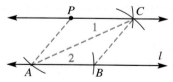

5. Draw a figure like the one shown. Without copying an angle, construct parallelogram *WXYZ*.

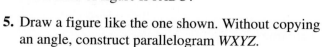

Thinking About Proof

Converses

On page 90 you learned that a *conditional* is a statement that can be written in *if-then* form. When you interchange the hypothesis and conclusion of a conditional statement, you get the *converse* of the original statement.

Example: **Write the converse of each conditional.**

    a. If two parallel lines are cut by a transversal, then each pair of alternate interior angles are congruent.

    b. A rectangle is a parallelogram.

Solution: **a.** If two lines are cut by a transversal so that each pair of alternate interior angles are congruent, then the lines are parallel.

    b. Rewrite the conditional in if-then form:
If a figure is a rectangle, then it is a parallelogram.

    Converse: If a figure is a parallelogram, then it is a rectangle.

Part (b) of the Example demonstrates that the converse of a true conditional does not have to be true. When a conditional and its converse are true, you can combine them and write the combined statement in *if-and-only-if* form:

Two lines cut by a transversal are parallel if and only if each pair of alternate interior angles are congruent.

Exercises

Write the converse of each conditional.

1. If two angles of a triangle are congruent, then the sides opposite those angles are congruent.
2. An equilateral triangle is also equiangular.
3. The three medians of a triangle are concurrent.
4. Vertical angles are congruent.

5. Which converses in Exercises 1–4 are true?
6. Combine Theorems 5.9 and 5.10 on page 138 in if-and-only-if form.
7. What is the converse of the converse of a conditional?

7.5 Interior Angle Measures of a Triangle

Objective: To use the sum of the interior angle measures of a triangle.

You know about the relationships among angles formed by parallel lines and a transversal. Important relationships also exist among the angles of a triangle.

EXPLORING

1. Use the *Geometric Supposer: Triangles* disk to draw an acute triangle, $\triangle ABC$. Through point C, draw $\overrightarrow{DE}$ so that it is parallel to side $\overline{AB}$.

2. Sketch the figure. Use the *Measure* option to measure all the angles formed. Record the measures on your sketch. What observations and conjectures can you make about the angles of $\triangle ABC$? Repeat using a right triangle, an isosceles triangle, and an equilateral triangle.

3. Explain why any of your conjectures must be true.

EXPLORING

Part A

1. Draw and cut out a scalene triangle. Fold the triangle as shown below so that all three vertices meet on one side of the triangle.

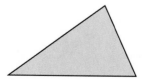 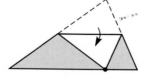 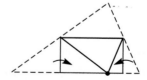

2. What is the sum of the measures of the three angles of the triangle?

Part B

1. What is $m\angle 5 + m\angle 3 + m\angle 4$?

2. Why is $\angle 5 \cong \angle 1$? Why is $\angle 4 \cong \angle 2$?

3. What is $m\angle 1 + m\angle 3 + m\angle 2$?

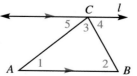

The angles of a triangle are sometimes called the *interior angles* of the triangle. The **EXPLORING** activities demonstrate the following theorem.

THEOREM 7.5 (Triangle Angle-Sum Theorem): The sum of the measures of the interior angles of any triangle is 180°.

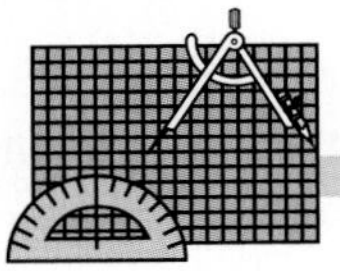

EXPLORING

Part A

1. What kind of triangle is *XYZ*?

2. What is $m\angle X + m\angle Y + m\angle Z$? $m\angle Y$? $m\angle X + m\angle Z$?

3. Describe the relationship of $\angle X$ and $\angle Z$.

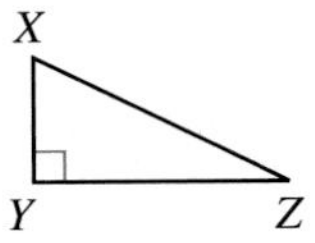

Part B

1. Draw an equilateral $\triangle RST$ and name the parts that are congruent.

2. What is the measure of each interior angle of $\triangle RST$?

The Triangle Angle-Sum Theorem has the following corollaries.

COROLLARY 1: The acute angles of any right triangle are complementary.

COROLLARY 2: The measure of each interior angle of an equilateral triangle is 60°.

Example: Find the measure of each numbered angle.

a.

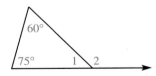

b.

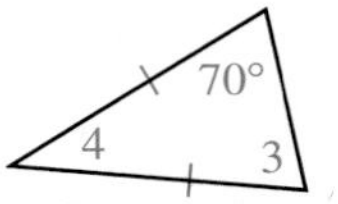

Solution: **a.** $m\angle 1 + 75° + 60° = 180°$

$$m\angle 1 + 135° = 180°$$
$$m\angle 1 = 180° - 135°$$
$$m\angle 1 = 45°$$
$$m\angle 2 = 180° - 45°$$
$$m\angle 2 = 135°$$

b. $m\angle 3 = 70°$

$$m\angle 4 + 70° + 70° = 180°$$
$$m\angle 4 + 140° = 180°$$
$$m\angle 4 = 180° - 140°$$
$$m\angle 4 = 40°$$

1. Is Part A of the second Exploring activity on page 201 an example of inductive reasoning or deductive reasoning? Which type of reasoning is Part B?

2. Explain why the following statements are true.
   a. A triangle must have at least one angle that measures at least 60°.
   b. Every triangle must have at least two acute angles.

Class Exercises

Find the measure of each numbered angle.

1.

2.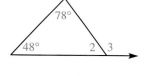

3.

4.

5.

6.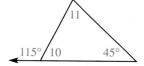

Exercises

True or false? Give a reason or example to support each answer.

1. A triangle can have two 89° angles.

2. The acute angles of any right triangle are supplementary.

3. An equiangular triangle is a special kind of acute triangle.

4. A triangle can have angles of measure 74°, 43°, and 62°.

5. A right triangle can be equilateral.

6. An isosceles triangle can be obtuse.

7. A triangle can have no obtuse angles.

8. A triangle can have two right angles.

9. A right triangle can have an obtuse angle.

10. A triangle can have exactly one acute angle.

Find the measure of each numbered angle.

11.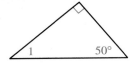

12.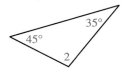

13.

14.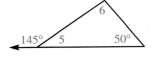

15.

16.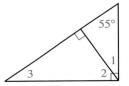

17.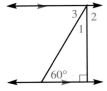

18.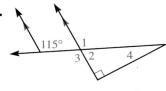

19.

20.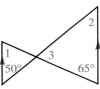

21.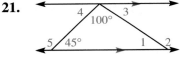

22.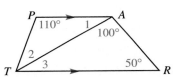

23. a. ∠1
   b. ∠DAB
   c. ∠DCB
   d. ∠2

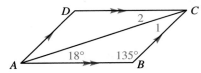

24. a. ∠1
   b. ∠PTR
   c. ∠2
   d. ∠3

25. One base angle of an isosceles triangle has a measure of 35°.

   a. Find the measure of the other base angle.

   b. Find the measure of the vertex angle.

26. The vertex angle of an isosceles triangle has a measure of 150°.
Find the measure of each base angle.

27. One acute angle of a right triangle has a measure of 28°.
Find the measure of the other acute angle.

28. △PQR is isosceles. One of the base angles, ∠P, has a measure
of 25°, and $\overline{PQ} \cong \overline{QR}$.

   a. Find $m\angle R.$ **b.** Find $m\angle Q.$ **c.** Classify △ PQR by its angles.

29. a. Find $m\angle 1$.

  b. Find $m\angle 2$.

  c. Classify $\triangle ABC$ by its sides. Give a reason for your answer.

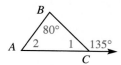

30. One angle of an obtuse isosceles triangle has a measure of 40°. Find the measures of the other two angles.

31. To be Proven: If two angles of one triangle are congruent to two angles of another triangle, then the third pair of angles are congruent (Theorem 6.1).

  Given: $\angle A \cong \angle D$; $\angle B \cong \angle E$
  Prove: $\angle C \cong \angle F$

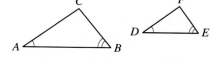

  a. Since $\angle A \cong \angle D$ and $\angle B \cong \angle E$, $m\angle A = \blacksquare$ and $m\angle B = \blacksquare$.

  b. $m\angle A + m\angle B = \blacksquare$

  c. What are $m\angle C$ and $m\angle F$? Why?

  d. Substitute your answer from part (b) into one of your equations in part (c). Is $m\angle C = m\angle F$?

  e. Why is $\angle C \cong \angle F$?

APPLICATIONS

32. Algebra Two angles of a triangle are congruent. The third angle has measure equal to the sum of the measures of the other two. Find the measures of all three angles.

33. Algebra The measure of one acute angle of a right triangle is five times the measure of the other acute angle. Find the measure of each acute angle.

34. Algebra and Mental Math The measure of the largest angle of a triangle is five times the measure of the smallest angle. The measure of the third angle is three times the measure of the smallest angle. Find the measures of all three angles.

Computer

Use the *Your Own* option on the *Geometric Supposer: Triangles* disk. Try to make a triangle such that $m\angle BAC = 120°$, $AB = 5$, and $m\angle CBA = 60°$. What happens? Explain.

7.6 Exterior Angle Measures of a Triangle

Objective: To use the relationship between an exterior angle of a triangle and its remote interior angles.

An *exterior angle* of a polygon is formed by one side of the polygon and the extension of the adjacent side. For △ABC, ∠1 is an exterior angle.

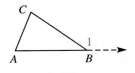

Every triangle has six exterior angles. The exterior angles of △DEF are ∠1, ∠2, ∠3, ∠4, ∠5, and ∠6. At each vertex, the two exterior angles are congruent. Why?

∠1 ≅ ∠2   ∠3 ≅ ∠4   ∠5 ≅ ∠6

Each exterior angle of a triangle has one *adjacent interior angle* and two *remote interior angles*.

Example: **For each exterior angle shown, name one adjacent interior angle and two remote interior angles.**

Solution:

Exterior angle	Adjacent interior angles	Remote interior angles
∠1	∠7	∠8, ∠9
∠4	∠8	∠7, ∠9

EXPLORING

1. For each triangle, one exterior angle is shown. Find the measure of each numbered angle.

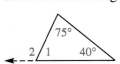

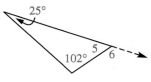
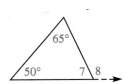

2. Compare the measure of each exterior angle shown with the measure of each of its remote interior angles. Are they the same? If not, which is greater?

3. Compare the measure of each exterior angle shown with the sum of the measures of its remote interior angles. Are they the same?

The **EXPLORING** activity demonstrates the following.

THEOREM 7.6: **In a triangle, the measure of each exterior angle is equal to the sum of the measures of its two remote interior angles.**

COROLLARY: **In a triangle, the measure of each exterior angle is greater than the measure of either of its remote interior angles.**

Thinking Critically

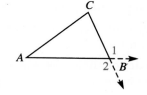

1. Do ∠1 and ∠2 have the same adjacent interior angle and remote interior angles?

2. Why is ∠1 ≅ ∠2?

3. What is the relationship between an exterior angle of a triangle and its adjacent interior angle?

4. Classify the exterior angles for an acute triangle, a right triangle, and an obtuse triangle.

Class Exercises

1. Name the exterior angle adjacent to ∠2.

2. Name the remote interior angles for ∠4.

3. Name an angle that is supplementary to ∠2.

4. $m\angle 1 + m\angle 3 = m$ ▨

5. $m\angle 4 = m$ ▨ $+ m$ ▨

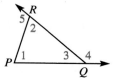

Draw a triangle similar to, but larger than, each one shown. Then draw and label an exterior angle at vertex *A*.

6.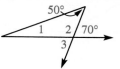

7.

8.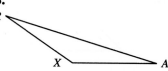

Find the measure of each numbered angle.

9.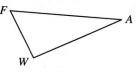

10.

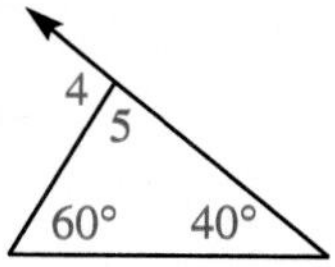

11.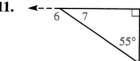

Exercises

Name the adjacent interior angle and the two remote interior angles for each exterior angle.

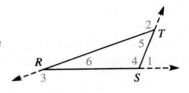

1. $\angle 1$ **2.** $\angle 2$ **3.** $\angle 3$

Name each of the following.

4. three interior angles

5. six exterior angles

6. three angles that are neither exterior nor interior angles

7. the adjacent interior angle and two remote interior angles for $\angle 4$

8. two exterior angles for which $\angle 3$ and $\angle 5$ are the two remote interior angles

9. six pairs of vertical angles

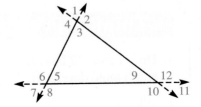

Draw a triangle similar to, but larger than, the one shown. Then draw and label an exterior angle at vertex A.

10. 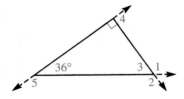   **11.**    **12.**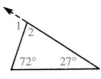

Find the measure of each numbered angle.

13.    **14.** 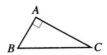   **15.**

16. 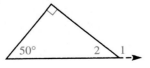   **17.** 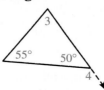   **18.**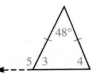

19. 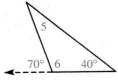   **20.**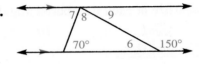

True or false?

21. All exterior angles of an acute triangle are obtuse angles.

22. All exterior angles of an equiangular triangle are congruent.

23. All exterior angles of an obtuse triangle are acute angles.

24. All exterior angles of a right triangle are right angles.

Find the measure of each numbered angle.

25.   **26.**   **27.** 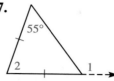   **28.**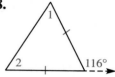

29. In $\triangle ABC$, $m\angle A = 45°$ and $m\angle B = 80°$. Find the measure of $\angle C$ and of an exterior angle at C.

30. In $\triangle PQR$, $m\angle P = 35°$ and the measure of an exterior angle at Q is 105°. Find the measures of $\angle R$ and $\angle Q$.

31. Explain why $\overline{AB}$ and $\overline{AC}$ could not both be perpendicular to $\overleftrightarrow{BD}$.

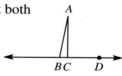

APPLICATION

32. Calculator   The measure of an exterior angle of a triangle is twenty-three times the measure of its adjacent interior angle. Find the measures of both angles.

Calculator

The measures of two angles of a triangle are 28° and 94°. Andy and Christine used different methods to find the measure of the third angle with their calculators:

Andy's Method
28 ⊞ 94 ⊟ 122
180 ⊟ 122 ⊟ 58

Christine's Method
180 ⊟ 28 ⊟ 94 ⊟ 58

Which method do you think was better? Give a reason for your answer.

7.7 Angle Measures of a Polygon

Objective: To find the sum of angle measures of any polygon.

You can use the sum of the interior angle measures of a triangle to determine angle measures in other polygons.

━━━━━━━━━━━ **EXPLORING**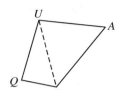

1. Draw a quadrilateral like the one shown. Use a protractor to measure $\angle Q$, $\angle QUA$, $\angle A$, and $\angle ADQ$. Use a calculator to find the sum.

2. Use the Triangle Angle-Sum Theorem to find the sum of the measures of the interior angles of *DQUA*. Compare your results with Step 1.

The **EXPLORING** activity demonstrates the following theorem.

THEOREM 7.7: **The sum of the measures of the interior angles of any quadrilateral is 360°.**

To find the sum of the measures of the interior angles of any polygon, draw all possible diagonals from one vertex. Then find the sum of the measures of the interior angles of the triangles formed.

Polygon	Sides	Figure	Triangles Formed	Sum of Interior Angle Measures
Pentagon	5		3	$3 \times 180° = 540°$
Hexagon	6		4	$4 \times 180° = 720°$

These results and the **EXPLORING** activity suggest the following theorem.

THEOREM 7.8 (Interior Angle Sum Theorem): The sum of the measures of the interior angles of a polygon with *n* sides is $(n - 2)180°$.

Example 1: Find $m\angle 1$.

Solution:
$$m\angle 1 + 98° + 115° + 90° + 121° = (5 - 2)\,180°$$
$$m\angle 1 + 424° = 540°$$
$$m\angle 1 = 540° - 424°$$
$$= 116°$$

EXPLORING

1. What is $m\angle 1 + m\angle 2$? What is the sum of the measures of each exterior angle and adjacent interior angle?

2. What is the sum of the measures of all the numbered angles shown?

3. What is the sum of the measures of the interior angles?

4. Subtract the sum of the measures of the interior angles (Step 3) from the sum in Step 2 to find the sum of the measures of the exterior angles, one at each vertex.

5. Draw a triangle or quadrilateral. Draw one exterior angle at each vertex. Number these exterior angles and the interior angles. Repeat Steps 1–4 for your figure.

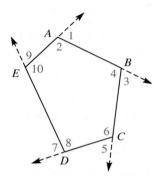

This **EXPLORING** activity demonstrates the following theorem.

THEOREM 7.9 (Exterior Angle Sum Theorem): The sum of the measures of the exterior angles, one at each vertex, of any polygon is 360°.

Example 2: The measure of each exterior angle of a regular polygon is 18°. Find the number of sides.

Solution: Let n = the number of sides.
$$18° \cdot n = 360°$$
$$n = 20$$

There are 20 sides.

1. Why are the exterior angles of a regular polygon congruent?
2. For what kind of polygon does the sum of the interior angles equal the sum of the exterior angles, one exterior angle at each vertex?

Class Exercises

Find the sum of the measures of the interior angles of a polygon with the given number of sides.

1. 7 sides **2.** 9 sides **3.** 10 sides

4. The measure of each interior angle of a regular polygon is 135°.
    a. Find the measure of each exterior angle.   **b.** Find the number of sides.

Find the measure of each numbered angle.

5.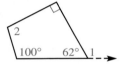
    6.
    7.

Exercises

Copy and complete the table.

	Number of Sides of Polygon	Sum of the Interior Angle Measures	Sum of the Exterior Angle Measures
1.	4	■	360°
2.	6	4 • 180 = 720°	■
3.	■	7 • 180 = 1,260°	■
4.	16	■	■
5.	24	■	■

Find the measure of each numbered angle.

6.
    7.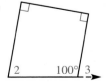
    8.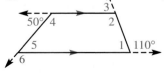
    9. regular pentagon

Copy and complete the table.

	Number of Sides of Regular Polygon	Measure of Each Interior Angle	Measure of Each Exterior Angle
10.	5	▦	▦
11.	9	▦	▦
12.	▦	▦	9°
13.	▦	144°	▦
14.	▦	160°	▦

The sum of the measures of the interior angles is given for some polygons. Find the number of sides of each polygon.

15. 540° **16.** 900° **17.** 1,440° **18.** 1,800° **19.** 2,340° **20.** 3,240°

Draw and label a figure for each polygon described, if possible. If it is not possible, explain why.

21. triangle: exterior angles—obtuse angles

22. triangle: exterior angles—2 right angles, 4 obtuse angles

23. quadrilateral: exterior angles—4 obtuse angles, 4 acute angles

24. quadrilateral: interior angles—4 acute angles

25. quadrilateral: interior angles—2 obtuse angles, 1 right angle, 1 acute angle

26. pentagon: interior angles—5 acute angles

APPLICATION

27. Sports Research the shape of the home plate used in baseball. Sketch home plate and give the measures of its interior angles.

Everyday Geometry

Bees construct honeycombs in the shape of regular hexagons. Do library research to find out what advantages this structure offers the bees.

7.8 Problem Solving Application: Interpreting Contour Maps

<u>Objective:</u> To solve problems involving contour maps.

If a plane intersects a portion of Earth's surface, the intersection is a closed curve. In the figure below, parallel planes L and M intersect the mountain at elevations of 7,000 ft and 8,000 ft respectively. The resulting *contour lines* show the shape of the mountain at the two elevations.

A map depicting Earth's surface at different cross sections is a *contour map*. The closer the lines are to each other, the steeper the terrain. Hikers, mountain climbers, and highway engineers use contour maps to plan routes between places.

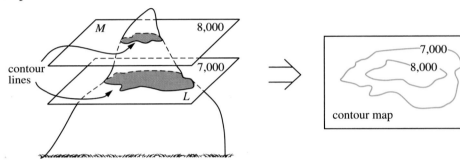

Example: Find the elevations of points A, B, and C on the contour map.

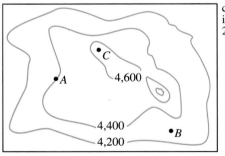

contour interval: 200 ft

Solution: Point A lies on the 4,400-ft contour line, so its elevation is exactly 4,400 ft. Point B lies between 4,200 ft and 4,400 ft. Point C lies inside the 4,600-ft curve, so its elevation is greater than 4,600 ft but less than 4,800 ft.

Chapter 8
Properties of Quadrilaterals

A special kind of quadrilateral — the rectangle — is commonly seen in the walls of buildings and containers. A less familiar sight is another kind of quadrilateral shaped like the wall of this dam, with only one pair of parallel sides.

Chapter 8

Focus on Skills

ALGEBRA

Solve.

1. $y + 14 = 5y - 7$
2. $z + 8 = \frac{1}{2}(6z - 9)$
3. $7b - 8 = 42 - 3b$
4. $3t - 5 = 2(t + 1)$
5. $2(x - 6) = 8x - 24$
6. $2x + 7 = 13x - 81$

Find the value of *x* and *y* for each pair of equations.

7. $5x - 3 = 12$
$y = 2x - 1$

8. $7x + 5 = 47$
$y = \frac{1}{2}(x - 4)$

9. $2x - 1 = 11$
$3y = 4x - 6$

GEOMETRY

Trace each figure and draw all the lines of symmetry.

10.

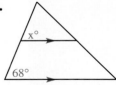

11.

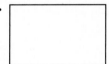

12.

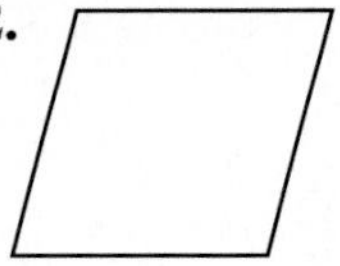

13.

True or false?

14. Every parallelogram is a rectangle.
15. Every square is a rhombus.
16. Every rectangle is a square.
17. Every rhombus is a square.
18. Every rectangle is a parallelogram.
19. Every square is a rectangle.

Find *x*.

20.

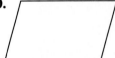

21.

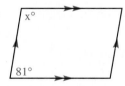

22.

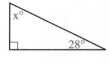

Complete.

23. If $\overleftrightarrow{AB}$ is perpendicular to $\overline{CD}$ at its midpoint, then $\overleftrightarrow{AB}$ is the ■ of $\overline{CD}$.

24. If $\overrightarrow{XY}$ divides $\angle AXB$ into two congruent angles, then $\overrightarrow{XY}$ is the ■ of $\angle AXB$.

25. A ■ is a quadrilateral that is both equilateral and equiangular.

8.1 Properties of Parallelograms

Objective: To use the properties of parallelograms.

Recall that a parallelogram is a quadrilateral with two pairs of parallel sides. We often use the symbol ▱ to represent a parallelogram.

EXPLORING

1. What is true of consecutive angles of ▱*ABCD* such as ∠*DAB* and ∠*CDA*? Why?

2. Does ▱*ABCD* have line symmetry? Does it have point symmetry?

3. When ▱*ABCD* is rotated 180° about point *X*, what is the image of $\overline{AD}$? of ∠*ADC*? What is true of $\overline{AD}$ and its image? of ∠*ADC* and its image?

4. When the figure is rotated 180° about point *X*, what is the image of $\overline{AX}$? of $\overline{DX}$? What can you conclude?

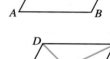

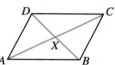

The **EXPLORING** activity demonstrates the following theorems describing the properties of any parallelogram.

THEOREM 8.1: **The opposite sides of a parallelogram are congruent.**

THEOREM 8.2: **The opposite angles of a parallelogram are congruent.**

THEOREM 8.3: **The consecutive angles of a parallelogram are supplementary.**

THEOREM 8.4: **The diagonals of a parallelogram bisect each other.**

Example 1: **Find each angle measure or segment length.**

    a. *m*∠*B*   **b.** *m*∠*C*

    c. *BC*   **d.** *AB*

Solution: **a.** *m*∠*B* = 130°   **b.** *m*∠*C* = 50°

            c. *BC* = 10   **d.** *AB* = 16

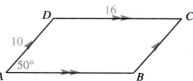

Example 2: In ▱ *RSTW*, *m*∠*RWT* = 73°, *WS* = 20, and
RX = 8. Find each angle measure or
segment length.

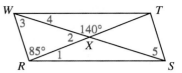

    a. *m*∠*WRS*   **b.** *m*∠1   **c.** *m*∠2

    d. *m*∠3   **e.** *m*∠4   **f.** *m*∠5

    g. *XT*   **h.** *RT*   **i.** *XS*

Solution: **a.** *m*∠*WRS* = 107° **b.** *m*∠1 = 22° **c.** *m*∠2 = 40°

         d. *m*∠3 = 55°   **e.** *m*∠4 = 18°   **f.** *m*∠5 = 55°

         g. *XT* = 8   **h.** *RT* = 16   **i.** *XS* = 10

Thinking Critically

True or false? **Give a reason to support your answer.**

1. If the measure of only one angle of a parallelogram is known, it is possible to find the measures of the other three angles without measuring.

2. If the length of only one side of a parallelogram is known, it is possible to find the lengths of the other three sides without measuring.

3. Are Theorems 8.1–8.4 also true for any rectangle, rhombus, or square? Why or why not?

4. Must any of the following be true for all parallelograms? Use Example 2 to explain.

    a. The diagonals are congruent.

    b. The diagonals are perpendicular.

    c. Each diagonal bisects two angles of the parallelogram.

Class Exercises

Find each angle measure or segment length.

  1. *m*∠*DAB*   **2.** *m*∠*ABC*

  3. *m*∠*BCD*   **4.** *m*∠*CDA*

  5. *DC*   **6.** *BC*

  7. *OB*   **8.** *AO*

  9. *DB*   **10.** *AC*

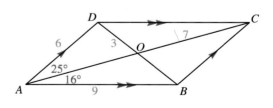

11. For ⊐*PQRS* name four pairs of congruent segments.

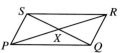

12. For ⊐*EFGH* name all angles congruent to each indicated angle.

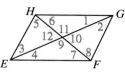

    a. ∠*HEF*   **b.** ∠*EFG*   **c.** ∠1

    d. ∠2   **e.** ∠5   **f.** ∠10

Exercises

True or false? Use the figure to answer 7–12.

    1. All parallelograms have line symmetry.

    2. All parallelograms have point symmetry.

    3. Opposite angles of any parallelogram are supplementary.

    4. Consecutive angles of any parallelogram are supplementary.

    5. The diagonals of any parallelogram are congruent.

    6. The diagonals of any parallelogram bisect each other.

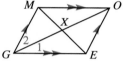

    7. ∠1 ≅ ∠2   **8.** $\overline{ME}$ ≅ $\overline{GO}$ **9.** $\overline{GX}$ ≅ $\overline{OX}$

  10. $\overline{MX}$ ≅ $\overline{EX}$ **11.** $\overline{MX}$ ≅ $\overline{GX}$ **12.** $\overline{ME}$ ⊥ $\overline{GO}$

13. a. Name two pairs of congruent angles.

    b. Name four pairs of supplementary angles.

    c. Name two pairs of congruent segments.

Find the measure of each angle or the length of each segment.

14. a. ∠*C*   **b.** ∠*D*   **15. a.** ∠*F*   **b.** ∠*FGH*   **16. a.** ∠*S*   **b.** ∠1

    c. *CD*   **d.** *AD*   **c.** ∠1   **d.** ∠2   **c.** ∠2   **d.** ∠3

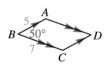 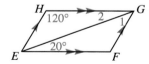 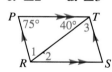

Find the measures of the other three angles of ⊐*NOPQ*.

  17. *m*∠*N* = 32°   **18.** *m*∠*O* = 90°   **19.** *m*∠*P* = 138°   **20.** *m*∠*Q* = 42°

21. If *DX* = 4 and *AX* = 6, find:

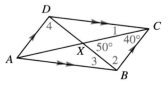

    a. *BX*   **b.** *BD*   **c.** *AC*

22. *m*∠*ABC* = 120°. Find the measure of each angle.

    a. ∠*ADC*   **b.** ∠*DAB*   **c.** ∠1

    d. ∠2   **e.** ∠3   **f.** ∠4

Tell whether *ABCD* is a parallelogram. Give a reason for your answer.

23.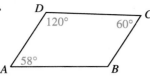

24.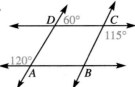

25.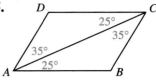

26. Draw two noncongruent parallelograms with congruent corresponding sides.

27. ▱*RSTW* is a rhombus.

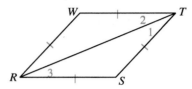

   a. Why is △*RST* ≅ △*RWT*?

   b. Why is ∠1 ≅ ∠2?

   c. Why is ∠1 ≅ ∠3?

   d. If *m*∠*WRT* = 22°, find the measures of the numbered angles.

APPLICATION

28. Lattices   A *lattice* is a structure of crossed strips or bars of metal, wood, plastic, or other material. These strips usually form a pattern of open spaces between the strips. Describe the types of quadrilaterals formed by the latticework shown.

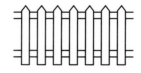

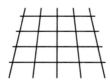

| Window | Picket Fence | Rose Trellis | Yard Fence | Gate |

Seeing in Geometry

***ABCD* is a parallelogram.**

   1. If you folded point *A* onto point *D*, which fold line would result?

   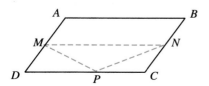

   2. If you folded point *A* onto point *C*, which fold line would result?

   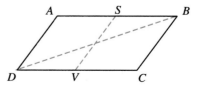

8.2 Properties of Rhombuses

Objective: To use the properties of rhombuses.

Recall that a rhombus is a parallelogram with all sides congruent. Thus a rhombus has all the properties of a parallelogram. A rhombus has additional properties that are not true for all parallelograms.

EXPLORING

Use a large rhombus that is not a square.

1. How many lines of symmetry does the rhombus have? Verify your answer by folding.

2. Fold the rhombus on either diagonal. Does the diagonal bisect the two angles whose vertices are its endpoints?

3. Label your rhombus as shown. Fold the rhombus along $\overline{AC}$. Then fold along $\overline{DO}$. What is true of ∠1 and ∠2? What is true of diagonals $\overline{AC}$ and $\overline{BD}$?

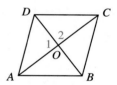

THEOREM 8.5: **Each diagonal of a rhombus bisects a pair of opposite angles.**

THEOREM 8.6: **Each diagonal of a rhombus is the perpendicular bisector of the other.**

Example: **a.** ***ABCD* is a rhombus. Find:**

$m\angle1$ $\quad$ $m\angle2$ $\quad$ $m\angle3$
$m\angle BCD$ $\quad$ $m\angle4$ $\quad$ $m\angle5$

b. ***MNOP* is a rhombus. Find:**

PO $\quad$ PM $\quad$ NQ
MQ $\quad$ MO

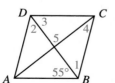

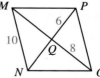

Solution: **a.** $m\angle1 = m\angle2 = m\angle3 = 55°$
$m\angle BCD = 70°$
$m\angle4 = 35°$
$m\angle5 = 90°$

b. $PO = PM = 10$
$NQ = 6$
$MQ = 8$
$MO = 16$

1. Would you have gotten the same results in the Exploring activity if you had used a rhombus that is a square?

2. Must the diagonals of a rhombus be congruent? If possible, draw a rhombus with diagonals that are not congruent and one with diagonals that are congruent.

Class Exercises

For each rhombus find each angle measure or segment length.

1. **a.** $m\angle 1$
   b. $m\angle 2$
   c. $m\angle 3$
   d. $m\angle 4$
   e. $m\angle JML$
   f. $m\angle MLK$

2. **a.** BX
   b. DB
   c. CX
   d. AC
   e. BC
   f. DC

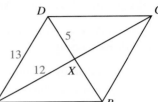

Exercises

True or false?

1. Every square is a rhombus.

2. Every rhombus is a square.

3. Every rhombus is a parallelogram.

4. Every parallelogram is a rhombus.

5. The opposite sides of any rhombus are parallel and congruent.

6. The opposite angles of any rhombus are congruent.

7. The consecutive angles of any rhombus are supplementary.

8. Every rhombus has point symmetry.

9. Every rhombus has at least two lines of symmetry.

ABCD is a rhombus. Classify each statement as true or false.

10. $\angle DAB \cong \angle BCD$

11. $\angle DAB \cong \angle ABC$

12. $\angle 1 \cong \angle 2$

13. $\angle 2 \cong \angle 3$

14. $m\angle DAB + m\angle ABC = 180°$

15. $\triangle ABC$ is isosceles.

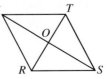

RSTU is a rhombus. Classify each statement as true or false.

16. $\angle UOT \cong \angle SOT$

17. $\angle SOT \cong \angle SOR$

18. $\overline{US} \perp \overline{TR}$

19. $\overline{UO} \cong \overline{SO}$

20. $\overline{RO} \cong \overline{TO}$

21. $\overline{US} \cong \overline{TR}$

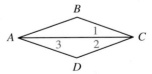

For each rhombus find each angle measure or segment length.

22.
a. $m\angle 1$
b. $m\angle 2$
c. $m\angle 3$
d. $m\angle 4$
e. $m\angle 5$
f. $m\angle 6$
g. $m\angle ADC$
h. $m\angle DCB$

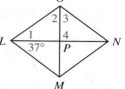

23.
a. LN
b. OP
c. PN
d. MP
e. $m\angle 1$
f. $m\angle 2$
g. $m\angle 3$
h. $m\angle 4$

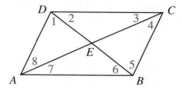

$LP = 4$
$MO = 6$

List all angles or segments, *if any*, that must be congruent to each indicated angle or segment.

24. *ABCD* is a rhombus.

a. $\angle 1$ **b.** $\angle 3$ **c.** $\angle DEC$
d. $\overline{DE}$ **e.** $\overline{DB}$ **f.** $\overline{DC}$

25. *ABCD* is a parallelogram.

a. $\angle 1$ **b.** $\angle 3$ **c.** $\angle DEC$
d. $\overline{DE}$ **e.** $\overline{DB}$ **f.** $\overline{DC}$

26. Draw two noncongruent rhombuses with congruent corresponding sides.

27. Draw two noncongruent rhombuses with congruent corresponding angles.

APPLICATIONS

28. Algebra *ABCD* is a rhombus, with $m\angle ABC = (3x - 5)°$ and $m\angle DBC = (x + 15)°$. Find x and $m\angle ABC$.

29. Algebra *RSTW* is a rhombus, with $m\angle RSW = 2x°$ and $m\angle SRT = (3x - 10)°$. Find x, $m\angle RSW$, and $m\angle SRT$.

Thinking in Geometry

Larry wanted to explore the properties of rhombuses. He drew rhombus *ABCD* and measured the angles and diagonals. His measurements led him to draw these false conclusions about rhombuses:

1. All the angles are right angles.

2. The diagonals are congruent.

What error did Larry make?

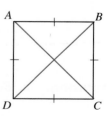

8.3 Properties of Rectangles and Squares

Objective: To use the properties of rectangles and squares.

Recall that a rectangle is a parallelogram with right angles. Thus all properties of parallelograms are also properties of rectangles. Rectangles have additional properties.

EXPLORING

1. Use LOGO to enter and run the procedure at the right. Sketch the result.

2. Write a procedure called **REFL.CORNER** that draws the reflection image of the figure in Step 1 over the vertical line through the home position.

3. Run the procedure at the right. Sketch the result. In a different color, draw all the lines of symmetry for the figure.

4. What is true of the diagonals? Why?

```
To corner
pu fd 30 pd
lt 90 fd 50 lt 90 fd 30 bk 30 home
end


To rectangle
corner refl.corner
rt 180 corner
rt 180 refl.corner
end
```

The **EXPLORING** activity demonstrates the following theorem.

THEOREM 8.7: The diagonals of a rectangle are congruent.

Example:
a. *ABCD* is a rectangle. Find:
BC AB DB
AO CO DO

b. *PQRS* is a rectangle. Find:
$m\angle 1$ $m\angle 2$ $m\angle 3$
$m\angle 4$ $m\angle 5$ $m\angle 6$

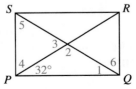

Solution:
a. $BC = 12$ $AB = 16$ $DB = 20$
$AO = CO = DO = 10$

b. $m\angle 1 = 32°$ $m\angle 2 = 116°$ $m\angle 3 = 64°$
$m\angle 4 = m\angle 5 = m\angle 6 = 58°$

A square is a parallelogram that is both a rectangle and a rhombus. Therefore, any property of a parallelogram, rectangle, or rhombus will also be a property of a square.

Thinking Critically

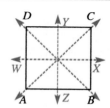

1. How many lines of symmetry does a square have? Which of the lines of symmetry for square *ABCD* at the right result from *ABCD* being a rhombus? Which result from *ABCD* being a rectangle?

2. Combine the properties of parallelograms, rhombuses, and rectangles developed in this chapter to make a list of properties for any square.

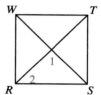

3. *RSTW* is a square. What are the measures of ∠1 and ∠2?

4. Draw a parallelogram with diagonals that satisfy each of the given conditions. Give the best name for the parallelogram.

   a. The diagonals are congruent but not perpendicular.

   b. The diagonals are perpendicular but not congruent.

   c. The diagonals are congruent and perpendicular.

Class Exercises

For each rectangle find each length or angle measure.

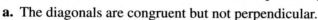

1. **a.** *XB*
   b. *AX*
   c. *DB*
   d. *AC*
   e. *AB*
   f. *BC*

2. **a.** *m∠1*
   b. *m∠2*
   c. *m∠3*
   d. *m∠4*
   e. *m∠5*
   f. *m∠6*

ABCD is a square such that *AD* = 12 and *DX* = 8.5. Find each length or angle measure.

3. *BX*   4. *AX*   5. *DB*   6. *AC*

7. *AB*   8. *BC*   9. *m∠AXB*   10. *m∠XAB*

Exercises

True or false?

1. The diagonals of a rectangle must be congruent.

2. The diagonals of a rectangle must bisect each other.

3. The diagonals of a rectangle must be perpendicular.

4. The diagonals of a square must bisect each other.

5. The diagonals of a square must be congruent.

6. If the diagonals of a parallelogram are congruent, then the parallelogram must be a square.

7. The diagonals of a square must be perpendicular.

8. If the diagonals of a parallelogram are perpendicular, then the parallelogram must be a square.

9. A rhombus can be a square.

10. Every square has exactly four lines of symmetry.

11. Every rhombus has exactly two lines of symmetry.

12. Every rectangle has exactly two lines of symmetry.

MNOP **is a rectangle such that** *MQ* = 5. **Find each length or angle measure.**

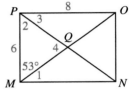

13. *QO*   14. *MN*   15. *ON*   16. *QN*

17. *m∠1*   18. *m∠2*   19. *m∠3*   20. *m∠4*

The diagonals of rectangle *RSTU* **intersect at** *P.* **Name each of the following.**

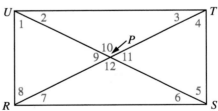

21. segments congruent to $\overline{UR}$

22. segments congruent to $\overline{RP}$

23. segments congruent to $\overline{UT}$

24. segments congruent to $\overline{RT}$

25. segments congruent to $\overline{US}$

26. angles congruent to ∠7

27. angles congruent to ∠8

28. triangles congruent to △RPS

29. triangles congruent to △UPR

30. right triangles

31. isosceles triangles

32. supplements of ∠10

The diagonals of square *RSTU* intersect at *P*. Name each of the following.

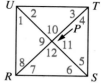

33. segments congruent to $\overline{UR}$

34. segments congruent to $\overline{RP}$

35. segments congruent to $\overline{RT}$

36. angles congruent to $\angle 7$

37. triangles congruent to $\triangle RPS$

38. right triangles

39. isosceles triangles

40. supplements of $\angle 10$

Name the quadrilaterals that have each property. Choose from parallelogram, rhombus, rectangle, and square.

41. All angles are congruent.

42. The diagonals are congruent.

43. The diagonals are perpendicular.

44. The diagonals bisect each other.

45. The diagonals are perpendicular bisectors of each other.

46. Consecutive angles are supplementary.

47. Each diagonal bisects two angles of the quadrilateral.

48. Each diagonal determines a line of symmetry.

Draw a parallelogram to fit each description.

49. The diagonals are congruent but not perpendicular.

50. The diagonals are perpendicular but not congruent.

51. There are exactly four lines of symmetry.

52. There are exactly two lines of symmetry, each determined by a diagonal.

APPLICATIONS

53. Algebra   The diagonals of rectangle *RSTW* intersect at point *O*. If $WO = x + 6$ and $RT = 5x + 3$, find the length of $\overline{RT}$.

54. Computer   Write a LOGO procedure that draws a rectangle and its lines of symmetry. Make the rectangle 120 units by 60 units.

Computer

A square with its four lines of symmetry is shown. Notice that the large square is composed of smaller squares. Write a LOGO procedure that draws one of the smaller squares and its diagonal. Then use that procedure in a procedure that draws the larger figure.

8.4 Properties of Trapezoids

Objective: To use the properties of trapezoids and isosceles trapezoids.

Recall that a trapezoid is a quadrilateral with exactly one pair of parallel sides. We use the following terms to describe trapezoids.

Bases: the parallel sides

Legs: the nonparallel sides

Base angles: a pair of consecutive angles whose included side is a base (Each trapezoid has two pairs of base angles.)

Median: the segment that joins the midpoints of the legs

Example 1: *ABCD* is a trapezoid.

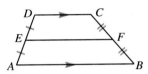

Bases: $\overline{AB}$ and $\overline{DC}$

Legs: $\overline{AD}$ and $\overline{BC}$

Base angles: $\angle A$ and $\angle B$; $\angle D$ and $\angle C$

Median: $\overline{EF}$

An *isosceles trapezoid* is one whose legs are congruent.

Trapezoid *RSTW* is an isosceles trapezoid because $\overline{WR} \cong \overline{TS}$.

EXPLORING

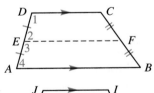

Use two large trapezoids, one of which is isosceles. Draw the median for each trapezoid. Label the figures as shown. Do the following for each trapezoid.

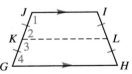

1. Name both pairs of base angles.

2. Measure each pair of base angles. Are any pairs congruent?

3. Measure both bases and the median to the nearest millimeter.

4. Add the lengths of the bases. How does this sum compare to the length of the median?

5. Each leg is a transversal for the bases and median. Measure $\angle 1$, $\angle 2$, $\angle 3$, and $\angle 4$. Are any pairs congruent? Are any pairs supplementary?

6. Why is the median parallel to both bases?

The **EXPLORING** activity demonstrates the following theorems. Notice that Theorem 8.8 is true for all trapezoids; Theorem 8.9 is true only for isosceles trapezoids.

THEOREM 8.8: **The median of any trapezoid is parallel to the bases and has a length equal to half the sum of the base lengths.**

THEOREM 8.9: **Base angles of an isosceles trapezoid are congruent.**

Example 2: Find $m\angle X$ and RS.

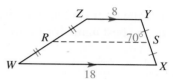

Solution: $m\angle X = 70°$ since $\overline{RS} \parallel \overline{WX}$.
$$RS = \frac{1}{2}(8 + 18) = \frac{1}{2}(26) = 13$$

If you think of $\triangle ABC$ as a trapezoid with one base of length zero, you will see how the following theorem is related to Theorem 8.8.

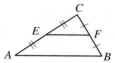

THEOREM 8.10: **A segment joining the midpoints of two sides of a triangle is parallel to the third side and half its length.**

Thinking Critically

1. $\triangle ABC$ is isosceles. X and Y are midpoints of $\overline{AB}$ and $\overline{AC}$.
   a. What kind of triangle is $\triangle AXY$?
   b. Why is $\angle 1 \cong \angle 2$ and $\angle B \cong \angle C$?
   c. Why is $\overline{XY} \parallel \overline{BC}$?
   d. What kind of figure is $XBCY$?

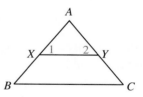

2. Does a trapezoid that is not isosceles have line symmetry? Does an isosceles trapezoid have line symmetry?

Class Exercises

1. Name the following for the trapezoid.
   a. the bases
   b. the legs
   c. two pairs of base angles

2. a. Find $m\angle 1$.
   b. Find $m\angle 2$.
   c. Find YZ.

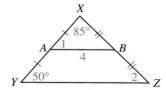

For each trapezoid find (a) *EF*, (b) *m∠*1, (c) *m∠*2, (d) *m∠*3, (e) *m∠*4, and (f) *m∠*5.

3.
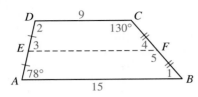

4.
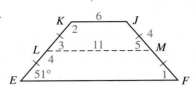

5. For the trapezoid in Exercise 4 find the following.
  a. the best name for *EFJK*   **b.** *FM*   **c.** *EK*

Exercises

For each trapezoid name (a) the bases, (b) the legs, and (c) two pairs of base angles.

1.

2.
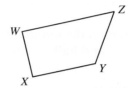

3.

Find the measure of each angle or the length of each segment.

4. a. ∠1
  b. ∠2
  c. ∠3
  d. $\overline{MN}$

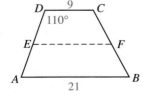

5. a. $\overline{RS}$ **b.** $\overline{TS}$
  c. $\overline{ZY}$ **d.** ∠1
  e. ∠2   **f.** ∠3
  g. ∠4   **h.** ∠5

XY = 5
XZ = 4
RT = 14

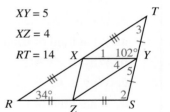

For each trapezoid find the measure of each angle or the length of each segment. The dashed segments shown are medians.

6. a. $\overline{EF}$ **b.** ∠A

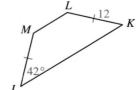

7. a. $\overline{MJ}$ **b.** ∠K
  c. ∠L   **d.** ∠M

8. a. ∠D   **b.** ∠B
  c. ∠C   **d.** ∠1
  e. $\overline{EF}$ **f.** $\overline{EA}$

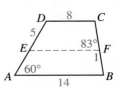

9. a. $\overline{TQ}$ **b.** $\overline{TS}$
c. $\overline{RS}$ **d.** $\angle W$
e. $\angle 1$ **f.** $\angle 2$

10. a. $\angle E$ **b.** $\angle C$
c. $\angle D$ **d.** $\overline{ED}$
e. $\overline{JC}$ **f.** $\overline{CD}$

11. a. $\overline{AB}$ **b.** $\overline{ZY}$
c. $\angle W$ **d.** $\angle Z$
e. $\angle 1$ **f.** $\angle 2$

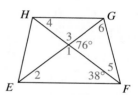

12. a. $\angle JKL$ **b.** $\angle 1$
c. $\angle 2$ **d.** $\angle 3$

13. a. $\angle DCB$ **b.** $\angle 1$
c. $\angle 2$ **d.** $\angle DAB$
e. $\angle 3$ **f.** $\angle B$

14. $m\angle EFG = 82°$
a. $\angle 1$ **b.** $\angle 2$
c. $\angle 3$ **d.** $\angle 4$
e. $\angle 5$ **f.** $\angle 6$

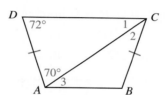

ABCD is an isosceles trapezoid with one base $\overline{AB}$. One angle measure is given. Find the measures of the other three angles.

15. $m\angle A = 70°$ **16.** $m\angle B = 50°$ **17.** $m\angle D = 120°$

18. The measures of two angles of a trapezoid are 65° and 100°. What are the measures of the other two angles?

19. *ABCD* is an isosceles trapezoid with $m\angle A = 75°$. Find all possible measures of $\angle B$.

WXYZ is an isosceles trapezoid, with $\overline{ZY} \parallel \overline{WX}$. Give a reason for each statement.

20. $\angle 1 \cong \angle 2$ **21.** $\angle WZY \cong \angle XYZ$

22. $\angle 3 \cong \angle 4$ **23.** $\overline{WZ} \cong \overline{XY}$

24. $m\angle 4 = m\angle 5 + m\angle 6$

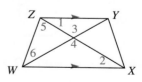

Draw a trapezoid to fit each description, if possible. If it is not possible, give a reason.

25. two right angles   **26.** four acute angles   **27.** exactly one right angle

28. three congruent sides   **29.** non-isosceles but with two congruent sides

APPLICATIONS

30. Visualization $\overline{DC} \parallel \overline{EF} \parallel \overline{AB}$.
Name all trapezoids shown.

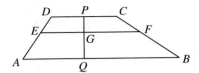

31. Algebra Find x, AB, and ST.

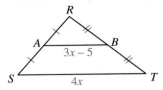

Test Yourself

Find the measure of each angle.

1. *ABCD* is a parallelogram.
  a. $\angle ABC$ **b.** $\angle C$
  c. $\angle 1$ **d.** $\angle 2$

2. *MNOP* is a rhombus.
  a. $\angle MNO$ **b.** $\angle 1$
  c. $\angle 2$ **d.** $\angle 3$

3. *RMCT* is a rectangle.
  a. $\angle 1$ **b.** $\angle 2$
  c. $\angle 3$ **d.** $\angle 4$

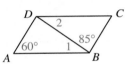

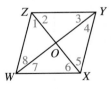

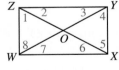

Name each of the following for each figure.
a. all segments congruent to $\overline{WO}$
b. all segments congruent to $\overline{WZ}$
c. all angles congruent to $\angle 1$

4. $\square WXYZ$ **5.** rhombus *WXYZ*   **6.** rectangle *WXYZ*   **7.** square *WXYZ*

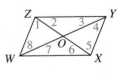

8. *ABCD* is a trapezoid, with $AD = 6$. Find each of the following.
  a. $m\angle A$ **b.** $m\angle D$ **c.** $m\angle B$
  d. AE **e.** BC **f.** AB

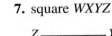

9. The measures of two angles of a trapezoid are 45° and 110°.
What are the measures of the other angles?

8.5 Proving Properties of Parallelograms

Objective: To prove theorems and other statements relating to parallelograms and trapezoids.

You can use the definition of a parallelogram and what you know about congruent triangles and parallel lines to prove the theorems of this chapter and other properties of parallelograms and trapezoids. In the following Exploring activity, you also may use any theorem you have learned in this chapter as a reason.

EXPLORING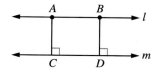

Given: $l \parallel m$;

A and B are any points on line l;

$\overline{AC}$ and $\overline{BD}$ are perpendicular segments to line m.

Prove: $\overline{AC} \cong \overline{BD}$

Give a reason for each statement.

1. $l \parallel m$ (so $\overline{AB} \parallel \overline{CD}$)

2. $\overline{AC} \perp m$ and $\overline{BD} \perp m$

3. $\overline{AC} \parallel \overline{BD}$

4. $ABDC$ is a parallelogram.

5. $\overline{AC} \cong \overline{BD}$

You have just proved the following theorem.

THEOREM 8.11: **Parallel lines are everywhere equidistant.**

In the **EXPLORING** activity you used the definition of a parallelogram to show that since both pairs of opposite sides of the quadrilateral are parallel, the quadrilateral is a parallelogram. Notice how the following Example uses the definition of a parallelogram in a different way.

Example: **To be Proven:** A diagonal of a parallelogram separates it into two congruent triangles.

Given: *ABCD* is a parallelogram.

Prove: $\triangle ABC \cong \triangle CDA$

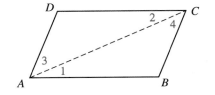

Give a reason for each statement.

a. *ABCD* is a parallelogram.

b. $\overline{AB} \parallel \overline{CD}$ and $\overline{BC} \parallel \overline{DA}$

c. $\angle 1 \cong \angle 2$ and $\angle 4 \cong \angle 3$

d. $\overline{AC} \cong \overline{CA}$

e. $\triangle ABC \cong \triangle CDA$

Solution:

a. Given

b. Definition of a parallelogram

c. Alternate Interior Angles Postulate

d. A segment is congruent to itself.

e. ASA Postulate

Thinking Critically

1. Explain how to modify the proof in the Example to prove this theorem.

   To be Proven: The opposite sides of a parallelogram are congruent.

   Given: *ABCD* is a parallelogram.

   Prove: $\overline{AB} \cong \overline{CD}$ and $\overline{BC} \cong \overline{DA}$

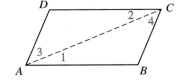

2. Explain how to prove the following theorem.

   To be Proven: The opposite angles of a parallelogram are congruent.

Class Exercises

Supply a reason for each conclusion.

1. **Given:** *PQRS* is a rhombus.
   Conclusion: *PQRS* is a parallelogram.

2. **Given:** *EFGH* is a rectangle.
   Conclusion: $\angle F$ and $\angle G$ are right angles.

3. **Given:** *ABCD* is a parallelogram.
   Conclusion: $\overline{AB} \parallel \overline{DC}$

4. **Given:** *ABCD* is a parallelogram.
   Conclusion: $\overline{AB} \cong \overline{DC}$

5. Given: *WXYZ* is an isosceles trapezoid with legs $\overline{WZ}$ and $\overline{XY}$.

   Conclusion: $\overline{WZ} \cong \overline{XY}$

6. Given: *ABCD* is a quadrilateral such that $\overline{AB} \parallel \overline{DC}$ and $\overline{AD} \parallel \overline{BC}$.

   Conclusion: *ABCD* is a parallelogram.

7. Given: *EFGH* is a rectangle.

   Conclusion: $\overline{EG} \cong \overline{FH}$

8. Given: $\overrightarrow{LP}$ bisects $\angle KLM$.

   Conclusion: $\angle KLP \cong \angle PLM$

Exercises

Give a reason for each statement.

1. To be Proven: Each diagonal of a rhombus bisects a pair of opposite angles (Theorem 8.5).

   Given: *ABCD* is a rhombus.

   Prove: $\overline{AC}$ bisects $\angle DAB$ and $\angle DCB$.

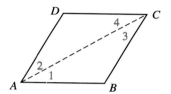

   a. *ABCD* is a rhombus.
   b. $\overline{AB} \cong \overline{AD}$ and $\overline{BC} \cong \overline{DC}$
   c. $\overline{AC} \cong \overline{AC}$
   d. $\triangle ABC \cong \triangle ADC$
   e. $\angle 1 \cong \angle 2$ and $\angle 3 \cong \angle 4$
   f. $\overline{AC}$ bisects $\angle DAB$ and $\angle DCB$.

2. To be Proven: The diagonals of a parallelogram bisect each other (Theorem 8.4).

   Given: *ABCD* is a parallelogram.

   Prove: $\overline{AC}$ and $\overline{BD}$ bisect each other.

   Note: You may use any theorem stated before Theorem 8.4 on page 221.

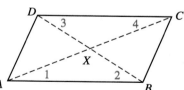

   a. *ABCD* is a parallelogram.
   b. $\overline{AB} \parallel \overline{DC}$
   c. $\angle 1 \cong \angle 4$ and $\angle 2 \cong \angle 3$
   d. $\overline{AB} \cong \overline{DC}$
   e. $\triangle ABX \cong \triangle CDX$
   f. $\overline{AX} \cong \overline{CX}$ and $\overline{BX} \cong \overline{DX}$
   g. $\overline{AC}$ and $\overline{BD}$ bisect each other.

3. **To be Proven:** The diagonals of an isosceles trapezoid are congruent.

Given: *ABCD* is an isosceles trapezoid with legs $\overline{AD}$ and $\overline{BC}$.

Prove: $\overline{DB} \cong \overline{CA}$

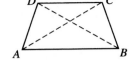

a. *ABCD* is an isosceles trapezoid with legs $\overline{AD}$ and $\overline{BC}$.

b. $\overline{AD} \cong \overline{BC}$

c. $\overline{AB} \cong \overline{BA}$

d. $\angle DAB \cong \angle CBA$

e. $\triangle DAB \cong \triangle CBA$

f. $\overline{DB} \cong \overline{AC}$

4. **To be Proven:** The diagonals of a rectangle are congruent (Theorem 8.7).

Given: *PQRS* is a rectangle.

Prove: $\overline{SQ} \cong \overline{RP}$

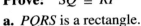

a. *PQRS* is a rectangle.

b. $\angle SPQ$ and $\angle RQP$ are right angles.

c. $\triangle SPQ$ and $\triangle RQP$ are right triangles.

d. *PQRS* is a parallelogram.

e. $\overline{PS} \cong \overline{QR}$

f. $\overline{PQ} \cong \overline{QP}$

g. $\triangle SPQ \cong \triangle RQP$

h. $\overline{SQ} \cong \overline{RP}$

5. **To be Proven:** Each diagonal of a rhombus is the perpendicular bisector of the other diagonal (Theorem 8.6).

Given: *ABCD* is a rhombus.

Prove: $\overline{BD}$ is the perpendicular bisector of $\overline{AC}$.

a. *ABCD* is a rhombus.

b. $\overline{AB} \cong \overline{CB}$

c. *ABCD* is a parallelogram.

d. $\overline{BD}$ bisects $\overline{AC}$.

e. $\overline{AX} \cong \overline{CX}$

f. $\overline{XB} \cong \overline{XB}$

g. $\triangle AXB \cong \triangle CXB$

h. $\angle AXB \cong \angle CXB$

i. $\angle AXB$ and $\angle CXB$ are right angles.

j. $\overline{BD}$ is the perpendicular bisector of $\overline{AC}$.

6. To be Proven: Base angles of an isosceles trapezoid are congruent (Theorem 8.9).

Note: To help prove the theorem, you can construct $\overleftrightarrow{WX}$ and $\overleftrightarrow{TY}$, the perpendiculars from W and T to $\overleftrightarrow{RS}$.

Given: $RSTW$ is an isosceles trapezoid with legs $\overline{RW}$ and $\overline{ST}$; $\overline{WX} \perp \overline{RS}$ and $\overline{TY} \perp \overline{RS}$.

Prove: $\angle R \cong \angle S$

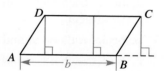

a. $RSTW$ is an isosceles trapezoid with legs $\overline{RW}$ and $\overline{ST}$.

b. $\overline{WT} \parallel \overline{RS}$ (so $\overline{WT} \parallel \overline{XY}$)

c. $\overline{WX} \perp \overline{RS}$ and $\overline{TY} \perp \overline{RS}$

d. $\overline{WX} \parallel \overline{TY}$

e. $XYTW$ is a parallelogram.

f. $\overline{WX} \cong \overline{TY}$

g. $\overline{RW} \cong \overline{ST}$

h. $\angle RXW$ and $\angle SYT$ are right angles.

i. $\triangle RXW$ and $\triangle SYT$ are right triangles.

j. $\triangle RXW \cong \triangle SYT$

k. $\angle R \cong \angle S$

APPLICATIONS

7. Area The area of a parallelogram is the product of the base length b and the height h. Explain why all the segments shown in red for $\square ABCD$ have the same length h.

8. Logic If a diagonal of a quadrilateral separates the quadrilateral into two congruent triangles, must the quadrilateral be a parallelogram? Why or why not?

Seeing in Geometry

Point M is the midpoint of $\overline{DE}$. What must be true of DM, EM, and CM? Why? (*Hint:* Consider rectangle $CDFE$.)

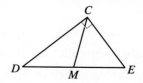

8.6 Finding the Quadrilaterals that are Parallelograms

Objective: To determine whether a quadrilateral must be a parallelogram, rectangle, rhombus, or square.

As you will discover in the Exploring activity, the diagonals of a quadrilateral determine whether the quadrilateral is a parallelogram and whether it is a special parallelogram.

EXPLORING

1. Use the *Geometric Supposer: Quadrilaterals* disk. Construct *Your Own* quadrilateral using the *Diagonals* option. Make the diagonals congruent, perpendicular to each other, and bisecting each other.

2. Measure to determine what kind of quadrilateral you made. Record your data.

3. Repeat Steps 1–2 for diagonals that:
   a. bisect each other, but are not perpendicular and are not congruent
   b. bisect each other and are congruent, but are not perpendicular
   c. bisect each other and are perpendicular, but are not congruent

The **EXPLORING** activity demonstrates the following theorems.

THEOREM 8.12: If the diagonals of a quadrilateral bisect each other, then the quadrilateral is a parallelogram.

THEOREM 8.13: If the diagonals of a quadrilateral bisect each other and:

a. are congruent, then the quadrilateral is a rectangle.

b. are perpendicular, then the quadrilateral is a rhombus.

c. are perpendicular and congruent, then the quadrilateral is a square.

To prove that a quadrilateral is a parallelogram, you could show that two pairs of opposite sides are parallel and use the definition of a parallelogram. Or you could show that the diagonals bisect each other (Theorem 8.12).

The following theorems provide additional ways to prove that a quadrilateral is a parallelogram. Theorem 8.14 and Theorem 8.15 are *converses* of theorems you saw earlier in the chapter. Theorem 8.16 may surprise you.

THEOREM 8.14: **If both pairs of opposite sides of a quadrilateral are congruent, then the quadrilateral is a parallelogram.**

THEOREM 8.15: **If both pairs of opposite angles of a quadrilateral are congruent, then the quadrilateral is a parallelogram.**

THEOREM 8.16: **If two sides of a quadrilateral are both parallel and congruent, then the quadrilateral is a parallelogram.**

Example: **Must each quadrilateral be a parallelogram? Explain.**

a.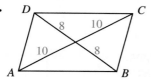

b.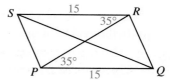

Solution: **a.** Yes, because the diagonals bisect each other.

b. Yes. $\overline{SR}$ and $\overline{PQ}$ are both congruent and parallel (since alternate interior angles are congruent).

Thinking Critically

1. Name five ways to show that a quadrilateral is a parallelogram.

2. Explain why Theorem 8.13c follows from Theorem 8.13a and b.

3. **a.** What is $2a° + 2b°$? Why?
   b. What is $a° + b°$?
   c. Why is $\overline{PQ} \parallel \overline{SR}$ and $\overline{SP} \parallel \overline{RQ}$?
   d. Why is $PQRS$ a parallelogram?
   e. What theorem have you just proved?

Class Exercises

Suppose you are given the following information about quadrilateral *MNOP*, whose diagonals intersect at *Q*. Give all the correct names (parallelogram, rhombus, rectangle, square) for *MNOP*.

1. $\overline{MQ} \cong \overline{OQ}$ and $\overline{PQ} \cong \overline{NQ}$
2. $\overline{MQ} \cong \overline{OQ}$, $\overline{PQ} \cong \overline{NQ}$, and $\overline{MO} \perp \overline{NP}$
3. $\overline{MQ} \cong \overline{OQ} \cong \overline{PQ} \cong \overline{NQ}$
4. $\overline{MO} \cong \overline{PN}$, $\overline{MO} \perp \overline{PN}$, $\overline{PQ} \cong \overline{NQ}$, and $\overline{MQ} \cong \overline{OQ}$

Must each quadrilateral be a parallelogram? If yes, explain.

5.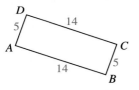

6.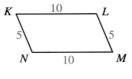

7.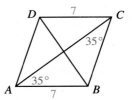

Exercises

Must each quadrilateral be a parallelogram? Explain.

1.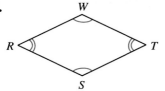

2.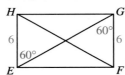

3.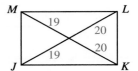

4.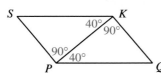

5.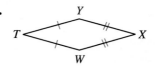

6.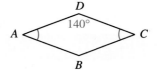

Draw a quadrilateral to fit each given description.

7. a parallelogram that is *not* a rhombus and that has diagonals of length 6 cm and 4 cm

8. a rectangle that is *not* a square and that has diagonals of length 6 cm

9. a rhombus with diagonals of length 6 cm and 4 cm

10. a square with both diagonals of length 6 cm

11. a quadrilateral with perpendicular diagonals that do not bisect each other

12. a quadrilateral with congruent diagonals that do not bisect each other

Give a reason for each statement.

13. **To be Proven:** If both pairs of opposite sides of a quadrilateral are congruent, then the quadrilateral is a parallelogram (Theorem 8.14).

   Given: $\overline{AB} \cong \overline{CD}$ and $\overline{CB} \cong \overline{AD}$

   Prove: $ABCD$ is a parallelogram.

   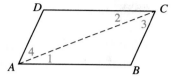

   a. $\overline{AB} \cong \overline{CD}$ and $\overline{CB} \cong \overline{AD}$

   b. $\overline{AC} \cong \overline{CA}$

   c. $\triangle ABC \cong \triangle CDA$

   d. $\angle 1 \cong \angle 2$ and $\angle 3 \cong \angle 4$

   e. $\overline{AB} \parallel \overline{DC}$ and $\overline{BC} \parallel \overline{AD}$

   f. $ABCD$ is a parallelogram.

14. **To be Proven:** If two sides of a quadrilateral are both parallel and congruent, then the quadrilateral is a parallelogram (Theorem 8.16).

   Given: $\overline{AB} \parallel \overline{CD}$ and $\overline{AB} \cong \overline{CD}$

   Prove: $ABCD$ is a parallelogram.

   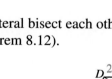

   a. $\overline{AB} \parallel \overline{CD}$ and $\overline{AB} \cong \overline{CD}$

   b. $\angle 1 \cong \angle 2$

   c. $\overline{AC} \cong \overline{CA}$

   d. $\triangle ABC \cong \triangle CDA$

   e. $\angle 3 \cong \angle 4$

   f. $\overline{BC} \parallel \overline{AD}$

   g. $ABCD$ is a parallelogram.

15. **To be Proven:** If the diagonals of a quadrilateral bisect each other, then the quadrilateral is a parallelogram (Theorem 8.12).

   Given: $\overline{AO} \cong \overline{CO}$ and $\overline{BO} \cong \overline{DO}$

   Prove: $ABCD$ is a parallelogram.

   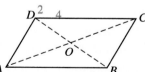

   a. $\overline{AO} \cong \overline{CO}$ and $\overline{BO} \cong \overline{DO}$

   b. $\angle AOB \cong \angle COD$; $\angle DOA \cong \angle BOC$

   c. $\triangle AOB \cong \triangle COD$; $\triangle DOA \cong \triangle BOC$

   d. $\angle 1 \cong \angle 2$ and $\angle 3 \cong \angle 4$

   e. $\overline{AB} \parallel \overline{DC}$ and $\overline{AD} \parallel \overline{BC}$

   f. $ABCD$ is a parallelogram.

16. Explain why it is not possible for a trapezoid to have congruent bases.

17. Explain why it is not possible for the diagonals of a trapezoid to bisect each other.

18. Refer to Exercises 13 and 14. Explain how you could change the proof in Exercise 14 so that it uses as a reason the theorem proved in Exercise 13.

19. Refer to Exercises 14 and 15. Explain how you could change the proof in Exercise 15 so that it uses as a reason the theorem proved in Exercise 14.

APPLICATIONS

20. Mechanical Drawing   Artists sometimes use *parallel rulers* to draw parallel lines. In the diagram, $WZ = XY$ and $ZY = WX$. Explain why the rulers are always parallel.

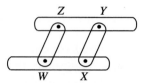

21. Manufacturing   Many ironing boards are made so that the intersection of the legs is the midpoint of both legs. In the diagram, $DX = BX$ and $AX = CX$. Explain why $\overline{DC}$ must be parallel to $\overline{AB}$.

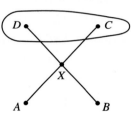

22. Computer   A *kite* is a kind of quadrilateral. Use the *Geometric Supposer: Quadrilaterals* disk to draw a kite and its diagonals. Measure to determine the properties of a kite. Record your data. Is a kite a parallelogram? Explain your answer.

Computer

1. Use the *Geometric Supposer: Quadrilaterals* disk to create a rectangle. Draw segments to connect the midpoints of the sides of the rectangle.

2. Draw the diagonals of the new quadrilateral. Label their intersection.

3. Measure to determine the properties of the new figure. Is it a parallelogram? If it is, what kind?

4. Repeat Steps 1–3 for a rhombus, a square, and a trapezoid.

Thinking About Proof

Auxiliary Lines

On page 238 you saw how the diagram shown below is involved in the proofs of the following three statements.

A diagonal of a parallelogram separates it into two congruent triangles.

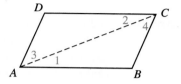

The opposite sides of a parallelogram are congruent.

The opposite angles of a parallelogram are congruent.

Notice that the first statement mentions a diagonal of the parallelogram. Although the other two statements do not mention the diagonal, it was very helpful in their proofs.

When you add an **auxiliary** line or segment to a diagram to help you prove a theorem, you must be sure there is such a figure.

Exercises

Look at each of the following sets of exercises. Name each exercise that uses an auxiliary line or segment and name the auxiliary figure.

  1. Exercises 3–6, pages 240–241

  2. Exercises 13–15, page 245

  3. the Class Exercise and Exercises 1–3, page 173.

Tell which of the following best describes each figure.
a. There is exactly one such figure.
b. There is more than one such figure.
c. There is no figure that satisfies all the conditions.

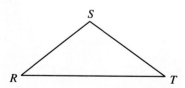

  4. midpoint X of $\overline{RS}$

  5. angle bisector $\overrightarrow{SY}$ of $\angle RST$

  6. median $\overline{TX}$ of $\triangle RST$

  7. altitude $\overline{TD}$ of $\triangle RST$

  8. perpendicular bisector $\overleftrightarrow{SZ}$ of $\overline{RT}$

  9. $\angle A$, congruent to $\angle S$ and complementary to $\angle R$

10. $\overrightarrow{SZ}$ so that $m\angle RSZ + m\angle ZST = m\angle RST$

11. Look at the Seeing in Geometry feature at the bottom of page 241. Describe how auxiliary lines help you answer the question.

8.7 Problem Solving Strategy: Finding a Pattern

Objective: To solve problems by finding a pattern.

Sometimes the numbers in a problem form a pattern.
By finding the pattern you can solve the problem.

Example: The drawing shows three rows of steel beams
stacked at a building site. Suppose the stack
had twelve rows of beams. What would be the
total number of beams in the stack?

Solution: You could draw a model and count the total
number of beams. Or you could
look for a pattern.

  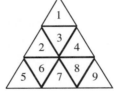

How is the total number of beams in each stack related
to the number of rows in the stack?

Look at the pattern. The number of rows multiplied
by itself equals the total number of beams.

You can extend the pattern to a stack of
12 rows.

$$12 \cdot 12 = 144$$

There would be 144 beams in the stack.

Number of Rows	1	2	3	4
Beams	1	4	9	16

Class Exercise

1. Suppose the measures of the vertex angles of three different isosceles triangles are 2°, 4°, and 6° respectively.

   a. Copy and complete the table for the measures of the vertex angles and the base angles in the next isosceles triangles in the pattern.

Triangle	1	2	3	4	5	6
Vertex angle	2°	4°	6°	■	■	■
Base angles	■	■	■	■	■	■

   b. What is the measure of the *vertex* angle of isosceles triangle number 25?

   c. What is the pattern for the measures of the base angles?

   d. What is the measure of each base angle of triangle number 37?

Exercises

1. The diagram shows three rows of boxes of toothpaste tubes stacked in a window display. Describe the pattern in the display. How many boxes will be in a 15-row stack?

2. One year Margo planted a peach tree in the field behind her house. The next year she planted three more trees in the pattern shown. The third year she planted five more trees. Margo continued the pattern each year.

   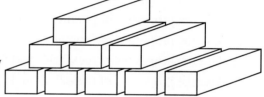

   a. If Margo continues the pattern, how many trees will she plant in year 14?

   b. How many trees are there in each row at the end of year 19?

   c. In which year will Margo plant 51 new trees?

3. The figures show the first four triangular numbers and their geometric representations.

   a. Find the fifth and sixth triangular numbers. Draw their geometric representations.

   b. What is the tenth triangular number?

   1   3   6   10

Vocabulary and Symbols

You should be able to write a brief description, draw a picture, or give an example to illustrate the meaning of each of the following terms.

Vocabulary

auxiliary (p. 247)
base angles of a trapezoid (p. 232)
bases of a trapezoid (p. 232)
isosceles trapezoid (p. 232)

legs of a trapezoid (p. 232)
median of a trapezoid (p. 232)

Symbol

▱ (parallelogram) (p. 221)

Summary

The following list indicates the major skills, facts, and results you should have mastered in this chapter.

8.1 Use the properties of parallelograms. (pp. 221–224)

8.2 Use the properties of rhombuses. (pp. 225–227)

8.3 Use the properties of rectangles and squares. (pp. 228–231)

8.4 Use the properties of trapezoids and isosceles trapezoids. (pp. 232–236)

8.5 Prove theorems and other statements relating to parallelograms and trapezoids. (pp. 237–241)

8.6 Determine whether a quadrilateral must be a parallelogram, rectangle, rhombus, or square. (pp. 242–246)

8.7 Solve problems by finding a pattern. (pp. 248–249)

Exercises

For each parallelogram find each length or angle measure.

1. a. $m\angle D$ **b.** $m\angle A$ **2. a.** $m\angle EGF$ **b.** $m\angle HFG$ **3.** *PQRS* is a rectangle.

    c. *AD*   **d.** *AB*   **c.** $m\angle EFG$ **d.** $m\angle HKG$ **a.** *PR* **b.** *SQ* **c.** $m\angle STR$

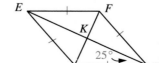 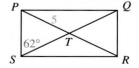

Name the quadrilaterals that have each property. Choose from parallelogram, rhombus, rectangle, and square.

4. All angles are congruent.

5. The diagonals are perpendicular.

6. Opposite angles are congruent.

7. The diagonals bisect each other.

Find each length or angle measure for each trapezoid.

8. a. $m\angle Z$
   b. $m\angle ZAB$
   c. AB

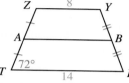

9. a. $m\angle S$
   b. $m\angle A$
   c. $m\angle P$

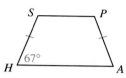

10. To be Proven: The opposite sides of a parallelogram are congruent.

Given: WXYZ is a parallelogram.

Prove: $\overline{WX} \cong \overline{YZ}$ and $\overline{WZ} \cong \overline{YX}$

Give a reason for each statement.

a. WXYZ is a parallelogram.

b. $\overline{WX} \parallel \overline{ZY}$ and $\overline{WZ} \parallel \overline{XY}$

c. $\angle XWY \cong \angle ZYW$ and $\angle XYW \cong \angle ZWY$

d. $\overline{WY} \cong \overline{WY}$

e. $\triangle XWY \cong \triangle ZYW$

f. $\overline{WX} \cong \overline{YZ}$ and $\overline{WZ} \cong \overline{YX}$

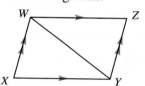

Must each quadrilateral be a parallelogram? Explain.

11.

12.

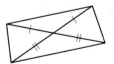

13.

14.

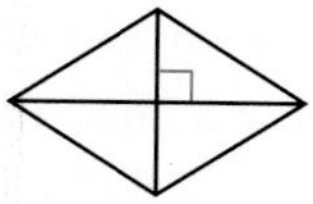

15. Complete: If the diagonals of a quadrilateral are congruent and are perpendicular bisectors of each other, then the quadrilateral must be a ▪.

PROBLEM SOLVING

16. The diagram shows the first five rows of an arrangement of numbers known as Pascal's Triangle. Look for a pattern. Then find the numbers that belong in the next two rows.

```
              1
           1     1
        1     2     1
     1     3     3     1
  1     4     6     4     1
```

Find *x* and *y*.

1. square

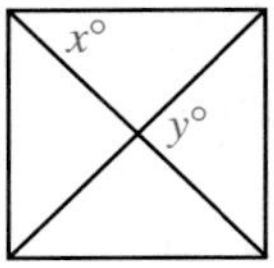

2. parallelogram

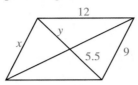

3. parallelogram

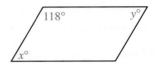

4. rectangle

5. rhombus

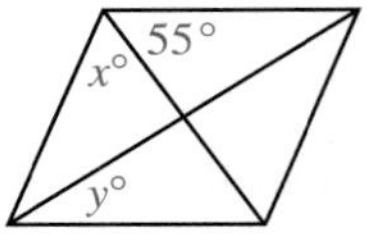

6. trapezoid

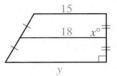

Supply a reason for each conclusion.

7. Given: *ABCD* is a parallelogram.

    Conclusion: ∠*A* ≅ ∠*C*

8. Given: *RSTW* is an isosceles trapezoid with legs $\overline{RW}$ and $\overline{ST}$.

    Conclusion: ∠*R* ≅ ∠*S*

Must each quadrilateral be a parallelogram? Explain.

9.

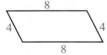

10.

11.

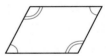

12. Complete: If the diagonals of a quadrilateral are the perpendicular bisectors of each other, then the quadrilateral must be a ▓.

13. Draw a quadrilateral with diagonals that bisect each other and are congruent but are not perpendicular.

PROBLEM SOLVING

14. The figures show the first four rectangular numbers and their geometric representations.

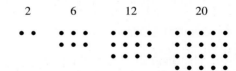

    a. Find the fifth and sixth rectangular numbers. Draw their geometric representations.

    b. What is the tenth rectangular number?

Chapter 9

Perimeter and Area

The trees planted as a windbreak around the edge of a farmer's field define a region whose size influences the farmer's estimates of such quantities as crop yield and amount of fertilizer required.

Chapter 9
Focus on Skills

ARITHMETIC

Solve.

1. $$\begin{array}{r} 6.8 \\ 13.5 \\ 8.0 \\ +\,11.8 \\ \hline \end{array}$$

2. $$\begin{array}{r} 15 \\ 7\frac{3}{4} \\ +\,4\frac{1}{2} \\ \hline \end{array}$$

3. $$\begin{array}{r} 6.9 \\ \times\,4.8 \\ \hline \end{array}$$

4. $$\begin{array}{r} 62.4 \\ 62.4 \\ 115.3 \\ +\,115.3 \\ \hline \end{array}$$

5. $$\begin{array}{r} 7\frac{3}{5} \\ \times\,6\frac{1}{2} \\ \hline \end{array}$$

6. $7.25 + 8.3 + 16.4 + 11 + 5.05$

7. $3\frac{5}{8} + 4\frac{3}{4} + 8\frac{1}{4} + 6 + 2\frac{1}{2}$

8. $19\frac{2}{3} \cdot 15\frac{1}{4}$

9. $75 \cdot 34$

10. $88.6 \cdot 102$

11. $16.6 \cdot 18.2$

ALGEBRA

Solve.

12. $49 = \frac{1}{2} \cdot 7(b + 5)$

13. $2x + 38 = 92$

14. $\frac{1}{2}d(13) = 65$

Multiply.

15. 5^2

16. 8^2

17. 10^2

18. $5 \cdot 3^2$

GEOMETRY

19. Find the radius of a circle with diameter 328 mm.

20. Find the diameter of a circle with radius 61 in.

Find the length of the indicated side for each figure.

21. square

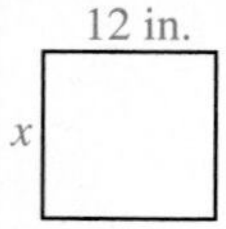

22. rectangle

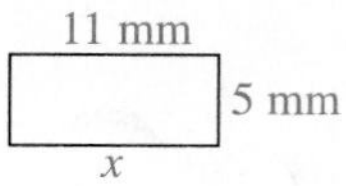

23. parallelogram

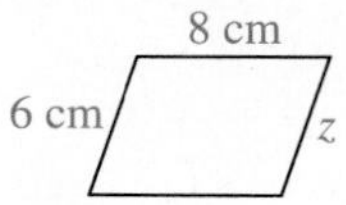

Complete.

24. All four sides of a rhombus are ■.

25. Opposite sides of a parallelogram are both ■ and ■.

26. An isosceles trapezoid has ■ pair(s) of congruent sides.

27. A parallelogram that includes one right angle is a ■.

9.1 Perimeter

Objective: To determine the perimeter of a concave or convex polygon.

You can determine the amount of baseboard molding needed for a room by finding its *perimeter*. The **perimeter (P)** is the distance around a region. You can find the perimeter of a polygon by adding the lengths of its sides.

Example 1: **Find the perimeter of the parallelogram.**

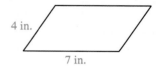

Solution: $P = 7 + 7 + 4 + 4$

$\qquad = 2 \cdot 7 + 2 \cdot 4$

$\qquad = 22$ in.

Example 2: **Use a calculator to find the perimeter of the regular pentagon.**

Solution: 5 ☒ 1.26 ▤ 6.3 cm

1.26 cm

EXPLORING

1. Write an equation for the perimeter of each polygon.

a.

Square

b.

Pentagon

c.

Regular octagon

d.

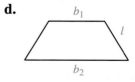

Isosceles trapezoid

2. The figures in parts (a) and (c) above are regular polygons. Use *n* for the number of sides and *s* for the length of each side. Write a formula for the perimeter, *P*, of any regular polygon.

1. You know that two quadrilaterals have the same perimeter. Can you conclude that each side of one is congruent to a side of the other? Explain.

2. Two quadrilaterals are different in shape but have the same length sides. Must they have the same perimeter? Explain.

Class Exercises

Find the perimeter of each polygon.

1.
7 cm
3 cm

2.
5 cm
5 cm

3.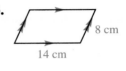
14 cm
8 cm

4.
4 in.
$3\frac{1}{2}$ in.
5 in.
3 in.

Find the lengths of the indicated sides.

5. $P = 32$ cm
  a. $a = $ ▨
  b. $b = $ ▨

12 cm
b b
a

6. **a.** $a = $ ▨
  b. $b = $ ▨
  c. $P = $ ▨

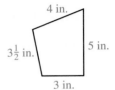
4 in.
16 in.
b
a
16 in.
6 in.

7. Find the perimeter of a regular hexagon with sides 8 cm long.

8. Find the number of sides of a regular polygon with perimeter 180 cm and sides 15 cm long.

9. $\triangle RST$ is an equiangular triangle. $RS = 8$ in. Find the perimeter of the triangle.

Exercises

Use a metric ruler to find the perimeter of each figure.

1.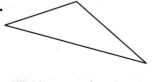

2.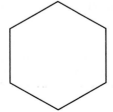

3.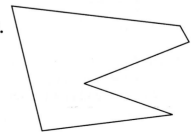

Find the perimeter of each polygon.

4.
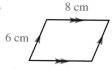
8 cm
6 cm

5.
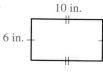
10 in.
6 in.

6.
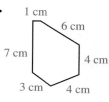
1 cm
6 cm
7 cm
4 cm
3 cm
4 cm

7.
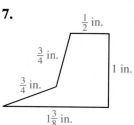
$\frac{1}{2}$ in.
$\frac{3}{4}$ in.
1 in.
$\frac{3}{4}$ in.
$1\frac{3}{8}$ in.

Find the length of each indicated side. A calculator may be helpful.

8. $P = 10$ cm
$x =$ ■
3.3 cm
x
2.9 cm
1.7 cm

9. $P = 83$ in.
$r =$ ■
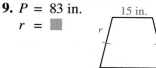
15 in.
r r
28 in.

10. $P = 96$ mm
a. $a =$ ■
b. $b =$ ■

32 mm
b
a

11. $P = 20$ cm
$x =$ ■
x
x
x

Find the perimeter of a polygon with sides of the given length.

12. regular pentagon; 6 cm

13. regular hexagon; 4 cm

14. regular octagon; 5 cm

15. equilateral triangle; 6 cm

Find the number of sides of a regular polygon with the given perimeter and with sides of the given length. A calculator may be helpful.

16. $P = 104$ cm
$s = 13$ cm

17. $P = 84$ cm
$s = 7$ cm

18. $P = 9.6$ in.
$s = 1.6$ in.

19. $P = 25.63$ in.
$s = 2.33$ in.

Use graph paper to draw polygons to fit each description.

20. a square with perimeter 12 units

21. a rectangle that is not a square with perimeter 12 units

22. three noncongruent rectangles, each with perimeter 14 units

23. a 6-sided figure with perimeter 8 units

Find (a) the length of each indicated side and (b) the perimeter of each figure. You may assume that angles that appear to be right angles are right angles.

24.
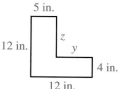
5 in.
12 in.
z
y
4 in.
12 in.

25.
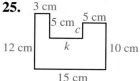
3 cm
5 cm
5 cm
c
12 cm
k
10 cm
15 cm

26.
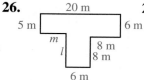
20 m
5 m
6 m
m
8 m
l
8 m
6 m

27.
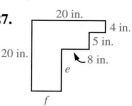
20 in.
4 in.
5 in.
20 in.
8 in.
e
f

28. Suppose you want to frame a 12 in. by 16 in. picture with a $1\frac{1}{2}$in.-wide frame. Draw a picture to represent the problem.

   a. Find the perimeter of the picture.

   b. Find the *outside* perimeter of the frame.

APPLICATIONS

29. Landscaping   A stockade fence costs $4.50 per meter. Find the cost of fencing a rectangular yard 18 m long and 7 m wide.

30. Construction

   a. How much baseboard molding would you need for the rectangular room pictured at the right? (Do *not* include the door or closet.)

   b. Suppose the closet is 3 ft deep. How much baseboard molding would you need for the interior of the closet?

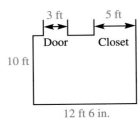

31. Gardening   How many fence sections, each 9 in. wide, would you need to make a border for the flower bed shown at the right?

Everyday Geometry

You can use the following method to approximate the perimeter of an irregularly shaped region such as a pond.

   1. Draw a polygon inside or around the outside of the given region.

   2. Measure the sides of the polygon.

   3. The perimeter of the polygon will be an approximation for the perimeter of the irregularly shaped region.

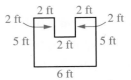

Trace each region. Approximate the perimeter of each region in centimeters.

1.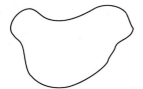

2.

3.

9.2 Area

Objective: To determine the area of a polygon by counting unit squares.

You can determine the amount of flooring needed by finding the *area* of the floor. *Area (A)* is the amount of surface in a region.

People often refer to the area of a rectangle or triangle when they mean the area of a rectangular region or triangular region. In this textbook, we will use *area of rectangle ABCD* to mean the area of the rectangular region.

Using the square unit shown to measure area, the area of the rectangle at the right is 8 square units.

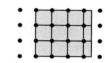

Example 1: **Find the area of the shaded region.**

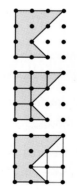

represents 1 unit of area.

Solution: There are 9 whole squares and 3 half squares. Area $= 10\frac{1}{2}$ square units

You can find the area of some regions in more than one way.

Example 2: **Find the area of the shaded region in two ways.**

represents 1 unit of area.

Solution (a): There are 4 whole shaded squares and 3 half shaded squares.
Area $= 5\frac{1}{2}$ square units

(b): Draw a rectangle around the region. Find the area of the region that is *unshaded*.

There are 2 whole unshaded squares and 3 half unshaded squares. Subtract the unshaded area from the area of the rectangle.
Area $= 9 - 3\frac{1}{2} = 5\frac{1}{2}$ square units

Part A

Count square units to find the area of each figure.

1. a.   **b.**   **c.**

2. a.   **b.**   **c.**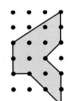

3. On dot paper or graph paper, draw three noncongruent figures with area five square units.

4. Does changing the shape of a figure change the area?

Part B

Copy each figure on dot paper. Demonstrate two different ways to find the area of each figure.

1.   **2.**   **3.**   **4.**

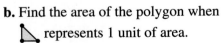 

Thinking Critically

1. a. Find the area of the polygon when represents 1 unit of area.

       b. Find the area of the polygon when represents 1 unit of area.

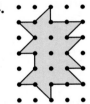

2. The figure is the same in parts (a) and (b). Is the area the same? Explain why your answers for parts (a) and (b) are different.

Class Exercises

Find the area of each shaded region. ⬜ **represents 1 unit of area.**

1.   2.   3.   4.   5.

Use dot paper or graph paper to draw a polygon to fit each description. A square on the paper represents 1 unit of area.

6. a rectangle with area 8 square units and perimeter 12 units

7. two noncongruent rectangles, each with area 18 square units

8. a triangle with area $4\frac{1}{2}$ square units

Exercises

Find the area of each shaded region. ⬜ **represents 1 unit of area.**

1.   2.   3.   4.   5.

6.   7.   8.   9.   10.

Use dot paper or graph paper to draw a polygon to fit each description. A square on the paper represents 1 unit of area.

11. a rectangle with area 20 square units and perimeter 24 units

12. a rectangle with area 20 square units and perimeter 18 units

13. a square with area 25 square units

14. a 6-sided figure with area 3 square units

15. a 6-sided figure with area 4 square units

16. a 6-sided figure with area 10 square units

Draw a triangle with the given area in square units.

17. 1   **18.** 2   **19.** 3   **20.** 4   **21.** 5   **22.** $7\frac{1}{2}$

23. a. Draw a triangle and a square having the same area.

   b. Which figure has the smaller perimeter?

   c. Try another example. Is the answer to part (b) the same?

24. a. Draw a rectangle and one of its diagonals.

   b. What is the area of the rectangle?

   c. What shape are the two parts of the rectangle?

   d. What is the area of each part?

APPLICATIONS

25. Construction   A ceiling is being installed in a basement that is shaped like the figure below. The ceiling tiles are 12 in. by 12 in. squares. How many tiles are needed?

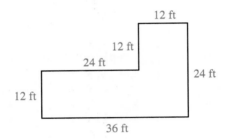

Thinking in Geometry

1. Find the area and perimeter of the large square at the right.

2. Draw regions that result from the following changes in the figure shown. For each part, try to find all ways in which you can do the problem.

   a. One fewer square; leave the perimeter the same.

   b. Two fewer squares; leave the perimeter the same.

   c. Three fewer squares; leave the perimeter the same.

   d. Four fewer squares; leave the perimeter the same.

   e. One fewer square; increase the perimeter by 2 units.

   f. Two fewer squares; increase the perimeter by 4 units.

   g. Three fewer squares; increase the perimeter by 6 units.

   h. Four fewer squares; increase the perimeter by 8 units.

9.3 Area of a Rectangle

Objective: To determine the area of a rectangle.

For a rectangle, the ***base*** can be any side. The base length is denoted by **b**. The corresponding height, **h,** is the length of each of the sides perpendicular to the base.

Example:

Base	Height
$\overline{WX}$	WZ or XY
$\overline{XY}$	YZ or WX

 EXPLORING

1. Make a table like the one shown. Assume that each square is 1 cm long on each side. To find the area of each figure below, count the squares in the figures. What is the base length, *b*, and the height, *h*, for each figure?

2. Calculate *b* • *h* for each figure.

3. Is the product of the base length and the height the same as the area of each figure?

Fig.	area (in cm²)	*b*	*h*	*b* • *h*
a.	■	■	■	■
b.	■	■	■	■
c.	■	■	■	■
d.	■	■	■	■
e.	■	■	■	■
f.	■	■	■	■

a.

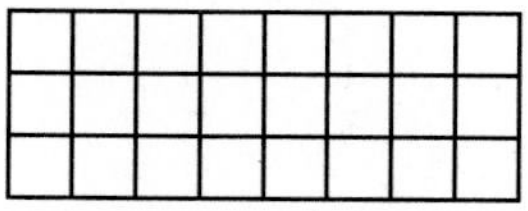

b.

c.

d.

e.

f.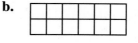

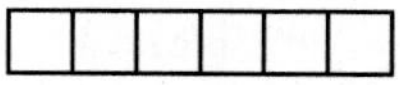

The **EXPLORING** activity demonstrates the following theorem.

THEOREM 9.1:   **The area (A) of a rectangle is the product of its base length (b) and its height (h). $A = bh$**

Example 1: **Find the area of the rectangle.**

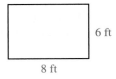

6 ft

8 ft

Solution: $A = bh$
$= 8 \cdot 6$
$= 48 \text{ ft}^2$

Example 2: **Find the area of the rectangle. A calculator may be helpful.**

50 cm

4.3 m

Solution: You must express the base length and height in the same unit of measure.

$50 \text{ cm} = \frac{50}{100} \text{ m}$
$= 0.5 \text{ m}$

$A = bh$
$= 4.3 \boxed{\times} 0.5 \boxed{=} 2.15$

The area is 2.15 m².

Thinking Critically

1. Suppose you double the length of one side of a rectangle. What happens to the area?

2. Write a simpler formula than $A = bh$ for the area of a square.

Class Exercises

Find the area of each rectangle.

1.

3 cm

3 cm

2.

42 ft

112 yd

3.

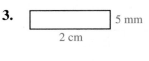

5 mm

2 cm

Find the area and perimeter of each rectangle.

4. $b = 15$ in., $h = 5$ in.   **5.** $b = 8$ cm, $h = 8$ cm   **6.** $b = 1$ ft, $h = 4$ in.

For each rectangle, find the length of the indicated side.

7. $A = 24$ cm^2
   $x = $
   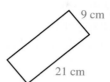 x
   6 cm

8. $A = 192$ cm^2
   $y = $
   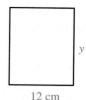 y
   12 cm

Exercises

Find the area of each rectangle.

1. 9 m   4 m

2. 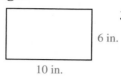 6 in.   10 in.

3. 9 cm   21 cm

4. 5 cm   9 cm

Copy and complete the table for each rectangle. A calculator may be helpful.

	Base length (b)	Height (h)	Area (A)	Perimeter (P)
5.	8 cm	7 cm	▩	▩
6.	4 ft	$2\frac{1}{2}$ ft	▩	▩
7.	8 cm	▩	16 cm^2	▩
8.	4 yd	9 ft	▩	▩
9.	150 cm	1 m	▩	▩
10.	5.5 cm	▩	▩	20 cm
11.	2 yd	▩	12 ft^2	▩

12. a. $P = $ ▩
   b. $A = $ ▩
   16 ft
   8 ft
   8 ft
   4 ft
   8 ft

13. a. $P = $ ▩
   b. $A = $ ▩
   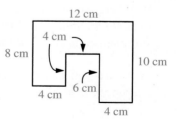
   12 cm
   4 cm
   8 cm
   10 cm
   4 cm   6 cm
   4 cm

Draw a rectangle or rectangles to fit each description.

14. $P = 12$ in.
   $A = 8$ in.^{2}

15. $P = 18$ cm
   $A = 8$ cm^2

16. $P = 14$ units
   $A = 12$ square units

17. $P = 16$ units
   $A = 12$ square units

18. three noncongruent rectangles having the same area but different perimeters

19. three noncongruent rectangles having the same perimeter but different areas

20. Two rectangular lots are for sale. The first lot is 100 ft wide and 150 ft long. The second lot is 20 ft wide and 750 ft long.

    a. Find the area of each lot.

    b. How do the areas compare?

    c. If you were to build a house, which lot would you prefer? Why?

21. A football field is 100 yd long and 160 ft wide.

    a. Find the area in square feet.

    b. Find the area in square yards.

    c. Is the area greater or less than 1 acre? (1 acre $= 4,840$ yd^2)

APPLICATIONS

22. Estimation About how many basketball courts would cover the same area as a baseball diamond? about how many baseball diamonds as a football field? about how many basketball courts as a football field?

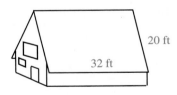

> **Baseball diamond: 90 ft by 90 ft**
> **Basketball court: 84 ft by 50 ft**
> **Football field: 300 ft by 160 ft**

23. Construction A rectangular shelf is 12 in. wide and 30 in. long. The rectangular base of a stereo turntable is 14 in. wide and 16 in. long.

    a. Find the area of the shelf and of the turntable base.

    b. Which has the greater area?

    c. Is it safe to place the turntable on the shelf? Draw a picture to represent the problem. Explain your answer.

24. Construction The picture at the right shows a cabin that needs a new roof.

    a. Find the area of the roof. (Include both sides.)

    b. Fiberglass shingles are sold by the "square" (enough to cover an area of 100 ft^2). How many squares of shingles are needed for the cabin roof?

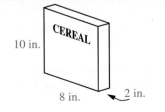

20 ft

32 ft

Seeing in Geometry

This cereal box is made of four rectangular sides, a rectangular top, and a rectangular bottom. Find the total area of the six rectangles of cardboard needed to construct the box.

10 in.

CEREAL

8 in.   2 in.

9.4 Area of a Parallelogram

Objective: To determine the area of a parallelogram.

Any side of a parallelogram can be called the **base.** An **altitude** is any segment perpendicular to the lines containing the base and the side opposite the base. The **height**, *h*, is the length of an altitude. Several altitudes are shown below.

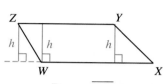

Base: $\overline{WX}$

Base: $\overline{RE}$

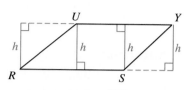

Base: $\overline{RS}$

EXPLORING

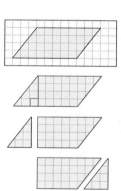

1. Draw a nonrectangular parallelogram on dot paper or graph paper. Find its area. A square on the paper represents 1 unit of area.

2. Draw an altitude as shown.

3. Cut out the parallelogram. Cut along the altitude, separating the parallelogram into two regions.

4. Fit the regions together as shown to form a rectangle. Find the area of the rectangle.

5. Compare the base length, height, and area of the parallelogram and the rectangle.

6. Repeat Steps 1–5 for a different parallelogram. Use your results to write a formula for the area of a parallelogram.

The **EXPLORING** activity demonstrates the following theorem.

THEOREM 9.2: **The area (*A*) of a parallelogram is the product of its base length (*b*) and its height (*h*).** $A = bh$

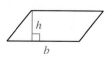

Example 1: **Find the area of the parallelogram.**

Solution: $A = bh$

$\qquad = 9 \cdot 7$

$\qquad = 63 \text{ in.}^2$

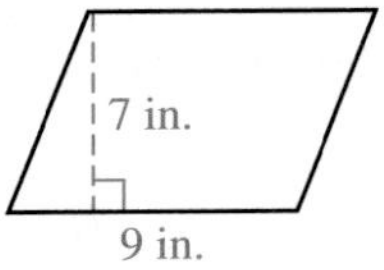

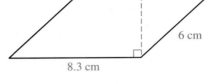

7 in.

9 in.

Example 2: **Find the indicated information.**
A calculator may be helpful.

$A = 33.2 \text{ cm}^2$

$h = \blacksquare$

6 cm

8.3 cm

Solution: $\quad A = bh$

$\qquad 33.2 = 8.3\, h$

$\qquad 33.2 \boxed{\div} 8.3 \boxed{=} 4$

$\qquad$ The height is 4 cm.

Thinking Critically

1. For any rectangle, the height is also the length of each side adjacent to the base. Is this true for all parallelograms? Explain.

2. Explain why the area computed for each figure is incorrect.

a. $A = bh$

$\quad = 1 \cdot 6$

$\quad = 6$

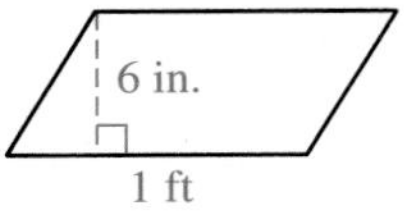

6 in.

1 ft

b. $A = bh$

$\quad = 8 \cdot 3$

$\quad = 24 \text{ square units}$

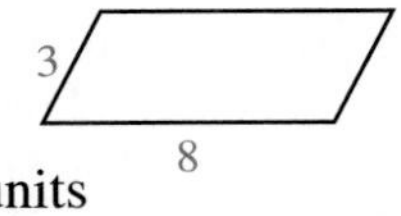

3

8

Class Exercises

Find the perimeter and area of each parallelogram.

1. 5 cm, 3 cm, 8 cm
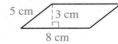

2.
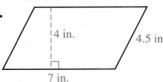
4 in.

4.5 in.

7 in.

3. 4 cm

15 cm

5 cm

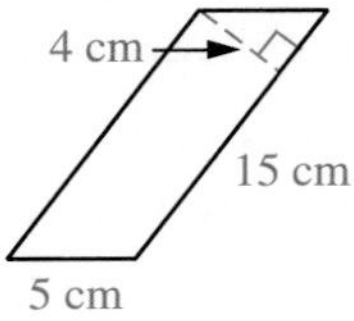

Find the indicated information.

4. $A = 60 \text{ cm}^2$
$\quad h = \blacksquare$
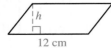
h
12 cm

5. $P = 26 \text{ cm}$
$\quad x = \blacksquare$

x
9 cm

Exercises

Find the area of each shaded region in square units .

1. a. 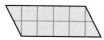 **b.** **c.** **d.**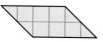

2. a. 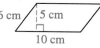 **b.**  **c.** **d.**

Find the perimeter and area of each parallelogram.

3. 6 cm 5 cm 10 cm **4.** 5 in. 8 in. **5.** 8 cm 10 cm 10 cm **6.** 5 cm 12 cm 6 cm

Find the area of a parallelogram with the given dimensions. A calculator may be helpful.

7. $b = 3.1$ cm
$h = 6.2$ cm

8. $b = 8$ ft
$h = 1.5$ ft

9. $b = 3$ yd
$h = 6$ ft

10. $b = 78.2$ cm
$h = 0.7$ m

11. $A = 50$ cm^2
$x = $ ▇

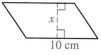

12. $A = $ ▇ in.^{2}

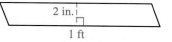

Draw a region or regions to fit each description.

13. a rectangle: $b = 8$, $h = 3$ **14.** three noncongruent parallelograms: $b = 4$, $h = 3$

For each parallelogram, name two altitudes and a corresponding base for each altitude.

15.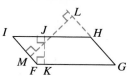

16.

17.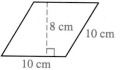

Find (a) the area of $\square ABCD$ and (b) the area of $\triangle ABC$.

18.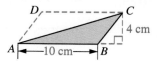

19.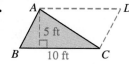

20.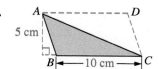

Find the area of the shaded region in each figure.

21.

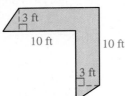

22.

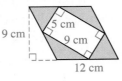

23.

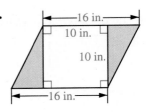

24. Draw a rectangle and a nonrectangular parallelogram, each of which has base length 6 units and height 2 units.

   a. How do the areas compare?

   b. How do the perimeters compare?

   c. Describe the relationship of the areas and perimeters for a rectangular and a nonrectangular parallelogram that have the same base length and same height.

APPLICATION

25. Landscaping

   a. Find the area of the yard at the right.

   b. One bag of fertilizer will cover 5,000 ft². How many bags would you need for the yard?

   c. How many square yards of sod would you need to cover the yard?

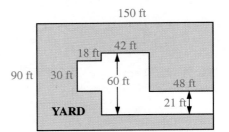

Test Yourself

Find the perimeter and area of each region described.

1. a square: side length = 8 cm

2. a rectangle: $b = 1$ yd, $h = 2$ ft

Find the indicated information.

3. $A = $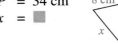

4. $P = 34$ cm
   $x = $ ▪

5. $A = 64$ cm²
   $h = $ ▪

6. $A = 50$ cm²
   a. $h = $ ▪
   b. $P = $ ▪

7. Draw a rectangle with area 36 square units and perimeter 24 units.

9.5 Area of a Triangle

Objective: To determine the area of a triangle.

Any side of a triangle can be the **base**. The height *h* is the length of the corresponding *altitude* from the opposite vertex to the base.

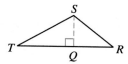

Base: $\overline{TR}$

Height: SQ

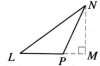

Base: $\overline{LP}$

Height: MN

EXPLORING

1. Use the *Geometric Supposer: Triangles* disk to draw an acute triangle and its three altitudes. Measure to find the length of each altitude and the length of the base to which it is drawn. Find the product of each base length and height.

2. Measure the area of the triangle. Compare the area to the products that you obtained in Step 1. What do you notice?

3. Repeat this process for a right triangle and for an obtuse triangle. How do your results compare to your results for an acute triangle?

The **EXPLORING** activity demonstrates the following theorem.

THEOREM 9.3: **The area (*A*) of any triangle is half the product of its base length (*b*) and height (*h*).** $A = \frac{1}{2}bh$

Example: Find each area.

a.

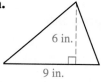

6 in.

9 in.

b.

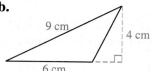

9 cm

4 cm

6 cm

Solution: **a.** $A = \frac{1}{2}bh$

$= \frac{1}{2} \cdot 9 \cdot 6$

$= 27$ in.^{2}

b. $A = \frac{1}{2}bh$

$= 0.5 \;\boxtimes\; 6 \;\boxtimes\; 4 \;\boxminus\; 12$ cm^2

Thinking Critically

Describe what happens to the area of a triangle under the following conditions.

1. You double its base length and leave its height the same.
2. You double the height and leave its base length the same.
3. You double both its base length and its height.
4. You double the height and halve its base length.

Class Exercises

Find the area of each shaded region.

1. 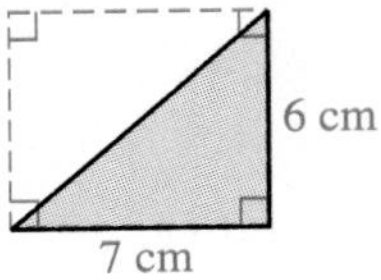
6 cm
7 cm

2. 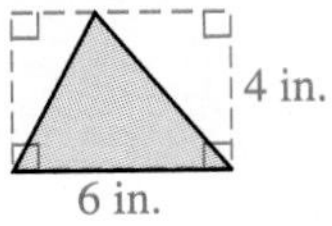
4 in.
6 in.

3. 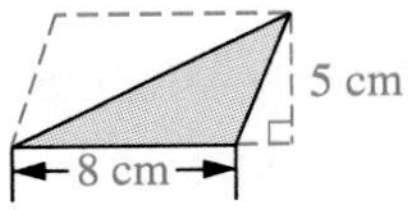
5 cm
8 cm

Find the area of each triangle with the given base length and height.

4. $b = 10$ cm, $h = 4$ cm
5. $b = 9$ ft, $h = 6$ ft
6. $b = 4$ cm, $h = 5$ mm

Find each indicated length.

7. $A = 30\text{ cm}^2$
   $h = ■$

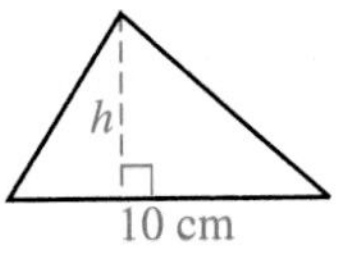
h
10 cm

8. $A = 16\text{ ft}^2$
   $x = ■$

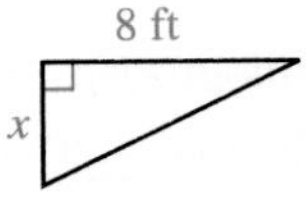
8 ft
x

Exercises

Find the area of each triangle with the given base length and height. A calculator may be helpful.

1. $b = 11$ cm, $h = 7$ cm
2. $b = 89.6$ cm, $h = 31.5$ cm
3. $b = 12$ ft, $h = 1\frac{1}{2}$ ft
4. $b = 43$ in., $h = 9$ ft
5. $b = 2.25$ yd, $h = 9$ ft
6. $b = 77$ cm, $h = 0.9$ m

Find the area of the triangle. Write each answer with an appropriate unit of area. A calculator may be helpful.

7. 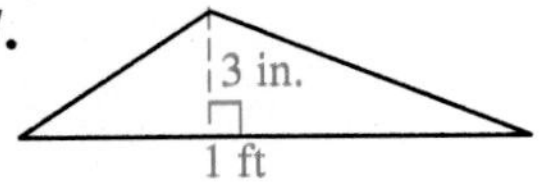
3 in.
1 ft

8. 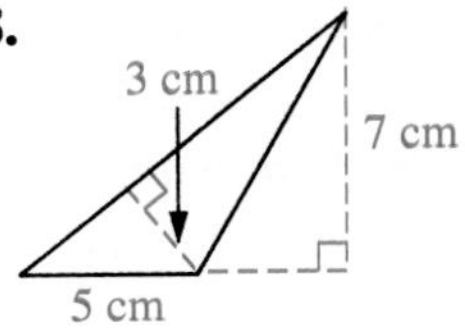
3 cm
7 cm
5 cm

9. 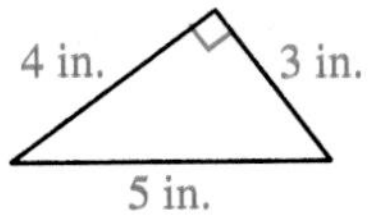
4 in.
3 in.
5 in.

10. 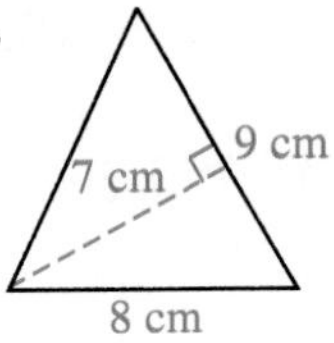
7 cm
9 cm
8 cm

Find each height.

11. $A = 32$ ft^2
$h = $ ■

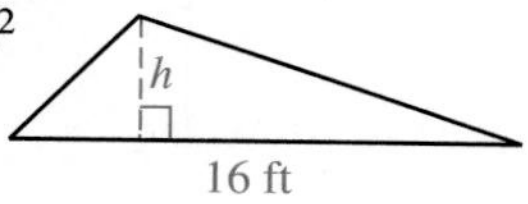

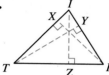

16 ft

12. $A = 35$ cm^2
$h = $ ■

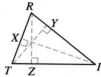

7 cm

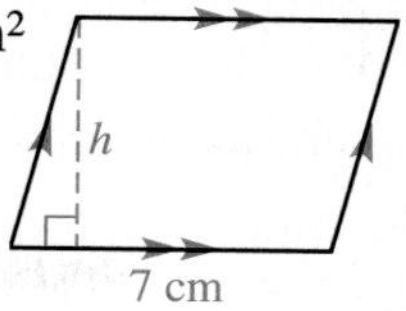

For $\triangle TRI$, **name three altitudes and the corresponding base for each altitude.**

13.

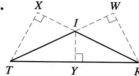

14.

15.

Draw a figure or figures to fit each description.

16. a right triangle: $A = 6$ cm^2

17. a right triangle: one leg $= 5$ cm; $A = 10$ cm^2

18. an obtuse triangle: $b = 4$ cm; $h = 3$ cm

19. an acute triangle: $b = 4$ cm; $h = 3$ cm

20. an obtuse triangle: $b = 6$ cm; $A = 9$ cm^2

21. an acute triangle: $b = 8$ in.; $A = 12$ in.^{2}

22. four noncongruent triangles: $A = 3$ cm^2

23. a triangle with the same area as a rectangle with $b = 4$ cm, $h = 3$ cm

24. Find the area of the following figures. Write each answer with an appropriate unit of area.
a. rectangle $ABCD$ **b.** $\triangle EAD$
c. $\triangle DCF$ **d.** $\triangle FBE$
e. the shaded region   **f.** $\triangle DEF$

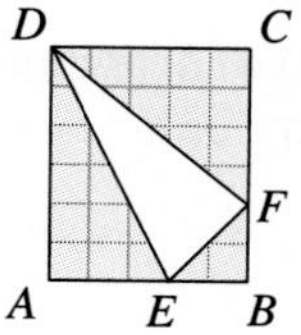

25. $\overline{AE}$ is an altitude for $\triangle ABC$. $\overline{AD}$ is a median for $\triangle ABC$. Find the area of each triangle. $AE = 6$ cm, $BC = 12$ cm.
a. $\triangle ABC$
b. $\triangle ABD$
c. $\triangle ACD$

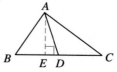

26. Four different triangles with base $\overline{XY}$ are shown.

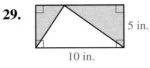

$\overleftrightarrow{AB} \parallel \overleftrightarrow{XY}$

    a. How does the altitude from A to $\overline{XY}$ compare to the altitude from B to $\overline{XY}$? from C to $\overline{XY}$? from D to $\overline{XY}$?

    b. Do the four triangles have the same area?

    c. Are the four triangles congruent?

Find the area of each shaded region. Write each answer with an appropriate unit of area.

27.

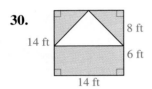

28.

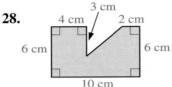

29.

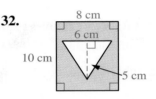

30.

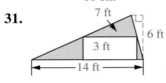

31.

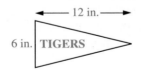

32.

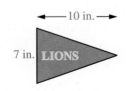

APPLICATIONS

33. Remodeling To remodel a kitchen you must consider perimeter and area in deciding the amount of material to buy. Which measurement would you need to consider for each job?

    a. installing new tile on floor   **b.** papering the walls

    c. painting the ceiling   **d.** installing new baseboard molding

    e. refinishing the cabinets   **f.** installing new laminate on counter tops

34. Design Which of the two pennants at the right has the greater area? How much greater?

Computer

Use the *Geometric Supposer: Triangles* disk to draw any triangle. Draw the altitude from vertex A to side $\overline{BC}$. Construct a rectangle around the triangle such that $\overline{BC}$ is a side of the rectangle and AD is the length of the other side. Explain why the area of the original triangle is half the area of the rectangle you constructed.

9.6 Area of Other Polygons

Objective: To determine the area of a polygon.

You can find the area of a polygon by dividing it into regions for which you know how to determine the area.

Example 1: **Find the area of the polygon.**

Solution: Area of triangle $= \frac{1}{2} \cdot 4 \cdot 3$

$\qquad\qquad\qquad = 6 \text{ m}^2$

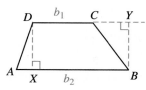

Area of rectangle $= 8 \cdot 3$

$\qquad\qquad\qquad = 24 \text{ m}^2$

Area of figure is $6 \text{ m}^2 + 24 \text{ m}^2 = 30 \text{ m}^2$.

EXPLORING

Part A

1. Copy trapezoid $ABCD$ and altitudes $\overline{DX}$ and $\overline{YB}$. If the length of $\overline{DX}$ is h, what is the length of $\overline{YB}$?

2. Draw diagonal $\overline{DB}$. How does the height of $\triangle ABD$ compare to the height of $\triangle BCD$?

3. Use b_1 and b_2 to describe the area of $\triangle ABD$ and $\triangle BCD$ in terms of height and base length.

4. Write a formula for the area of $ABCD$.

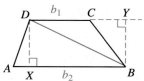

Part B

1. Copy rhombus $RSTV$ and diagonals $\overline{RT}$ and $\overline{VS}$.

2. In terms of d_1 and d_2, what is the height of $\triangle RST$? of $\triangle RVT$? In terms of d_1 and d_2, what is the base length of $\triangle RST$? of $\triangle RVT$?

3. In terms of d_1 and d_2, what is the area of $\triangle RST$? of $\triangle RVT$?

4. Write a formula for the area of $RSTV$.

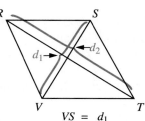

$$VS = d_1$$
$$RT = d_2$$

The **EXPLORING** activity leads to the following theorems.

THEOREM 9.4: **The area of a trapezoid is half the product of the height (h) and the sum of the base lengths ($b_1 + b_2$).**
$A = \frac{1}{2}h(b_1 + b_2)$

THEOREM 9.5: **The area of a rhombus is half the product of the lengths of the diagonals.** $A = \frac{1}{2}d_1 d_2$

Recall that a regular polygon is both equilateral and equiangular. The **center** of a regular polygon is the point equidistant from all vertices. The **apothem** of a regular polygon is the length of the perpendicular segment drawn from the center of the polygon to any side.

Center: O
Apothem: OA, OB, OC,
OD, or OE

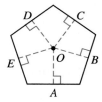

EXPLORING

1. Copy and label the regular hexagon as shown (R is the center of the hexagon, s is the length of a side, and a is the apothem).

2. What is the area of $\triangle ZRT$ in terms of s and a?

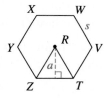

3. Draw $\overline{RV}$, $\overline{RW}$, $\overline{RX}$, and $\overline{RY}$. Are the triangles formed congruent? What is the area of each triangle?

4. Can you say that the area of $TVWXYZ = 6(\frac{1}{2}sa) = (\frac{1}{2}a)6s$?

5. What does $6s$ represent for $TVWXYZ$?

6. Use P to represent perimeter. Write a formula for the area of a regular hexagon.

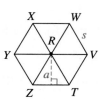

This **EXPLORING** activity demonstrates the following theorem.

THEOREM 9.6: **The area of a regular polygon is half the product of the apothem and the perimeter.** $A = \frac{1}{2}aP$

Example 2: Use a calculator to find the area of each polygon.

a.

b.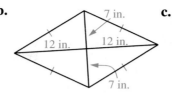

c.

Solution:

a. $A = \frac{1}{2}h(b_1 + b_2)$

10 ⊞ 12 ⊟ ⊠ 0.5 ⊠ 8

⊟ 88 cm^2

b. $A = \frac{1}{2}d_1d_2$

0.5 ⊠ 24 ⊠ 14

⊟ 168 in.^{2}

c. $A = \frac{1}{2}aP$

0.5 ⊠ 16 ⊠ 5 ⊠ 23.2

⊟ 928 m^2

Thinking Critically

1. Can you use the formula $A = \frac{1}{2}d_1d_2$ to find the area of a square? Why or why not?

2. Suppose two congruent copies of a trapezoid are arranged as shown. The figure formed is a parallelogram.

   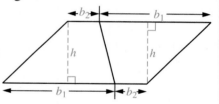

   a. What is the base length of the parallelogram?

   b. Write a formula for the area of the parallelogram in terms of $b_1, b_2,$ and h.

   c. Explain how you could use this figure to develop the formula for the area of a trapezoid.

Class Exercises

Find the area of each shaded figure.

1.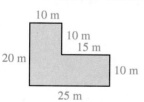

2.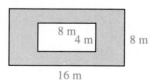

3.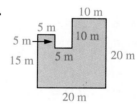

Find the area of each figure described.

4. a regular hexagon with $s = 12.0$ cm and $a = 10.4$ cm

5. a rhombus with diagonals of length 10 in. and 6 in.

6. a square with both diagonals of length 10 in.

7. a trapezoid with height 4 cm and bases of length 10 cm and 15 cm

Exercises

Find the area of each shaded figure.

1.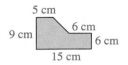
5 cm
6 cm
9 cm
6 cm
15 cm

2.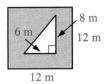
8 m
6 m
12 m
12 m

3.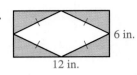
6 in.
12 in.

Find the area of each figure described. A calculator may be helpful.

4. a rhombus with diagonals of length 6 in. and 8 in.

5. a regular pentagon: $s = 6$ cm, $a = 4.1$ cm

6. a regular hexagon: $s = 10$ cm, $a = 8.7$ cm

7. a trapezoid with height 6 cm and bases of length 5 cm and 10 cm

Find the area of each rhombus.

8. $AC = 12$ in.
$BD = 20$ in.

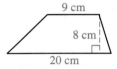
C
D   B
A

9. $RT = 8$ cm
$SU = 8$ cm

U   T
R   S

Find the area of each trapezoid.

10.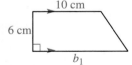
9 cm
8 cm
20 cm

11.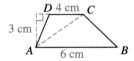
10 in.
6 in.
15 in.

12. a. Find the area of trapezoid *ABCD*.

b. Draw rectangular and nonrectangular parallelograms with height 5 cm that have the same area as trapezoid *ABCD*.

D 4 cm C
3 cm
A   6 cm   B

13. $A = 72$ cm²
$b_1 = $ ■

10 cm
6 cm
b_1

For each problem, (a) tell whether there is too much or too little information, (b) tell what information is or is not needed, and (c) solve if possible.

14. Find the area of a rectangle that is 5 ft by 12 ft and has a diagonal of length 13 ft.

15. How many 2 ft by 4 ft ceiling tiles do you need to cover a 12 ft wide ceiling?

16. How many tiles are needed to cover a floor that is 12 ft long and 10 ft wide?

17. Describe how the formula $A = bh$ for the area of a parallelogram, a rectangle, or a square is a special case of the formula $A = \frac{1}{2}h(b_1 + b_2)$ for the area of a trapezoid.

18. A *kite* is a quadrilateral for which exactly one diagonal is a line of symmetry. For the kite at the right, $\overleftrightarrow{QS}$ is a line of symmetry, $QM = 8$ cm, $MS = 12$ cm, and $PR = 12$ cm.

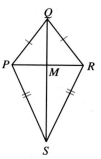

a. What is the relationship between $\overline{PR}$ and $\overline{QS}$?

b. Name three pairs of congruent triangles.

c. How could you prove each pair of triangles congruent?

d. Find the areas of $\triangle PQR$, $\triangle PRS$, and $PQRS$.

e. Use the formula $A = \frac{1}{2}d_1d_2$ to find the area of $PQRS$. Does the formula work?

APPLICATIONS

19. Home Repair   A hip roof has four sloping sides. The roof pictured at the right consists of two isosceles triangles and two isosceles trapezoids. Find the area of the roof shown.

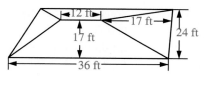

20. Remodeling   How many square feet of laminate do you need to cover the countertop at the right?

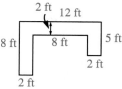

Calculator

Laurie used the formula $A = \frac{1}{2}h(b_1 + b_2)$ to find the area of the trapezoid. On her calculator, she entered the following sequence of keys:

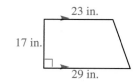

$$0.5 \; \boxed{\times} \; 17 \; \boxed{\times} \; 23 \; \boxed{+} \; 29 \; \boxed{=} \; 224.5$$

The area of the trapezoid is 442 in². Why did Laurie get the wrong answer?

9.7 Problem Solving Strategy: Using Formulas

Objective: To use formulas to solve problems.

In this chapter, you have learned formulas for the area of several geometric figures. To use a formula to solve a problem, follow these steps:

1. Choose a formula in which you know the values for all but one variable.
2. Substitute appropriate values for each variable.
3. Solve for the unknown variable.
4. Check your work.

Example: For the baseball field shown in the diagram, the infield is a square with sides measuring 90 ft. The outfield is the remaining area. Find the approximate area of the outfield.

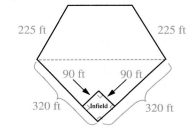

Solution: You can divide the field into two parts. The first part is half a regular hexagon with apothem of about 195 ft. The second part is an isosceles right triangle with base length and height both 320 ft.

1. Use a calculator to find the area of the hexagon.

   $A = \frac{1}{2}aP$

   0.5 ☒ 195 ☒ 225 ☒ 6 ▤ 131,625 ft^2

   The area of the half hexagon is half that of the hexagon.

   0.5 ☒ 131,625 ▤ 65,812.5 ft^2

2. Use a calculator to find the area of the triangle.

   $A = \frac{1}{2}bh$

   0.5 ☒ 320 ☒ 320 ▤ 51,200 ft^2

3. The infield is a square.

   $A = bh$

   $= 90 \cdot 90 = 8,100$ ft^2

4. Add the areas of the half hexagon and triangle. Subtract the area of the infield to find the area of the outfield.

   65,812.5 ⊞ 51,200 ⊟ 8,100 ▤ 108,912.5 ft^2

   The area is about 109,000 ft^2.

Class Exercises

Choose the formula(s) you would use to find the area of each figure. You may assume that the figures shown in Exercises 1 and 3 consist of regular polygons.

a. $A = \frac{1}{2}bh$

b. $A = \frac{1}{2}d_1d_2$

c. $A = \frac{1}{2}h(b_1 + b_2)$

d. $A = \frac{1}{2}aP$

e. $A = bh$

1.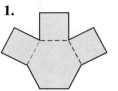

2.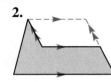

3.

4.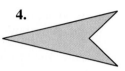

Exercises

Write the needed formula(s). Then solve. A calculator may be helpful.

1. The sheet metal pattern at the right consists of a regular pentagon, a trapezoid, and a rectangle. The apothem of the pentagon measures about 5.5 in. Find the total area of the pattern.

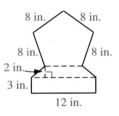

Use the formula $D = rt$, where D = distance, r = rate, and t = time.

2. A jet traveling at 900 mi/h takes 3 h to fly between two cities. How far apart are the cities?

3. How long will it take you to get to your aunt's house if she lives 525 mi away and you travel at 50 mi/h?

4. A cyclist traveled 25 mi in 2 h. How fast was she going?

5. The formula $w = \dfrac{11(h - 40)}{2}$, where w is weight in pounds and h is height in inches, relates weight and height for average adults. Use the formula to complete the table.

Height (in.)	Weight (lb)
60	▩
65	▩
▩	121
▩	170.5

9.8 Circumference of a Circle

Objective: To determine the circumference of a circle.

The *circumference* (*C*) of a circle is the distance around the circle. As a wheel makes a complete turn, it covers a distance equal to its circumference.

EXPLORING

1. Use LOGO to enter and run the procedure at the right.

2. Write a procedure that draws a circle. Run the procedure and determine the circumference of the circle.

3. Enter the procedure at the right. Type "var. circle 6."
   a. What is the circumference of the circle?
   b. Type "rt 90." Use forward moves to find the diameter of the circle.

4. Repeat Step 3 for var.circle 10 and then for var.circle 2.5.

5. For each of the circles in Steps 3 and 4, divide the circumference by the diameter. The quotients are close to what whole number?

```
To partcircle
repeat 120 [fd 1 rt 1]
end
```

```
To var.circle :side
repeat 36 [fd :side rt 10]
end
```

More than 2,000 years ago mathematicians proved that the circumference of a circle divided by the diameter is the same number for every circle. They named this constant value with the Greek letter *pi* (π). The most commonly used approximations for π are shown below, where ≈ is read as "is approximately equal to."

$$\pi \approx \tfrac{22}{7} \quad \pi \approx 3\tfrac{1}{7} \quad \pi \approx 3.14$$

In this textbook we will use 3.14 for π unless otherwise stated.

The **EXPLORING** activity demonstrates the following theorem.

THEOREM 9.7: **The circumference of any circle is the product of its diameter and π.** $C = \pi d = 2\pi r$

An answer stated in terms of π is an *exact* answer. When you use an approximation for π, such as 3.14, your answer is an approximation.

Example 1: Find (a) the exact circumference and (b) the approximate circumference. Round to the nearest hundredth. A calculator may be helpful.

Solution: **a.** $C = 2\pi \cdot 2.7$

$= 5.4\pi$ cm

2.7 cm

b. $C \approx 5.4 \;\boxed{\times}\; 3.14$

$\boxed{=}\; 16.956$

≈ 16.96 cm

Example 2: The circumference of a circle is approximately 25.125 ft. Find the diameter and the radius.

Solution: $C = \pi d$

$25.125 \approx 3.14d$

$\dfrac{25.125}{3.14} \approx d$

$8.002 \approx d$

The diameter is about 8 ft, and the radius is about 4 ft.

Thinking Critically

Describe what happens to the circumference of a circle under the following conditions.

1. You double its diameter.   **2.** You double its radius.

3. You halve its diameter.   **4.** You increase its radius by 1.

Class Exercises

Find the circumference for the circle with the given diameter or radius as (a) an exact answer and (b) an approximation. A calculator may be helpful.

1. $d = 12$ cm   **2.** $d = 9$ cm   **3.** $r = 8$ in.   **4.** $r = 10.6$ m   **5.** $d = 0.85$ m

6. The circumference of a circle is 100π ft. Find the diameter and the radius.

7. The circumference of a wheel on a bike is approximately 219.8 cm. Find the diameter and the radius.

Exercises

Use the given information to find the circumference as (a) an exact answer and (b) an approximation. Use the value of π that makes your calculations easiest. A calculator may be helpful.

1. $r = 8$ cm **2.** $r = 5$ cm **3.** $d = 15$ in. **4.** $d = 2.5$ cm **5.** $r = 28.2$ m

6. $d = 3\frac{1}{2}$ in. **7.** $r = 7$ in. **8.** $d = 14$ in. **9.** $r = 35$ cm **10.** $d = 0.75$ m

11. The circumference of a circle is 14π ft. Find the diameter and the radius.

12. The circumference of a pizza is approximately $37\frac{5}{7}$ in. Find the diameter and the radius. Use $\frac{22}{7}$ for π.

Find the distance around each object.

13.
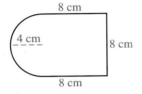
8 cm
4 cm
8 cm
8 cm

14.
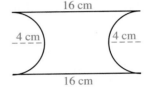
16 cm
4 cm
4 cm
16 cm

15. The minute hand on a watch is 8 mm long. How far does the tip of the minute hand travel in 1 h?

16. How many revolutions will a 26-in. bicycle wheel make in 1 mi? (1 mi = 5,280 ft)

APPLICATION

17. Aviation The diameter of Earth is about 8,000 mi. Suppose a jet is traveling around Earth at an altitude of 30,000 ft. How many miles will the jet travel in a complete circle around Earth? (1 mi = 5,280 ft)

Computer

Use LOGO to enter and run the procedure at right for several values of *r*. Then write a LOGO procedure to draw several circles with the same center.

```
To circle :r
pu fd :r rt 95 pd
repeat 36 [fd :r * 3.14159 / 18 rt 10]
pu home pd
end
```

Thinking About Proof

Using Properties from Algebra

When you solve an equation, you can justify each step with a property from algebra. For example, you can think of the display at the right as a proof that 3 is the solution of the original equation.

$$3(4x - 7) = 15$$
$$12x - 21 = 15 \quad \text{Distributive property}$$
$$12x = 36 \quad \text{Addition property}$$
$$x = 3 \quad \text{Division property}$$

Other properties from algebra that are commonly used are the subtraction property, the multiplication property, and the substitution property.

Exercises

Name the property from algebra that justifies each step.

1. $3h = 12$
$h = 4$

2. $5y - 9 = 6$
$5y = 15$

3. $4z + 2 = 6$
$4z = 4$

4. $\frac{1}{2}b = 10$
$b = 20$

5. $10 = 2(a + 3)$
$10 = 2a + 6$

6. $x + y = 5 + y$
$x = 5$

7. $4x + 5x = 18$
$9x = 18$

8. $A = \frac{1}{2}hb_1 + \frac{1}{2}hb_2$
$A = \frac{1}{2}h(b_1 + b_2)$

9. Theorem 8.15 is: If both pairs of opposite angles of a quadrilateral are congruent, then the quadrilateral is a parallelogram. Part of the proof of Theorem 8.15 is shown below. Give a reason for each step.

a. $2a° + 2b° = 360°$
b. $2(a° + b°) = 360°$
c. $a° + b° = 180°$

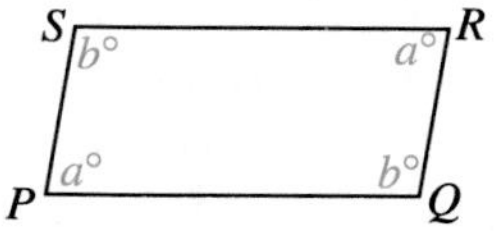

10. To be Proven: Angles that are congruent and supplementary are right angles.

Given: $\angle 1 \cong \angle 2$ and $m\angle 1 + m\angle 2 = 180°$
Prove: $\angle 1$ and $\angle 2$ are right angles.

Give a reason for each statement.

a. $\angle 1 \cong \angle 2$ and $m\angle 1 + m\angle 2 = 180°$
b. $m\angle 1 = m\angle 2$
c. $m\angle 1 + m\angle 1 = 180°$
d. $2 \cdot m\angle 1 = 180°$
e. $m\angle 1 = 90°$
f. $m\angle 2 = 90°$
g. $\angle 1$ and $\angle 2$ are right angles.

9.9 Area of a Circle

Objective: To determine the area of a circle.

You can use what you know about the area of a parallelogram to develop a formula for the area of a circle. In the Exploring activity that follows, the figure you form by rearranging the circle is not really a parallelogram. However, as you divide the circle into smaller and smaller congruent parts, the figure becomes more like a parallelogram.

EXPLORING

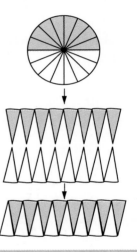

1. The diagrams show a circle separated into congruent parts and rearranged to form a new figure that resembles a parallelogram. How do the areas of the two figures compare?

2. What is the formula for the area of a parallelogram?

3. How does the base length (b) of the parallelogram relate to the circumference ($2\pi r$) of the circle?

4. How does the height (h) of the parallelogram relate to the radius (r) of the circle?

5. Use your answers from Steps 2–4 to write a formula for the area of a circle.

The **EXPLORING** activity demonstrates the following theorem.

THEOREM 9.8: **The area (A) of any circle is the product of π and the square of the radius (r).** $A = \pi r^2$

Example 1: **For a circle with radius 5 cm, find (a) the exact area and (b) the approximate area.**

Solution: **a.** $A = \pi r^2$

$= \pi \cdot 5^2$

$= \pi \cdot 25$

$= 25\pi$ cm^2

b. A calculator may be helpful.

$A \approx 25 \; \boxed{\times} \; 3.14 \; \boxed{=} \; 78.5$ cm^2

Example 2: **Find the area of the shaded region. Use $\frac{22}{7}$ for π.**

Solution: $A \approx \dfrac{\overset{11}{\cancel{22}}}{\underset{1}{\cancel{7}}} \cdot \dfrac{\overset{1}{\cancel{7}}}{\underset{1}{\cancel{2}}} \cdot \dfrac{7}{2}$

$ = \dfrac{11 \cdot 7}{2}$

$ = \dfrac{77}{2}$

$ = 38\dfrac{1}{2} \text{ ft}^2$

$\dfrac{1}{4} \cdot 38\dfrac{1}{2} = \dfrac{1}{4} \cdot \dfrac{77}{2}$

$\phantom{\dfrac{1}{4} \cdot 38\dfrac{1}{2}} = \dfrac{77}{8}$

$\phantom{\dfrac{1}{4} \cdot 38\dfrac{1}{2}} = 9\dfrac{5}{8} \text{ ft}^2$

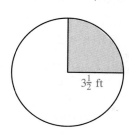

$3\frac{1}{2}$ ft

Thinking Critically

Describe what happens to the area of a circle under the following conditions.

1. You double its diameter.   **2.** You double its radius.

3. You halve its diameter.   **4.** You increase its radius by 1.

Class Exercises

For a circle with each given radius or diameter, find (a) the exact area and (b) the approximate area. Round to the nearest hundredth.

1. $r = 4$ cm   **2.** $r = 5.5$ cm   **3.** $d = 78$ cm   **4.** $d = 12.6$ cm   **5.** $r = 20$ ft

6. Find the area of the shaded region.

5 cm

Exercises

For a circle with each given radius or diameter, find (a) the exact area and (b) the approximate area. A calculator may be helpful.

1. $r = 22$ cm   **2.** $r = 3.5$ cm   **3.** $d = 1.6$ cm

4. $r = 1.2$ m   **5.** $r = 37$ ft   **6.** $d = 7.6$ m

For a circle with each given radius or diameter, find (a) the exact area and (b) the approximate area. Use $\frac{22}{7}$ for π.

7. $r = 7$ cm

8. $r = 14$ cm

9. $d = 20$ in.

10. $r = 28$ ft

11. $r = 1\frac{2}{5}$ ft

12. $d = 42$ ft

Use a calculator to complete the table. Round answers to the nearest hundredth.

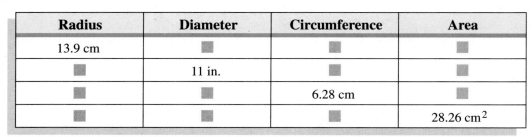

	Radius	Diameter	Circumference	Area
13.	13.9 cm	▨	▨	▨
14.	▨	11 in.	▨	▨
15.	▨	▨	6.28 cm	▨
16.	▨	▨	▨	28.26 cm^2

Find the area of each region.

17. dime, $r = 9$ mm

18. watch face, $r = 15$ mm

19. pie pan, $d = 8$ in.

20. pizza pan, $r = 7\frac{1}{2}$ in.

Find the indicated information for each shaded region.

21. A = ▨

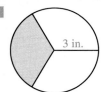

3 in.

22. a. A = ▨

b. P = ▨

9 cm

23. A = ▨

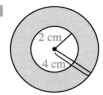

2 cm

4 cm

24. A = ▨

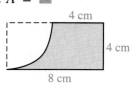

4 cm

4 cm

8 cm

25. A = ▨

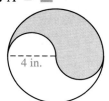

4 in.

26. a. A = ▨

b. P = ▨

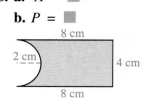

8 cm

2 cm

4 cm

8 cm

27. Let n be the measure in degrees of $\angle ACB$. Then you can use the following formula to find the area of the shaded region.

$$A = \frac{n}{360}(\pi r^2)$$

a. Find the area of the shaded region.

b. Explain why the formula works.

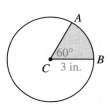

A

$60°$

C 3 in. B

28. a. Find the exact area of the circle at the right.

b. Find the area of the three circles formed when the radius of the region in part (a) is multiplied by 2, 3, and 4.

c. Compare your results in parts (a) and (b). What happens to the area of a circle when its radius is multiplied by some number n?

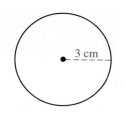
3 cm

APPLICATIONS

Calculator   A pound of grass seed costs $4.95 and covers about 675 ft². Find (a) the area and (b) the total cost to seed a circular lawn area with the given diameter. Round answers to the nearest cent.

29. 75 ft   **30.** 15 yd   **31.** 68.5 ft

32. Landscaping   A type of lawn sprinkler consists of a rigid "arm" that rotates about a point. The arm of the sprinkler is 15 ft long. Find the area of the circular region of grass that can be watered by the sprinkler.

33. Design   A round table has a diameter of 48 in. You want to make a round tablecloth that will hang 4 in. all the way around the table. Draw a picture to represent the situation.

a. What should the diameter of the tablecloth be?

b. What will the area of the tablecloth be in square inches? square yards?

c. Your fabric is 60 in. wide. Is the fabric wide enough for the tablecloth? Will $1\frac{3}{4}$ yd of fabric be enough for the tablecloth?

Historical Note

In 1706, the Greek letter π was first used to denote the ratio of the circumference of a circle to its diameter. One earlier symbol used was the letter "e."

In 1989, two Columbia University mathematicians used a computer to calculate the value of pi to 480 million decimal places. If printed linearly, this number would extend 600 mi.

In 470, the Chinese mathematician Tsu Ch'ung-Chih discovered that the fraction $\frac{355}{113}$ gives a remarkably close approximation of π. Use your calculator to convert $\frac{355}{113}$ to a decimal. To the nearest ten-millionth, $\pi = 3.1415927$. By how much do your calculator results differ from this value?

Vocabulary and Symbols

You should be able to write a brief description, draw a picture, or give an example to illustrate the meaning of each of the following terms.

Vocabulary

altitude of a parallelogram (p. 267)
apothem of a regular polygon (p. 276)
area (p. 259)
base of a parallelogram (p. 267)
base of a rectangle (p. 263)
base of a triangle (p. 271)

center of a regular polygon (p. 276)
circumference (p. 282)
height of a parallelogram (p. 267)
height of a rectangle (p. 263)
height of a triangle (p. 271)
perimeter (p. 255)
pi (p. 282)

Symbols

A (area) (p. 259)
b (base length) (p. 263)
C (circumference) (p. 282)
h (height) (p. 263)
P (perimeter) (p. 255)
π (pi) (p. 282)
$\approx$ (is approximately equal to) (p. 282)

Summary

The following list indicates the major skills, facts, and results you should have mastered in this chapter.

9.1 Determine the perimeter of a concave or convex polygon. (pp. 255–258)

9.2 Determine the area of a polygon by counting unit squares. (pp. 259–262)

9.3 Determine the area of any rectangle. (pp. 263–266)

9.4 Determine the area of any parallelogram. (pp. 267–270)

9.5 Determine the area of any triangle. (pp. 271–274)

9.6 Determine the area of a polygon. (pp. 275–279)

9.7 Use formulas to solve problems. (pp. 280–281)

9.8 Determine the circumference of any circle. (pp. 282–284)

9.9 Determine the area of any circle. (pp. 286–289)

Exercises

Find the perimeter of each polygon.

1.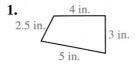
4 in.
2.5 in.
3 in.
5 in.

2.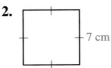
7 cm

3.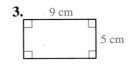
9 cm
5 cm

4.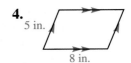
5 in.
8 in.

Find the length of the indicated side.

5. $P = 56$ m

$x = $ ▨

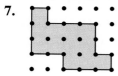

16 m

16 m

6. $P = 41$ cm

$s = $ ▨

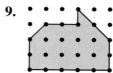

10 cm

15 cm

Find the area of each shaded region. ▨ **represents 1 unit of area.**

7.

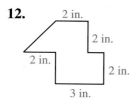

8.

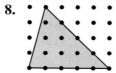

9.

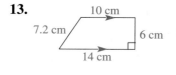

Copy and complete the table. A calculator may be helpful.

	Figure	Base Length (*b*)	Height (*h*)	Area (*A*)
10.	parallelogram	32 ft	18 ft	▨
11.	triangle	18 cm	10 cm	▨

Find the area of each region.

12.

2 in.

2 in.

2 in.

2 in.

3 in.

13.

10 cm

7.2 cm

6 cm

14 cm

14.

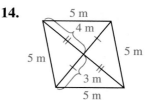

5 m

4 m

5 m

5 m

3 m

5 m

Find (a) the exact circumference and area and (b) the approximate circumference and area of the circle with the given diameter or radius.

15. $r = 12$ cm   **16.** $d = 8$ ft   **17.** $r = 7.5$ m

PROBLEM SOLVING

18. The formula $I = prt$ denotes the amount of interest paid on a principal amount (*p*) borrowed at a rate (*r*) for a given time (*t*). Use the formula to determine how much interest Jim paid on a principal amount of $4,500 borrowed for 3 years at a rate of 8.75%.

Find the perimeter of the polygon described.

1. a rectangle with length 5 cm and width 2.5 cm

2. a square with sides of length 2.3 mm

3. a parallelogram with two sides 12 cm long and two sides 8 cm long

4. a regular pentagon with sides of length 34 in.

Find the area of each figure.

5. a rectangle with base length $7\frac{1}{2}$ in. and height 4 in.

6. a parallelogram with base length 19 cm and height 9 cm

7. a square with sides of length 51.3 m

8. a triangle with base length 32 cm and height 18 cm

9. a rhombus with diagonals of length 15 cm and 17 cm

10. a trapezoid with bases of length 21 ft and 14 ft and height 6 ft

Complete.

11. A = ▨

   ▢ represents
   1 unit of area

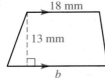

12. A = ▨

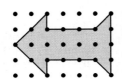

13. A = 286 mm²

   b = ▨

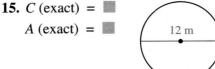

14. A = 27 ft²

   h = ▨

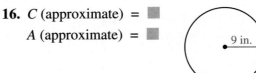

15. C (exact) = ▨

   A (exact) = ▨

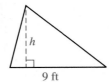

16. C (approximate) = ▨

   A (approximate) = ▨

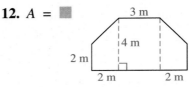

PROBLEM SOLVING

17. The formula $P = IE$ relates electrical power (P) in watts to the amount of
    current (I) in amperes , and the potential difference (E) in volts. How many
    amperes of current are used by a 1,200-watt hair dryer on a 110-volt line?

Similarity

Traditional woodcarvers make Russian stacking dolls hollow and similar in shape so the dolls can be placed inside one another to form a nested set.

Focus on Skills

ARITHMETIC

Express each fraction in simplest form.

1. $\frac{6}{9}$ **2.** $\frac{10}{30}$ **3.** $\frac{8}{10}$ **4.** $\frac{12}{16}$ **5.** $\frac{18}{30}$ **6.** $\frac{15}{45}$

7. $\frac{8}{12}$ **8.** $\frac{6}{12}$ **9.** $\frac{10}{16}$ **10.** $\frac{21}{24}$ **11.** $\frac{16}{20}$ **12.** $\frac{10}{25}$

13. $\frac{12}{60}$ **14.** $\frac{75}{100}$ **15.** $\frac{70}{100}$ **16.** $\frac{45}{100}$ **17.** $\frac{60}{360}$ **18.** $\frac{30}{360}$

Complete.

19. 12 cm = ▇ mm **20.** $2\frac{1}{2}$ ft = ▇ in. **21.** $1\frac{1}{2}$ h = ▇ min

22. 3.2 m = ▇ cm **23.** $1\frac{3}{4}$ lb = ▇ oz **24.** 12 min = ▇ h

GEOMETRY

The polygons shown are congruent. Complete.

25. a. $\triangle JKL \cong$ ▇   **b.** $\triangle JLK \cong$ ▇   **26. a.** $ABCD \cong$ ▇   **b.** $DCBA \cong$ ▇

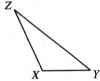
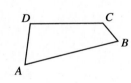
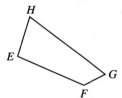

The polygons shown are congruent.

a. Name all pairs of corresponding angles.

b. Name all pairs of corresponding sides.

c. Write a congruence statement for the polygons.

27.
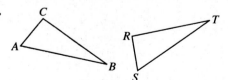

28.
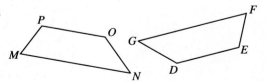

29. If $\triangle GHI \cong \triangle KLM$, then $\angle H$ and ▇ are corresponding angles and $\overline{MK}$ and ▇ are corresponding sides.

10.1 Ratios

Objective: To express ratios in simplest form.

A *ratio* is a comparison of two numbers by division. The ratio of *a* to *b* may be written as $\frac{a}{b}$ or $a : b$. Ratios are usually expressed in simplest form.

Example 1: **Express each ratio in simplest form.**

a. *AC* to *CB* **b.** *CB* to *AC* **c.** *AC* to *AB*

Solution: **a.** $\dfrac{AC}{CB} = \dfrac{3}{12}$ **b.** $\dfrac{CB}{AC} = \dfrac{12}{3}$ **c.** $\dfrac{AC}{AB} = \dfrac{3}{15}$

$\qquad\qquad = \dfrac{1}{4}$ $\qquad = \dfrac{4}{1}$ $\qquad = \dfrac{1}{5}$

EXPLORING

Express each ratio in simplest form.

1. Use the table at the right to determine the ratio of the number of free throws made (FTM) to the number of free throws attempted (FTA) for each basketball player.

2. The team has won 12 games and lost 8 games.
   a. Write the ratio of games won to games lost.
   b. Write the ratio of games lost to games won.
   c. Write the ratio of games won to total games.

Player	FTM	FTA
Beth	15	23
Maria	50	60
Kim	14	21

Example 2: **Express each ratio in simplest form.**

a. 3 cm to 15 mm **b.** 1 dime to 1 dollar

Solution: **a.** $\dfrac{3\,cm}{15\,mm} = \dfrac{30\,mm}{15\,mm} = \dfrac{2}{1}$ **b.** $\dfrac{1\,dime}{1\,dollar} = \dfrac{10\cancel{c}}{100\cancel{c}} = \dfrac{1}{10}$

Thinking Critically

1. Jon's paycheck is $180. He saves $\frac{2}{5}$ of it, or $72. What comparison is being made by the ratio $\frac{2}{5}$?

2. **a.** Name three ratios equal to 1 : 3. **b.** Name three ratios equal to 12 : 16.

Class Exercises

Make a drawing to represent each ratio.

1. shaded parts to all parts is 5 : 12

2. shaded parts to unshaded parts is 5 : 12

Express each ratio in simplest form.

3. $\dfrac{8}{36}$

4. $\dfrac{24}{36}$

5. 45 : 60

6. 16 : 20

7. 8 mm to 2 cm

8. 2 quarters to 3 dimes

Exercises

Express each ratio in simplest form.

1. shaded squares to unshaded squares

2. unshaded squares to shaded squares

3. shaded squares to all squares

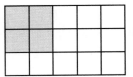

Make a drawing to represent each ratio.

4. shaded parts to all parts is 3 to 7

5. shaded parts to unshaded parts is 3 to 7

6. A room is 10 ft wide and 13 ft long. Find the ratio of length to width.

7. Rick has 3 hits in 5 times at bat. Find the ratio of hits to times at bat.

8. A geometry class has 17 girls and 13 boys. Find each ratio.

    a. girls to boys   **b.** boys to girls

    c. girls to all students   **d.** boys to all students

    e. all students to boys

Express each ratio in simplest form.

9. $\dfrac{9}{27}$

10. $\dfrac{15}{45}$

11. $\dfrac{18}{30}$

12. $\dfrac{24}{60}$

13. 6 m to 18 m

14. 10 cm to 15 cm

15. 15 cm to 10 cm

16. 10 mi to 4 mi

17. 8 : 20

18. 9 : 36

19. 24 : 32

20. 28 : 24

21. *RS* : *ST*

22. *ST* : *RS*

23. *RS* : *RT*

24. $\dfrac{AX}{XB}$

25. $\dfrac{XB}{AX}$

26. $\dfrac{AX}{AB}$

27. $\dfrac{BC}{XY}$

28. $\dfrac{AY}{YC}$

29. $\dfrac{AY}{AC}$

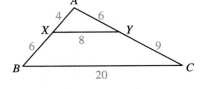

30. $\dfrac{DE}{RS}$ **31.** $\dfrac{FD}{TR}$ **32.** $\dfrac{TS}{FE}$

33. $\dfrac{RS}{DE}$ **34.** $\dfrac{TR}{FD}$ **35.** $\dfrac{FE}{TS}$

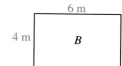

36. 2 ft to 6 in. **37.** 1 dime to 1 quarter **38.** 6 cm to 50 mm

39. 15 min to 1 h **40.** 7 oz to 1 lb **41.** 3 nickels to 2 quarters

42. the measure of a right angle to the measure of a straight angle

43. the length of a side of a square to the perimeter of the square

44. If $EF = \frac{2}{3} \cdot XY$, find $EF : XY$. **45.** If $RS = 3 \cdot AB$, find $AB : RS$.

46. The ratio of the measures of two consecutive angles of a parallelogram is $2 : 3$. Find the measure of each angle. (*Hint*: Represent the measures by $2x$ and $3x$.)

47. The ratio of the measures of the three interior angles of a triangle is $2 : 3 : 4$. Find the measure of each angle.

APPLICATION

48. Perimeter and Area Find the indicated ratios for rectangles A and B.

   a. base length of B to base length of A

   b. height of B to height of A

   c. perimeter of B to perimeter of A

   d. area of B to area of A

Everyday Geometry

The musical sound of a guitar or violin is produced by the vibration of the string. The number of vibrations each second is the *frequency* measured in Hertz (Hz). When the ratio of the frequencies of two sounds is $2 : 1$, the sounds are one *octave* apart. The higher sound has twice the frequency of the sound an octave lower.

  1. The lowest string on a guitar vibrates with a frequency of 82.5 Hz. Find the frequency of a sound:

    a. one octave higher **b.** two octaves higher

  2. The highest note on a piano is produced by a string vibrating with a frequency of 4,186 Hz. Find the frequency of a sound:

    a. one octave lower **b.** four octaves lower

10.2 Proportions

Objective: To solve proportions.

Since both $\frac{6}{8}$ and $\frac{9}{12}$ are equal to $\frac{3}{4}$, they are equal to each other. A statement that two ratios are equal is called a **proportion**. A proportion can be written in either of the following ways.

$$\frac{6}{8} = \frac{9}{12} \quad \text{or} \quad 6:8 = 9:12$$

EXPLORING

Choose the ratios that are equal to the given ratio.

1. $\frac{2}{5}$ **a.** $\frac{4}{10}$ **b.** $\frac{10}{4}$ **c.** $\frac{20}{50}$ **d.** $\frac{12}{30}$

2. $12:16$ **a.** $9:12$ **b.** $18:24$ **c.** $16:12$ **d.** $15:20$

The following terms are used to describe proportions.

1st term 3rd term   1st term 3rd term
$$\frac{a}{b} = \frac{c}{d} \qquad a:b = c:d$$
2nd term 4th term   2nd term 4th term

The first and fourth terms of a proportion are called the **extremes**. The second and third terms of a proportion are called the **means**. You can use the following property to solve a proportion for one term when the other three terms are known.

Means-Extremes Property

In a proportion, the product of the means equals the product of the extremes.

$$\text{If } \frac{a}{b} = \frac{c}{d}, \text{ then } ad = bc.$$
$$\text{If } a:b = c:d, \text{ then } ad = bc.$$

Example: Solve each proportion.

a. $\dfrac{4}{x} = \dfrac{12}{60}$ **b.** $\dfrac{n}{6} = \dfrac{7}{42}$ **c.** $\dfrac{3}{5} = \dfrac{5}{x}$

Solution:

a. $12 \cdot x = 4 \cdot 60$ **b.** $42 \cdot n = 6 \cdot 7$ **c.** $3 \cdot x = 5 \cdot 5$

$12x = 240$ $42n = 42$ $3x = 25$

$x = \dfrac{240}{12} = 20$ $n = \dfrac{42}{42} = 1$ $x = \dfrac{25}{3} = 8\dfrac{1}{3}$

Thinking Critically

1. In the proportion $a : b = c : d$, are there any restrictions on b and d? Explain.

2. A proportion like $\dfrac{x}{36} = \dfrac{5}{6} = \dfrac{20}{y}$ is called an *extended proportion*.

   a. Describe how you would solve for x and y.

   b. Solve for x and y.

Class Exercises

1. For the proportion $\dfrac{x}{2} = \dfrac{5}{7}$, name each of the following.

   a. the first term   **b.** the second term   **c.** the third term

   d. the fourth term   **e.** the means   **f.** the extremes

Solve each proportion.

2. $\dfrac{a}{15} = \dfrac{2}{5}$ **3.** $\dfrac{6}{16} = \dfrac{9}{x}$ **4.** $3 : x = 12 : 16$ **5.** $\dfrac{3}{5} = \dfrac{c}{8}$

Exercises

Solve each proportion.

1. $\dfrac{3}{2} = \dfrac{x}{4}$ **2.** $\dfrac{a}{12} = \dfrac{3}{4}$ **3.** $\dfrac{4}{y} = \dfrac{2}{5}$ **4.** $\dfrac{c}{3} = \dfrac{10}{5}$

5. $\dfrac{2}{7} = \dfrac{8}{z}$ **6.** $\dfrac{1}{2} = \dfrac{d}{9}$ **7.** $\dfrac{x}{10} = \dfrac{3}{4}$ **8.** $\dfrac{1}{4} = \dfrac{n}{5}$

9. $9 : x = 3 : 2$ **10.** $4 : 7 = 8 : y$ **11.** $n : 2 = 5 : 3$ **12.** $5 : a = 2 : 5$

13. $7.5 : 15 = b : 20$ **14.** $1 : 2.5 = 4 : x$ **15.** $4 : d = 5 : 7.5$ **16.** $1.5 : 4.5 = 2 : z$

17. The numbers 2, 3, and 6 are the first three terms (listed in order) of a proportion. Find the fourth term.

If possible, use the given numbers to complete the proportion.

18. 8, 16; $\dfrac{2}{4} = \dfrac{\blacksquare}{\blacksquare}$ **19.** 2, 8; $\dfrac{3}{12} = \dfrac{\blacksquare}{\blacksquare}$ **20.** 6, 2; $\dfrac{5}{15} = \dfrac{\blacksquare}{\blacksquare}$ **21.** 9, 12; $\dfrac{8}{6} = \dfrac{\blacksquare}{\blacksquare}$

22. 6, 8; $\dfrac{15}{225} = \dfrac{\blacksquare}{\blacksquare}$ **23.** 16, 18; $\dfrac{27}{24} = \dfrac{\blacksquare}{\blacksquare}$ **24.** 20, 8; $\dfrac{18}{45} = \dfrac{\blacksquare}{\blacksquare}$ **25.** 16, 21; $\dfrac{15}{20} = \dfrac{\blacksquare}{\blacksquare}$

Solve each proportion.

26. $\dfrac{3}{5} = \dfrac{x}{4 + x}$ **27.** $\dfrac{2x - 5}{2x + 5} = \dfrac{5}{7}$ **28.** $\dfrac{9}{12} = \dfrac{x}{20} = \dfrac{21}{y}$

APPLICATIONS

29. Elections   A candidate won an election by a 3 to 2 margin. Her opponent received 1,110 votes. How many votes did the winner receive?

30. Calculator   Use a calculator to solve the proportion $70 : 175 = 98 : x$.

31. Property Taxes   The property tax on the Johnsons' house, which is valued at $100,000, is $1,500. Next door, the Lees' house is valued at $110,000. What is the property tax on the Lees' house?

Seeing in Geometry

Complete.

1. is to as is to $\blacksquare$.

2. is to as is to $\blacksquare$.

3. is to as is to $\blacksquare$.

4. is to as is to $\blacksquare$.

10.3 Similar Figures

Objective: To recognize figures that appear to be similar and identify corresponding parts of similar polygons.

Figures that are the same shape but not necessarily the same size are called *similar* (~). A photograph and its enlargement are a familiar example of similar figures. In each case below, the rectangles are similar.

The red rectangle is an enlargement of the black rectangle.

The red rectangle is a reduction of the black rectangle.

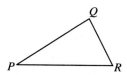

EXPLORING

Part A

1. Use the *Geometric Supposer: Triangles* disk to create an acute triangle. Sketch the triangle and record its measurements.

2. Use the *Scale change* option. Sketch the new figure. What do you notice about the two figures?

3. Repeat Steps 1 and 2 for a right triangle and an obtuse triangle.

4. Insert the *Quadrilaterals* disk and repeat Steps 1 and 2 for several different quadrilaterals.

Part B

1. The two triangles shown at the right are similar. Using "↔" for "corresponds to," write correspondences between six pairs of corresponding parts.

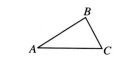

2. A *similarity statement* that indicates the corresponding order of the vertices is △ABC ~ △PQR, where "~" is read as "is similar to." Write five other similarity statements for the two triangles.

Example: Assume that *GHIJ ~ RSTW*. Complete.

    a. $\angle H \leftrightarrow$ ■   **b.** $\overline{HI} \leftrightarrow$ ■

    c. *IJGH ~* ■   **d.** *IHGJ ~* ■

Solution: **a.** $\angle H \leftrightarrow \angle S$ **b.** $\overline{HI} \leftrightarrow \overline{ST}$

    c. *IJGH ~ TWRS*   **d.** *IHGJ ~ TSRW*

Thinking Critically

1. Tell whether each pair of figures appear to be similar.

  a.   **b.**    **c.**

  d.    **e.**    **f.**

2. Are congruent figures also similar?

3. What appears to be true about corresponding angles of similar polygons?

Class Exercises

The polygons shown are similar. Name the side or angle that corresponds to each indicated part.

1. a. $\overline{ST}$ **2. a.** $\overline{HK}$

    b. $\overline{XY}$ **b.** $\overline{BP}$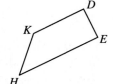

    c. $\angle R$ **c.** $\angle D$

    d. $\angle X$ **d.** $\angle P$

3. Complete each similarity statement for the triangles in Exercise 1.

    a. $\triangle XYZ \sim$ ■   **b.** $\triangle STR \sim$ ■   **c.** $\triangle ZYX \sim$ ■   **d.** $\triangle TSR \sim$ ■

4. Assume that $\triangle PTO \sim \triangle RMB$. Complete.

    a. $\angle O \leftrightarrow$ ■   **b.** $\overline{OT} \leftrightarrow$ ■   **c.** $\overline{MR} \leftrightarrow$ ■   **d.** $\angle BMR \leftrightarrow$ ■

Exercises

True or false?

1. Similar figures must be the same size.
2. Congruent figures must be similar.
3. If $\triangle ABC \sim \triangle WXY$, then $\overline{BC}$ corresponds to $\overline{WX}$.
4. If "$\triangle ABC \sim \triangle WXY$" is true, then "$\triangle CAB \sim \triangle YWX$" is also true.
5. Any two triangles are similar.
6. Any two rectangles are similar.

Select the figure that appears to be similar to the given figure.

7. 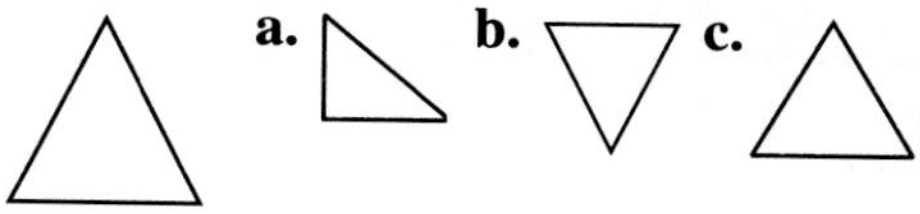   a. b. c. 8. 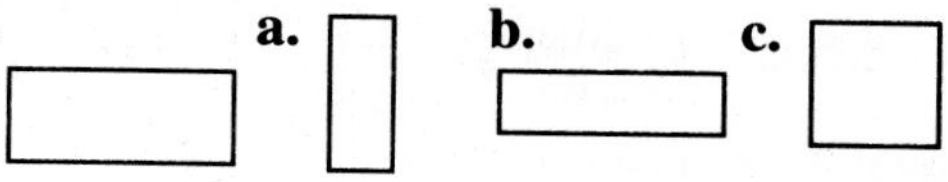 a. b. c.

9.   a. b. c. 10. 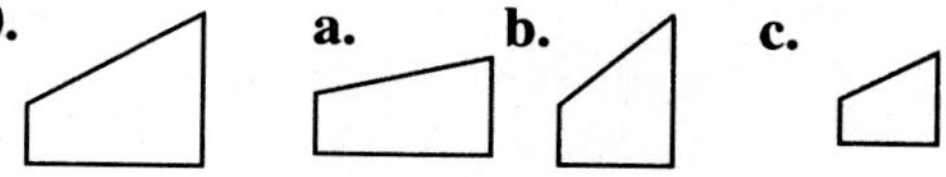 a. b. c.

Complete each similarity statement for the similar polygons.

11. **a.** $BQNST \sim$ ▧ **b.** $CDFGH \sim$ ▧   12. **a.** $TRPS \sim$ ▧ **b.** $AEBD \sim$ ▧
     c. $NQBTS \sim$ ▧ **d.** $GFDCH \sim$ ▧   **c.** $PRTS \sim$ ▧ **d.** $EADB \sim$ ▧

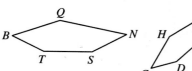

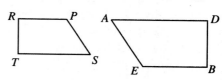

13. Use the similar polygons in Exercise 11.
     a. Name five pairs of corresponding angles.
     b. Name five pairs of corresponding sides.

14. **a.** Write a similarity statement for the two similar triangles shown.
     b. Name three pairs of corresponding angles.
     c. Name three pairs of corresponding sides.

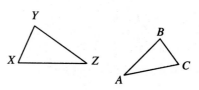

15. Assume that $\triangle RST \sim \triangle JKL$.
     a. Name three pairs of corresponding angles.
     b. Name three pairs of corresponding sides.

For each pair of similar triangles:
a. Name three pairs of corresponding angles.
b. Name three pairs of corresponding sides.
c. Write two other similarity statements for the two triangles.

16. $\triangle ABC \sim \triangle XYC$

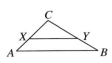

17. $\triangle AMN \sim \triangle AEF$

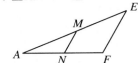

18. $\triangle ACD \sim \triangle BCD$

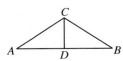

Draw and label similar polygons to fit each description.
19. $\triangle CDE \sim \triangle PLG$
20. rectangle $ABCD \sim$ rectangle $WXYZ$
21. $\triangle RST \sim \triangle RWY$
22. $\triangle ABC \sim \triangle ABD$

APPLICATION

23. Patterns

  a. The length of each side of the large rectangle is twice the length of the corresponding side of the other rectangle. How do the areas of the rectangles compare? (*Hint*: How many of the small rectangles can you place inside the large one?)

  b. The length of each side of the large rectangle is three times the length of the corresponding side of the other rectangle. How do the areas of the rectangles compare?

  c. The length of each side of a large rectangle is 10 times the length of the corresponding side of another rectangle. How do the areas of the rectangles compare?

Test Yourself

Express each ratio in simplest form.
  1. $\dfrac{24}{40}$
  2. $32 : 48$
  3. $\dfrac{18}{15}$
  4. 4 in. to 1 ft

Solve each proportion.
  5. $\dfrac{2}{5} = \dfrac{6}{y}$
  6. $5 : 6 = x : 24$
  7. $\dfrac{3}{c} = \dfrac{15}{35}$
  8. $a : 8 = 8 : 1$

10.4 Similar Polygons

Objective: To apply the definition of similar polygons.

In the previous lesson we defined similar figures as figures having the same shape but not necessarily the same size. We can develop a more precise definition of similarity for *polygons*.

When the ratios of the lengths of corresponding sides of two polygons are equal, we say that the lengths are ***proportional.*** The ***scale factor*** of the similarity is the ratio of the lengths of the corresponding sides.

EXPLORING

1. Use LOGO to enter and run the procedure at the right.

To similar.pentagons
pu lt 90 fd 70 lt 90 pd
repeat 5 [rt 72 fd 30]
pu rt 90 fd 80 lt 90 pd
repeat 5 [rt 72 fd 60]
end

2. The pentagons shown are similar. What measures are the same? How do their other measures compare? What is the scale factor of the first pentagon to the second pentagon?

3. Change the procedure so the sides of the second pentagon are three times as long as the sides of the first pentagon. Are the pentagons similar? What is the scale factor of the first pentagon to the second pentagon?

4. Write a procedure that draws four similar regular hexagons. Make the sides of the first hexagon 12 units. The scale factors of the first hexagon to each of the other three hexagons should be 1 : 3, 1 : 4, and 1 : 5. Find the scale factor of the second hexagon to the third hexagon.

The **EXPLORING** activity suggests the following definition of similarity for two polygons.

Similar polygons are polygons for which

1. corresponding angles are congruent, and
2. corresponding side lengths are proportional.

Example: **Given:** $\triangle RST \sim \triangle DEF$

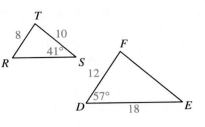

    a. Find the scale factor of $\triangle RST$ to $\triangle DEF$.

    b. Find the scale factor of $\triangle DEF$ to $\triangle RST$.

    c. $m\angle E = $ ▨   **d.** $m\angle R = $ ▨

    e. $m\angle T = $ ▨   **f.** $RS = $ ▨ $\bullet\ DE$

    g. $RS = $ ▨   **h.** $FE = $ ▨

Solution: **a.** Since $\dfrac{RT}{DF} = \dfrac{8}{12} = \dfrac{2}{3}$, the scale factor of $\triangle RST$ to $\triangle DEF$ is $\dfrac{2}{3}$.

    b. Since $\dfrac{DF}{RT} = \dfrac{12}{8} = \dfrac{3}{2}$, the scale factor of $\triangle DEF$ to $\triangle RST$ is $\dfrac{3}{2}$.

    c. $m\angle E = m\angle S = 41°$

    d. $m\angle R = m\angle D = 57°$

    e. $m\angle T = 180° - (41° + 57°) = 82°$

    f. $RS = \dfrac{2}{3} \bullet DE$, since the scale factor of $\triangle RST$ to $\triangle DEF$ is $\dfrac{2}{3}$.

    g. Two methods of solution are shown.

Using the scale factor:	**Using proportions:**

$$RS = \tfrac{2}{3} \bullet DE$$

$$RS = \tfrac{2}{3} \bullet 18 = \tfrac{36}{3} = 12$$

$$\dfrac{RS}{DE} = \dfrac{RT}{DF}$$

$$\dfrac{RS}{18} = \dfrac{8}{12}$$

$$12 \bullet RS = 8 \bullet 18$$

$$= 144$$

$$RS = 12$$

    h. $FE = \dfrac{3}{2} \bullet TS$

$$FE = \dfrac{3}{2} \bullet 10 = 15$$

Thinking Critically

1. Are similar polygons always congruent? Why or why not?

2. Are congruent polygons always similar? Why or why not?

3. Are all rectangles similar? Why or why not?

4. Are all rhombuses similar? Why or why not?

5. Are all squares similar? Why or why not?

Class Exercises

Given: $\triangle ABC \sim \triangle XYZ$

1. What is the scale factor of $\triangle ABC$ to $\triangle XYZ$?
2. What is the scale factor of $\triangle XYZ$ to $\triangle ABC$?

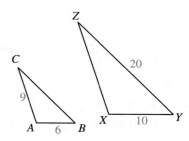

3. $\dfrac{BC}{YZ} = $ ▨
4. $BC = $ ▨ $\bullet\, YZ$
5. $BC = $ ▨

6. $\dfrac{XZ}{AC} = $ ▨
7. $XZ = $ ▨ $\bullet\, AC$
8. $XZ = $ ▨

9. If $m\angle C = 29°$, then $m\angle Z = $ ▨.

Exercises

The polygons shown are similar. Find each length or angle measure.

1. **a.** a **b.** $m\angle 1$

2. **a.** c **b.** d **c.** e **d.** $m\angle 2$

 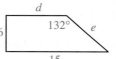

3. f

4. **a.** x **b.** y **c.** $m\angle 3$ **d.** $m\angle 4$

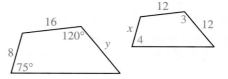

The triangles shown are similar.

5. **a.** What is the scale factor of $\triangle DEF$ to $\triangle XYZ$?

   b. $\dfrac{EF}{YZ} = $ ▨
   c. $EF = $ ▨ $\bullet\, YZ$

   d. $EF = $ ▨
   e. $\dfrac{DE}{XY} = $ ▨

   f. $DE = $ ▨ $\bullet\, XY$
   g. $XY = $ ▨

   h. If $m\angle X = 48°$, then $m\angle D = $ ▨.

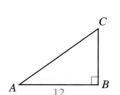

6. **a.** $\dfrac{AB}{RS} = $ ▨
   b. $\dfrac{BC}{ST} = $ ▨

   c. $BC = $ ▨ $\bullet\, ST$
   d. $BC = $ ▨

   e. $AC = $ ▨

   f. If $m\angle T = 53°$, then $m\angle A = $ ▨.

The triangles shown are similar.

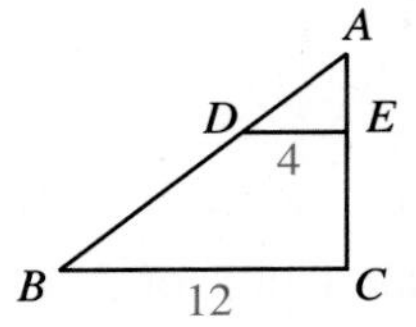

7. **a.** What is the scale factor of $\triangle ABC$ to $\triangle ADE$?
   b. If $AD = 5$, then $AB =$ ■.
   c. If $m\angle ADE = 35°$, then $m\angle ABC =$ ■.

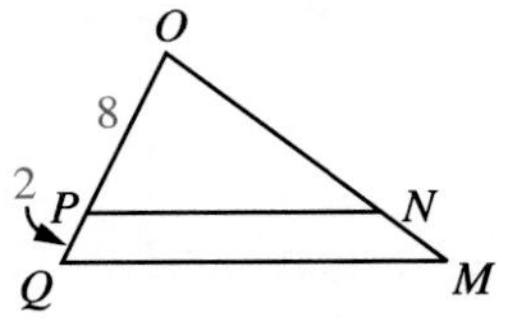

8. **a.** $OQ =$ ■
   b. What is the scale factor of $\triangle NOP$ to $\triangle MOQ$?
   c. If $NP = 12$, then $MQ =$ ■.
   d. If $m\angle M = 38°$, then $m\angle ONP =$ ■.

9. $\triangle ABC \sim \triangle PQR$ and $AB = 3 \cdot PQ$
   a. $\frac{AB}{PQ} =$ ■ **b.** $\frac{PQ}{AB} =$ ■
   c. $PQ =$ ■ $\cdot AB$
   d. If $PQ = 6$ cm, then $AB =$ ■.
   e. If $AB = 15$ cm, then $PQ =$ ■.
   f. $CA =$ ■ $\cdot RP$
   g. $\frac{CA}{RP} =$ ■
   h. If $m\angle P = 45°$, then $m\angle A =$ ■.
   i. What is the scale factor of $\triangle ABC$ to $\triangle PQR$?

10. $\triangle WLC \sim \triangle XYZ$ and $LC = \frac{2}{3} \cdot YZ$
   a. $\frac{LC}{YZ} =$ ■ **b.** $\frac{YZ}{LC} =$ ■
   c. $YZ =$ ■ $\cdot LC$
   d. If $YZ = 12$ cm, then $LC =$ ■.
   e. If $LC = 30$ cm, then $YZ =$ ■.
   f. $WL =$ ■ $\cdot XY$
   g. $\frac{WL}{XY} =$ ■
   h. If $m\angle Y = 30°$, then $m\angle L =$ ■.
   i. What is the scale factor of $\triangle WLC$ to $\triangle XYZ$?

Find the scale factor of the first polygon to the second polygon.

11. $\triangle ABC \sim \triangle PQR$, $AB = 5$ cm, and $PQ = 10$ cm
12. $\triangle ACE \sim \triangle MOT$, $AC = 6$ cm, and $MO = 2$ cm
13. $ABCD \sim JKLM$, $KL = 5$ cm, and $BC = 2$ cm
14. The lengths of the sides of the smaller of two similar triangles are 3 cm, 4 cm, and 5 cm. The shortest side of the larger triangle is 9 cm. Find the lengths of the remaining sides of the larger triangle.
15. The smaller of two similar rectangles has dimensions of 6 ft and 8 ft. If the ratio of a pair of corresponding sides is 2 to 5, find the dimensions of the larger rectangle.
16. The lengths of the sides of the larger of two similar triangles are 8 cm, 14 cm, and 18 cm. The longest side of the smaller triangle is 9 cm. Find the lengths of the remaining sides of the smaller triangle.

Draw and label a pair of polygons to fit each description. If the polygons described cannot be drawn, write "not possible."

17. rectangles that are similar

18. rectangles that are not similar

19. squares that are not similar

20. isosceles triangles that are similar

21. isosceles triangles that are not similar

22. parallelograms that are not similar

23. regular polygons that are similar

24. regular polygons that are not similar

APPLICATIONS

25. Photography A rectangular photo is 5 in. by 7 in. It is enlarged so that the longer dimension is 21 in. What is the shorter dimension of the enlargement?

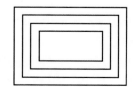

26. Perimeter and Area The rectangles shown are similar.

a. $\dfrac{AB}{EF} = $ ▪

b. $\dfrac{BC}{FG} = $ ▪

c. $\dfrac{\text{Perimeter of } ABCD}{\text{Perimeter of } EFGH} = $ ▪

d. $\dfrac{\text{Area of } ABCD}{\text{Area of } EFGH} = $ ▪

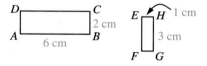

27. Computer Examine the two LOGO procedures. shown at the right. Are the triangles similar? Why or why not? If the triangles are not similar, change the procedures so that $\triangle ABC \sim \triangle XYZ$. What is the scale factor?

```
To ABC
fd 30 rt 120
fd 30 rt 120
fd 30
end
```

```
To XYZ
fd 75 rt 90
fd 75 rt 135
fd 75*sqrt 2
end
```

Computer

Packing boxes are often nested inside one another, as shown at the right. Write a LOGO procedure that will draw similar nested rectangles.

Using Definitions

The exercises in Lesson 10.4 applied the definition of similar polygons only one way—to show what is true of corresponding sides and angles if the polygons are known to be similar. To use the definition to show that two polygons are similar, you must show that there is a correspondence between the polygons such that corresponding angles are congruent *and* corresponding side lengths are proportional.

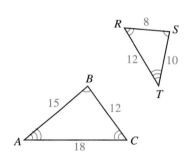

Example: **Are the triangles similar? If so, write a similarity statement.**

Solution: $\angle A \cong \angle T, \angle B \cong \angle S, \angle C \cong \angle R$

$\dfrac{AB}{TS} = \dfrac{15}{10} = \dfrac{3}{2}, \dfrac{BC}{SR} = \dfrac{12}{8} = \dfrac{3}{2},$

$\dfrac{CA}{RT} = \dfrac{18}{12} = \dfrac{3}{2},$ so $\dfrac{AB}{TS} = \dfrac{BC}{SR} = \dfrac{CA}{RT}$.

Therefore $\triangle ABC \sim \triangle TSR$.

Exercises

1. **Given:** $\triangle ABC \sim \triangle EDC$

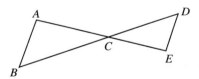

   a. Write congruence statements for three pairs of congruent angles.

   b. Complete: $\dfrac{AB}{\blacksquare} = \dfrac{BC}{\blacksquare} = \dfrac{AC}{\blacksquare}$

   c. If $AB = 8, BC = 15, AC = 12$, and $ED = 6$, find DC and CE.

Are the polygons described similar? Why or why not?

2. a rectangle 15 cm by 10 cm and a rectangle 12 cm by 8 cm

3. a rhombus with sides 12 cm long and one angle of measure 50° and a rhombus with sides 5 cm long and one angle of measure 130°

4. **a.** Explain why the triangles are similar.

   b. Write a similarity statement for the triangles.

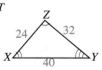

5. In $\triangle ABC$, $m\angle A = 50°$ and $m\angle B = 70°$. In $\triangle DEF$, $m\angle E = 50°$. As part of showing that the triangles are similar, what must you show about $\angle D$ and $\angle F$?

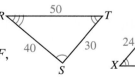

10.5 Scale Drawings

Objective: To find actual lengths represented on scale drawings.

Scale drawings are drawings or plans of objects that are either too large or too small to draw actual size on a sheet of paper. The relationship between the size of a scale drawing and the actual size of the object represented is indicated by the *scale*. The scale may be expressed in several ways, such as $\frac{1}{36}$, $\frac{1}{4}$ in. = 1 ft, or 1:48.

A scale such as $\frac{1}{36}$ means that for every unit of length on the scale drawing, there are 36 units of length on the actual object. For example, a length of 1 cm on the drawing would represent 36 cm on the actual object.

A scale such as $\frac{1}{4}$ in. = 1 ft means that every $\frac{1}{4}$ in. on the drawing represents a length of 1 ft on the actual object. This scale represents a ratio of 1:48 between lengths on the drawing and actual lengths.

EXPLORING

Measure the appropriate part of the scale drawing and use the given scale to find the following actual measurements.

1. the width of Bedroom 1

2. the length and width of Bedroom 2

3. the length and width of the patio

4. the length and width of the living room

5. the length and width of the garage

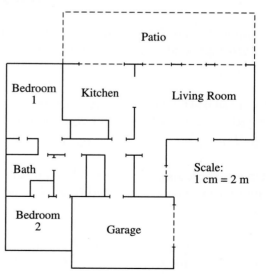

1. Using the scale 1 cm = 2 m, what would be the length on a drawing of a room that is 7 m long?

2. A scale drawing has a scale of $\frac{1}{8}$ in. = 6 ft. What would be the length on the drawing corresponding to each of the following?

   a. an actual length of 18 ft

   b. an actual length of 9 ft

Class Exercises

Suppose a scale drawing has a scale of 1 : 4. Find the actual length represented by each length on the drawing.

1. 2 cm   **2.** 2.5 cm   **3.** 6 cm   **4.** 0.5 cm

Suppose a scale drawing has a scale of $\frac{1}{4}$ in. = 6 ft. Find the actual length represented by each length on the drawing.

5. $\frac{1}{4}$ in.   **6.** $\frac{1}{2}$ in.   **7.** $\frac{5}{8}$ in.   **8.** 2 in.

Exercises

Suppose a scale drawing has a scale of $\frac{1}{2}$ in. = 8 ft. Find the actual length represented by each length on the drawing.

1. 1 in.   **2.** 3 in.   **3.** $4\frac{1}{2}$ in.   **4.** $\frac{1}{4}$ in.

Suppose a scale drawing has a scale of $\frac{1}{16}$ in. = 1 in. Find the actual length represented by each length on the drawing.

5. $\frac{1}{2}$ in.   **6.** 1 in.   **7.** 2 in.   **8.** $\frac{1}{4}$ in.

On a certain map, a distance of 50 mi is represented by $2\frac{1}{2}$ in. Find the number of miles represented by each length on the map.

9. 5 in.   **10.** 2 in.   **11.** 1 in.   **12.** 10 in.

13. On a scale drawing of a room, the dimensions are $2\frac{1}{2}$ in. by $2\frac{1}{4}$ in. Suppose the scale is $\frac{1}{4}$ in. = 2 ft. Find the following information.

    a. the actual dimensions of the room

    b. the actual area of the room in square feet

Minimum regulation sizes for various athletic fields are given. Using the indicated scale, make a scale drawing of each.

14. Soccer field: 91 m by 46 m
Scale: 1 mm = 1 m

15. Baseball diamond: 90 ft by 90 ft
Scale: $\frac{1}{4}$ in. = 10 ft

16. Basketball court: 84 ft by 50 ft
Scale: $\frac{1}{16}$ in. = 2 ft

17. Football field: 100 yd by 160 ft
Scale: $\frac{1}{4}$ in. = 20 ft

Each scale drawing is labeled with the actual dimensions being represented. Measure each drawing and find an appropriate scale.

18.
15 ft
45 ft

19.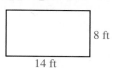
8 ft
14 ft

20.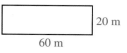
20 m
60 m

21.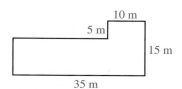
10 m
5 m
15 m
35 m

APPLICATIONS

22. Technical Drawing Choose an appropriate scale and make a scale drawing of your classroom.

23. Research Collect some examples of scale drawings from newspapers, magazines, and other sources. List the different scales used.

Calculator

To make using scales easier, you can store the scale in your calculator's memory. Find out how to use your calculator's memory.

A map has a scale of 1 cm = 2.5 km. Use a calculator to find the actual distance represented by each distance on the map.

1. 2.2 cm   **2.** 4.6 cm   **3.** 7 cm   **4.** 11 cm   **5.** 8.4 cm

6. 14.6 cm   **7.** 15 cm   **8.** 21 cm   **9.** 8.3 cm   **10.** 10.5 cm

Use the scale 1 cm = 2.5 km to find the map distance that represents each actual distance.

11. 12.5 km   **12.** 32.5 km   **13.** 15 km   **14.** 12 km   **15.** 29 km

16. 17 km   **17.** 8.75 km   **18.** 49 km   **19.** 55.5 km   **20.** 93.5 km

10.6 Problem Solving Application: Using Proportions to Estimate Distances

Objective: To solve problems by using proportions to estimate distances.

You can use proportions based on similar polygons to measure distances indirectly. For example, astronomers cannot measure directly the heights of mountains on the moon or other planets. However, they can estimate those heights by using a simple proportion based on similar triangles. One of the terms in this proportion is the distance of the mountain from the *terminator*, the line that separates the sunlit side of the moon or planet from the side facing away from the sun. The proportion is written:

$$\frac{\text{height of mountain}}{\text{length of shadow}} = \frac{\text{distance to terminator}}{\text{radius of moon}}$$

Example: A photograph of the moon shows a mountain casting a shadow that astronomers have calculated to be about 4.5 mi long. The distance of the mountain from the *terminator* is about 720 mi. The radius of the moon is 1,080 mi. How high is the mountain?

Solution: Substitute the known values in the proportion. Let h represent the unknown height.

$$\frac{h}{4.5} = \frac{720}{1,080}$$

$$1,080\,h = 720 \cdot 4.5$$

$$1,080\,h = 3,240$$

$$h = \frac{3,240}{1,080}$$

$$h = 3$$

The mountain is about 3 mi high.

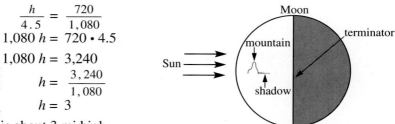

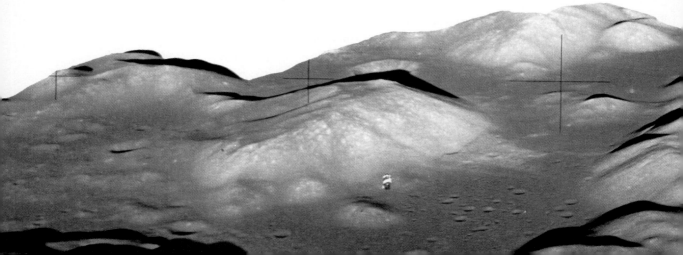

Astronomers used this method to discover that the borders of some of the huge craters on the moon are over 15,000 ft high. They also learned that some mountains exceed 30,000 ft, making them higher than Mount Everest, the highest point on Earth.

Class Exercises

1. Measurements from a photo of the moon show a mountain that casts a shadow 22.5 mi long. The mountain appears to be 204 mi from the terminator. How high is the mountain?

2. A mountain on the moon is 500 mi from the terminator. The mountain casts a shadow 2.6 mi long. How high is the mountain?

Exercises

1. Mars has a radius of 2,100 mi. Suppose a photograph of Mars shows a mountain 1,575 mi from the terminator. The shadow cast by the mountain is 5 mi long. How high is the mountain?

2. Astronomers have determined that the highest mountain on a certain planet is 24 mi high. A photograph of the planet shows that the mountain casts a shadow 32 mi long. The distance from the mountain to the terminator is 42,000 mi. What is the radius of the planet?

3. A botanist used a photograph of a redwood tree to measure the height of the tree. On the photo, the tree was $1\frac{3}{4}$ in. tall and cast a shadow $\frac{1}{2}$ in. long. When the photograph was taken, the botanist measured the actual length of the tree's shadow as 72 ft.

   a. Using the fact that the right triangle formed by the tree and its shadow is similar to the right triangle formed by the picture of the tree and its shadow on the photograph, write a proportion relating the length of the shadow and the height of the tree on the photograph to the actual length and height.

   b. Solve the proportion to find the height of the tree.

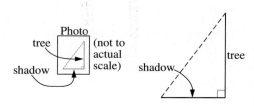

Vocabulary and Symbols

You should be able to write a brief description, draw a picture, or give an example to illustrate the meaning of each of the following terms.

Vocabulary

extremes (p. 298)
means (p. 298)
proportion (p. 298)
proportional (p. 305)
ratio (p. 295)
scale drawing (p. 311)
scale factor (p. 305)

scale of a scale drawing (p. 311)
similar figures (p. 301)
similar polygons (p. 305)
similarity statement (p. 301)

Symbols

~ (is similar to) (p. 301)
$\frac{a}{b}$ or $a : b$ (the ratio of a to b) (p. 295)

Summary

The following list indicates the major skills, facts, and results you should have mastered in this chapter.

10.1 Express ratios in simplest form. (pp. 295–297)

10.2 Solve proportions. (pp. 298–300)

10.3 Recognize figures that appear to be similar and identify corresponding parts of similar figures. (pp. 301–304)

10.4 Apply the definition of similar polygons. (pp. 305–309)

10.5 Find actual lengths represented on scale drawings. (pp. 311–313)

10.6 Solve problems by using proportions to estimate distances. (pp. 314–315)

Exercises

Express each ratio in simplest form.

1. the ratio of shaded triangles to unshaded triangles
2. the ratio of unshaded triangles to all triangles

Express each ratio in simplest form.

3. $\frac{13}{39}$

4. $\frac{12}{18}$

5. $\frac{15}{35}$

6. $25 : 10$

7. $18 : 24$

8. 12 m to 60 m

9. 2 cm to 25 mm

10. 4 in. to 2 ft

Solve each proportion.

11. $\dfrac{3}{y} = \dfrac{4}{24}$ **12.** $\dfrac{12}{18} = \dfrac{x}{6}$ **13.** $\dfrac{8}{12} = \dfrac{a}{6}$ **14.** $\dfrac{5}{12} = \dfrac{15}{z}$

15. $9 : 24 = x : 4$ **16.** $12 : 4 = 5 : n$ **17.** $c : 8 = 15 : 24$ **18.** $5 : 7 = z : 21$

Select the figure that appears to be similar to the given figure.

19. 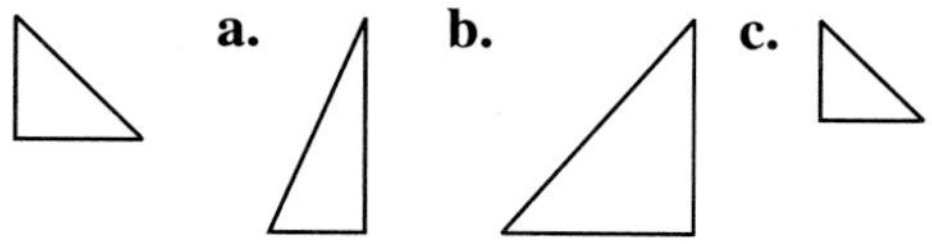   **a.** **b.** **c.**   **20.** 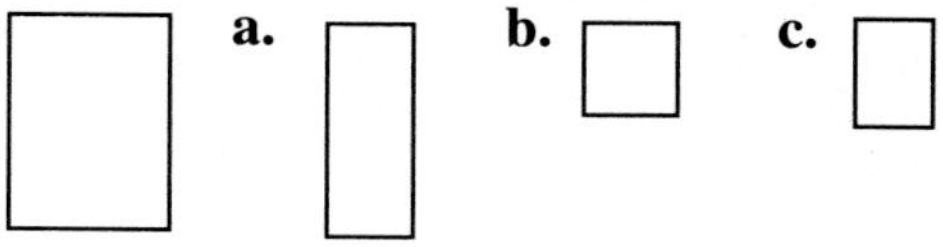 **a.** **b.** **c.**

The polygons shown are similar.

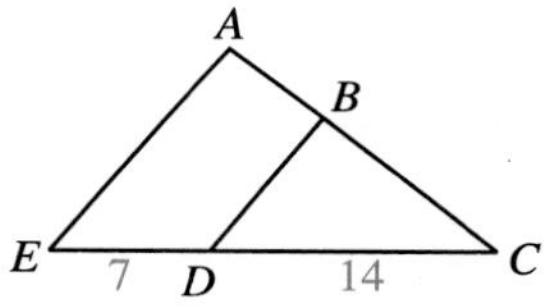

21. Name three pairs of corresponding sides.

22. Name three pairs of corresponding angles.

23. Write three similarity statements for the triangles.

24. What is the scale factor of the smaller triangle to the larger triangle?

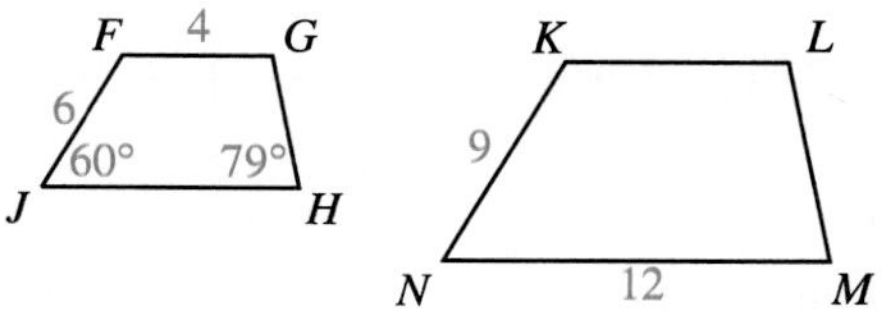

25. Name the side that corresponds to $\overline{GH}$.

26. Name the angle that corresponds to $\angle H$.

27. What is the scale factor of $FGHJ$ to $KLMN$?

28. Find $m\angle N$.

29. a. $\dfrac{JH}{NM} = $ ▦   **b.** $JH = $ ▦ $\cdot NM$ **c.** $JH = $ ▦

30. Find KL.

Suppose a scale drawing has a scale of 1 cm = 3.5 m. Find the actual length represented by each length on the drawing.

31. 2 cm   **32.** 3.5 cm   **33.** 5 cm   **34.** 0.5 cm   **35.** 4.6 cm

Suppose a scale drawing has a scale of $\frac{1}{4}$ in. = 3 ft. Find the actual length represented by each length on the drawing.

36. 1 in.   **37.** $\frac{1}{2}$ in.   **38.** $\frac{3}{4}$ in.   **39.** $1\frac{1}{2}$ in.   **40.** 3 in.

PROBLEM SOLVING

41. The moon has a radius of 1,080 mi. A photograph of the moon shows a mountain 195 mi from the terminator. The shadow cast by the mountain is 9.6 mi long. How high is the mountain?

Express each ratio in simplest form.

1. $\dfrac{20}{35}$ 2. $\dfrac{26}{10}$ 3. $16 : 20$ 4. $12 : 2$

5. $18 : 27$ 6. 16 cm to 4 cm   7. 8 in. to 3 ft   8. 2 dimes to 5 nickels

Solve each proportion.

9. $\dfrac{x}{8} = \dfrac{7}{2}$ 10. $\dfrac{36}{42} = \dfrac{6}{t}$ 11. $\dfrac{9}{6} = \dfrac{y}{4}$ 12. $\dfrac{5}{6} = \dfrac{n}{2}$

13. $3 : 4 = 5 : x$ 14. $7 : z = 12 : 18$ 15. $m : 12 = 4.5 : 8$ 16. $6 : 4 = y : 3$

17. The polygons shown are similar. Complete.

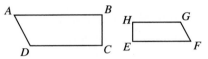

     a. $ABCD \sim$ ▨   **b.** $EFGH \sim$ ▨

     c. $BCDA \sim$ ▨   **d.** $GHEF \sim$ ▨

Given: $\triangle UVW \sim \triangle XYZ$

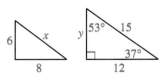

18. **a.** Name three pairs of corresponding sides.

     b. Name three pairs of corresponding angles.

19. **a.** What is the scale factor of $\triangle UVW$ to $\triangle XYZ$?

     b. What is the scale factor of $\triangle XYZ$ to $\triangle UVW$?

20. **a.** Find $m\angle Y$.   **b.** Find UV.   **c.** Find YZ.

21. Find each of the following for the similar triangles shown.

     a. x **b.** y **c.** $m\angle 1$

Suppose a scale drawing has a scale of $\frac{1}{2}$ in. $= 5$ ft. **Find the actual length represented by each length on the drawing.**

22. 1 in.   23. $\dfrac{3}{4}$ in.   24. $1\dfrac{1}{2}$ in.   25. 4 in.   26. $\dfrac{1}{4}$ in.

PROBLEM SOLVING

27. The moon has a radius of 1,080 mi. A photograph of the moon shows a mountain 354 mi from the terminator. The shadow cast by the mountain is 17.6 mi long. How high is the mountain?

Similar Triangles

As the repairer climbs to the top of this tower, he is framed by ever smaller triangles, all of which appear to be similar. The triangles are formed by beams that provide stability and support for the tower.

Focus on Skills

ARITHMETIC

Match each ratio on the left with an equal ratio on the right.

1. $\dfrac{5}{10}$

2. $\dfrac{10}{15}$

3. $\dfrac{16}{20}$

4. $\dfrac{9}{12}$

5. $\dfrac{15}{18}$

a. $\dfrac{6}{8}$

b. $\dfrac{1}{2}$

c. $\dfrac{10}{12}$

d. $\dfrac{4}{6}$

e. $\dfrac{12}{15}$

ALGEBRA

Solve.

6. $\dfrac{4}{x} = \dfrac{3}{12}$

7. $\dfrac{6}{9} = \dfrac{4}{y}$

8. $\dfrac{12}{27} = \dfrac{t}{18}$

9. $\dfrac{n}{45} = \dfrac{2}{15}$

10. $\dfrac{z}{9} = \dfrac{5}{6}$

GEOMETRY

For each pair of similar triangles (a) write congruence statements for three pairs of congruent angles and (b) write two ratios of side lengths equal to $\dfrac{RS}{RT}$.

11. $\triangle RSW \sim \triangle RTZ$

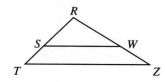

12. $\triangle RSX \sim \triangle RTY$

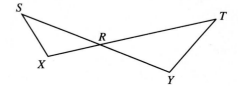

13. Draw an acute $\angle A$. Construct $\angle B \cong \angle A$.

14. Draw $\overleftrightarrow{XY}$ and point Z not on $\overleftrightarrow{XY}$. Construct the line through Z that is parallel to $\overleftrightarrow{XY}$.

15. Draw $\overline{CD}$ about 8 cm long.

   a. Construct the perpendicular bisector of $\overline{CD}$.

   b. Construct $\overline{EF}$ with length $3 \cdot CD$.

   c. Construct $\overline{RS}$ with length $\frac{1}{2} \cdot CD$.

11.1 AA Postulate

Objective: To use the AA Postulate to show that two triangles are similar.

You can use the definition of similar polygons to show that two triangles are similar. In this lesson and the next one you will develop some additional methods for determining that two triangles are similar.

EXPLORING

Part A

In the two triangles shown, $\angle C \cong \angle F$ and $\angle B \cong \angle E$.

1. Measure the sides to the nearest millimeter.
   Compare the following ratios:
   $\dfrac{FE}{CB}$, $\dfrac{DE}{AB}$, and $\dfrac{DF}{AC}$

2. Why is $\angle A \cong \angle D$?

3. Why is $\triangle DEF \sim \triangle ABC$?

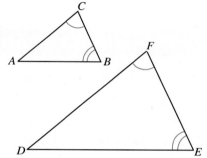

Part B

1. Draw $RS = 4$ cm and $XY = 6$ cm.

2. Draw any $\triangle XYZ$ that has side $\overline{XY}$. Construct $\angle R \cong \angle X$ and $\angle S \cong \angle Y$. Extend the sides to form $\triangle RST$.

3. Measure the sides to the nearest millimeter.
   Compare the following ratios: $\dfrac{RT}{XZ}$, $\dfrac{ST}{YZ}$, and $\dfrac{RS}{XY}$

4. Why is $\triangle RST \sim \triangle XYZ$?

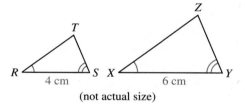

(not actual size)

The **EXPLORING** activity demonstrates the following postulate, which can be used to prove that two triangles are similar.

POSTULATE 11 (AA Postulate): If two angles of one triangle are congruent to two angles of another triangle, then the triangles are similar.

Example: Must the triangles be similar? If yes, write a similarity statement.

a.

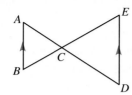

b.

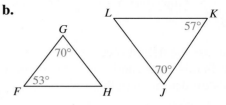

Solution: **a.** $\angle A \cong \angle D$ and $\angle B \cong \angle E$, since $\overline{AB} \parallel \overline{ED}$.
$\triangle ABC \sim \triangle DEC$ by the AA Postulate.

b. $m\angle H = 180° - (53° + 70°) = 57°$
$\triangle FGH \sim \triangle LJK$ by the AA Postulate.

Thinking Critically

Complete each statement with *always*, *sometimes*, or *never*. Explain your answer.

1. Two isosceles triangles are ▨ similar.

2. Two equilateral triangles are ▨ similar.

3. Draw two nonsimilar quadrilaterals whose corresponding angles are congruent.

4. Describe how to construct $\triangle DEF$ so that $\triangle DEF \sim \triangle ABC$ with scale factor 3 : 1.

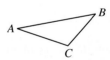

Class Exercises

1. Find each angle measure.
   a. $m\angle ACB$　　**b.** $m\angle B$　　**c.** $m\angle E$

2. $\triangle ABC \sim$ ▨

3. Find each length.
   a. CD　　　　**b.** CE

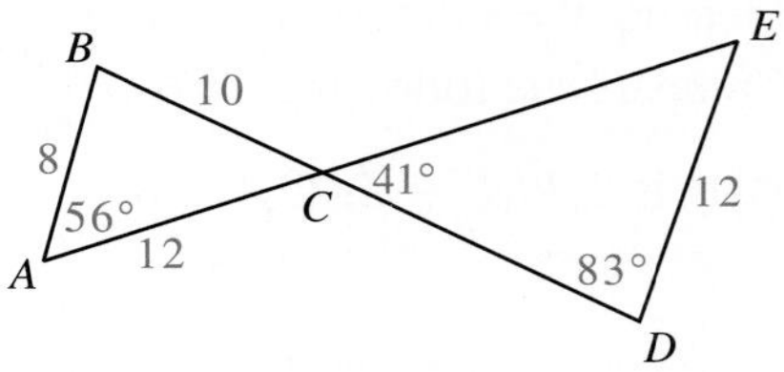

4. Name three pairs of congruent angles.

5. $\triangle RST \sim$ ▨

6. $\dfrac{RS}{▨} = \dfrac{RT}{▨} = \dfrac{▨}{▨}$

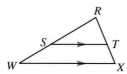

Exercises

Must the triangles be similar? If yes, write a similarity statement.

1.

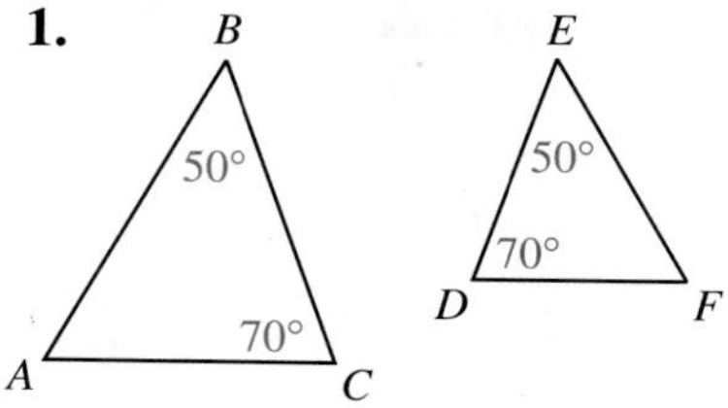

2.

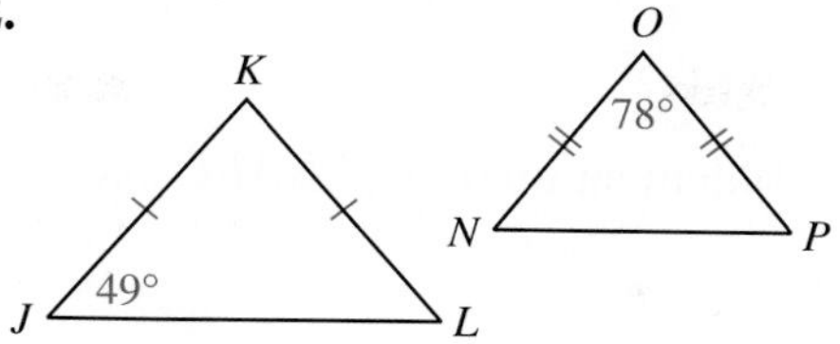

3.

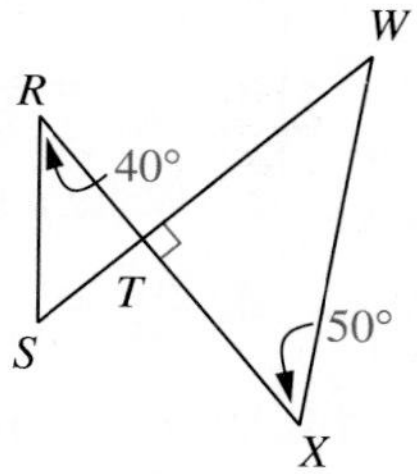

Complete.

4. a. $\triangle AXY \sim$ ■

b. $\frac{AX}{AB} = \frac{AY}{■} = \frac{XY}{■}$

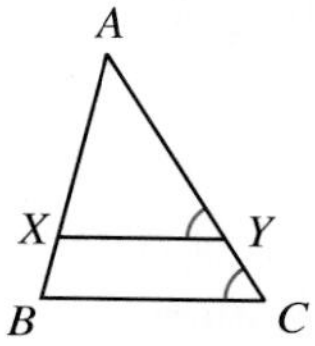

5. a. $\triangle RWZ \sim$ ■

b. $\frac{RW}{RS} = \frac{■}{ST} = \frac{RZ}{■}$

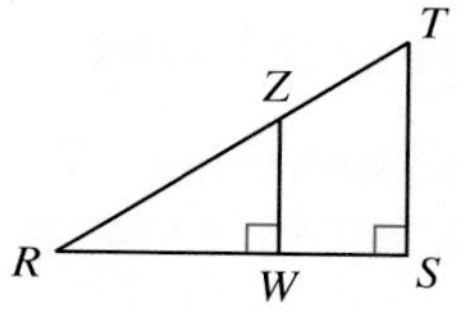

6. a. $m\angle T =$ ■

b. $\triangle RST \sim$ ■

c. $x =$ ■

d. $y =$ ■

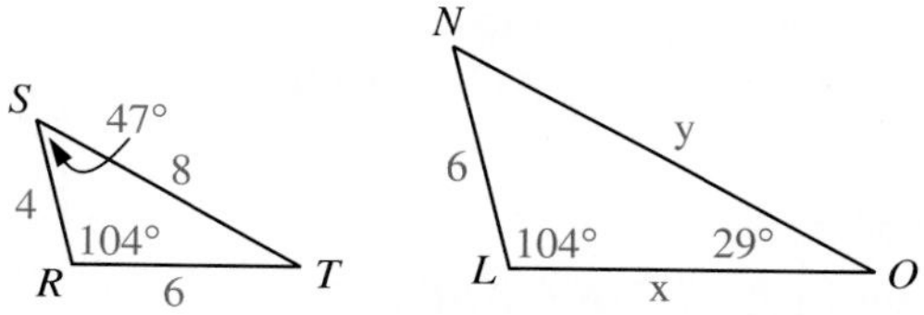

7. a. $m\angle B =$ ■

b. $\triangle ABC \sim$ ■

c. $DE =$ ■

d. $EF =$ ■

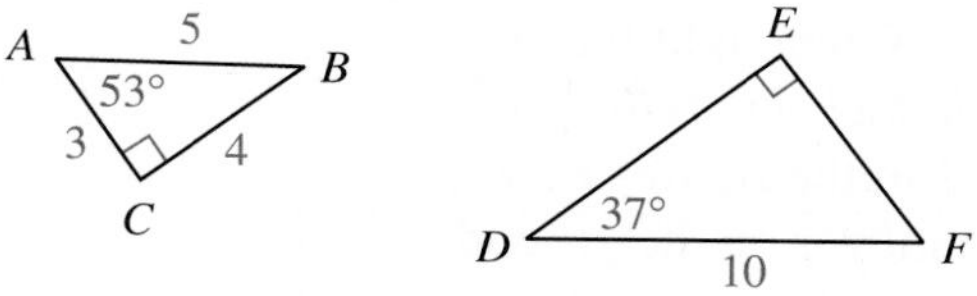

Must the parallelograms be similar?

8.

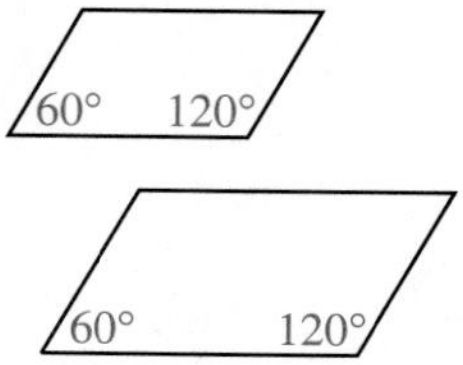

9.

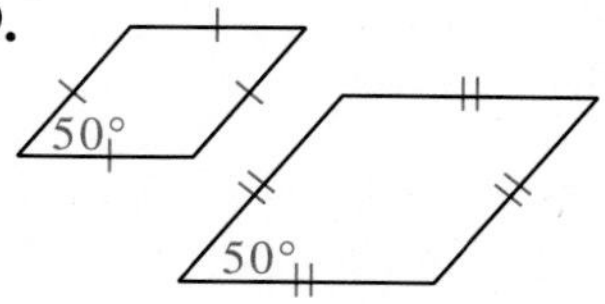

10.

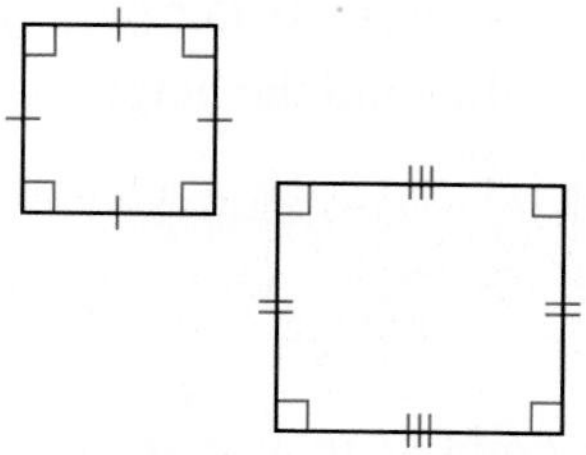

Draw a large $\triangle ABC$. Construct $\triangle DEF$ so that $\triangle DEF \sim \triangle ABC$ with each scale factor.

11. $\frac{1}{2}$

12. 2 : 1

13. $\angle ACB$ and $\angle BDC$ are right angles. Notice that $\angle B$ is an angle of both $\triangle ABC$ and $\triangle BCD$.

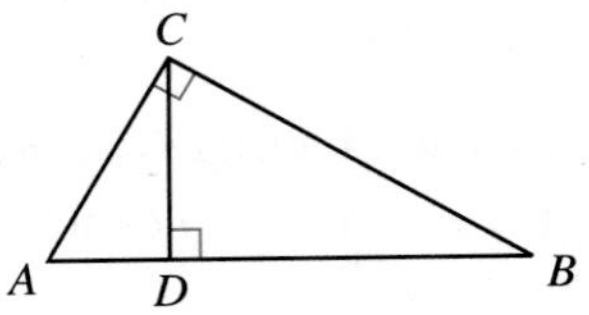

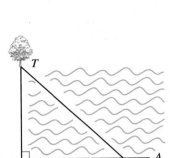

  a. Write a similarity statement involving $\triangle ABC$ and $\triangle BCD$.

  b. $\dfrac{AC}{\blacksquare} = \dfrac{CB}{\blacksquare} = \dfrac{AB}{\blacksquare}$

  c. Write a similarity statement involving $\triangle ABC$ and $\triangle ADC$.

  d. $\dfrac{AC}{\blacksquare} = \dfrac{CB}{\blacksquare} = \dfrac{AB}{\blacksquare}$

14. Write a similarity statement involving $\triangle ADC$ and $\triangle BDC$ in the figure for Exercise 13. Explain why the triangles are similar.

APPLICATIONS

15. Surveying To measure the width of a river, a surveyor used tree T as a reference point. On the other side of the river she placed stakes at points S, A, B, and C so that $\angle TSA$ and $\angle SAB$ were right angles. Measuring, she found that $SC = 200$ ft, $AC = 50$ ft, and $AB = 40$ ft.

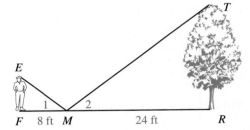

  a. Why is $\triangle STC \sim \triangle ABC$?

  b. What is the scale factor of $\triangle STC$ to $\triangle ABC$?

  c. How long is $\overline{ST}$?

16. Indirect Measurement When light hits a mirror, the angles of *incidence* and *reflection* are congruent ($\angle 1 \cong \angle 2$ in the figure at the right). Rick has placed a mirror on the ground so that the top of the tree is reflected in the mirror. Rick's eyes are 6 ft above the ground.

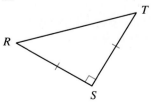

  a. Why is $\triangle EFM \sim \triangle TRM$?

  b. Find the height of the tree.

Seeing in Geometry

Complete in as many correct ways as possible: $\triangle ABC \sim \blacksquare$

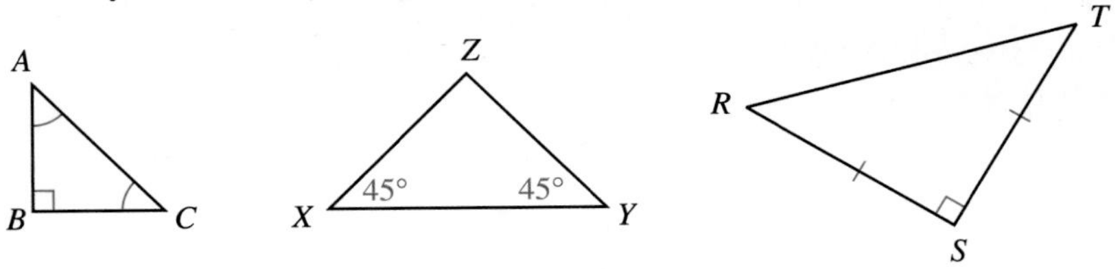

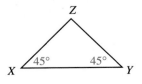

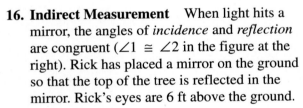

11.2 SSS and SAS Similarity Postulates

Objective: To use the SSS and SAS Similarity Postulates to show that two triangles are similar.

In this lesson you will discover two additional ways to show that two triangles are similar.

 EXPLORING

1. Use the *Geometric Supposer*: *Triangles* disk to draw a triangle. Measure the angles and side lengths. Record your results.

2. Use the *Your Own* option and Side-Side-Side to draw a triangle. Make the sides twice as long as the sides of the triangle in Step 1. (If the new triangle would be too large to fit on the screen, make the sides half as long as the sides of the triangle in Step 1 instead.)

3. Measure the angles of the triangle you drew in Step 2. Why are the triangles in Steps 1 and 2 similar?

4. Use the *Your Own* option and Side-Angle-Side to draw a triangle. Make the two sides twice as long (or half as long) as two sides of the triangle in Step 1. Make the included angle the same measure as the corresponding included angle in Step 1.

5. Measure the other two angles of the triangle you drew in Step 4. Why are the triangles in Steps 1 and 4 similar?

The **EXPLORING** activity demonstrates the following postulates.

POSTULATE 12 (SSS Similarity Postulate): **If the lengths of the corresponding sides of two triangles are proportional, then the triangles are similar.**

POSTULATE 13 (SAS Similarity Postulate): **If the lengths of two pairs of corresponding sides of two triangles are proportional and the corresponding included angles are congruent, then the triangles are similar.**

Example: Which postulate, if any, could you use to prove that the triangles are similar?

a.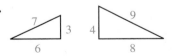

b.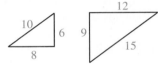

c.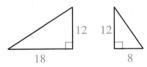

d.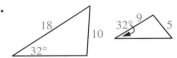

Solution: **a.** $\dfrac{18}{12} = \dfrac{3}{2}$; $\dfrac{12}{8} = \dfrac{3}{2}$
SAS Similarity Postulate

b. $\dfrac{10}{15} = \dfrac{2}{3}$; $\dfrac{6}{9} = \dfrac{2}{3}$; $\dfrac{8}{12} = \dfrac{2}{3}$
SSS Similarity Postulate

c. $\dfrac{6}{8} = \dfrac{3}{4}$; $\dfrac{7}{9} \ne \dfrac{3}{4}$
none

d. $\dfrac{18}{9} = \dfrac{2}{1}$ and $\dfrac{10}{5} = \dfrac{2}{1}$, but the *included* angles are not congruent.
none

Thinking Critically

1. Draw two nonsimilar quadrilaterals whose corresponding sides have proportional lengths.
2. Describe how to construct $EFGH$ so that $EFGH \sim ABCD$ with scale factor $2 : 1$.

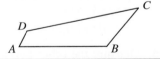

Class Exercises

Which postulate, if any, could you use to prove that the triangles are similar?

1.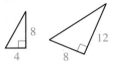

2.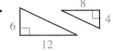

3.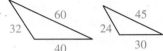

4.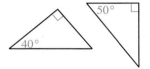

5.

6.

20. a. Write a similarity statement involving $\triangle ABD$ and $\triangle BCD$.

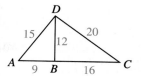

b. Which similarity postulate did you apply in part (a)?

c. Write three congruence statements involving angles of $\triangle ABD$ and $\triangle BCD$.

d. Why is $m\angle ABD = 90°$?

e. Write a similarity statement involving $\triangle ABD$ and $\triangle ACD$.

f. Which similarity postulate did you apply in part (e)? Explain.

21. $\triangle ABC \sim \triangle DEF$;
$\overline{AX}$ and $\overline{DY}$ are altitudes.

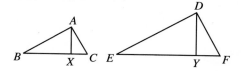

a. Why is $\triangle AXC \sim \triangle DYF$?

b. $\dfrac{AX}{DY} = \dfrac{AC}{\blacksquare}$

c. What is true of corresponding altitudes of similar triangles?

22. $\triangle ABC \sim \triangle DEF$;
$\overline{AM}$ and $\overline{DN}$ are medians.

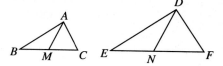

a. Why is $\triangle AMC \sim \triangle DNF$?

b. $\dfrac{AM}{DN} = \dfrac{\blacksquare}{\blacksquare}$

c. What is true of corresponding medians of similar triangles?

APPLICATION

23. Photography When light enters the pinhole P of a pinhole camera, an upside-down image is formed on the film at the opposite end of the box. For the camera shown, $PX = 20$ cm. Suppose you are photographing a segment ($\overline{AB}$) 12 cm high and you want its image ($\overline{A'B'}$) to be 3 cm high on the film.

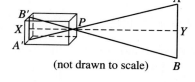

(not drawn to scale)

a. The distance from the pin-hole to the segment is PY. How is $\overline{PY}$ related to $\triangle PAB$?

c. How far should the box be placed from the segment?

b. What is $\dfrac{A'B'}{AB}$? $\dfrac{PX}{PY}$?

d. If the segment is 8 cm wide, how wide is its image?

Computer

Use the *Geometric Supposer: Triangles* disk to draw a triangle and the bisector of $\angle BAC$. Compare $BD : DC$ and $AB : AC$. Repeat for a different triangle and a different angle of the triangle. What seems to be true?

11.3 Triangles and Proportional Segments

Objective: To use the Triangle Proportionality Theorem.

Two segments are said to be *divided proportionally* when the ratios of corresponding lengths are equal.

Example 1: Since $\frac{AB}{BC} = \frac{2}{3}$ and $\frac{DE}{EF} = \frac{4}{6} = \frac{2}{3}$, $\frac{AB}{BC} = \frac{DE}{EF}$.

Therefore $\overline{AC}$ and $\overline{DF}$ are divided proportionally by points B and E.

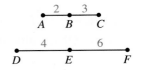

EXPLORING

1. Use the *Geometric Supposer*: *Triangles* disk to draw a large triangle. Label a random point on $\overline{AB}$.

2. Through point D draw a segment parallel to $\overline{BC}$ that intersects $\overline{AC}$.

3. Measure each of the following ratios.

   a. $\frac{AD}{AB}$ **b.** $\frac{AE}{AC}$ **c.** $\frac{ED}{CB}$

4. Why is $\triangle ADE \sim \triangle ABC$?

5. Measure and compare the following ratios.

   a. $\frac{AD}{DB}$ **b.** $\frac{AE}{EC}$

The **EXPLORING** activity demonstrates the following theorem.

THEOREM 11.1 (Triangle Proportionality Theorem): If a line is parallel to one side of a triangle and intersects the other two sides, then the triangle formed is similar to the original triangle and the sides of the original triangle are divided proportionally.

$\frac{a}{g} = \frac{c}{h} = \frac{e}{f}$ and $\frac{a}{b} = \frac{c}{d}$

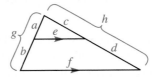

Example 2: a. Find x.   **b. Find y.**

Solution: $\angle RWZ \cong \angle RST$. Therefore $\overline{WZ} \parallel \overline{ST}$. (Why?)

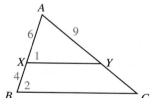

a. $\overline{RS}$ and $\overline{RT}$ are divided
proportionally.

$$\frac{x}{8} = \frac{3}{4}$$

$$4x = 24$$

$$x = 6$$

b. $\triangle RWZ \sim \triangle RST$

$$\frac{RZ}{RT} = \frac{WZ}{ST}$$

$$\frac{3}{7} = \frac{y}{14}$$

$$7y = 42$$

$$y = 6$$

The following theorem is also true. You will justify it in the Thinking
Critically questions.

THEOREM 11.2: **If a line divides two sides of a triangle
proportionally, then the line is parallel to the third side
of the triangle.**

Thinking Critically

1. **Given:** $\overline{XY}$ divides $\overline{AB}$ and $\overline{AC}$ proportionally.

   a. Do any angles of $\triangle AXY$ and $\triangle ABC$ have to
   be congruent? Explain.

   b. Find YC.

   c. Find $\dfrac{AX}{AB}$ and $\dfrac{AY}{AC}$.

   d. Why is $\triangle AXY \sim \triangle ABC$?

   e. Why is $\angle 1 \cong \angle 2$?

   f. Why is $\overline{XY} \parallel \overline{BC}$?

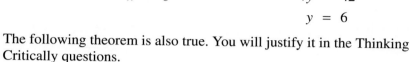

2. **a.** Find AY and AC.   **b.** Complete: $\dfrac{b}{g} = \dfrac{\blacksquare}{\blacksquare}$

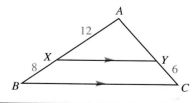

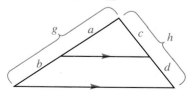

Class Exercises

Complete.

1. $\dfrac{PX}{XQ} = \dfrac{PY}{\blacksquare}$

2. $\dfrac{PX}{PQ} = \dfrac{PY}{\blacksquare}$

3. $\dfrac{XY}{QR} = \dfrac{PX}{\blacksquare}$

4. $\dfrac{PY}{PR} = \dfrac{XY}{\blacksquare}$

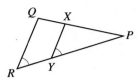

Find the value of each ratio.

5. $\dfrac{FA}{AD}$

6. $\dfrac{FB}{BE}$

7. $\dfrac{FA}{FD}$

8. $\dfrac{AB}{DE}$

9. $\dfrac{FB}{FE}$

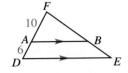

10. If $FB = 15$ and $AB = 20$, find:

    a. BE **b.** DE

Exercises

Complete.

1. $\dfrac{RA}{AS} = \dfrac{RB}{\blacksquare}$

2. $\dfrac{RA}{RS} = \dfrac{\blacksquare}{RT}$

3. $\dfrac{AB}{ST} = \dfrac{RA}{\blacksquare}$

4. $\dfrac{RB}{RT} = \dfrac{AB}{\blacksquare}$

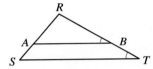

Find the value of each ratio.

5. $\dfrac{XM}{MY}$

6. $\dfrac{XN}{NZ}$

7. $\dfrac{XM}{XY}$

8. $\dfrac{XN}{XZ}$

9. $\dfrac{MN}{YZ}$

10. $\dfrac{NZ}{XZ}$

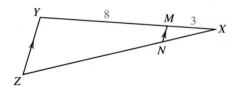

Find each length.

11. **a.** c
   b. d

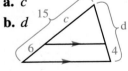

12. **a.** x
   b. y

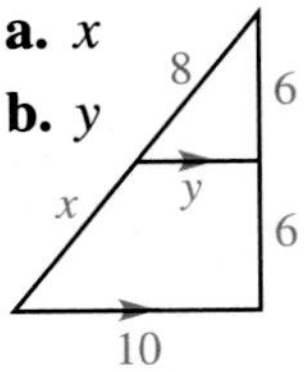

13. x

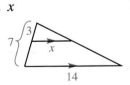

14. x

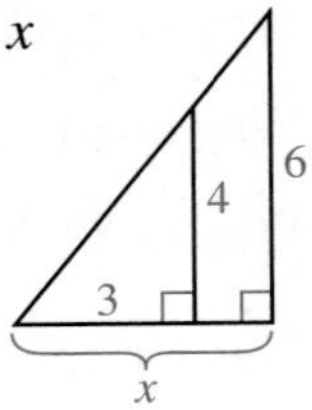

15. Find:

   a. x
   b. y

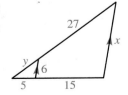

16. Find:

   a. $m\angle B$
   b. $m\angle A$
   c. EC
   d. BC

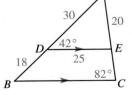

Is $\overline{DE} \parallel \overline{BC}$? Explain.

17.

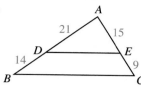

18.

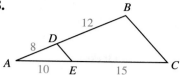

19. Given: $\overline{MN}$ is the median of trapezoid $ABCD$.
Give a reason for each statement.

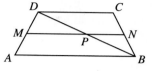

a. $\overline{MN} \parallel \overline{AB}$

b. $DM = MA$

c. $DP = PB$

d. What have you proven about the median of
a trapezoid and a diagonal of the trapezoid?

APPLICATIONS

20. Surveying Find the length of the
lake.

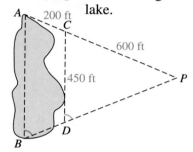

21. Indirect Measurement Maria stood
so that the tip of her shadow coincided
with the tip of the tree's shadow. Find
the height of the tree.

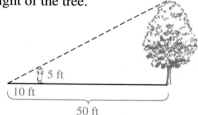

Test Yourself

**Given the information, which postulate could you use
to prove $\triangle ABC \sim \triangle RST$?**

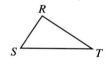

1. $\dfrac{AB}{RS} = \dfrac{BC}{ST} = \dfrac{AC}{RT}$ **2.** $\angle A \cong \angle R$ and $\dfrac{AC}{RT} = \dfrac{AB}{RS}$

**A congruence statement is given. What additional information would
enable you to use the given postulate to prove that $\triangle DEF \sim \triangle GHK$?**

3. $\angle F \cong \angle K$; SAS Similarity Postulate   **4.** $\angle D \cong \angle G$; AA Postulate

Find each length.

5. x **6.** y

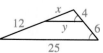

Problem Solving Application: Measuring Indirectly

Objective: To find unknown measurements using similar triangles.

In this chapter you have seen some ways that you can use similar triangles to calculate a length that is difficult to measure directly. The following Example shows how you can use the lengths of nearby shadows to find the height of a tall object.

Example: A person 1.6 m tall casts a shadow 2 m long. At the same time, a nearby tree casts a shadow 15 m long. Find the height of the tree.

Solution: Draw a diagram.

Because the sun is so far away, we can consider its rays to be parallel. Thus, $\angle CAB \cong \angle FDE$. Since $\angle ABC$ and $\angle DEF$ are both right angles, $\angle ABC \cong \angle DEF$. By the AA Postulate, $\triangle ABC \sim \triangle DEF$.

$$\frac{AB}{DE} = \frac{CB}{FE}$$

$$\frac{2}{15} = \frac{1.6}{FE}$$

$$2 \cdot FE = 15 \cdot 1.6$$

$$FE = 15 \boxtimes 1.6 \boxdiv 2 \boxminus 12$$

The tree is 12 m high.

Class Exercise

1. Mike stood so that he could just see the top of a tree over a wall. Mike's eyes (point E in the diagram at the right) are 5 ft above the ground. The wall ($\overline{WG}$) is 6 ft tall. Mike is 10 ft from the wall, and the wall is 20 ft from the tree.

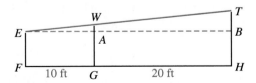

   a. Write a similarity statement involving two triangles.

   b. What do you know about lengths of sides of these two triangles?

   c. Write a proportion you can solve to find TB.

   d. Find TB and then TH, the height of the tree.

Exercises

1. When a meter stick 1 m long is held vertical to the ground, it casts a shadow 0.75 m long. At the same time, a nearby telephone pole casts a shadow 4.5 m long. How tall is the telephone pole?

2. The length of a shadow cast by a fence post 1.2 m high is 0.8 m. Find the height of a nearby building that casts a shadow 11 m long at the same time of day.

3. A ladder rests against the top of a wall. The head of a person 5 ft tall just touches the ladder. The person is 9 ft from the wall and 6 ft from the foot of the ladder. Find *DC*, the height of the wall.

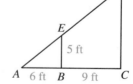

4. To find the height of a building, Kim held a meter stick vertically in front of her so that the two lines of sight past the ruler (the dashed lines) hit the top and bottom of the building. The meter stick was 0.6 m from her eyes and 4.2 m from the building. Find the height of the building. (*Note:* The ratio of the heights of similar triangles is equal to the scale factor of the triangles.)

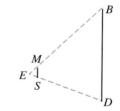

5. The diagram at the right shows a cross section of the Great Pyramid of Giza in Egypt. Point *E* represents the entrance to the pyramid. The top of the pyramid is missing now, but you can use the dimensions *AN* = 43 ft, *NB* = 333 ft, and *EN* = 55 ft to estimate the original height, *TB*, of the pyramid. Estimate the original height to the nearest foot.

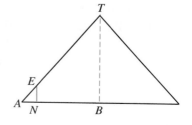

Thinking About Proof

Alternative Proofs

How would you show that the triangles at the right are similar?

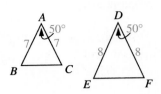

You could use either the AA Postulate or the SAS Similarity Postulate.

Using the SAS Similarity Postulate:

$\angle A \cong \angle D$

$\dfrac{AB}{DE} = \dfrac{7}{8}$ and $\dfrac{AC}{DF} = \dfrac{7}{8}$, so $\dfrac{AB}{DE} = \dfrac{AC}{DF}$

Thus $\triangle ABC \sim \triangle DEF$ by the SAS Similarity Postulate.

Using the AA Postulate:

Since $\overline{AB} \cong \overline{AC}$, $\angle B \cong \angle C$.

Since $\overline{DE} \cong \overline{DF}$, $\angle E \cong \angle F$.

$m\angle B + m\angle C + m\angle A = 180°$

$m\angle B + m\angle B + 50° = 180°$

$2m\angle B = 180° - 50° = 130°$

$m\angle B = 65°$

Similarly, $m\angle E = 65°$

Since $\angle B \cong \angle E$ and $\angle A \cong \angle D$,

$\triangle ABC \sim \triangle DEF$ by the AA Postulate.

Exercises

Name *all* the postulates you could use to prove that the triangles are similar. If more than one postulate can be used, explain.

1.

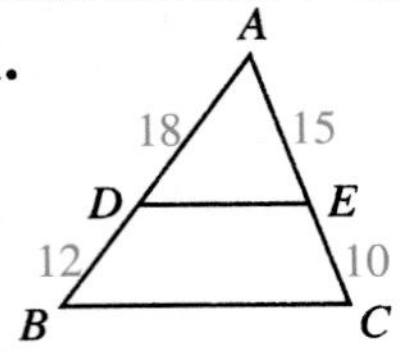

2.

3.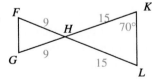

Using the congruence marks, name *all* of the following that you could use to prove that the triangles are congruent: SSS, SAS, ASA, AAS, HL. If more than one can be used, explain.

4.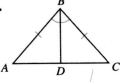

5.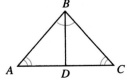

6.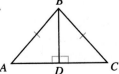

11.5 Parallel Lines and Proportional Segments

Objective: To use the fact that parallel lines divide transversals proportionally.

Suppose you wish to divide a sheet of paper into evenly spaced columns, but you do not have a ruler. What you learn in this lesson will enable you to divide the paper using a second sheet of lined paper and a straightedge.

EXPLORING

Part A

1. Draw two or three segments on a sheet of lined paper. (These are transversals for the parallel segments.)

2. Do the following for each transversal.

   a. Measure the segments intercepted (cut off) by the parallel lines.

   b. Compare the lengths of these segments.

Part B

Refer to $\triangle AEF$ and $\triangle ABF$ in the figure. Give a reason for each statement.

1. $\dfrac{AC}{CE} = \dfrac{AG}{GF}$ 2. $\dfrac{AG}{GF} = \dfrac{BD}{DF}$ 3. $\dfrac{AC}{CE} = \dfrac{BD}{DF}$

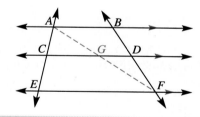

The **EXPLORING** activity demonstrates the following theorem and corollary.

THEOREM 11.3: **If three or more parallel lines intersect two or more transversals, then the segments intercepted on the transversals are proportional.**

COROLLARY: **If three or more parallel lines intercept congruent segments on one transversal, then they intercept congruent segments on every transversal.**

Example: **a.** Find x.

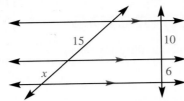

b. If $AC = 20$ and $AD = 8$, find AB, BC, and DE.

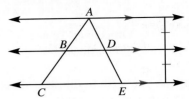

Solution: **a.** $\dfrac{15}{x} = \dfrac{10}{6}$

$10\,x = 15 \cdot 6$

$x = 9$

b. $AB = BC = 10$

$DE = 8$

The corollary to Theorem 11.3 enables you to divide a given segment into any number of congruent segments.

CONSTRUCTION 12

Divide $\overline{AB}$ into three congruent parts.

1. Draw a ray with endpoint A. Use a compass to mark off three congruent segments on this ray.

$$\overline{AC} \cong \overline{CD} \cong \overline{DE}$$

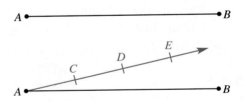

2. Draw $\overline{EB}$.

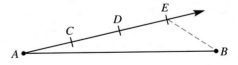

3. Construct lines through C and D that are parallel to $\overline{EB}$:

$$\overleftrightarrow{CX} \parallel \overleftrightarrow{DY} \parallel \overline{EB}.$$

Then $\overline{AX} \cong \overline{XY} \cong \overline{YB}$ by the corollary to Theorem 11.3.

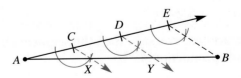

---

1. Would the steps of Construction 12 be different if the measure of ∠A were different? if the length of $\overline{AB}$ were different? if the length of $\overline{AC}$ were different?

2. How would you change the construction to divide $\overline{AB}$ into five congruent parts? into six congruent parts?

3. Suppose you wish to divide $\overline{AB}$ into two or four congruent parts. What construction could you use instead of Construction 12?

4. Given $\overline{AB}$, describe how you would construct a segment with each length.

   a. $\frac{2}{3} \cdot AB$ **b.** $\frac{4}{3} \cdot AB$ **c.** $\frac{5}{3} \cdot AB$

Class Exercises

Find the value of each ratio.

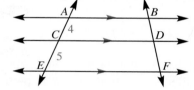

1. $\dfrac{AC}{CE}$ 2. $\dfrac{BD}{DF}$ 3. $\dfrac{AC}{AE}$

4. $\dfrac{BD}{BF}$ 5. $\dfrac{CE}{AE}$ 6. $\dfrac{DF}{BF}$

Find each length.

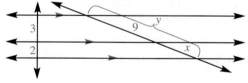

7. x

8. y

Draw a segment, $\overline{CD}$, at least 8 cm long.

9. Use Construction 12 to divide $\overline{CD}$ into four congruent parts.

10. Construct a segment with length $\frac{3}{4} \cdot CD$.

Exercises

Find the value of each ratio.

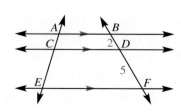

1. $\dfrac{BD}{DF}$ 2. $\dfrac{AC}{CE}$ 3. $\dfrac{BD}{BF}$

4. $\dfrac{AC}{AE}$ 5. $\dfrac{DF}{BF}$ 6. $\dfrac{CE}{AE}$

Use the given information to find each length.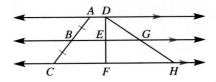

7. $AC = 10$; $BC = $

8. $DF = 8$; $EF = $ ▨

9. $DG = 7$; $GH = $ ▨

10. a. a　**b.** b　**c.** c　**d.** d　　　**11. a.** w　**b.** x　**c.** y　**d.** z

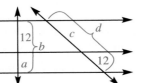 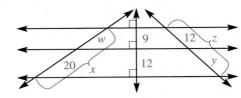

Draw a segment $\overline{XY}$ at least 9 cm long. Use Construction 12 to divide $\overline{XY}$ into the indicated number of congruent parts.

12. 3　　　　**13.** 4　　　　　**14.** 5

Draw a segment, $\overline{CD}$, at least 8 cm long. Construct a segment with each length.

15. $\frac{2}{3} \cdot CD$　　**16.** $\frac{5}{3} \cdot CD$

17. $\overline{AB}$ has been drawn on a sheet of lined paper. Why is $\overline{AB}$ divided into five congruent parts?

Draw a triangle similar to, but larger than, $\triangle ABC$. Construct $\triangle DEF \sim \triangle ABC$ with each scale factor.

18. $\frac{1}{3}$　　　**19.** $\frac{2}{3}$　　　**20.** $\frac{8}{5}$

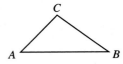

APPLICATIONS

21. Drawing Ann wanted to divide a 5 in. by 7 in. sheet of paper into evenly-spaced columns. She did not have a ruler, but she did have a sheet of lined paper with evenly-spaced lines. Using the edge of the lined paper diagonally, she put the first and eleventh lines on the edges of her paper. Then she marked the blank paper every two lines. She slid the paper down and marked the paper again. Finally, she drew lines through each pair of points.

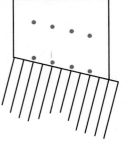

a. How many lines would Ann use to make six columns?

b. Use the same method to divide a sheet of paper into seven columns

Algebra   Find *x*, *AB*, and *BC*.

22.

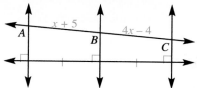

23.

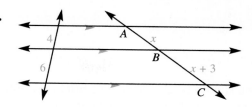

Everyday Geometry

One special rectangle, the *golden rectangle*, has fascinated mathematicians and visual artists for centuries. The ancient Greeks thought such a rectangle was most pleasing to the eye.

A **golden rectangle** is one with length and width that satisfy the equation $\frac{l}{w} = \frac{l+w}{l}$. When *l* and *w* satisfy this equation, the ratio $\frac{l}{w}$ is called the **golden ratio**. It has a value of about 1.6.

The golden ratio was known as early as the fifth century B.C. Several famous works of architecture and art, including the Greek Parthenon (shown at the right) and some of Leonardo da Vinci's paintings, appear to be framed in the golden rectangle. Present-day architects and artists have also used the golden ratio in their designs.

1. Begin with a square *ABCD*. (If necessary, construct the square.) Follow the steps below to construct a golden rectangle, *AEFD*.

   a. Locate the midpoint *M* of $\overline{AB}$ by constructing the perpendicular bisector of $\overline{AB}$.

   b. Draw an arc with *M* as center and *MC* as radius, intersecting $\overrightarrow{AB}$ at *E*.

   c. Construct the perpendicular to $\overleftrightarrow{AB}$ at *E*. Extend $\overline{DC}$ to intersect the perpendicular at *F*.

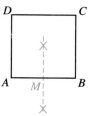

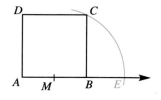

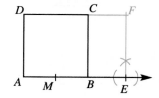

2. Research the history and uses of the golden rectangle.

Vocabulary and Symbols

You should be able to write a brief description, draw a picture, or give an example to illustrate the meaning of the following.

Vocabulary
segments divided proportionally (p. 330)

Summary

The following list indicates the major skills, facts, and results you should have mastered in this chapter.

11.1 Use the AA Postulate to show that two triangles are similar. (pp. 321–324)

11.2 Use the SSS and SAS Similarity Postulates to show that two triangles are similar. (pp. 325–329)

11.3 Use the Triangle Proportionality Theorem. (pp. 330–333)

11.4 Find unknown measurements using similar triangles. (pp. 334–335)

11.5 Use the fact that parallel lines divide transversals proportionally. (pp. 337–341)

Exercises

Which postulate, if any, could you use to show that the triangles are similar?

1.

2.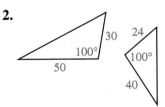

3.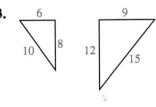

4.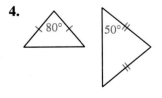

5.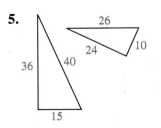

6.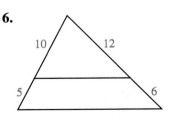

Given the information, which postulate, if any, could you use to prove that $\triangle GHK \sim \triangle LMN$?

7. $\angle G \cong \angle L$ and $\angle H \cong \angle M$

8. $\dfrac{HK}{MN} = \dfrac{GK}{LN}$ and $\angle K \cong \angle N$

9. $\dfrac{GK}{LN} = \dfrac{HG}{ML}$ and $\angle H \cong \angle M$

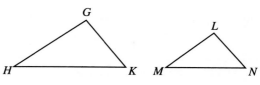

Suppose you wish to prove that $\triangle PQR \sim \triangle STU$.

10. What must you show to use the SSS Similarity Postulate?

11. If $\angle P \cong \angle S$, what must you show to use the SAS Similarity Postulate?

12. If $\angle R \cong \angle U$, what must you show to use the AA Postulate?

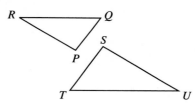

13. Complete.

   a. $\dfrac{AB}{BC} = \dfrac{AE}{\blacksquare}$

   b. $\dfrac{BE}{CD} = \dfrac{AE}{\blacksquare}$

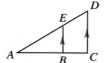

14. Find each length.

   a. x

   b. y

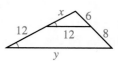

Find each length.

15. a

16. b

17. c

18. d

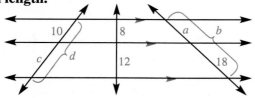

Draw a segment, $\overline{SE}$, at least 10 cm long. Use Construction 12 to divide $\overline{SE}$ into the given number of segments.

19. 3   **20.** 4   **21.** 5

PROBLEM SOLVING

22. A person 6 ft tall casts a shadow 8 ft long. Find the height of a nearby lighthouse that casts a shadow 76 ft long at the same time.

Which postulate, if any, could you use to prove that the triangles are similar?

1.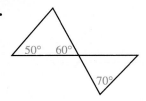

2.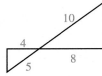

3.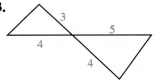

4.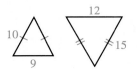

5.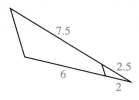

6.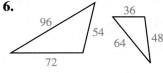

Complete.

7. If $\dfrac{AB}{ED} = \dfrac{AC}{EF} = \dfrac{BC}{DF}$, then $\triangle ABC \sim \triangle \blacksquare$ by the $\blacksquare$ Postulate.

8. If $\angle A \cong \angle E$ and $\angle B \cong \angle D$, then the triangles are similar by the $\blacksquare$ Postulate.

9. If $\angle C \cong \angle F$ and $\dfrac{\blacksquare}{\blacksquare} = \dfrac{\blacksquare}{\blacksquare}$, then the triangles are similar by the $\blacksquare$ Postulate.

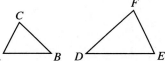

Find each length.

10. a. PR
  b. TS
  c. PS
  d. RS

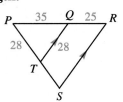

11. a. u
  b. v
  c. w
  d. x

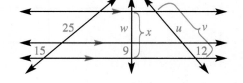

12. Draw a segment about 15 cm long. Use Construction 12 to divide the segment into 3 congruent parts.

PROBLEM SOLVING

13. Rosa is 5 ft tall. When her shadow is 3 ft long, the shadow of a nearby building is 45 ft long. How tall is the building?

Chapter 12

Square Roots and Right Triangles

The unusual shape of this solar observatory, with a vertical wing connected at the top to a wing extending out from it at an angle, creates a right triangle with the ground.

Focus on Skills

ARITHMETIC

Multiply. In Exercises 7–10 write answers in simplest form.

1. $72 \cdot 81$ **2.** $83 \cdot 36$ **3.** $17 \cdot 17$ **4.** $124 \cdot 124$ **5.** $65 \cdot 65$

6. $70 \cdot 70$ **7.** $\dfrac{3}{4} \cdot \dfrac{2}{5}$ **8.** $\dfrac{5}{13} \cdot \dfrac{13}{15}$ **9.** $\dfrac{2}{3} \cdot \dfrac{6}{9}$ **10.** $\dfrac{7}{8} \cdot \dfrac{8}{14}$

Round to the nearest tenth.

11. 8.77 **12.** 1.73205 **13.** 19.25 **14.** 21.045 **15.** 1.4142135

16. 16.449 **17.** 0.08743 **18.** 47.0075 **19.** 19.957 **20.** 25.980762

ALGEBRA

Solve.

21. $15n = 225$ **22.** $x + 16 = 25$ **23.** $\dfrac{2}{3}y = \dfrac{4}{9}$ **24.** $49 = 40 + z$

GEOMETRY

Can the given lengths be the lengths of the sides of a triangle?

25. 9 m, 41 m, 40 m   **26.** 5 cm, 5 cm, 10 cm   **27.** 2 cm, 4 cm, 8 cm

28. 2 mm, 3 mm, 4 mm   **29.** 8 m, 15 m, 12 m   **30.** 15 cm, 25 cm, 20 cm

Find the area of each region.

31.   **32.**   **33.**   **34.**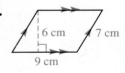

Find the measure of each indicated angle.

35. $m\angle B = $ ▓   **36.** $m\angle Z = $ ▓   **37.** $m\angle T = $ ▓

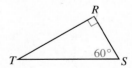

38. Name the legs and hypotenuse of $\triangle XYZ$ in Exercise 36.

39. Name the legs and hypotenuse of $\triangle RST$ in Exercise 37.

12.1 Squares and Square Roots

Objective: To find squares and square roots.

The area of a square having sides of length 5 is 25 square units. The symbol 5^2 represents $5 \cdot 5$. It is read as "five squared" or "the *square* of 5."

$$5^2 = 5 \cdot 5 = 25$$
$$\text{Area} = 25 \text{ square units}$$

The length of a side of a square with area 9 square units is 3 because $3^2 = 9$. The number 3 is the *square root* of 9.

$$\text{Area} = 9 \text{ square units}$$
$$x^2 = 9$$
$$x = 3$$

The *square root* of a number n is the number that when squared (multiplied by itself) gives n as the product. The symbol $\sqrt{}$ is read as "the square root of" and is called a *radical sign*.

Example 1: $\sqrt{36} = 6$, because $6^2 = 6 \cdot 6 = 36$.

$\sqrt{81} = 9$, because $9^2 = 9 \cdot 9 = 81$.

A number that can be written as the product of two equal factors is called a *perfect square*. For example, 36 and 81 are perfect squares.

EXPLORING

1. Make a table like the one at the right. In the first column list the whole numbers from 1 to 12. Complete the table. (Notice that each entry in the third column is the square root of the corresponding number in the second column.)

Number (n)	Square (n^2)	$\sqrt{n^2}$
1	1	1
2	4	2

2. List the first ten whole numbers that are perfect squares.

Example 2: The area of a square is 15 square units.

    a. What is the length of a side?

    b. Estimate the length of a side by finding the two consecutive whole numbers between which the length lies.

Solution: **a.** $\sqrt{15}$ **b.** Since $\sqrt{9} = 3$ and $\sqrt{16} = 4$, $3 < \sqrt{15} < 4$.

Thinking Critically

1. Complete: If n is a nonnegative number, then $\sqrt{n^2} = $ ▮.

2. In this book, "square root" will mean the *positive* square root, because we are dealing with numbers that represent lengths or distances. However, every positive number has two square roots. Name *two* square roots for each of the following.

    a. 25   **b.** 121   **c.** $\dfrac{1}{25}$ **d.** $\dfrac{1}{4}$

3. Is there any number that has exactly one square root? Explain.

Class Exercises

Find the value of each expression.

  1. square of 4   **2.** 4 squared   **3.** square root of 4   **4.** 4^2 **5.** $\sqrt{16}$

Estimate each square root by naming the two consecutive whole numbers between which the square root is located.

  6. $\sqrt{19}$ **7.** $\sqrt{63}$ **8.** $\sqrt{48}$ **9.** $\sqrt{80}$

Exercises

Find the length of a side of a square with the given area. If necessary, express the length using a radical sign.

  1. Area $= 9\ \text{cm}^2$ **2.** Area $= 36\ \text{cm}^2$ **3.** Area $= 21\ \text{cm}^2$ **4.** Area $= 33\ \text{cm}^2$

Complete.

  5. Since $32^2 = 1{,}024$, $\sqrt{1{,}024} = $ ▮.   **6.** Since $4.5^2 = 20.25$, $\sqrt{20.25} = $ ▮.

  7. State whether each number is a perfect square.

    a. 2   **b.** 4   **c.** 8   **d.** 10

    e. 25   **f.** 40   **g.** 49   **h.** 100

Find the value of each expression.

8. square of 10   **9.** square of 9   **10.** square root of 100

11. square root of 81   **12.** 15 squared   **13.** square root of 9

14. 11^2 **15.** 20^2 **16.** $\sqrt{49}$ **17.** $\sqrt{64}$ **18.** $\sqrt{144}$ **19.** $\sqrt{400}$

Estimate each square root by naming the two consecutive whole numbers between which the square root is located.

20. $\sqrt{3}$ **21.** $\sqrt{21}$ **22.** $\sqrt{30}$ **23.** $\sqrt{33}$ **24.** $\sqrt{69}$ **25.** $\sqrt{125}$

APPLICATIONS

26. Gardening   A square flower bed has an area of 64 ft². What are the dimensions of the flower bed?

27. Calculator   Many calculators have a square key [x²] or in some other way allow you to square a number without entering it twice. Find out how your calculator works. Use it to find each square.

    a. 56^2 **b.** 0.56^2 **c.** 0.056^2 **d.** $5,600^2$

    e. 123^2 **f.** 12.3^2 **g.** 1.23^2 **h.** 0.123^2

    i. 17.7^2 **j.** 2.54^2 **k.** 9.25^2 **l.** 795^2

Everyday Geometry

Illumination is the amount of light that falls on a unit area, such as 1 ft². Illumination depends on both the intensity of the source of light (watts) and the distance from the source.

The diagram shows the effect of distance on illumination. Light rays spread out as they travel away from a bulb. The same amount of light that passes through 1 ft² at a distance of 1 ft from the bulb passes through 4 ft² at 2 ft and 9 ft² at 3 ft. So, compared with the illumination at 1 ft, the illumination is $\frac{1}{4}$ as bright at 2 ft and $\frac{1}{9}$ as bright at 3 ft.

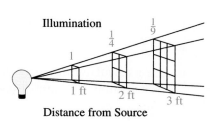

Distance from Source

The desk lamp described below has one bulb. How many bulbs of the same wattage would you need in the ceiling fixture to give the same illumination as the desk lamp?

   1. The desk lamp is 2 ft above the desk. The ceiling is 6 ft above the desk.

   2. The desk lamp is 0.5 m above the desk. The ceiling is 2 m above the desk.

12.2 Simplifying and Approximating Square Roots

Objective: To simplify square roots and to find an exact or approximate square root with the aid of a table or calculator.

The following property is used for simplifying square root expressions.

PRODUCT PROPERTY OF SQUARE ROOTS

For any nonnegative numbers a and b, it is true that $\sqrt{a} \cdot \sqrt{b} = \sqrt{a \cdot b}$.

A square root expression can be written in *simpler form* if the number under the radical sign has at least one perfect square factor other than 1. For example, $\sqrt{2}$ and $\sqrt{21}$ are in simplest form. The expressions $\sqrt{8}$ and $\sqrt{80}$ can be written in simpler form.

Example: **Simplify.**

    a. $\sqrt{8}$ **b.** $\sqrt{80}$

Solution: **a.** $\sqrt{8} = \sqrt{4 \cdot 2}$ **b.** $\sqrt{80} = \sqrt{16 \cdot 5}$

                        $= \sqrt{4} \cdot \sqrt{2}$ $= \sqrt{16} \cdot \sqrt{5}$

                        $= 2\sqrt{2}$ $= 4\sqrt{5}$

 EXPLORING

1. Use the table on page 566 to find the value of each expression.

    a. $\sqrt{27}$ **b.** $\sqrt{30}$ **c.** $\sqrt{41}$ **d.** $\sqrt{102}$

2. **a.** Use a calculator to square the answer to each part of Step 1. Compare each result to the number under the radical sign. Are they equal?

    b. Is the answer to any part of Step 1 an exact square root?

3. Use the table to find the value of each expression.

    a. 89^2 **b.** 132^2 **c.** $\sqrt{1,369}$ **d.** $\sqrt{5,776}$

4. Use the table to name the two consecutive whole numbers between which each square root is located.

    a. $\sqrt{282}$ **b.** $\sqrt{1,250}$ **c.** $\sqrt{8,000}$

Decimal approximations commonly are used in answers to applications. For example, 10.1 m is a better description of a distance or length than is $\sqrt{102}$ m. However, since the radical form represents an exact value, it is usually used when computing with square roots or to give an exact answer.

Some calculators are useful in finding approximate square roots. To find a decimal approximation for $\sqrt{12}$, enter 12 and press the $\boxed{\sqrt{}}$ key.

Thinking Critically

1. In which of the three columns (n, n^2, $\sqrt{n}$) of the table on page 566 is every number a perfect square?

2. Name the numbers from 1 through 150 that are perfect squares.

3. Use a calculator to find the square root of 1,530,169. Is 1,530,169 a perfect square?

Class Exercises

State whether each expression is in simplest form. If it isn't, simplify it.

1. $\sqrt{10}$　　**2.** $\sqrt{64}$　　**3.** $\sqrt{18}$　　**4.** $\sqrt{7}$　　**5.** $\sqrt{32}$　　**6.** $\sqrt{99}$

Find the value of each of the following.

7. 72^2　　**8.** $\sqrt{441}$　　**9.** $\sqrt{23}$　　**10.** $\sqrt{62}$　　**11.** $\sqrt{2,916}$　　**12.** $\sqrt{14,400}$

Exercises

Use the table on page 566 when needed. State whether each number is a perfect square.

1. 200　　**2.** 225　　**3.** 250　　**4.** 400　　**5.** 576　　**6.** 625

State whether each expression is in simplest form. If it isn't, simplify it.

7. $\sqrt{14}$　　**8.** $\sqrt{50}$　　**9.** $\sqrt{24}$　　**10.** $\sqrt{30}$　　**11.** $3\sqrt{12}$　　**12.** $9\sqrt{15}$

Simplify each expression.

13. $\sqrt{20}$　　**14.** $\sqrt{27}$　　**15.** $\sqrt{48}$　　**16.** $\sqrt{49}$　　**17.** $\sqrt{75}$　　**18.** $\sqrt{300}$

Find the value of each expression. Use a calculator or the table on page 566 as needed.

19. $\sqrt{289}$　　**20.** $\sqrt{67}$　　**21.** $\sqrt{86}$　　**22.** $\sqrt{99}$　　**23.** $\sqrt{121}$　　**24.** $\sqrt{400}$

Find the value of each expression. Use a calculator or the table on page 566 as needed.

25. $\sqrt{625}$ **26.** $\sqrt{2,500}$ **27.** 117^2 **28.** 54^2

29. 25^2 **30.** $\sqrt{25^2}$ **31.** $\sqrt{1,936}$ **32.** $\sqrt{3,969}$

33. $\sqrt{10,000}$ **34.** 32 squared   **35.** square of 68   **36.** square root of 68

Use a calculator to find each square root to the nearest tenth.

37. $\sqrt{12.5}$ **38.** $\sqrt{42.25}$ **39.** $\sqrt{92.16}$ **40.** $\sqrt{85.5}$

41. $\sqrt{0.25}$ **42.** $\sqrt{0.8}$ **43.** $\sqrt{175}$ **44.** $\sqrt{425}$

45. $\sqrt{750}$ **46.** $\sqrt{360}$ **47.** $\sqrt{245}$ **48.** $\sqrt{1,000}$

49. $\sqrt{1,900}$ **50.** $\sqrt{2,000}$ **51.** $\sqrt{7,000}$ **52.** $\sqrt{50,000}$

53. Find the value of each expression.

    a. $\sqrt{9}$ **b.** $\sqrt{16}$ **c.** $\sqrt{9} + \sqrt{16}$ **d.** $\sqrt{9 + 16}$

True or false?

54. $\sqrt{3} + \sqrt{5} = \sqrt{8}$ **55.** $\sqrt{\dfrac{4}{9}} = \dfrac{2}{3}$ **56.** $\sqrt{5} \cdot \sqrt{6} = \sqrt{30}$

57. $\sqrt{8} = 2\sqrt{4}$ **58.** $\sqrt{72} = 6\sqrt{12}$ **59.** $\sqrt{10} = \sqrt{5} + \sqrt{5}$

60. Show by example that $\sqrt{a + b}$ is not necessarily equal to $\sqrt{a} + \sqrt{b}$.

APPLICATIONS

61. Landscaping   State the dimensions of the square patio that can be built with 256 concrete patio stones that are 1 ft by 1 ft squares.

62. Gardening   The area of a square garden is about 40 m^2. Find the length of each side of the garden to the nearest tenth of a meter.

Everyday Geometry

When an object falls, the distance that it falls is given by the formula $d = 16t^2$, where d is the distance in feet and t is the time in seconds.

Find the distance an object falls in each given time period.

1. 2 s   **2.** 5 s   **3.** 12 s   **4.** 20 s

5. A rock that falls from the top of a cliff takes 7 s to reach the ground. How high is the cliff?

6. A rock falls from the top of a 1,600-ft cliff. How long does it take the rock to reach the ground?

Thinking Critically

1. For each right triangle write an equation relating the lengths of the sides.

   a.

   b.

2. a. You know that a triangle with sides having lengths 9 cm, 12 cm, and 15 cm is a right triangle. If the lengths of the sides of a triangle are 9 cm, 12 cm, and c cm, where $12 < c < 15$, what is true of the angle opposite the longest side? Why?

   b. If the lengths of the sides of a triangle are 9 cm, 12 cm, and c cm, where $15 < c$, what is true of the angle opposite the longest side? Why?

   c. If a, b, and c are the lengths of the sides of a triangle, with c being the length of the longest side, what is true if $a^2 + b^2 > c^2$? if $a^2 + b^2 < c^2$?

Class Exercises

Determine whether the given lengths can be the lengths of the sides of a right triangle.

1. 5 cm, 12 cm, 13 cm
2. 5 cm, 9 cm, 7 cm
3. 10 cm, 6 cm, 8 cm

Find each missing length. Simplify any square root expressions.

4.

5.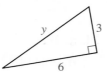

6.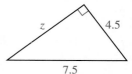

Exercises

For each right triangle write an equation that can be used to find the missing length.

1.

2.

3.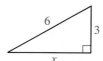

4.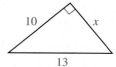

Determine whether the given lengths can be the lengths of the sides of a right triangle.

5. 5 ft, 6 ft, 7 ft   **6.** 11 in., 60 in., 61 in.   **7.** 9 cm, 40 cm, 41 cm

8. 16 cm, 63 cm, 65 cm   **9.** 8 ft, 5 ft, 6 ft   **10.** 12 cm, 37 cm, 35 cm

Find each missing length.

11.

12.

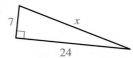

13.

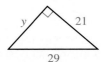

14.

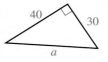

15.

16.

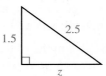

The lengths of two sides of a right triangle are given. Find the third length if c is the length of the hypotenuse. Simplify any square root expressions in each answer.

17. $a = 4, b = 2, c = $ ■   **18.** $a = 10, b = 20, c = $ ■

19. $a = 6, c = 12, b = $ ■   **20.** $b = 1, c = 2, a = $ ■

Find each of the following. Simplify any square root expressions in your answers.

21. h

22. x

23. a. h **b.** area of $ABCD$

24. a. WP
     b. area of $XYZP$
     c. area of $\triangle WXP$
     d. area of $WXYZ$
     e. perimeter of $WXYZ$

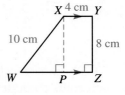

25. $ABCD$ is a rhombus, with $AB = 10$ cm and $AC = 16$ cm.
     a. AX
     b. BX
     c. area of $\triangle AXB$
     d. area of $ABCD$

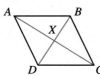

APPLICATIONS

26. **Engineering** How many meters of cable are needed to reach from point A, 15 m high on a pole, to a point B on the ground 8 m from the base of the pole?

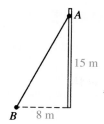

27. **Calculator** The length of the hypotenuse of a right triangle is 15.5 m. The length of one leg is 11.5 m. Use a calculator to find the length of the other leg to the nearest tenth of a meter.

28. **Sports** Softball and baseball diamonds are both in the shape of a square: 60 ft by 60 ft and 90 ft by 90 ft, respectively. To the nearest foot, find the distance from home plate to second base for each field.

29. **Area** Find the area of a triangular region if the hypotenuse is 25 m long and one leg is 20 m long.

30. **Area** The longest side of a triangle is 13 cm long. A second side is 12 cm long. The area of the triangular region is 24 cm². Is the triangle a right triangle? Explain.

31. **Home Repair** How far up on a building will a 12-foot ladder reach if the ladder's base is 4 ft from the building? Express the answer in the following three ways.

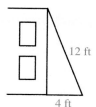

    a. as a simplified square root expression

    b. to the nearest tenth of a foot

    c. to the nearest foot

Thinking in Geometry

Three whole numbers such as 3, 4, and 5 that can be the lengths of the sides of a right triangle are called *Pythagorean triples.*

1. Two sets of Pythagorean triples are given below. For each set, multiply each number by 2, by 3, and by 4. Are these sets also Pythagorean triples?

    a. 3, 4, 5   **b.** 5, 12, 13

2. There are 50 sets of Pythagorean triples in which each number is less than 100. List as many as you can.

12.4 Products of Square Roots

Objective: To simplify products of square root expressions.

The property $\sqrt{a} \cdot \sqrt{b} = \sqrt{ab}$, used to simplify square root expressions, is also used to find the product of any two square root expressions.

Example: **a.** $\sqrt{3} \cdot \sqrt{2} = \sqrt{3 \cdot 2} = \sqrt{6}$

b. $\sqrt{3} \cdot 2\sqrt{5} = 2 \cdot \sqrt{3} \cdot \sqrt{5} = 2\sqrt{3 \cdot 5} = 2\sqrt{15}$

c. $\sqrt{3} \cdot \sqrt{6} = \sqrt{18} = \sqrt{9 \cdot 2} = \sqrt{9} \cdot \sqrt{2} = 3\sqrt{2}$

d. $3\sqrt{5} \cdot 2\sqrt{5} = 3 \cdot 2 \cdot \sqrt{5} \cdot \sqrt{5} = 6\sqrt{25} = 6 \cdot 5 = 30$

e. $5\sqrt{2} \cdot 3\sqrt{3} = 5 \cdot 3 \cdot \sqrt{2} \cdot \sqrt{3} = 15\sqrt{6}$

f. $(2\sqrt{3})^2 = 2\sqrt{3} \cdot 2\sqrt{3} = 2 \cdot 2 \cdot \sqrt{3} \cdot \sqrt{3} = 4\sqrt{9} = 4 \cdot 3 = 12$

EXPLORING

Simplify each product.

1. $(\sqrt{2})^2$ 　　　　　　　**2.** $(\sqrt{3})^2$ 　　　　　　　**3.** $(\sqrt{5})^2$

4. $(\sqrt{7})^2$ 　　　　　　　**5.** $(\sqrt{8})^2$ 　　　　　　　**6.** $(\sqrt{10})^2$

7. $2(\sqrt{5})^2$ 　　　　　　**8.** $(2\sqrt{5})^2$ 　　　　　　**9.** $5(\sqrt{2})^2$

10. $(5\sqrt{2})^2$ 　　　　　**11.** $3(\sqrt{2})^2$ 　　　　　**12.** $(3\sqrt{2})^2$

13. Complete: $(\sqrt{n})^2 = $ ▧

14. Explain why the expressions are not equal.
　　a. $6(\sqrt{5})^2$ and $(6\sqrt{5})^2$ 　　　　　**b.** $(3\sqrt{5})^2$ and $(5\sqrt{3})^2$

Thinking Critically

1. Can $\sqrt{5}$, $\sqrt{2}$, and $\sqrt{3}$ be the lengths of the sides of a right triangle?

2. Find x.

Class Exercises

Simplify each product.

1. $\sqrt{3} \cdot \sqrt{7}$ 2. $\sqrt{5} \cdot \sqrt{15}$ 3. $(\sqrt{6})^2$ 4. $(4\sqrt{3})^2$ 5. $4(\sqrt{3})^2$

Exercises

Simplify each product.

1. $\sqrt{6} \cdot \sqrt{7}$ 2. $\sqrt{7} \cdot \sqrt{5}$ 3. $\sqrt{5} \cdot \sqrt{10}$ 4. $\sqrt{2} \cdot \sqrt{10}$

5. $\sqrt{3} \cdot \sqrt{8}$ 6. $5\sqrt{6} \cdot \sqrt{2}$ 7. $5\sqrt{3} \cdot 2\sqrt{3}$ 8. $\sqrt{6} \cdot 2\sqrt{3}$

9. $(\sqrt{5})^2$ 10. $(\sqrt{3})^2$ 11. $(4\sqrt{2})^2$ 12. $4(\sqrt{2})^2$

13. $6(\sqrt{3})^2$ 14. $(6\sqrt{3})^2$ 15. $(3\sqrt{5})^2$ 16. $(7\sqrt{3})^2$

Determine whether the given lengths can be the lengths of the sides of a right triangle.

17. $\sqrt{3}, \sqrt{5}, \sqrt{8}$ 18. $\sqrt{9}, 5, 4$ 19. $\sqrt{3}, \sqrt{9}, \sqrt{4}$ 20. $\sqrt{2}, 5, \sqrt{23}$

Find each missing length. Simplify each answer.

21.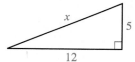

22.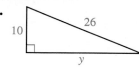

23.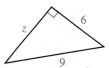

24.

25.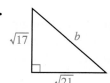

26.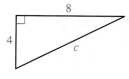

Solve. Simplify any square root expressions in your answer.

27. A diagonal of a square is 10 cm long. Find the length of a side of the square.

28. The length of a side of a square is 10 cm. Find the length of a diagonal.

29. Find x.

30. $ABCD$ is a rectangle. Find:

   a. x

   b. area of $\triangle ABD$

   c. area of $ABCD$

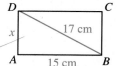

31. Find:

  a. *KL*

  b. area of $\triangle JKL$

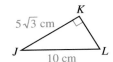
5√3 cm K
J ———— L
  10 cm

32. Find:

  a. *x*

  b. area of $\triangle DEF$

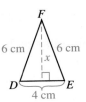
F
6 cm   6 cm
   x
D ——— E
  4 cm

33. *RHOM* is a rhombus, with $RO = 4\sqrt{3}$ cm and $MH = 4$ cm.
Find:

  a. *RX*

  b. *MX*

  c. *RM*

  d. area of $\triangle RXM$

34. For the rectangular prism find *x* and *d*.

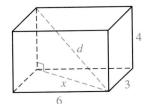

35. For the rectangular prism find *x* in terms of *a* and *b*. Then find *d* in terms of *a*, *b*, and *c*.

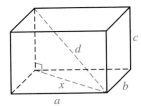

APPLICATION

36. Surveying   Stakes are placed at points *A* and *B* on opposite sides of a small pond. Another stake is placed at point *C* so that $\angle B$ is a right angle. If $BC = 128$ m and $AC = 160$ m, how long is the pond? (Find *AB*.)

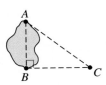

Thinking in Geometry

Find the value of *a*, *b*, *c*, *d*, and *e*.

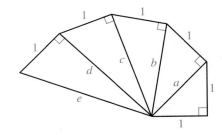

12.5 Quotients of Square Roots

Objective: To write square root expressions, including expressions with a square root in the denominator, in simplest radical form.

To simplify a fraction whose denominator contains a radical, you can use the fact that the product of a number and 1 is equal to the original number.

For example, since $\dfrac{\sqrt{2}}{\sqrt{2}}$ is equivalent to 1, $\dfrac{1}{\sqrt{2}} = \dfrac{1}{\sqrt{2}} \cdot \dfrac{\sqrt{2}}{\sqrt{2}}$.

Example: Simplify.

a. $\dfrac{\sqrt{3}}{\sqrt{2}}$

b. $\dfrac{6}{\sqrt{3}}$

Solution:

a. $\dfrac{\sqrt{3}}{\sqrt{2}} = \dfrac{\sqrt{3}}{\sqrt{2}} \cdot \dfrac{\sqrt{2}}{\sqrt{2}}$

$= \dfrac{\sqrt{3} \cdot \sqrt{2}}{\sqrt{2} \cdot \sqrt{2}}$

$= \dfrac{\sqrt{6}}{2}$

b. $\dfrac{6}{\sqrt{3}} = \dfrac{6}{\sqrt{3}} \cdot \dfrac{\sqrt{3}}{\sqrt{3}}$

$= \dfrac{6\sqrt{3}}{\sqrt{3} \cdot \sqrt{3}}$

$= \dfrac{6\sqrt{3}}{3} = 2\sqrt{3}$

EXPLORING

1. What fraction equivalent to 1 would you multiply by in order to simplify each expression?

a. $\dfrac{\sqrt{5}}{\sqrt{6}}$

b. $\dfrac{\sqrt{8}}{\sqrt{2}}$

c. $\dfrac{2\sqrt{7}}{\sqrt{3}}$

d. $\dfrac{\sqrt{13}}{\sqrt{5}}$

2. Simplify each expression in Step 1.

An expression having a square root radical is said to be in **simplest radical form** when the following is true.

1. The number under the radical has no perfect-square factors other than 1.

2. The denominator does not contain a radical, and there is no fraction under the radical sign.

When a and b are both whole numbers and $b \neq 0$, then:

$$\sqrt{\frac{a}{b}} = \frac{\sqrt{a}}{\sqrt{b}}$$

Write each of the following in simplest radical form.

1. $\sqrt{\frac{1}{2}}$ **2.** $\sqrt{\frac{3}{8}}$

Class Exercises

State whether each expression is in simplest radical form. If it isn't, write it in simplest radical form.

1. $\sqrt{30}$ **2.** $\sqrt{44}$ **3.** $\dfrac{3}{\sqrt{5}}$ **4.** $\dfrac{\sqrt{5}}{\sqrt{2}}$ **5.** $\dfrac{3\sqrt{10}}{5}$ **6.** $\sqrt{\frac{2}{3}}$

Exercises

State whether each expression is in simplest radical form. If it isn't, write it in simplest radical form.

1. $\sqrt{10}$ **2.** $\sqrt{35}$ **3.** $\sqrt{27}$ **4.** $\sqrt{98}$ **5.** $\dfrac{\sqrt{7}}{\sqrt{2}}$

6. $\dfrac{6}{\sqrt{5}}$ **7.** $\dfrac{\sqrt{5}}{3}$ **8.** $\dfrac{\sqrt{6}}{3}$ **9.** $\dfrac{6\sqrt{2}}{5}$ **10.** $\sqrt{\frac{5}{3}}$

Express in simplest radical form.

11. $\dfrac{8}{\sqrt{3}}$ **12.** $\dfrac{\sqrt{5}}{\sqrt{3}}$ **13.** $\dfrac{\sqrt{6}}{\sqrt{2}}$ **14.** $\dfrac{\sqrt{10}}{\sqrt{2}}$ **15.** $\dfrac{\sqrt{7}}{\sqrt{5}}$

16. $\dfrac{3\sqrt{5}}{\sqrt{7}}$ **17.** $\dfrac{2}{\sqrt{2}}$ **18.** $\dfrac{3}{\sqrt{3}}$ **19.** $\sqrt{\frac{5}{6}}$ **20.** $\sqrt{\frac{6}{20}}$

Solve. Express answers in simplest radical form.

21. For $\triangle ABC$ find the following.

     a. CD

     b. AC

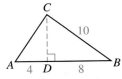

22. Find the length of the diagonal of a rectangle 10 in. wide and 20 in. long.

23. a. Find x.

  b. What are $m\angle R$ and $m\angle S$? Why?

24. In right $\triangle ABC$, hypotenuse $\overline{AB}$ is $\sqrt{5}$ times as long as $\overline{AC}$.

  a. If $AC = 2$, find AB and BC.

  b. If $AB = 2$, find AC and BC.

APPLICATION

25. Distance   A person travels 5 mi north, then 4 mi east, then 4 mi north, then 2 mi east.

  a. Draw the path on graph paper.

  b. Draw a segment from the starting point to the finish point.

  c. Use the Pythagorean Theorem to find the person's distance from the starting point. Give the distance in both simplest radical form and to the nearest tenth of a mile.

Test Yourself

Find the value of each expression. Use a calculator or the table on page 566 as needed.

  1. $\sqrt{61}$ **2.** $\sqrt{1{,}024}$ **3.** 83^2 **4.** square of 18   **5.** square root of 18

Simplify.

  6. $\sqrt{32}$ **7.** $\sqrt{63}$ **8.** $\sqrt{250}$ **9.** $\sqrt{6} \cdot \sqrt{8}$

  10. $(\sqrt{13})^2$ **11.** $7(\sqrt{3})^2$ **12.** $\dfrac{2}{\sqrt{3}}$ **13.** $\dfrac{\sqrt{2}}{\sqrt{5}}$

Determine whether the given lengths can be the lengths of a right triangle.

  14. 5 in., 6 in., $\sqrt{11}$ in.   **15.** 16 cm, 30 cm, 34 cm   **16.** 3 ft, 3 ft, $3\sqrt{2}$ ft

Find each missing length. Simplify each answer.

  17.
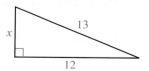
  18.
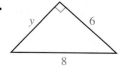

12.6 Problem Solving Application: Using the Pythagorean Theorem

Objective: To use the Pythagorean Theorem to find the shortest distance across surfaces of a room.

In a plane the shortest path between two points is a straight line. When the path between two points cannot lie in one plane, the problem is a little more complicated.

Example: The room shown at the right is 20 ft long, 8 ft wide, and 8 ft high. A spider is in a corner of the room, where two walls intersect the ceiling. A fly is in the center of the floor. Find the length of the shortest path from the spider to the fly. (Assume that the fly does not move.)

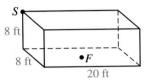

Solution: **1.** One path is down the edge where the walls meet and then across the floor. The path across the floor would be the hypotenuse of a right triangle with legs 10 ft and 4 ft long.

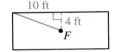

$$10^2 + 4^2 = 100 + 16$$
$$= 116$$

Use the table on page 566 or a calculator:
$$\sqrt{116} \approx 10.770$$

This path is about 8 ft + 10.77 ft, or 18.77 ft long.

2. If you unfold the room as shown at the right, $\overline{SF}$ is the path across one wall and the floor. $\overline{SF}$ is the hypotenuse of a right triangle with legs 10 ft and 12 ft long.

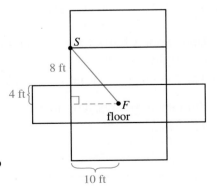

$$10^2 + 12^2 = 100 + 144$$
$$= 244$$
$$\sqrt{244} \approx 15.620$$

Thus the shortest path is about 15.6 ft long (to the nearest tenth of a foot).

Class Exercises

Find the lengths of the legs of each right triangle.

1.

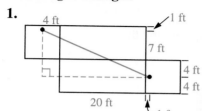

2.

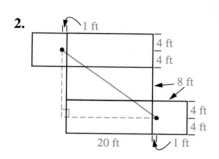

Exercises

1. The room shown at the right is 20 ft long, 8 ft wide, and 8 ft high. A spider is on one wall, 1 ft from the ceiling and equidistant (4 ft) from the adjacent walls. A fly is on the opposite wall, 1 ft from the floor and equidistant from the adjacent walls.

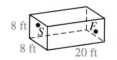

The room is unfolded in three different ways below. Use a calculator to find each distance from *S* to *F* to the nearest tenth of a foot. Which path provides the shortest distance?

a.

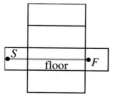

b.

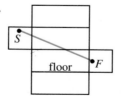

c.

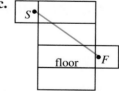

2. Repeat Exercise 1 for a room that is 30 ft long, 12 ft wide, and 12 ft high.

3. Repeat Exercise 1 for a room that is 18 ft long, 8 ft wide, and 8 ft high.

12.7 Special Right Triangles

Objective: To find the lengths of two sides of a special right triangle, given the length of the third side.

You can use the Pythagorean Theorem to find the relationships among the lengths of the sides of certain special right triangles.

================= **EXPLORING** =================

Part A

$\triangle RST$ is called either an isosceles right triangle or a 45°-45°-90° triangle.

1. If $RS = 1$, what is ST? Why? What is RT?

2. If $RS = 5$, what is ST? What is RT?

3. If $RT = 4$, what is RS? What is ST?

4. Describe the relationships among the lengths of the sides of a 45°-45°-90° triangle.

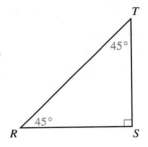

Part B

$\triangle ABC$ is called a 30°-60°-90° triangle.

1. $\triangle DBC$ is the reflection of $\triangle ABC$ over $\overline{BC}$. Describe the figure determined by the points A, B, D, and C.

2. If $AB = 2$, what is AD? AC? BC?

3. If $AB = 10$, what is AC? BC?

4. If $AC = 2$, what is AB? BC?

5. Describe the relationships among the lengths of the sides of a 30°-60°-90° triangle.

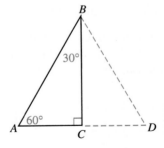

The **EXPLORING** activity leads to the following theorems.

THEOREM 12.3: In every 45°-45°-90° triangle,

a. **both legs are congruent, and**

b. **the hypotenuse is $\sqrt{2}$ times as long as each leg.**

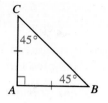

$$AC = AB$$
$$BC = \sqrt{2} \cdot AC$$
$$= \sqrt{2} \cdot AB$$

THEOREM 12.4: In every 30°-60°-90° triangle,

a. **the hypotenuse is twice as long as the side opposite the 30° angle (the shorter leg),**

b. **the side opposite the 30° angle (the shorter leg) is half as long as the hypotenuse, and**

c. **the side opposite the 60° angle (the longer leg) is $\sqrt{3}$ times as long as the side opposite the 30° angle (the shorter leg).**

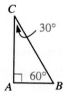

$$BC = 2 \cdot AB$$
$$AB = \frac{1}{2} \cdot BC$$
$$AC = \sqrt{3} \cdot AB$$
$$= \frac{\sqrt{3}}{2} \cdot BC$$

Example 1: a. **Find a.**

Solution: a. $a = 3\sqrt{2}$

b. **Find d.**

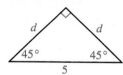

b. $5 = \sqrt{2} \cdot d$

$$d = \frac{5}{\sqrt{2}} = \frac{5}{\sqrt{2}} \cdot \frac{\sqrt{2}}{\sqrt{2}} = \frac{5\sqrt{2}}{2}$$

Example 2: a. **Find x and y.**

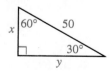

Solution: a. $x = 25$

$$y = 25\sqrt{3}$$

b. **Find r and s.**

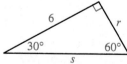

b. $6 = \sqrt{3} \cdot r$

$$r = \frac{6}{\sqrt{3}} = \frac{6}{\sqrt{3}} \cdot \frac{\sqrt{3}}{\sqrt{3}} = \frac{6\sqrt{3}}{3} = 2\sqrt{3}$$

$$s = 2 \cdot 2\sqrt{3} = 4\sqrt{3}$$

1. Could you use the Pythagorean Theorem instead of Theorems 12.3 and 12.4 to solve Examples 1 and 2?

2. $\triangle ABC$ is a 45°-45°-90° triangle with hypotenuse $\overline{AC}$. $\triangle ADC$ is the reflection of $\triangle ABC$ over $\overline{AC}$. Draw a figure and describe $ABCD$.

Class Exercises

Find the indicated information.

1. Identify the shorter leg, the longer leg, and the hypotenuse of $\triangle DEF$.

2. If $EF = 10$, then $FD = $ ■ and $DE = $ ■.

3. If $FD = 8$, then $EF = $ ■ and $DE = $ ■.

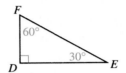

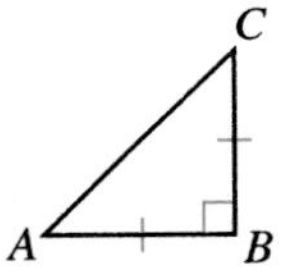

4. $m\angle A = $ ■

5. $m\angle C = $ ■

6. If $AB = 6$, then $BC = $ ■ and $AC = $ ■.

7. If $AC = 16$, then $AB = $ ■ and $BC = $ ■.

Exercises

Find each length or angle measure. Express all lengths in simplest radical form.

1. **a.** AB **b.** BC

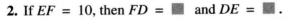

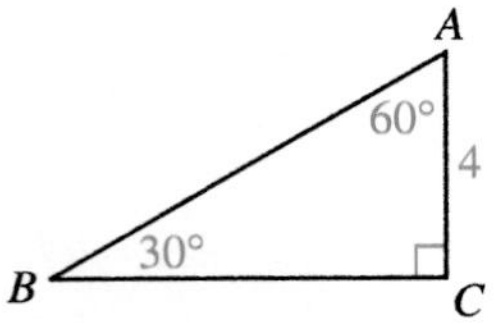

2. **a.** YZ **b.** XY

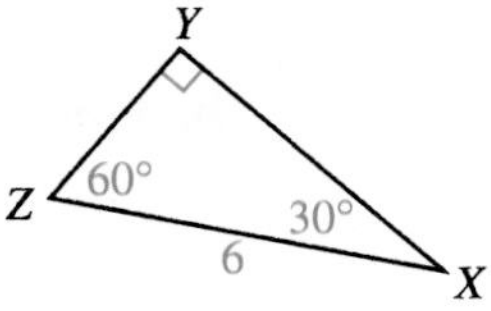

3. **a.** RX
   b. $m\angle R$
   c. $m\angle RTS$
   d. $m\angle RTX$

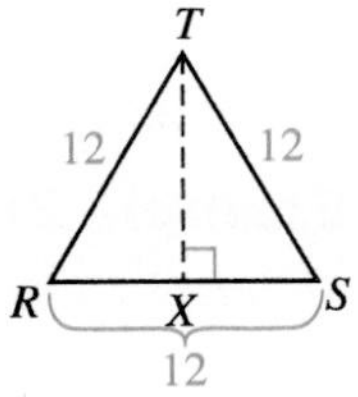

4. **a.** AC **b.** BC

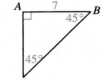

5. **a.** NO **b.** NM

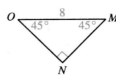

6. **a.** DE
   b. DF
   c. $m\angle D$
   d. $m\angle F$

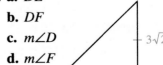

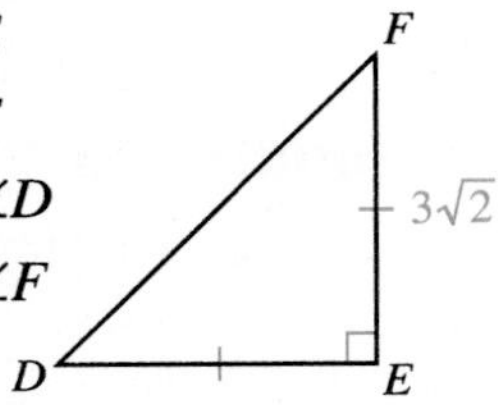

Find the indicated information.

7. a. length of a side of the square
   b. area of the square region

10 cm

8. a. *MX*   **b.** area of □*PGRM*

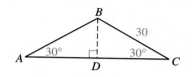

9. a. *m∠1*   **b.** *m∠2*   **c.** *TR*
   d. *RE*   **e.** area of *RECT*

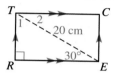

10. a. *BD*   **b.** *CD*   **c.** *AC*

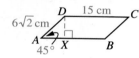

11. a. *DX*
   b. *AX*
   c. area of □*ABCD*
   d. perimeter of □*ABCD*

12. a. *AX*
   b. *DX*
   c. area of □*ABCD*
   d. perimeter of □*ABCD*

13. a. *AD*
   b. *DC*
   c. *BC*
   d. area of △*ABC*

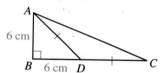

APPLICATIONS

14. Sports Ski-lift cables are strung at an angle of 30° to the top of a 5,000-ft mountain. How long are the cables?

15. Calculator Use a calculator to find each of the following to the nearest tenth.
   a. $\sqrt{2}$ **b.** $\sqrt{3}$ **c.** $5\sqrt{2}$ **d.** $5\sqrt{3}$ **e.** $50\sqrt{2}$

Computer

Use LOGO to enter the procedure at the right. Run the procedure for several values of *s* by typing, for example, SQUARE 40. Edit the procedure so that it draws a square and both diagonals.

```
To square :s
repeat 4 [fd :s rt 90]
rt 45 fd :s * sqrt 2
end
```

12.8 Similar Right Triangles

Objective: To use the relationships involving the altitude to the hypotenuse of a right triangle.

If r, s, and t are positive numbers such that $\frac{r}{s} = \frac{s}{t}$, then s is called the **geometric mean** of r and t.

Example 1: **Find the geometric mean of the numbers.**

        a. 16 and 9   **b.** 2 and 10

Solution: **a.** $\dfrac{16}{x} = \dfrac{x}{9}$ **b.** $\dfrac{2}{x} = \dfrac{x}{10}$

$$x^2 = 144 \qquad\qquad\qquad x^2 = 20$$

$$x = \sqrt{144} \qquad\qquad\quad x = \sqrt{20}$$

$$= 12 \qquad\qquad\qquad\quad = \sqrt{4 \cdot 5}$$

$$= \sqrt{4} \cdot \sqrt{5} = 2\sqrt{5}$$

EXPLORING

1. Use the *Geometric Supposer: Triangles* disk to create a right triangle. Draw the altitude from vertex A. Measure the angles of the three triangles shown.

2. Use your angle measures from Step 1 to write a similarity statement for each pair of triangles.

3. One of your similarity statements in Step 2 should involve two triangles that have $\overline{AD}$ as a side. Use that statement to complete the following proportion.

$$\frac{\blacksquare}{AD} = \frac{AD}{\blacksquare}$$

4. Use two of your similarity statements from Step 2 to complete the following proportions.

$$\frac{\blacksquare}{AC} = \frac{AC}{\blacksquare} \qquad\qquad\qquad \frac{\blacksquare}{AB} = \frac{AB}{\blacksquare}$$

The **EXPLORING** activity demonstrates the following theorem and corollaries.

THEOREM 12.5: The altitude to the hypotenuse of a right triangle forms two triangles that are similar to the original triangle and to each other.

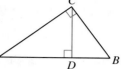

$$\triangle ADC \sim \triangle ACB \sim \triangle CDB$$

COROLLARY 1: The length of the altitude drawn to the hypotenuse of a right triangle is the geometric mean between the lengths of the segments of the hypotenuse.

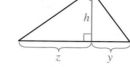

$$\frac{z}{h} = \frac{h}{y}$$

COROLLARY 2: When the altitude is drawn to the hypotenuse of a right triangle, the length of each leg is the geometric mean between the length of the hypotenuse and the length of the adjacent segment of the hypotenuse.

$$\frac{c}{b} = \frac{b}{z} \quad \text{and} \quad \frac{c}{a} = \frac{a}{y}$$

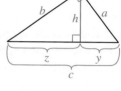

Example 2:   **a.** Find h.   **b.** Find d.

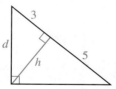

Solution: **a.** $\dfrac{3}{h} = \dfrac{h}{5}$ **b.** $\dfrac{8}{d} = \dfrac{d}{3}$

$$h^2 = 3 \cdot 5 \qquad\qquad d^2 = 8 \cdot 3$$

$$h = \sqrt{15} \qquad\qquad d = \sqrt{24}$$

$$= \sqrt{4 \cdot 6}$$

$$= \sqrt{4} \cdot \sqrt{6}$$

$$= 2\sqrt{6}$$

Thinking Critically

1. 10 is the geometric mean of 5 and what number?

2. For which of r, s, and t can you find the value? Explain.

Class Exercises

Find the geometric mean of the numbers.

1. 2 and 8   **2.** 5 and 7   **3.** 4 and 7   **4.** 5 and 10

Find each length.

5. x

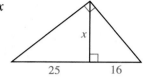

6. a. r
   b. s

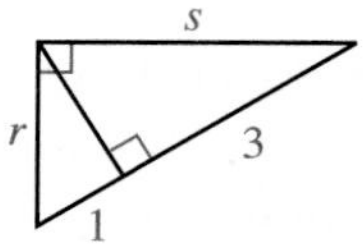

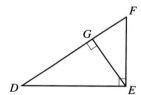

7. a. h
   b. a
   c. b

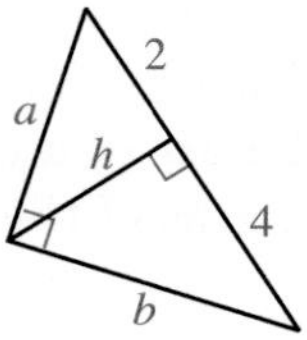

Exercises

Find the geometric mean of the numbers.

1. 4 and 16   **2.** 6 and 5

3. 2 and 20   **4.** 15 and 10

Complete.

5. $\triangle RSW \sim$ ▦ $\sim$ ▦   **6.** $\triangle DEG \sim$ ▦ $\sim$ ▦

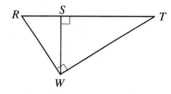

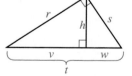

Find each length.

7. $v = 9, w = 4, h =$ ▦   **8.** $v = 8, w = 5, h =$ ▦

9. $v = 11, w = 5, s =$ ▦   **10.** $v = 3, w = 6, r =$ ▦

11. $r = 12, v = 9, t =$ ▦   **12.** $s = \sqrt{6}, w = \sqrt{2}, t =$ ▦

Given the following values for w and v, find lengths r, h, and s.

13. $v = 16, w = 9$

14. $v = 6, w = 3$

Find all other lengths for the figure at the right.

15. $b = 7, c = 9$ **16.** $a = 2, c = 11$

17. $a = 2, c = 4$ **18.** $c = 12, d = 6$

19. $a = 3, h = \sqrt{39}$ **20.** $a = \sqrt{3}, d = 2\sqrt{3}$

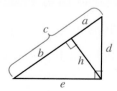

21. The lengths of the sides of a right triangle are 3 cm, 4 cm, and 5 cm. Find the length of the altitude to the hypotenuse.

22. You can use algebra to prove the Pythagorean Theorem.

Given: a right triangle with the lengths shown
Prove: $c^2 = a^2 + b^2$

Complete. Give a reason for each statement.

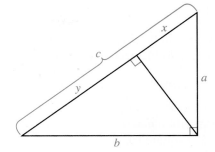

a. $\dfrac{c}{\blacksquare} = \dfrac{\blacksquare}{x}$ and $\dfrac{c}{\blacksquare} = \dfrac{\blacksquare}{y}$

b. $cx = \blacksquare$ and $cy = \blacksquare$

c. $cx + cy = \blacksquare + \blacksquare$

d. $\blacksquare(\blacksquare + \blacksquare) = \blacksquare + \blacksquare$

e. $\blacksquare \cdot \blacksquare = \blacksquare + \blacksquare$

APPLICATIONS

23. Computer Use LOGO to enter the procedure at the right. Then run the procedure for the values of *a* and *b* shown below by typing, for example, SPECIAL.TRI 80 45. Sketch each result.

a. 80 45   **b.** 45 80

c. 20 40   **d.** 50 32

```
To special.tri :a :b
bk sqrt :a * :b
lt 90 fd :a home
bk sqrt :a * :b
rt 90 fd :b home
end
```

24. Computer What theorem or corollary of this lesson does the procedure in Exercise 23 use?

25. Computer Write a procedure that draws the triangle shown at the right.

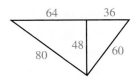

Seeing in Geometry

Some of the right triangles that have $\overline{AB}$ as the hypotenuse are shown at the right.

1. How many right triangles do you think have hypotenuse $\overline{AB}$?

2. The set of all points that satisfy a certain condition is called a *locus*. Describe the locus of all points *C* such that $\triangle ABC$ is a right triangle.

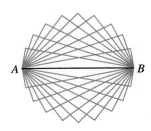

Vocabulary and Symbols

You should be able to write a brief description, draw a picture, or give an example to illustrate the meaning of each of the following terms.

Vocabulary

45°-45°-90° triangle (p. 366)
geometric mean (p. 370)
perfect square (p. 347)
Pythagorean triple (p. 357)
radical sign (p. 347)

square (p. 347)
square root (p. 347)
simplest radical form (p. 361)
30°-60°-90° triangle (p. 366)

Symbol

$\sqrt{}$ (the square root of) (p. 347)

Summary

The following list indicates the major skills, facts, and results you should have mastered in this chapter.

12.1 Find squares and square roots. (pp. 347–349)

12.2 Simplify square roots and find an exact or approximate square root with the aid of a table or calculator. (pp. 350–352)

12.3 Use the Pythagorean Theorem and its converse. (pp. 353–357)

12.4 Simplify products of radicals. (pp. 358–360)

12.5 Write square root expressions, including expressions with a square root in the denominator, in simplest radical form. (pp. 361–363)

12.6 Use the Pythagorean Theorem to find the shortest distance across surfaces of a room. (pp. 364–365)

12.7 Find the lengths of two sides of a special right triangle, given the length of the third side. (pp. 366–369)

12.8 Use the relationships involving the altitude to the hypotenuse of a right triangle. (pp. 370–373)

Exercises

Find the value of each expression.

1. 12 squared   **2.** square root of 16   **3.** square of 13

4. $\sqrt{169}$ **5.** 8^2 **6.** $\sqrt{400}$ **7.** 25^2 **8.** $\sqrt{25}$ **9.** $\sqrt{625}$

10. Name five numbers that are perfect squares.

Simplify each expression

11. $\sqrt{75}$ **12.** $\sqrt{96}$ **13.** $\sqrt{200}$ **14.** $\sqrt{108}$

Find the value of each expression. Use a calculator or the table on page 566 as needed.

15. $\sqrt{6,561}$ **16.** 73^2 **17.** 133^2 **18.** $\sqrt{7}$

Determine whether the given lengths can be the lengths of the sides of a right triangle.

19. 10 m, 26 m, 24 m   **20.** 0.3 km, 0.4 km, 0.5 km   **21.** 4, $2\sqrt{5}$, 6

Find each missing length. Simplify each answer.

22.   **23.** 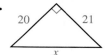   **24.**

25. 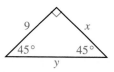   **26.** 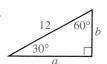   **27.**

Express in simplest radical form.

28. $\sqrt{6} \cdot \sqrt{2}$ **29.** $\sqrt{3} \cdot \sqrt{12}$ **30.** $(\sqrt{17})^2$ **31.** $(5\sqrt{8})^2$

32. $\dfrac{5}{\sqrt{2}}$ **33.** $\dfrac{3\sqrt{2}}{\sqrt{3}}$ **34.** $\sqrt{\dfrac{4}{5}}$ **35.** $\dfrac{\sqrt{24}}{\sqrt{6}}$

36. Find the geometric mean of 2 and 50.

Find each length.

37. 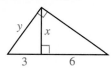   **38.**

PROBLEM SOLVING

39. Point A is on one wall of a room, 2 ft from the ceiling and 6 ft from an adjacent wall. Point B is on the adjacent wall, 7 ft from the ceiling and 5 ft from the intersection of the walls. Find the length of the shortest path from point A to point B.

Find the length of a side of a square with the given area. If necessary, express the length using a radical sign.

1. Area = 144 m²   **2.** Area = 25 cm²   **3.** Area = 47 cm²

Find the value of each expression. Use a calculator or the table on page 566 as needed.

4. square of 71   **5.** square root of 8   **6.** 31 squared   **7.** $\sqrt{196}$

8. 103^2 **9.** $\sqrt{48}$ **10.** $\sqrt{74^2}$ **11.** $\sqrt{1,225}$

Determine whether the given lengths can be the lengths of the sides of a right triangle.

12. 5 cm, 6 cm, 11 cm   **13.** 5 ft, 13 ft, 12 ft   **14.** 2, 3, $\sqrt{13}$

Find each missing length. Simplify each answer.

15.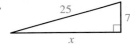

16.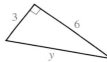

17.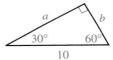

18.

19.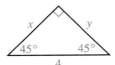

20.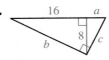

Express in simplest radical form.

21. $\sqrt{28}$ **22.** $\sqrt{7} \cdot \sqrt{14}$ **23.** $\sqrt{3} \cdot \sqrt{20}$ **24.** $4(\sqrt{5})^2$

25. $(4\sqrt{5})^2$ **26.** $\dfrac{2}{\sqrt{2}}$ **27.** $\dfrac{3\sqrt{3}}{\sqrt{5}}$ **28.** $\sqrt{\dfrac{3}{4}}$

29. area = ▓

PROBLEM SOLVING

30. The room shown at the right is 16 ft long, 12 ft wide, and 9 ft high. Point *P* is in a corner of the room, where two walls intersect the ceiling. Find the length of the shortest path from *P* to point *C*, the center of the floor.

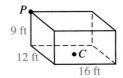

Select the best response.

1. Judging by appearance, which is *not* a correct name for the figure?

   a. rectangle

   b. square

   c. rhombus

   d. trapezoid

2. Which of the following reverse orientation?

   I. reflection
   II. translation
   III. rotation

   a. I only   **b.** I and II

   c. I and III   **d.** I, II, and III

3. An isosceles triangle has:

   a. no congruent angles

   b. exactly two congruent angles

   c. at least two congruent angles

   d. three congruent angles

4. Which of the following will *always* intersect at a point inside a triangle?

   a. the altitudes

   b. the angle bisectors

   c. the perpendicular bisectors of the sides

   d. none of the above

5. Which postulate or theorem could you use to prove that $\triangle ABC \cong \triangle DEF$ if you know that $\overline{AC} \cong \overline{DF}, \overline{AB} \cong \overline{DE},$ and $\angle B \cong \angle E$?

   a. SSS   **b.** SAS

   c. ASA   **d.** none of the above

6. In $\triangle ARC$, $m\angle A = 68°$ and $m\angle R = 45°$. Find $m\angle C$.

   a. 113°   **b.** 67°   **c.** 112°   **d.** 23°

7. Find $m\angle 1$ in the figure.

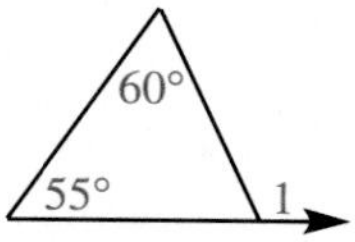

   a. 115°   **b.** 125°

   c. 120°   **d.** 65°

8. Find the sum of the measures of the interior angles of a 22-sided polygon.

   a. 3,600° **b.** 3,960° **c.** 7,200° **d.** 7,920°

9. *RHMB* is a rhombus with diagonals intersecting at *O*. Choose the statement that is not necessarily true.

   a. $\overline{RO} \cong \overline{OM}$

   b. $\overline{RM} \perp \overline{BH}$

   c. $\overrightarrow{RM}$ is an angle bisector.

   d. $\overline{RM} \cong \overline{BH}$

10. *RECT* is a rectangle. Choose the statement that must be true.

    a. $\overline{RC} \cong \overline{ET}$ **b.** $\overline{RC} \perp \overline{ET}$

    c. $\overline{RE} \cong \overline{EC}$ **d.** $\angle TRC \cong \angle CRE$

11. *WXYZ* is a trapezoid with bases $\overline{WX}$ and $\overline{ZY}$. If $WX = 24$ and $ZY = 36$, find the length of the median of *WXYZ*.

    a. 60   **b.** 30   **c.** 12   **d.** 6

12. Which of the following would allow you to conclude that *READ* is a parallelogram?

    a. $\overline{RA} \perp \overline{ED}$

    b. $\overline{RA} \cong \overline{ED}$

    c. $\overline{RE} \cong \overline{RD}$ and $\overline{EA} \cong \overline{DA}$

    d. $\angle R \cong \angle A$ and $\angle E \cong \angle D$

13. Find the area of the parallelogram.

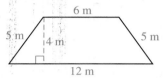

    a. 40 cm²

    b. 32 cm²

    c. 26 cm²

    d. 20 cm²

14. Find the area of the trapezoid.

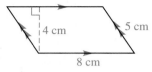

    a. 36 m²

    b. 45 m²

    c. 48 m²

    d. 28 m²

15. Find the exact area of a circle with diameter 14 ft.

    a. 14π ft²   **b.** 196π ft²

    c. 7π ft²   **d.** 49π ft²

16. A rectangle is 25 cm wide and 1.25 m long. Find the ratio of the width to the length.

    a. 20 : 1   **b.** 1 : 20

    c. 1 : 5   **d.** 5 : 1

17. $\triangle ABC \sim \triangle DEF$ with scale factor 2 : 1. Choose the true statement.

    a. $m\angle A = 2 \cdot m\angle D$

    b. $DE = 2 \cdot AB$

    c. $m\angle E = 2 \cdot m\angle B$

    d. $BC = 2 \cdot EF$

18. A scale drawing has scale 1 cm = 5 m. Find the actual length represented by 2.5 cm on the drawing.

    a. 6.5 m   **b.** 0.5 m

    c. 12.5 m   **d.** 7.5 m

19. Find z.

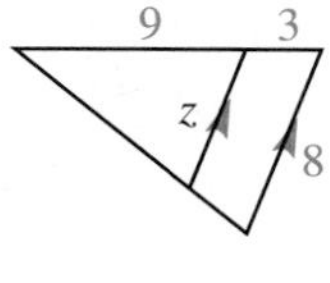

    a. 24   **b.** $13\frac{1}{2}$

    c. 6   **d.** $2\frac{2}{3}$

Use the diagram for Exercises 20 and 21.

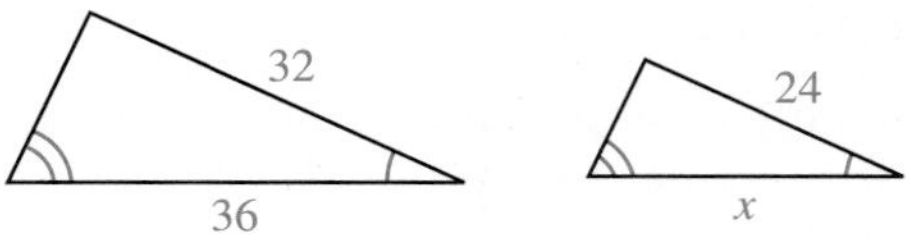

20. Which of the following could be used to prove that the triangles are similar?

    a. AA Postulate

    b. ASA Postulate

    c. SAS Similarity Postulate

    d. SSS Similarity Postulate

21. Find x in the diagram above.

    a. 27   **b.** 32   **c.** 28   **d.** 48

22. A right triangle has one leg 20 cm long and hypotenuse 29 cm long. Find the length of the other leg.

    a. 3 cm   **b.** 9 cm

    c. 21 cm   **d.** 441 cm

23. Find y.

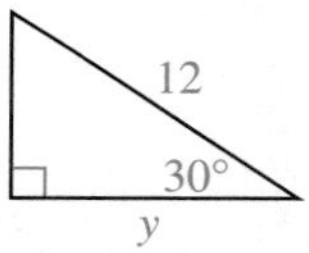

    a. 6

    b. $6\sqrt{3}$

    c. $6\sqrt{2}$

    d. $12\sqrt{2}$

24. Find t.

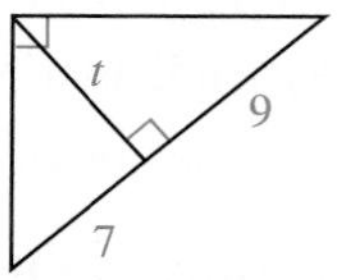

    a. $7\sqrt{3}$

    b. 4

    c. $3\sqrt{7}$

    d. 63

25. Draw a line j and a point Q not on j. Construct the line through Q parallel to j.

26. Draw a segment at least 10 cm long. Use Construction 12 to divide the segment into 5 congruent parts.

Chapter 13

Circles

At carnivals around the world people riding on ferris wheels get a bird's eye view of the fairgrounds. The invention of the wheel, essential to transportation and industry, was critical to the development of modern civilization.

Focus on Skills

Multiply.

1. $\frac{1}{2} \cdot 182$

2. $\frac{1}{2} \cdot 47$

3. $\frac{1}{2}(58 + 112)$

4. $\frac{1}{2}(220 - 140)$

GEOMETRY

Draw each figure described. Construct the indicated line or ray.

5. $\overline{AB}$; the perpendicular bisector of $\overline{AB}$

6. line m and Q on m; the perpendicular to m at Q

7. any acute angle DEF; the bisector of $\angle DEF$

In Exercises 8 and 12 use 3.14 for π.

8. The diameter of a circle is 20 cm. Find (a) the radius, (b) the circumference, and (c) the area of the circle.

9. Complete: 1 turn = ■°

10. What fractional part of the circle is each section shown?

11. Find the measure of $\angle RPT$ in amount of turn and in degrees.

12. Find the area of the shaded region.

4 cm

13. Find the area of a triangle with sides of length 15 m, 12 m, and 9 m.

Complete and give a reason for each statement.

14.

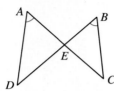

a. $\angle AED \cong$ ■
b. $\triangle AED \sim$ ■
c. $\dfrac{AE}{■} = \dfrac{DE}{■}$

15.

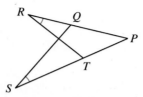

a. $\angle P \cong$ ■
b. $\triangle PRT \sim$ ■
c. $\dfrac{PR}{■} = \dfrac{PT}{■}$

16.

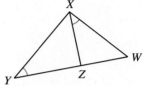

a. $\angle W \cong$ ■
b. $\triangle WXY \sim$ ■
c. $\dfrac{WY}{■} = \dfrac{WX}{■}$

13.1 Arcs and Central Angles

Objective: To determine the measures of arcs of circles.

A diameter of a circle divides the circle into two **semicircles**. In the figure shown, $\overset{\frown}{ACB}$ (read as "*arc ACB*") and $\overset{\frown}{ADB}$ are semicircles of $\odot O$.

Minor arcs are shorter than semicircles.
Major arcs are longer than semicircles.
Minor arcs: $\overset{\frown}{AC}$, $\overset{\frown}{CB}$, $\overset{\frown}{BD}$, $\overset{\frown}{DA}$, $\overset{\frown}{DC}$
Major arcs: $\overset{\frown}{CBD}$, $\overset{\frown}{BDC}$, $\overset{\frown}{DCB}$, $\overset{\frown}{ACD}$, $\overset{\frown}{CBA}$

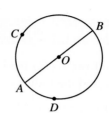

Notice that we name a minor arc by its endpoints. We use three points to name semicircles and major arcs.

The **measure** of a semicircle is 180°. The **measure** of a minor arc is defined as the measure of its central angle. The **measure** of a major arc is 360° minus the measure of the associated minor arc.

Example : For $\odot O$, $m\angle XOY = 110°$. Find each of the following.

      a. $m\overset{\frown}{XY}$ **b.** $m\overset{\frown}{XYZ}$

      c. $m\overset{\frown}{YZX}$ **d.** $m\overset{\frown}{YZ}$

Solution: **a.** $m\overset{\frown}{XY} = 110°$ **b.** $m\overset{\frown}{XYZ} = 180°$

              c. $m\overset{\frown}{YZX} = 250°$ **d.** $m\overset{\frown}{YZ} = 70°$

EXPLORING

$\odot O$ **has radius 1 cm. The circles with center** P **have radii 1 cm and 1.5 cm.**

1. Name the arcs shown that have measure 60°.

2. The **length** of an arc is a fractional part of the *circumference* of a circle. **a.** What fractional part of the circle is $\overset{\frown}{GH}$? $\overset{\frown}{EF}$? **b.** Find the lengths of $\overset{\frown}{GH}$ and $\overset{\frown}{EF}$. **c.** Must arcs that have the same measure also have the same length?

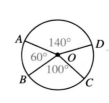

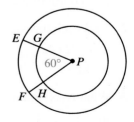

3. How would you define *congruent arcs*?

Circles that have the same radii are called ***congruent circles***. The **EXPLORING** activity suggests the following definition and theorem.

Congruent arcs are arcs of the same circle (or congruent circles) that have the same measure.

THEOREM 13.1: **In the same circle (or congruent circles), two minor arcs are congruent when their central angles are congruent.**

Thinking Critically

1. Suggest a formula for the length l of an arc in terms of the measure $n°$ of the arc and the radius r of the circle.

2. The shaded region is called a ***sector*** of the circle.
   a. How could you find the area of the shaded region?
   b. If the radius of the circle is 3 cm and the measure of the arc is 80°, find the area of the shaded region.

Class Exercises

1. Name each of the following for $\odot O$.
   a. two minor arcs   **b.** two major arcs   **c.** a semicircle

2. Find each measure.
   a. $m\angle YOZ$ **b.** $m\widehat{XY}$ **c.** $m\widehat{YZ}$
   d. $m\widehat{XYZ}$ **e.** $m\widehat{YZX}$ **f.** $m\widehat{YXZ}$

Exercises

1. Name each of the following for $\odot Q$.
   a. two minor arcs   **b.** two major arcs   **c.** a semicircle

2. Find each measure.
   a. $m\angle LQC$ **b.** $m\angle LQP$ **c.** $m\widehat{LP}$
   d. $m\widehat{CLP}$ **e.** $m\widehat{LPC}$ **f.** $m\widehat{LCP}$

3. What fractional part of a circle is an arc with the given measure?
   a. 30°   **b.** 45°   **c.** 120°

Find each indicated measure for ⊙O.

4. $m\angle AOB = 45°$
$m\widehat{AB} = $ ▨

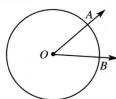

5. $m\widehat{CD} = 120°$
$m\angle COD = $ ▨

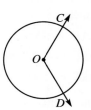

6. $m\widehat{XY} = 110°$
  a. $m\angle XOY = $ ▨
  b. $m\widehat{XAY} = $ ▨

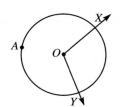

7. $m\widehat{XB} = 55°$
  a. $m\widehat{AX} = $ ▨
  b. $m\widehat{ABX} = $ ▨

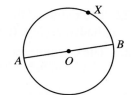

8. **a.** $m\angle 3$
  b. $m\widehat{BC}$
  c. $m\angle 1$
  d. $m\widehat{AD}$
  e. $m\angle 2$
  f. $m\widehat{ACB}$

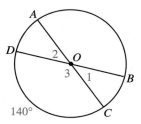

9. **a.** $m\angle ROS$
  b. $m\widehat{ST}$
  c. $m\widehat{PT}$
  d. $m\widehat{SP}$
  e. $m\widehat{RST}$
  f. $m\widehat{PRT}$

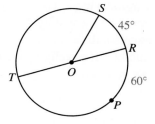

10. **a.** $m\angle AOB$
  b. $m\widehat{BC}$
  c. $m\angle AOC$
  d. $m\widehat{AC}$
  e. $m\widehat{ADC}$
  f. $m\widehat{BCA}$

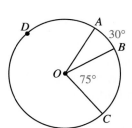

11. **a.** $m\angle XOY$
  b. $m\angle YOZ$
  c. $m\angle XOZ$
  d. $m\widehat{XZ}$
  e. $m\widehat{XWZ}$
  f. $m\widehat{YWZ}$

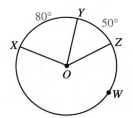

12. The three circles have center O. Each circle is divided into sixteen congruent arcs.

  a. $m\angle AOB = $ ▨   **b.** $m\angle COD = $ ▨   **c.** $m\angle EOF = $ ▨
  d. $m\widehat{AB} = $ ▨   **e.** $m\widehat{CD} = $ ▨   **f.** $m\widehat{EF} = $ ▨

  g. Are the arcs on the largest circle the same length as the arcs on the smaller circles?

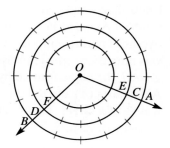

13. Draw two congruent circles.

14. Draw two circles that have center X and radii 2 cm and 5 cm.

For each exercise draw a circle having an arc or arcs with the indicated measure(s).

15. $m\widehat{AB} = 75°$

16. $m\widehat{CDE} = 210°$

17. $m\widehat{FG} = 80°, m\widehat{JK} = 80°$

18. $m\widehat{XY} = 110°, m\widehat{YZ} = 70°$

19. $m\widehat{RS} = 100°, m\widehat{TW} = 35°$

20. $m\widehat{AB} = 125°, m\widehat{ABC} = 300°$

21. The radius of a circle is 14 in. Find the length of an arc whose measure is 45°. (Use $\frac{22}{7}$ for π.)

22. The radius of the circle shown at the right is 6 cm. Find the area of the shaded region. (Use 3.14 for π.)

23. The circumference of a circle is 72π cm. If the length of one arc of the circle is 49π cm, what is the measure of the arc?

APPLICATIONS

Engineering The positions of the five lugs on an automobile wheel could be located by drawing five congruent central angles. For each exercise draw a circle with a radius of at least 5 cm. Locate the given number of equally-spaced points around the circle.

24. 5 points   **25.** 3 points   **26.** 6 points   **27.** 8 points

Historical Note

About 200 B.C. Eratosthenes, a Greek geographer, estimated the circumference of Earth. His calculations were based on assumptions that Earth is round and that since the sun is so far from Earth, the sun's rays are virtually parallel when they reach Earth.

Eratosthenes noticed that in Syene (*S*), a town south of Alexandria, Egypt, the sun was directly overhead at noon on a certain day. When the sun was directly overhead in Syene, he measured $\angle OAB$, the angle formed by a vertical post in Alexandria (*A*) and the tip of its shadow. The angle was $\frac{1}{50}$ of a circle. Thus Eratosthenes also knew the measure of $\angle AOS$.

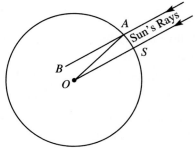

1. Why were $\angle OAB$ and $\angle AOS$ congruent?

2. Eratosthenes knew the distance between Alexandria and Syene. How did he estimate the circumference of Earth?

13.2 Inscribed Angles

Objective: To use the relationship between the measures of an inscribed angle and its intercepted arc.

An *inscribed angle* of a circle is an angle with vertex on the circle and sides containing chords of the circle. In the figure at the right, $\angle RST$ is an inscribed angle. $\overset{\frown}{RT}$ is its *intercepted* arc.

EXPLORING

1. Use the *Geometric Supposer: Circles* disk to draw a circle with radius 5. Label moveable points B, C, D, and E on the circle. Position point E so that $\overset{\frown}{BCE}$ is a major arc.

2. Draw $\overline{CB}$, $\overline{CE}$, $\overline{DB}$, $\overline{DE}$, $\overline{AB}$, and $\overline{AE}$.

3. Measure the ratio of $m\angle BCE$ to $m\angle BAE$. Measure the ratio of $m\angle BDE$ to $m\angle BAE$.

4. What is the ratio of the measure of each inscribed angle to $m\overset{\frown}{BE}$? Why?

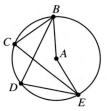

The **EXPLORING** activity demonstrates the following theorem and its corollary for the special case of an inscribed angle that intercepts a semicircle.

THEOREM 13.2: **The measure of an inscribed angle is equal to half the measure of its intercepted arc.**

COROLLARY: **An inscribed angle that intercepts a semicircle is a right angle.**

Thinking Critically

1. If $\overset{\frown}{CD}$ is a minor arc of $\odot P$, how many central angles intercept $\overset{\frown}{CD}$? How many inscribed angles intercept $\overset{\frown}{CD}$?

2. If two inscribed angles of a circle intercept the same arc or congruent arcs, what must be true of the inscribed angles?

Class Exercises

For ⊙C find each measure.

1. $m\widehat{XY}$
2. $m\angle WCX$
3. $m\widehat{WX}$
4. $m\angle ZWY$
5. $m\widehat{WZY}$
6. $m\widehat{WZ}$
7. $m\widehat{XZ}$
8. $m\widehat{XZY}$

Exercises

For each exercise draw a figure like ⊙P. Use the points shown to draw each of the following.

1. a central angle that intercepts $\widehat{AB}$
2. an inscribed right angle
3. two inscribed angles that intercept $\widehat{AB}$
4. an obtuse central angle
5. an acute inscribed angle
6. an obtuse inscribed angle

7. **a.** $m\widehat{CFE}$ = ▪
   b. $m\widehat{CE}$ = ▪

8. **a.** $m\widehat{RWT}$ = ▪
   b. $m\angle RST$ = ▪

9. **a.** $m\angle XBY$ = ▪
   b. $m\widehat{XBY}$ = ▪

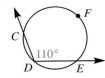

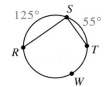

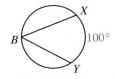

10. $m\widehat{AB}$ = 60°
    a. $m\angle 1$ = ▪
    b. $m\angle 2$ = ▪

11. $\widehat{AX} \cong \widehat{BX}$; $m\angle 1$ = 25°
    a. $m\angle 2$ = ▪
    b. $m\widehat{AB}$ = ▪

12. **a.** $m\widehat{UD}$ = ▪
    b. $m\widehat{AD}$ = ▪
    c. $m\widehat{UQD}$ = ▪

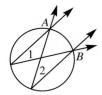

13. $\overline{AZ}$ is a diameter of $\odot O$.

$m\widehat{YZA} = 260°$, $m\widehat{XY} = 40°$

a. $m\widehat{AX} = $ ▨

b. $m\widehat{ABZ} = $ ▨

c. $m\widehat{YZ} = $ ▨

d. $m\angle XAY = $ ▨

e. $m\angle YAZ = $ ▨

f. $m\widehat{XAZ} = $ ▨

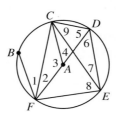

14. $\overline{AC}$ is a diameter of $\odot O$.

$m\widehat{AB} = 50°$, $m\widehat{DC} = 80°$

a. $m\angle 1 = $ ▨

b. $m\angle 2 = $ ▨

c. $m\angle 3 = $ ▨

d. $m\angle 4 = $ ▨

e. $m\angle 5 = $ ▨

f. $m\widehat{AD} = $ ▨

15. $\overline{DF}$ is a diameter of $\odot A$, $m\widehat{BF} = 70°$, $m\widehat{CD} = 50°$, and $m\widehat{DE} = 90°$. Find the measure of each numbered angle.

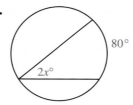

16. If all four vertices of a quadrilateral are on the same circle, what must be true of opposite angles of the quadrilateral? Draw a diagram and use it to explain.

17. Given a circle but not its center, describe how you could use the corner of a sheet of paper and a straightedge to locate each of the following.

a. a diameter   **b.** the center of the circle

APPLICATIONS

Algebra   Find the value of *x*.

18.

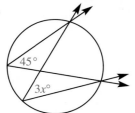

45°

3x°

19.

80°

2x°

20.

50°

(6x+10)°

Computer

Write LOGO procedures that draw a circle and an inscribed angle with each of the following measures. You can use the CIRCLE procedure on page 284.

1. 45°   **2.** 90°   **3.** 120°

13.3 Tangents

Objective: To construct tangents and apply theorems about tangents.

A line is *tangent* to a circle if it is in the same plane as the circle and
intersects it in exactly one point. The point of intersection is called the
point of tangency. In the figure at the right, line *l* is tangent to ⊙R at
point *S*. *S* is the point of tangency.

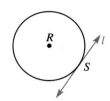

A line tangent to two or more circles is a *common tangent* to the circles.

Example: Line *t* is a common *internal*
tangent of ⊙P and ⊙Q.

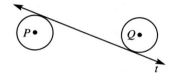

Line *m* is a common *external*
tangent of ⊙R and ⊙S.

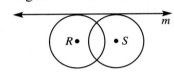

EXPLORING

1. **a.** Draw three or four circles and one tangent for each circle. For each
    circle draw the radius to the point of tangency.
    b. Describe the relationship of each radius to the tangent.

2. **a.** Draw three or four circles and one radius for each circle. For each
    circle draw the line that is perpendicular to the radius and passes
    through its endpoint on the circle.
    b. Describe the relationship of each line to the circle.

The **EXPLORING** activity demonstrates the following theorems.

THEOREM 13.3   If a line is tangent to a circle, then the line is
perpendicular to the radius drawn to the point of tangency.

THEOREM 13.4:   If a line is in the same plane as a circle and is
perpendicular to a radius at its endpoint on the circle, then the line is
tangent to the circle.

Construction 13 is based on Theorem 13.4.

CONSTRUCTION 13

Construct a tangent to ⊙O at point A on the circle.

1. Draw $\overrightarrow{OA}$.

2. Construct the line through A that is perpendicular to $\overrightarrow{OA}$.

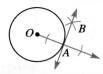

CONSTRUCTION 14

Construct the tangents to ⊙O from point A outside the circle.

1. Draw $\overline{OA}$. Locate the midpoint M of $\overline{OA}$ by constructing the perpendicular bisector of $\overline{OA}$.

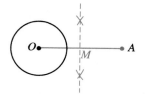

2. Construct a circle with center M and radius MA. Label the points where the circles intersect as B and C.

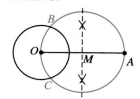

3. Draw $\overleftrightarrow{AB}$ and $\overleftrightarrow{AC}$.

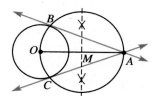

You can show that $\overline{AB} \cong \overline{AC}$ in Construction 14.

THEOREM 13.5: **The two segments tangent to a circle from a point outside the circle are congruent.**

Thinking Critically

1. When will a segment or ray be tangent to a circle?

2. Look at Construction 13. Why is $\overleftrightarrow{AB}$ tangent to ⊙O?

3. $\overline{OB}$ and $\overline{OC}$ have been added to the completed figure for Construction 14.

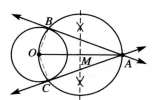

  a. ∠OCA and ∠OBA are inscribed angles of ⊙M. Name the intercepted arc for each angle.

  b. $m\angle OCA = $ ▇ and $m\angle OBA = $ ▇

  c. Why are $\overleftrightarrow{AB}$ and $\overleftrightarrow{AC}$ tangent to ⊙O?

Class Exercises

Draw circles similar to those shown. Draw all common tangents for each pair of circles.

1.  **2.** 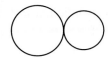 **3.**

Draw a figure similar to, but larger than, the one shown.

4. Construct the line tangent to $\odot C$ at point A.

5. Construct two lines through B that are tangent to $\odot C$.

Exercises

1. Draw a circle and select any point outside the circle. Construct two lines through this point that are tangent to the circle.

2. Draw a circle and select any point on the circle. Construct the line tangent to the circle at this point.

$\overrightarrow{CX}$ **and** $\overrightarrow{CY}$ **are tangent to** $\odot D$ **at points** X **and** Y.

3. $m\angle DXC =$ ▦ and $m\angle DYC =$ ▦

4. If $DX = 8$, then $DY =$ ▦.

5. If $CX = 20$, then $CY =$ ▦

6. Name two isosceles triangles.

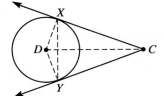

Draw circles similar to those shown. Draw all common tangents for each pair of circles.

7.   **8.**  **9.** 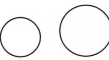 **10.**   **11.**

Find each length. O **and** P **are centers of circles. Tangents are shown in Exercises 13-17.**

12. OP

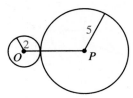

13. a. PB **b.** PD

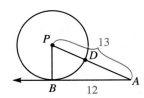

14. a. GO **b.** FE

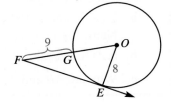

15. a. *AB*
   b. Area of △*ABO*

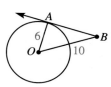

16. a. *RS*   **b.** *ST*
   c. *RT*
   d. Perimeter of △*RST*
   e. *m∠S*   (Why?)
   f. Area of △*RST*

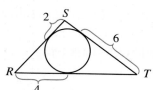

17. a. *AW*   **b.** *AD*
   c. *YD*   **d.** *XC*
   e. Perimeter of *ABCD*
   f. Area of *ABCD*

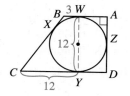

18. a. $\overrightarrow{PA}$ and $\overrightarrow{PB}$ are tangents to ⊙*C*. Name a supplement of ∠*APB*. Justify your answer.
   b. △*PDA* ≅ ▨ by ▨
   c. *m∠ADP* = ▨
   d. Why is △*PDA* ~ △*ADC*?
   e. Name four complements of ∠*CAD*.

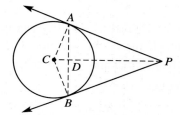

APPLICATION

19. Distance to the Horizon   The diagram at the right represents a cross section of Earth. The radius of Earth is about 6,400 km. Use a calculator to find the distance *d* that a person could see on a clear day from each of the following heights *h* above Earth. Round your answer to the nearest tenth of a kilometer.

   a. 100 m   **b.** 200 m   **c.** 300 m

Seeing in Geometry

A nickel, a dime, and a quarter are tangent as shown. Tangents are drawn to the points of tangency from point *A*. If *AB* = 1 in., how long is $\overline{AE}$? Give a reason for your answer.

Thinking About Proof

Using Definitions

The following statement, which you can use as a reason in proofs, follows directly from the definitions of a circle, congruent circles, and congruent segments.

Radii of a circle (or congruent circles) are congruent.

Exercise

To be Proven: The two segments tangent to a circle from a point outside the circle are congruent (Theorem 13.5).

Given: $\overline{AB}$ and $\overline{AC}$ are tangent to $\odot O$.

Prove: $\overline{AB} \cong \overline{AC}$

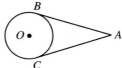

Note: To prove Theorem 13.5, you can add $\overline{OA}$ and radii $\overline{OB}$ and $\overline{OC}$ to the diagram.

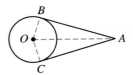

The statements and corresponding reasons for a proof of Theorem 13.5 are given below in a scrambled order. Put the statements on the left in a logical order and match each with an appropriate reason chosen from the list on the right.

(1) $\overline{AB}$ and $\overline{AC}$ are tangent to $\odot O$.

(2) $\overline{AB} \cong \overline{AC}$

(3) $\triangle OBA$ and $\triangle OCA$ are right triangles.

(4) $\angle OBA$ and $\angle OCA$ are right angles.

(5) $\triangle OBA \cong \triangle OCA$

(6) $\overline{AB} \perp \overline{OB}$ and $\overline{AC} \perp \overline{OC}$

(7) $\overline{OB} \cong \overline{OC}$

(8) $\overline{OA} \cong \overline{OA}$

(a) Definition of a right angle

(b) Definition of a right triangle

(c) A segment is congruent to itself.

(d) Given

(e) CPCTC

(f) HL Theorem

(g) Radii of a circle are congruent.

(h) If a line is tangent to a circle, then the line is perpendicular to the radius drawn to the point of tangency.

13.4 Chords and Arcs

Objective: To apply theorems about chords of circles.

The following Exploring activity uses the fact that all radii of a circle are congruent.

EXPLORING

Part A

In ⊙O, $\overline{AB} \cong \overline{CD}$. **Give a reason for each statement.**

1. $\triangle AOB \cong \triangle DOC$ **2.** $\angle AOB \cong \angle DOC$ **3.** $\overparen{AB} \cong \overparen{DC}$

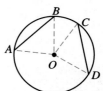

Part B

In ⊙X, $\overparen{EF} \cong \overparen{GH}$. **Give a reason for each statement.**

1. $\angle EXF \cong \angle HXG$ **2.** $\triangle EXF \cong \triangle HXG$ **3.** $\overline{EF} \cong \overline{HG}$

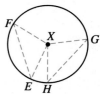

Part C

In ⊙P, $\overline{PD} \perp \overline{AB}$. **Give a reason for each statement.**

1. $\triangle APC \cong \triangle BPC$ **2.** $\overline{AC} \cong \overline{BC}$

3. $\angle APC \cong \angle BPC$ **4.** $\overparen{AD} \cong \overparen{BD}$

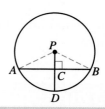

The **EXPLORING** activity demonstrates the following theorems.

THEOREM 13.6: In a circle (or congruent circles), congruent chords determine congruent arcs and congruent arcs determine congruent chords.

THEOREM 13.7: In a circle, a diameter that is perpendicular to a chord bisects the chord and its arc.

Example 1: **Given:** $\overline{AB} \cong \overline{CD}$; $\overline{PX} \perp \overline{AB}$ and $\overline{PY} \perp \overline{CD}$

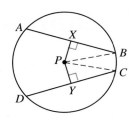

    a. Name three segments congruent to $\overline{XB}$.

    b. Why is $\triangle PXB \cong \triangle PYC$?

    c. Why is $\overline{PX} \cong \overline{PY}$?

Solution: **a.** $\overline{AX}$, $\overline{DY}$, and $\overline{YC}$

    b. $\overline{XB} \cong \overline{YC}$ and $\overline{PB} \cong \overline{PC}$

       $\angle PXB$ and $\angle PYC$ are right angles.

       Thus $\triangle PXB \cong \triangle PYC$ by HL.

    c. CPCTC

Example 2: **Given:** $\overline{PX} \cong \overline{PY}$; $\overline{PX} \perp \overline{AB}$ and $\overline{PY} \perp \overline{CD}$

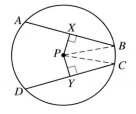

    a. Why is $\triangle PXB \cong \triangle PYC$?

    b. Why is $\overline{XB} \cong \overline{YC}$?

    c. Why is $\overline{AB} \cong \overline{DC}$?

Solution: **a.** HL

    b. CPCTC

    c. Since $\overline{PX}$ and $\overline{PY}$ are perpendicular to $\overline{AB}$ and $\overline{CD}$,
       respectively, they bisect $\overline{AB}$ and $\overline{CD}$. Since $\overline{XB} \cong \overline{YC}$,
       $\overline{AB} \cong \overline{CD}$.

The Examples demonstrate the following theorem.

THEOREM 13.8: **In a circle, congruent chords are equidistant from the center and chords which are equidistant from the center are congruent.**

Thinking Critically

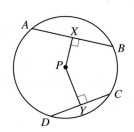

1. If the radius of $\odot P$ is 5, $AB = 8$, and $CD = 6$, what can you conclude about PX and PY?

2. If $AB > CD$, what can you conclude about PX and PY? Justify your answer.

3. If $PX < PY$, what can you conclude about AB and CD?

Class Exercises

Give a reason for each statement about ⊙C.

1. $\overline{DF} \cong \overline{EF}$ 2. $\widehat{DA} \cong \widehat{EA}$
3. $\overline{DA} \cong \overline{EA}$ 4. $\angle ADB$ is a right angle.
5. If $EF = 6$, then $DE = $ ■.
6. If $m\widehat{DE} = 140°$, then $m\widehat{AD} = $ ■.
7. If $m\angle ABD = 36°$, then $m\widehat{AD} = $ ■ and $m\widehat{DE} = $ ■.
8. If $BC = 5$ and $BD = 8$, then $AD = $ ■.

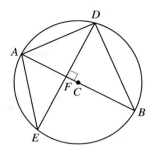

Exercises

1. For ⊙C, if $CA = 9$, then $CB = $ ■ and $RS = $ ■.
2. If $AB = 16$, then $AD = $ ■.
3. If $m\angle ACS = 64°$, then $m\widehat{AS} = $ ■ and $m\widehat{AB} = $ ■.
4. If $m\widehat{AS} = 60°$, then $m\widehat{AR} = $ ■.
5. If $m\widehat{AB} = 130°$, then $m\widehat{SB} = $ ■ and $m\angle BCS = $ ■.
6. If $AC = 13$ and $CD = 5$, then $AB = $ ■.
7. If $AB = 30$ and $CD = 8$, then $AC = $ ■.

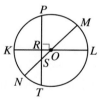

$\overline{KL}$ and $\overline{MN}$ are diameters of ⊙O. Name the following.

8. the midpoint of $\overline{PT}$
9. an arc congruent to $\widehat{PK}$
10. an arc congruent to $\widehat{LT}$
11. an arc congruent to $\widehat{LM}$
12. a segment congruent to $\overline{TR}$
13. a segment congruent to $\overline{MN}$

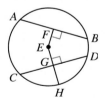

Give a reason for each statement about ⊙E.

14. If $\overline{EF} \cong \overline{EG}$, then $\overline{AB} \cong \overline{CD}$.
15. If $\overline{AB} \cong \overline{CD}$, then $\widehat{AB} \cong \widehat{CD}$.
16. If $\overline{AB} \cong \overline{CD}$, then $\overline{EF} \cong \overline{EG}$.
17. $\widehat{CH} \cong \widehat{DH}$ 18. $\overline{AF} \cong \overline{BF}$

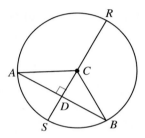

For ⊙O can you conclude that $\widehat{AB} \cong \widehat{BC}$? Explain.

19.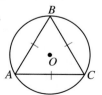
20.
21.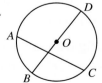
22.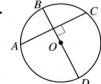

For ⊙O find each length or angle measure.

23. $AO = AB = 8$
   a. AD **b.** OD

24. a. OD **b.** OE
   c. OG **d.** GE
   e. FG

25. a. AB **b.** AC
   c. $m\angle OAC$ **d.** $m\angle AOC$
   e. OC

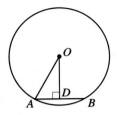
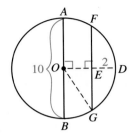
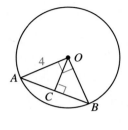

APPLICATIONS

Algebra   For ⊙O find x and y.

26.

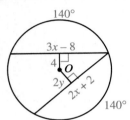

27.

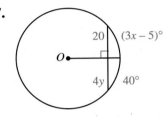

Test Yourself

For ⊙O find each measure or length. In Exercise 3, $\overline{PR}$ and $\overline{PS}$ are tangents.

1. a. $m\widehat{AB}$
  b. $m\angle BOC$
  c. $m\angle BAC$

2. a. $m\widehat{FG}$
  b. $m\widehat{EG}$
  c. XG

3. a. PS
  b. PO

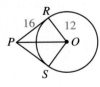

Draw a figure to fit each description.

4. all common tangents for two circles like those shown

5. an obtuse angle inscribed in a circle

For each exercise draw a figure similar to the one shown.

6. Construct a line tangent to ⊙C at point A.

7. Construct two lines tangent to ⊙C from point B.

13.5 Inscribed and Circumscribed Polygons and Circles

Objective: To draw and construct inscribed and circumscribed polygons and circles.

If every vertex of a polygon is on the same circle, we say that the circle is *circumscribed about the polygon* and the polygon is *inscribed in the circle*.

If every side of a polygon is tangent to the same circle, we say that the polygon is *circumscribed about the circle* and the circle is *inscribed in the polygon*.

Example: $\odot M$ is circumscribed about *ABCD*.

ABCD is inscribed in $\odot M$

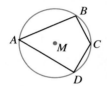

EFGH is circumscribed about $\odot P$

$\odot P$ is inscribed in *EFGH*.

Constructions 15 and 16 are based on the following theorems, which were stated on page 139:

The perpendicular bisectors of the three sides of a triangle intersect in a point that is equidistant from the vertices of the triangle.

The three angle bisectors of a triangle intersect in a point that is equidistant from the sides of the triangle.

CONSTRUCTION 15

Construct a circle circumscribed about △*ABC*.

1. Construct the perpendicular bisectors of any two sides of the triangle. Label the point of intersection as *O*.

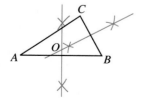

2. Construct a circle using *O* as the center and either *OA*, *OB*, or *OC* as the radius.

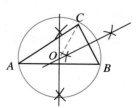

CONSTRUCTION 16

Construct a circle inscribed in △ABC.

1. Construct the bisectors of any two angles of the triangle. Label the point of intersection as *O*.

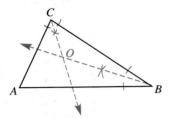

2. Construct a perpendicular from *O* to any side of the triangle. Label the point of intersection as *X*. Construct a circle using *O* as the center and *OX* as the radius.

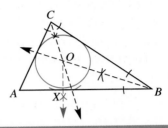

You can use either a compass or a protractor to inscribe certain polygons in a given circle.

EXPLORING

Part A

1. Use a compass to construct a circle. Using the same compass setting, construct arcs to mark off six points on the circle.

2. Use a straightedge to connect the points, forming the polygon shown.

3. Draw dashed lines to connect the vertices of the polygon to the center of the circle. What kind of triangles are formed? Why?

4. What is the measure of each angle of each triangle? Why?

5. Describe the inscribed polygon.

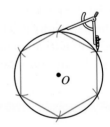

Part B

1. If *ABCDE* is a regular pentagon inscribed in ⊙*O*, what is the measure of each central angle shown?

2. Construct a ⊙*O*. Use a protractor and straightedge to inscribe a regular pentagon in ⊙*O*.

3. How could you use a protractor and straightedge to inscribe a regular octagon in a circle?

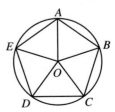

Thinking Critically

1. How could you modify Part A of the Exploring activity to construct an inscribed equilateral triangle?

2. In the figure at the right, diameters $\overline{AB}$ and $\overline{CD}$ are perpendicular.
   a. Why is *ACBD* a square?
   b. Describe how you could modify this figure to construct an inscribed regular octagon.

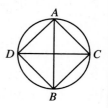

3. Describe how to construct a circle passing through any three noncollinear points.

4. a. Must the perpendicular bisector of a chord of a circle contain the center of the circle? Why or why not?
   b. Suppose you were given an arc of a circle, such as the one at the right. How could you use two chords to find the center of the circle?

Class Exercises

1. Which of the four figures show(s) a circle circumscribed about a triangle? Which show(s) a circle inscribed in a triangle?

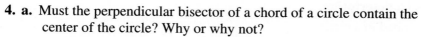

2. Draw a large triangle. Use Construction 15 to construct a circle circumscribed about the triangle.

3. Draw a large triangle. Use Construction 16 to construct a circle inscribed in the triangle.

4. Use a protractor to draw an inscribed regular 9-sided polygon.

Exercises

Name the figures (A–F) that fit each description.

1. a triangle inscribed in a circle
2. a triangle circumscribed about a circle
3. a circle inscribed in a triangle
4. a circle circumscribed about a triangle

  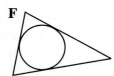

For each exercise begin with a circle. Use a protractor and straightedge to draw each inscribed polygon described.

5. a regular hexagon

6. an equilateral triangle

7. a square

8. a regular octagon

Construct each of the following.

9. a circle circumscribed about an obtuse triangle

10. a circle circumscribed about a right triangle

11. a circle inscribed in an acute triangle

12. a circle inscribed in an obtuse triangle

Draw a figure to fit each description.

13. a right triangle inscribed in a circle

14. an obtuse triangle inscribed in a circle

15. an obtuse triangle circumscribed about a circle

16. a rectangle inscribed in a circle

17. a trapezoid circumscribed about a circle

18. a rectangle circumscribed about a circle

For each exercise begin with a circle. Use a compass and straightedge to construct each inscribed polygon described.

19. an equilateral triangle

20. a square

21. a regular octagon

22. a regular dodecagon

23. Find each of the following.

  a. AC

  b. $m\angle ABC$

  c. AD

  d. area of $ABCD$

  e. area of $\odot O$

  f. area of shaded region

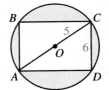

24. Draw a figure like the one shown at the right. Locate the center of the circle.

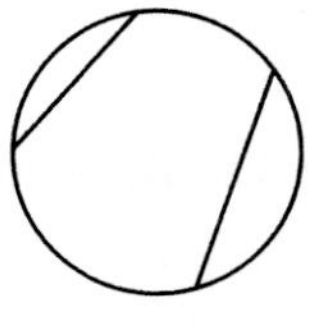

25. Construct an equilateral triangle circumscribed about a circle. (*Hint:* Locate the vertices for an inscribed equilateral triangle. Construct a tangent to the radius drawn to each point.)

26. Construct a rectangle circumscribed about a circle.

APPLICATIONS

27. Archaeology An archaeologist found part of a circular plate, as shown at the right. Draw a similar figure, locate the center of the circle, and complete the circle.

28. Design Use a compass to construct the design shown. (*Hint:* Construct a circle. With the compass point on the circle and the same radius, construct an arc. Use the endpoints of that arc as centers for the next arcs.)

29. Computer The following LOGO procedures use the CIRCLE procedure on page 284. Edit each procedure so that it draws a square inscribed in a circle.

a. To inscr.sq1 :r
  circle :r
  pu fd :r pd
  rt 90
  repeat 4[fd :r * sqrt 2 rt 90]
  pu home pd
  end

b. To inscr.sq2 :r
  circle :r
  pu lt 45 fd :r pd
  rt 135
  repeat 4[fd :r rt 90]
  pu home pd
  end

Computer

Write LOGO procedures to draw each of the following. You can use the CIRCLE procedure on page 284.

1. a regular hexagon inscribed in a circle

2. a square circumscribed about a circle

3. an equilateral triangle inscribed in a circle

13.6 Angles Formed by Chords, Tangents, and Secants

Objective: To determine the measures of angles formed by chords, tangents, and secants.

A *secant* of a circle is a line that contains a chord. In the figure at the right, $\overleftrightarrow{AB}$ and $\overleftrightarrow{CD}$ are secants. Notice that a secant intersects a circle in exactly two points.

In the Exploring activities you will investigate the relationship between certain angles and the arcs they intercept.

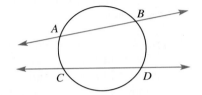

EXPLORING

Part A

1. In the figures below, notice that as point C moves closer to point B, $\overrightarrow{BC}$ becomes more like a tangent. What do you think is the relationship between $m\angle 1$ and $m\widehat{AC}$ in the last figure, where B and C coincide?

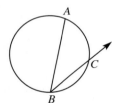
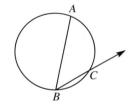
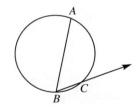
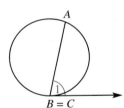

2. Begin with three or four circles. In each circle draw an angle formed by a chord and a tangent. Measure each angle and its intercepted arc. How do the measures compare?

Part B

1. Name the arcs intercepted by $\angle 1$ and its vertical angle.

2. Is $\angle 1$ a central angle of the circle? What is the relationship of $m\angle 1$, $m\angle 2$, and $m\angle 3$? Why?

3. What are $m\angle 2$ and $m\angle 3$? Why? What is $m\angle 1$?

4. What is the relationship between $m\angle 1$, $m\widehat{BC}$, and $m\widehat{AD}$?

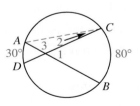

This **EXPLORING** activity demonstrates the following theorems.

THEOREM 13.9: The measure of an angle formed by a tangent and a chord is half the measure of the intercepted arc.

THEOREM 13.10: The measure of an angle formed by two chords that intersect inside a circle is half the sum of the measures of the intercepted arcs.

EXPLORING

1. Name the arcs intercepted by $\angle A$.

2. Find $m\angle 1$ and $m\angle 2$.

3. Find $m\angle A$.

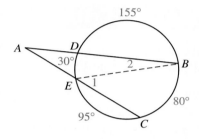

4. Repeat Steps 1–3 for $\odot O$. $\overline{AB}$ is a tangent. (*Hint*: Apply one of the theorems above.)

5. Repeat Steps 1–3 for $\odot P$. $\overline{AB}$ and $\overline{AC}$ are tangents.

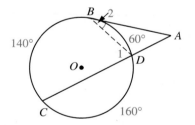

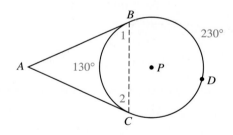

6. What is the relationship in each case between $m\angle A$ and the measures of the intercepted arcs?

This **EXPLORING** activity demonstrates the following theorem.

THEOREM 13.11: The measure of an angle formed by two secants, a tangent and a secant, or two tangents drawn from a point outside the circle is half the difference of the measures of the intercepted arcs.

Example: Find the measure of each numbered angle.

Solution: $m\angle 1 = \frac{1}{2}(60° + 100°)$

$= \frac{1}{2}(160°) = 80°$

$m\angle 2 = \frac{1}{2}(100° - 40°)$

$= \frac{1}{2}(60°) = 30°$

$m\angle 3 = \frac{1}{2}(100° + 90°)$

$= \frac{1}{2}(190°) = 95°$

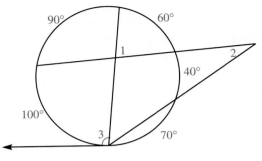

Thinking Critically

1. Describe how Theorem 13.3 on page 388 can be considered to be a special case of Theorem 13.9.

2. Theorem 13.11 deals with the angle formed by two secants that intersect outside a circle.

   a. What theorem deals with the angle(s) formed by two secants that intersect inside a circle? (*Hint:* Recall the definition of a secant.)

   b. What theorem deals with the angle formed by two secants that intersect on a circle?

Class Exercises

Find the measure of each arc or angle. In Exercise 1, $\overleftrightarrow{BE}$ is a tangent.

1. a. $\angle CAD$ **b.** $\angle BAC$ **2. a.** $\angle 1$ **b.** $\angle 2$ **3. a.** $\angle 1$

   c. $\overarc{DA}$ **d.** $\angle DAE$ **c.** $\angle 3$ **d.** $\overarc{RW}$ **b.** $\angle 2$

   e. $\angle CAE$

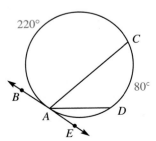

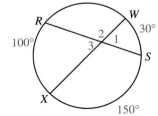

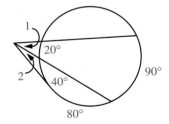

Exercises

Find the measure of each arc or angle. When point *O* is shown, it is the center of the circle. You may assume that segments that look like tangents are tangents.

1. a. ∠1   **b.** ∠2

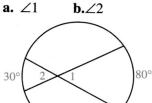

2. ∠1

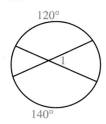

3. a. $\widehat{AB}$ **b.** ∠1

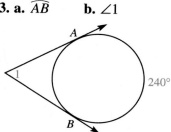

4. ∠1

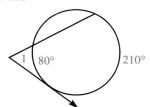

5. a. ∠1   **b.** $\widehat{AB}$

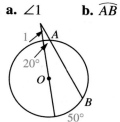

6. a. $\widehat{ACB}$ **b.** ∠1

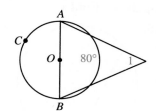

7. a. ∠1   **b.** ∠2

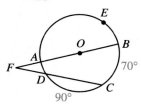

8. a. ∠1   **b.** ∠2

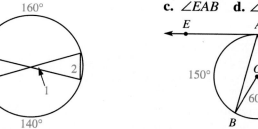

9. a. $\widehat{BC}$ **b.** ∠*BAC*
c. ∠*EAB*   **d.** ∠*EAC*

10. a. $\widehat{AD}$ **b.** ∠*F*
c. $\widehat{AEB}$

11. a. ∠*R*   **b.** ∠*RTY*
c. ∠*RZS*   **d.** ∠*TSZ*
e. ∠*TXZ*

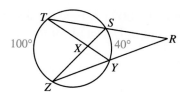

12. a. $\widehat{BC}$ **b.** ∠*C*
c. ∠*CEB*

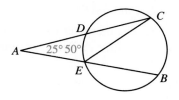

Find the measure of each arc or angle. In Exercise 14, *O* is the center of the circle. Segments and rays that look like tangents are tangents.

13. a. $\overset{\frown}{BD}$ **b.** $\overset{\frown}{AD}$

14. a. $\overset{\frown}{DE}$ **b.** $\overset{\frown}{FE}$ **c.** $\overset{\frown}{DF}$
   d. $\overset{\frown}{DEF}$ **e.** $\angle B$ **f.** $\angle A$

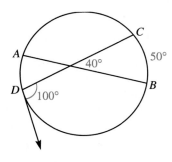

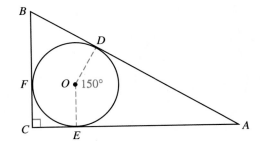

APPLICATIONS

Algebra   Find *x*.

15.

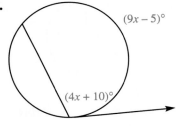

16.

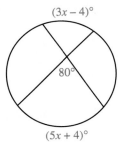

17.

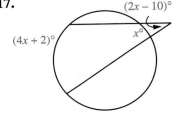

Everyday Geometry

A communications satellite *S* is in orbit 22,000 mi above Earth's equator. The circle shown at the right represents the equator. *A* and *B* are the farthest points on the equator that are in a direct line of communication with the satellite. $m\angle S = 18°$.

1. Find $m\angle SAB$.

2. Find $m\overset{\frown}{AB}$.

3. The diameter of Earth is about 8,000 mi. Find the distance on the surface of Earth between *A* and *B*. Use 3.14 for π.

13.7 Segments of Chords, Secants, and Tangents

Objective: To determine the lengths of segments of chords, secants, and tangents.

You may find the result developed in the following Exploring activity a little surprising.

 EXPLORING

Give a reason for each statement.

1. $m\angle A = \frac{1}{2}m\widehat{BC}$ and $m\angle D = \frac{1}{2}m\widehat{BC}$

2. $\angle A \cong \angle D$

3. $\angle AXC \cong \angle DXB$

4. $\triangle AXC \sim \triangle DXB$

5. $\dfrac{AX}{DX} = \dfrac{CX}{BX}$

6. $AX \cdot BX = CX \cdot DX$

7. If $BX = 15$, $CX = 18$, and $DX = 10$, use the equation in Step 6 to find AX.

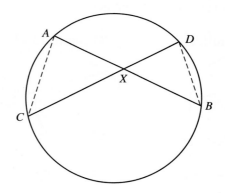

The **EXPLORING** activity demonstrates the following theorem.

THEOREM 13.12: **If two chords of a circle intersect, then the product of the lengths of the segments of one chord is equal to the product of the lengths of the segments of the other chord.**

The following terms are illustrated in the figure at the right.

$\overline{PA}$ is a *tangent segment*.

$\overline{PC}$ is a *secant segment*.

$\overline{PB}$ is an *external secant segment*.

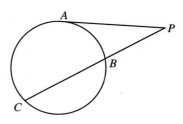

You will justify the following theorems in the Thinking Critically questions.

THEOREM 13.13:   If two secants intersect in the exterior of a circle, then the product of the lengths of one secant segment and its external segment is equal to the product of the lengths of the other secant segment and its external segment.

THEOREM 13.14:   If a secant and a tangent intersect in the exterior of a circle, then the product of the lengths of the secant segment and its external segment is equal to the square of the length of the tangent segment.

The following diagrams illustrate the theorems. Notice that e, g, and s are lengths of external secant segments.

Theorem 13.12

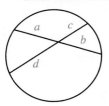

$$a \cdot b = c \cdot d$$

Theorem 13.13

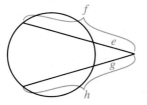

$$f \cdot e = h \cdot g$$

Theorem 13.14

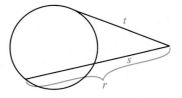

$$r \cdot s = t^2$$

Example: Find the value of x.

a.

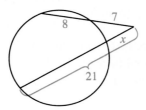

b.

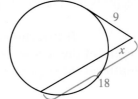

Solution: **a.** $21 \cdot x = (8 + 7)7$

$21x = 15 \cdot 7$

$x = \frac{105}{21}$

$x = 5$

b. $18 \cdot x = 9^2$

$18x = 81$

$x = \frac{81}{18}$

$x = 4\frac{1}{2}$

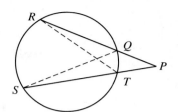

1. **a.** Name two angles whose measures equal $\frac{1}{2} m\widehat{QT}$.
   b. Name two pairs of congruent angles for $\triangle PRT$ and $\triangle PSQ$.
   c. Why is $\triangle PRT \sim \triangle PSQ$?
   d. Complete and give a reason: $\dfrac{PR}{\blacksquare} = \dfrac{PT}{\blacksquare}$
   e. Complete and give a reason: $PR \cdot \blacksquare = PT \cdot \blacksquare$
   f. What have you just shown to be true?

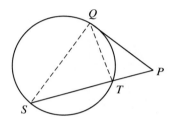

2. **a.** Write two congruence statements between angles of $\triangle PQS$ and $\triangle PTQ$. Explain why each congruence is true.
   b. Write a similarity statement between the triangles. Why is it true?
   c. Write a proportion that involves PQ and PS.
   d. Apply the means-extremes property to your proportion in part (c). What have you just shown to be true?

Class Exercises

Find the value of x.

1.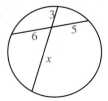

2.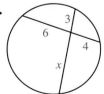

3.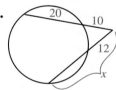

4.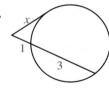

Exercises

Find the value of x.

1.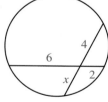

2.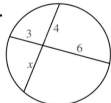

3.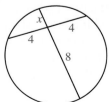

4.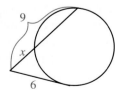

Find the value of x.

5.

6.

7.

8.
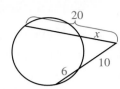

9.
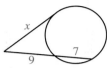

10.

11.
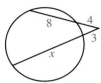

12.
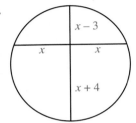

13.
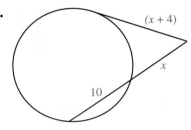

APPLICATION

14. Estimation You can estimate the diameter of a circular swimming pool by holding the ends of a yardstick against the inside wall of the pool and measuring the distance from the midpoint of the yardstick to the wall. To the nearest foot, what diameter of the pool corresponds to each of the following distances from the midpoint to the wall? (*Hint*: Use Theorem 13.12.)

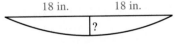

a. 1 in.   **b.** $\frac{7}{8}$ in.   **c.** $1\frac{1}{8}$ in.   **d.** $1\frac{1}{2}$ in.

Everyday Geometry

A train wheel consists of the main wheel, which rides on the rail, and behind it a larger flange wheel, which holds the train on the track. In the figure at the right, $MN = 1$ in. and $AB = 12$ in. Find the radius of each wheel.

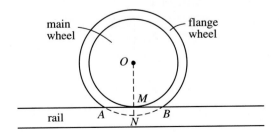

13.8 Locus of Points

Objective: To describe the set of points that satisfy a given set of conditions.

A *locus* is the set of all points (and only those points) that satisfy a given set of conditions. For example, the locus of points in a plane that are 5 cm from a given point *O* in the plane is the circle with center *O* and radius 5 cm. Without the restriction "in a plane" the locus would be the *sphere* with center *O* and radius 5 cm. In this lesson you may assume, unless otherwise stated, that all solutions are in a plane.

EXPLORING

1. Describe the locus of points equidistant from two points *A* and *B*.

2. Describe the locus of points equidistant from the sides of ∠*CDE*.

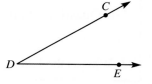

Drawing a diagram will often help you solve a locus problem. Sometimes a diagram helps you to describe a locus.

Example: **Describe the locus of points equidistant from two parallel lines, *l* and *m*.**

Solution: Find three or four points that satisfy the conditions.

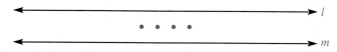

Think of the location of other points satisfying the conditions. Draw the figure formed by these points and describe the locus.

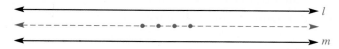

The locus is the line halfway between *l* and *m* and parallel to both lines.

1. **a.** For which types of quadrilaterals (parallelogram, rectangle, rhombus, square, trapezoid) will a diagonal always follow the same path as the locus of points equidistant from any two adjacent sides?

   b. Describe the locus of points equidistant from two adjacent sides of a square.

   c. Describe the locus of points equidistant from sides $\overline{AB}$ and $\overline{AD}$ of rhombus $ABCD$. Use a diagram to help you describe the locus.

   d. Describe the locus of points equidistant from sides $\overline{AB}$ and $\overline{BC}$ of rhombus $ABCD$. Use a diagram to help you describe the locus.

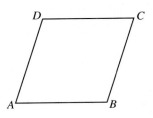

For Questions 2 and 3 do *not* limit your answers to a plane.

2. What is the locus of points equidistant from two parallel planes?

3. What is the locus of points equidistant from two points?

Class Exercises

If AB = 6 cm, describe each locus.

1. the locus of points that are 3 cm from A and 3 cm from B

2. the locus of points that are 3 cm from A and 4 cm from B

3. the locus of points that are 3 cm from A and 2 cm from B

Exercises

Describe each locus.

1. the locus of points 3 cm from a given line l

2. the locus of points equidistant from two perpendicular lines

3. the locus of midpoints of all radii of a circle that has center O and radius 4 cm

4. the locus of points equidistant from all four vertices of a rectangle

5. the locus of points 3 cm from a given segment, $\overline{AB}$

6. the locus of points equidistant from sides $\overline{AD}$ and $\overline{BC}$ of rectangle $ABCD$

7. the locus of the centers of all circles tangent to a line l at a given point P on l

8. the locus of points equidistant from $\overrightarrow{AB}$ and $\overrightarrow{AC}$, where $\overrightarrow{AB}$ and $\overrightarrow{AC}$ are legs of isosceles $\triangle ABC$.

9. R and S are any two points. Find the locus of points that are 3 cm from R and 4 cm from S. (*Hint:* Consider three cases.)

10. X, Y, and Z are any three noncollinear points. Find the locus of points that are 2 cm from X and equidistant from Y and Z.

11. Describe the locus of the center of a wheel with diameter 15 in. as it rolls along a straight path on a flat surface. (Note that this problem is not restricted to a plane.)

APPLICATION

12. Distance A pizza company advertises free delivery within a 5-mi radius of its store. That is, free deliveries are made to any location inside or on a circle with the pizza store as center and a radius of 5 mi. Will the distance traveled round trip on a single delivery always be 10 mi or less? Why or why not?

Thinking in Geometry

The tree on the treasure map is 30 paces from the river. The instructions for finding the buried treasure say that it can be found 40 paces from the river and 15 paces from the tree. In how many locations might you have to dig in order to find the treasure? Give a reason for your answer.

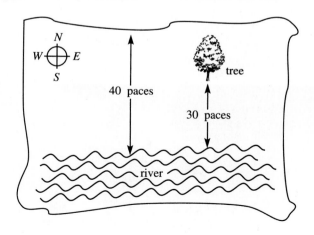

Problem Solving Application: Circle Graphs

Objective: To display data using a circle graph.

Sometimes circle graphs are used to display information.

Example: The table shows the results of a poll of skiers at the Ponderosa Mountain Ski Resort. Display the results in a circle graph.

Ponderosa Skiers by Age	
Age	Percent of all skiers
0-10	10%
11-20	15%
21-30	25%
31-40	32%
Over 40	18%

Solution:
1. Write each percent as a decimal and multiply by 360° to find the measure of the central angle. Round to the nearest degree.

$$0–10 \quad 0.1 \times 360° = 36°$$
$$11–20 \quad 0.15 \times 360° = 54°$$
$$21–30 \quad 0.25 \times 360° = 90°$$
$$31–40 \quad 0.32 \times 360° = 115.2° \approx 115°$$
$$\text{over } 40 \quad 0.18 \times 360° = 64.8° \approx 65°$$

2. Draw a circle. Use a protractor to draw each central angle.

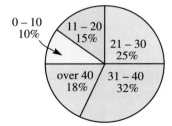

3. Label each section and write a title for the graph.

Ponderosa Skiers by Age

0 – 10
10%
11 – 20
15%
21 – 30
25%
31 – 40
32%
over 40
18%

Notice that the relative sizes of the categories can be gauged simply by comparing the central angles in a circle graph. The smallest category is clearly the "0–10" category. The largest is "31–40."

Other relationships can be seen by comparing groups of categories. Notice that half of all skiers come from the "0–30" age group. One quarter are in the "0–20" age group.

Class Exercises

Answer these questions about the continents of origin of the animals at Central Zoo.

1. Which continent has the largest representation at the zoo? Which has the smallest representation?

2. Which two continents together with South America produced half of the zoo population?

3. Find the measure of each central angle.

a. North America (25%)   **b.** Europe (8%)   **c.** Oceania (2%)

Animals' Continents of Origin

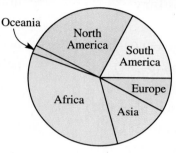

Exercises

Answer these questions about commuting methods in one metropolitan area.

1. What is the most popular method of commuting?

2. What methods together with riding the subway represent half of all commuting methods?

3. Find the measure of each central angle.

a. subway (35%)   **b.** car (42%)   **c.** bus (8%)

Methods of commuting

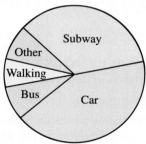

Make a circle graph for the given data.

4. Preference for Mayor

Candidate	Percent
Smith	55%
Jones	30%
Green	10%
Other	5%

5. Elements in Earth's Crust

Element	Percent
Oxygen	48%
Silicon	27%
Aluminum	8%
Other	17%

6. Family Budget

Category	Percent
Rent	30%
Food	25%
Utilities	12%
Clothing	10%
Transportation	10%
Miscellaneous	13%

Chapter *13 Review*

Vocabulary and Symbols

You should be able to write a brief statement, draw a picture, or give an example to illustrate the meaning of each term or symbol.

Vocabulary

arcs

  congruent (p. 382)

  length (p. 381)

  major (p. 381)

  measure of (p. 381)

  minor (p. 381)

  semicircle (p. 381)

circles

  circumscribed (p. 397)

  congruent (p. 382)

  inscribed (p. 397)

inscribed angle (p. 385)

locus (p. 411)

point of tangency (p.388)

polygons

  circumscribed (p. 397)

  inscribed (p. 397)

secant (p. 402)

segments

  external secant (p. 407)

  secant (p. 407)

  tangent (p. 407)

tangent

  common (p. 388)

  internal (p. 388)

  external (p. 388)

  line (p. 388)

Symbols

$\overset{\frown}{AB}$ minor arc *AB* (p. 381)

$\overset{\frown}{PQR}$ major arc *PQR* or semicircle *PQR* (p. 381)

Summary

The following list indicates the major skills, facts, and results you should have mastered in this chapter.

13.1 Determine the measures of arcs of circles. (pp. 381–384)

13.2 Use the relationship between the measures of an inscribed angle and its intercepted arc. (pp. 385–387)

13.3 Construct tangents and apply theorems about tangents. (pp. 388–391)

13.4 Apply theorems about chords of circles. (pp. 393–396)

13.5 Draw and construct inscribed and circumscribed polygons and circles. (pp. 397–401)

13.6 Determine the measures of angles formed by chords, tangents, and secants. (pp. 402–406)

13.7 Determine the lengths of segments of chords, secants, and tangents. (pp. 407–410)

13.8 Describe the set of points that satisfy a given set of conditions.(pp. 411–413)

13.9 Display data using a circle graph (pp. 414–415)

Exercises

Find each measure for ⊙C. $\overrightarrow{BF}$ is a tangent.

1. $m\angle ACD$ 2. $m\widehat{ADB}$ 3. $m\angle DBE$
4. $m\widehat{BE}$ 5. $m\angle EBF$ 6. $m\angle DBA$

Find each measure or length. Tangents are shown in Exercises 7 and 10.

7. ST and SO 8. $m\angle 1$ 9. z

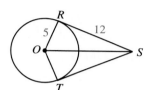

10. x 11. x 12. t

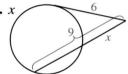

In ⊙O, $OX = 17$ and $OC = OZ = 8$.

13. Find $m\widehat{XY}$ if $m\angle XOZ = 62°$.

14. Find XY.

15. Name two congruent chords and two congruent arcs.

For each exercise, begin with ⊙P or △ ABC.

16. Draw Q on ⊙P and construct the tangent to ⊙P at Q.

17. Draw X not on ⊙P and construct two tangents to ⊙P from X.

18. Construct a circle circumscribed about $\triangle ABC$.

19. Inscribe a square in ⊙P.

20. Given diameter $\overline{JK}$ of ⊙O, describe the locus of points L such that $\angle JLK$ is a right angle.

PROBLEM SOLVING

21. Make a circle graph to show the distribution of the world's population:
North America, 7%; Latin America, 10%; Europe, 12%; U.S.S.R., 7%; Asia,
48%; Africa, 15%; Oceania (some of the islands in the Pacific Ocean), 1%.

Find each measure or length. In Exercises 1, 5, and 6, *O* is the center of the circle. Tangents are shown in Exercises 2, 6, and 8.

1. $m\widehat{AB}$ and $m\angle 1$

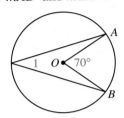

2. $m\angle DEF$ and $m\angle 2$

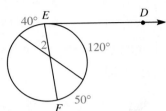

3. x

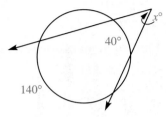

4. y

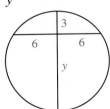

5. MP

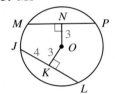

6. OB and CB

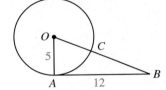

7. y

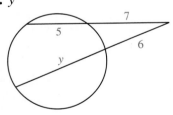

8. x

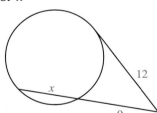

9. $m\angle 3$ and $m\angle 4$

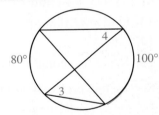

10. Draw an acute $\triangle XYZ$ and inscribe a circle in $\triangle XYZ$.

11. Draw an obtuse $\triangle PQR$ and circumscribe a circle about $\triangle PQR$.

12. Draw a circle and inscribe a regular hexagon in the circle.

13. Draw $\odot J$ and point Z on $\odot J$. Construct the tangent to $\odot J$ at Z.

14. Describe the locus of points equidistant from the vertices of a triangle.

PROBLEM SOLVING

15. Make a circle graph to show the distribution of the U.S. population by age: under age 5, 7%; ages 5–19, 25%; ages 20–44, 37%; ages 45–62, 20%; over age 62, 11%.

Chapter 14

Area and Volume of Prisms

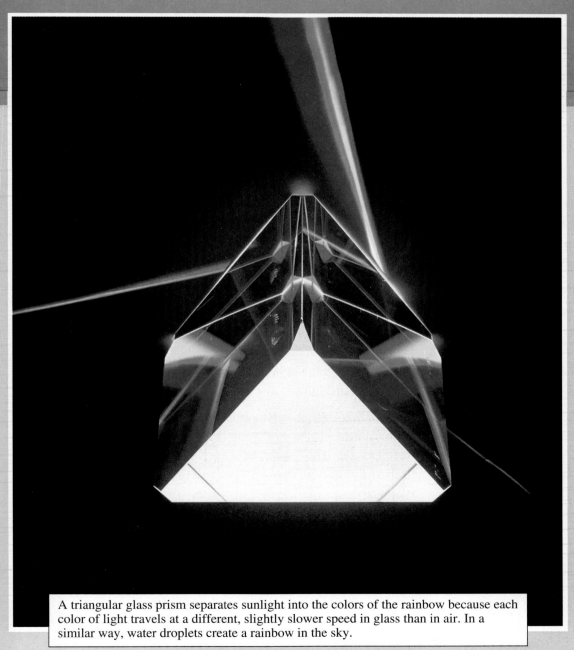

A triangular glass prism separates sunlight into the colors of the rainbow because each color of light travels at a different, slightly slower speed in glass than in air. In a similar way, water droplets create a rainbow in the sky.

Focus on Skills

Use a calculator where appropriate.

ARITHMETIC

Multiply.

1. $72 \cdot 28$ 2. $18 \cdot 65$ 3. $42 \cdot 80$ 4. $35 \cdot 52$

5. $105 \cdot 88$ 6. $66 \cdot 12$ 7. $14(45 + 68)$ 8. $16(21 + 33)$

Complete.

9. $85 \text{ mm} = \blacksquare \text{ cm}$ 10. $152 \text{ cm} = \blacksquare \text{ m}$ 11. $4.2 \text{ km} = \blacksquare \text{ m}$

12. $72 \text{ in.} = \blacksquare \text{ ft}$ 13. $144 \text{ in.} = \blacksquare \text{ yd}$ 14. $712 \text{ ft} = \blacksquare \text{ in.}$

GEOMETRY

The lengths of two sides of a right triangle are given. Find the third length if c is the length of the hypotenuse.

15. $a = 20$ cm, $b = 21$ cm   16. $a = 24$ m, $b = 32$ m

17. $c = 17$ ft, $a = 8$ ft   18. $b = 60$ in., $c = 65$ in.

Find the area of each figure.

19. a square with sides of length 20.5 m

20. a rectangle with base length 42 in. and height 7.5 in.

21. a triangle with base length 48 cm and height 16 cm

22. a trapezoid with bases of length 62 ft and 40 ft and height 9 ft

23. a triangle with sides of length 4.5 m, 6.0 m, and 7.5 m

For each prism (a) state the number of faces, (b) describe the shape of the bases, and (c) classify the prism.

24.   25.   26.   27.

28. The measure of the vertex angle of an isosceles triangle is three times that of each base angle. Use *Guess and Check* to find the measure of each angle.

14.1 Unfolding Prisms

Objective: To describe the size and shape of the faces of a prism.

If you unfold the rectangular prism on the left below,
one pattern you might get is shown on the right.

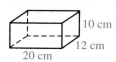 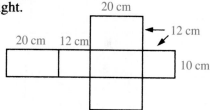

Example 1: **Describe the size and shape of the faces of the rectangular prism shown above.**

Solution: 2 rectangular faces – each 20 cm × 10 cm

2 rectangular faces – each 12 cm × 10 cm

2 rectangular faces – each 20 cm × 12 cm

The result of unfolding a triangular prism is shown at the right.

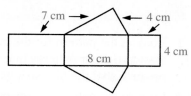

Example 2: **Draw the faces of each prism, label the dimensions, and indicate the number of congruent faces.**

a. 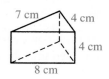   b.

Solution: **a.** Front/Back (2)   Ends (2)   Bases (2)

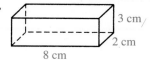

b. Front Face (1)   Back Faces (2)   Bases (2)

Describe the size and shape of the faces of each prism.

1.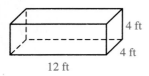
12 ft, 4 ft, 4 ft

2.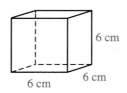
6 cm, 6 cm, 6 cm

3.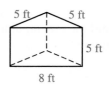
5 ft, 5 ft, 5 ft, 8 ft

4.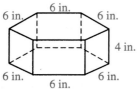
6 in., 6 in., 6 in., 4 in., 6 in., 6 in., 6 in.

5.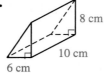
8 cm, 10 cm, 6 cm

6.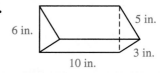
6 in., 5 in., 3 in., 10 in.

7. Classify each prism shown in parts 1–6.

Thinking Critically

1. Look at the pattern for the rectangular prism on page 421. Draw at least two other patterns that could be folded into the rectangular prism. Label the dimensions.

2. Look at the pattern for the triangular prism on page 421.

   a. Describe the size and shape of the faces

   b. Draw at least one other pattern that could be folded into the triangular prism. Label the dimensions.

Class Exercises

Describe the size and shape of the faces of each prism.

1.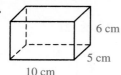
6 cm, 5 cm, 10 cm

2.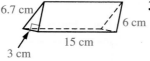
6.7 cm, 6 cm, 15 cm, 3 cm

3.
14 ft, 6 ft, 10 ft, 10 ft

4.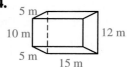
5 m, 10 m, 12 m, 5 m, 15 m

5. Classify the prisms shown in Exercises 1–4.

Exercises

Describe the size and shape of the faces of each prism.

1. 6 cm, 6 cm, 6 cm

2. 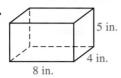 5 in., 4 in., 8 in.

3. 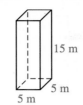 15 m, 5 m, 5 m

4. 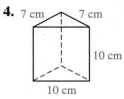 7 cm, 7 cm, 10 cm, 10 cm

5. 8 in., 9 in., 6 in., 4 in.

6. 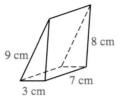 9 cm, 8 cm, 7 cm, 3 cm

7. 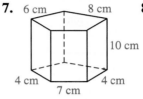 6 cm, 8 cm, 10 cm, 4 cm, 7 cm, 4 cm

8. 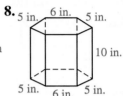 5 in., 6 in., 5 in., 10 in., 5 in., 6 in., 5 in.

9. Classify the prisms shown in Exercises 1–8.

Select the patterns that you can fold into the prism shown.

10.   **a.**   **b.**   **c.**   **d.**   **e.**

11.   **a.**   **b.**   **c.**   **d.**   **e.**

Two copies of a pattern are given for each prism shown. In each case the bottom face is labeled _B_. State the number of the face that becomes the top when the pattern is folded.

12.

a.

			5
1	B	3	4
	2		

b.

			5
1	2	3	4
	B		

13.

a.

	5		
B	2	3	4
	1		

b.

	5		
2	B	3	4
	1		

Two copies of a pattern are given for the prism shown. In each case the bottom face is labeled _B_. State the number of the face that becomes the top when the pattern is folded.

14.

a.

b.

		B		
2	3	4	5

For each prism three faces are shown. Find the value of each variable.

15.

Front
8 cm × 4 cm

Top
8 cm × 3 cm

Side
3 cm × 4 cm

16.

Front
5 cm × 12 cm

Side
5 cm × 12 cm

Top
5 cm × 5 cm

17.

Top
9 cm
7 cm   7 cm

Back
9 cm × 8 cm

Right Side
7 cm × 8 cm

APPLICATIONS

Packaging   You are packing boxes for mailing. You want to tape the boxes twice around in each direction as shown. Answer each of the following questions for Exercises 18–22.

a. How much tape do you need for each box?

b. How many boxes can you tape with one roll of tape?

For Exercises 18–20 assume a 2 in. overlap for each strip of tape. One roll of $\frac{3}{4}$ in. strapping tape has 40 yd (120 ft) of tape.

18.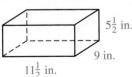

$5\frac{1}{2}$ in.

9 in.

$11\frac{1}{2}$ in.

19.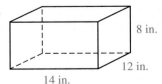

8 in.

12 in.

14 in.

20.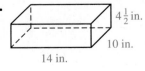

$4\frac{1}{2}$ in.

10 in.

14 in.

In Exercises 21–22 assume a 5 cm overlap for each strip of tape. One roll of 19 mm strapping tape has 36.6 m (3,660 cm) of tape.

21.

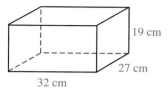

19 cm

27 cm

32 cm

22.

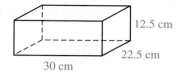

12.5 cm

22.5 cm

30 cm

Seeing in Geometry

A *crystal* is a solid that is composed of atoms that repeat in a pattern. Most nonliving substances are made up of crystals. For example, metals, rocks, snowflakes, salt, and sugar consist of crystals.

Many minerals have a prism-shaped crystal structure. Galena, cobalt, and halite occur naturally in the form of cubes. Calcite is found in rectangular prism form. Quartz sometimes forms triangular prisms and aragonite forms hexagonal prisms.

Look at the photos of crystals and identify as many different kinds of prisms as you can.

Beryl

Tourmaline

Halite

Galena

14.2 Total Area of a Prism

Objective: To find the total area of a right prism.

The *total area* of a prism is the sum of the areas of all faces.

Example: **Find the total area of the rectangular prism.**

Solution: The prism has six faces, as shown.

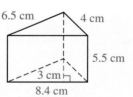

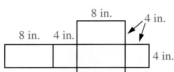

Number	Size	Area	
2	4 in. by 4 in.	2 ☒ 4 ☒ 4 ▤	32 in.²
4	8 in. by 4 in.	4 ☒ 8 ☒ 4 ▤	128 in.²
		Total Area:	160 in.²

EXPLORING

1. Describe the size and shape of all faces of the prism shown at the right.

2. Use a calculator to find the area of each face of the prism.

3. Find the total area to the nearest square centimeter.

Thinking Critically

1. One base of a triangular prism is shown at the right. How would you find the area of this base?

2. Each base of a certain triangular prism has sides of length 30 cm, 40 cm, and 50 cm. Can you find the area of a base? Why or why not?

Class Exercises

State the length of each edge.

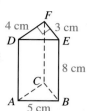

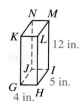

1. $\overline{FC}$ 2. $\overline{AC}$

3. $\overline{AD}$ 4. $\overline{KG}$

5. $\overline{KN}$ 6. $\overline{LH}$

7. $\overline{JI}$ 8. $\overline{MN}$

State the dimensions (length and width) of each face of the prisms above.

9. *ABED*   10. *ADFC*   11. *BCFE*   12. *GHIJ*   13. *GHLK*   14. *KGJN*

15. Find the total area of each prism shown.

Exercises

State the length of each edge.

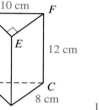

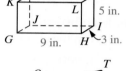

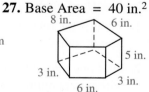

1. $\overline{EF}$ 2. $\overline{AD}$ 3. $\overline{AC}$

4. $\overline{DE}$ 5. $\overline{BE}$ 6. $\overline{JI}$

7. $\overline{KL}$ 8. $\overline{KG}$ 9. $\overline{KN}$

10. $\overline{NJ}$ 11. $\overline{GJ}$ 12. $\overline{QT}$

13. $\overline{ST}$ 14. $\overline{SW}$ 15. $\overline{QR}$

State the dimensions (length and width) of each face.

16. *ABED*   17. *ACFD*   18. *KGHL*   19. *KLMN*

20. *KGJN*   21. *PRWS*   22. *PSTQ*   23. *RWTQ*

Describe the size and shape of all faces of each prism. Then find the total area of the prism. A calculator may be helpful.

24.

25.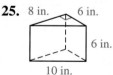

26.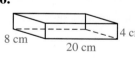

27. Base Area = 40 in.²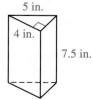

Find the total area of each prism. A calculator may be helpful.

28.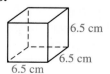

29.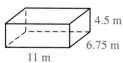

30.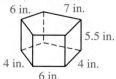

31. Base Area = 50 in.²

32. Find the area of the five faces of this open-topped box.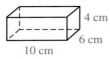

33. Find the total area of a cube that is 5 in. on each edge.

34. A cube has a total area of 294 ft². What is the area of each face of the cube? What is the length of an edge of the cube?

35.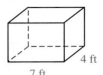

Total area: 166 ft²
Length: 7 ft
Width: 4 ft
Height: ▓

Find the total area of each figure.

36.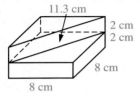

37.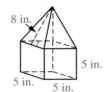

38. Metalworking   Suppose you want to make a copper flower box in the shape shown.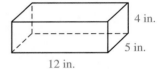

  a. Ignoring waste, how many square inches of sheet copper are needed for the five faces?

  b. A square piece of sheet copper 14 in. on each edge has an area of 196 in.². Could you cut a pattern for the flower box out of this sheet? Why or why not?

Seeing in Geometry

The length of each side of the cube shown is 3 units.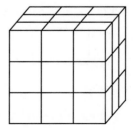

1. What is the total area of the cube in square units?

2. Find one unit cube that may be removed without changing the total area. How many cubes have this property? Describe the location(s) of these cubes.

3. Find one unit cube whose removal will increase the total area by 2 square units. How many cubes have this property? Describe the location(s) of these cubes.

4. Find one unit cube whose removal will increase the total area by 4 square units. How many cubes have this property? Describe the location(s) of these cubes.

14.3 Lateral Area of a Prism

Objective: To find the lateral area of a right prism.

Recall that total area means the sum of the areas of *all* faces. The *lateral area* of a prism is the sum of the areas of its faces *not including* the bases. The following series of pictures shows the relationship between the total area and lateral area for two prisms.

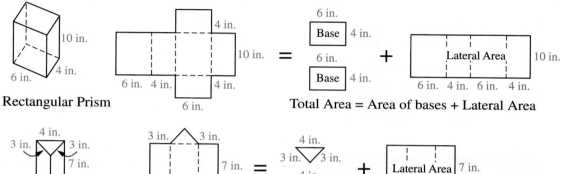

Rectangular Prism

Total Area = Area of bases + Lateral Area

Triangular Prism

Total Area = Area of bases + Lateral Area

EXPLORING

Refer to the rectangular and triangular prisms shown above.

1. Find the lateral area of the rectangular prism.

2. Find the lateral area of the triangular prism.

3. **a.** Find the perimeter (*P*) of a base of each prism.

   b. The length of a lateral edge is called the **height** (*h*) of a right prism. What is the height of each of the prisms?

   c. For each prism find the product of the perimeter of a base and the height.

4. Suggest a formula for the lateral area of a prism.

The **EXPLORING** activity demonstrates the following theorem.

THEOREM 14.1: The lateral area (L.A.) of a prism is the product of the perimeter (*P*) of a base and the height (*h*). **L.A. = *Ph***

Example: Find the lateral area (L.A.) and total area (T.A.) of the prism. A calculator may be helpful.

Solution: By the Pythagorean Theorem, the length of the third side of a base is 13 cm.

L.A. = *Ph*

5 ⊞ 12 ⊞ 13 ⊟ ⊠ 18 ⊟ 540

Base Area: 0.5 ⊠ 5 ⊠ 12 ⊟ 30

T.A.: 540 ⊞ 30 ⊞ 30 ⊟ 600

The lateral area is 540 cm² and the total area is 600 cm².

Thinking Critically

1. a. The first rectangular prism shown at the right has total area 136 cm² and lateral area 88 cm². Does the second rectangular prism have the same total area? the same lateral area? Why or why not?

  b. Do the two triangular prisms have the same total area? the same lateral area? Why or why not?

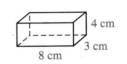

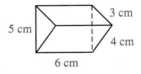

2. Compare the lateral areas and total areas of the prisms in each part. What can you conclude?

  a.

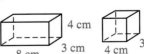

  b.

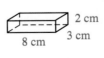

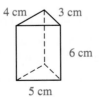

3. Suggest a formula for the total area (T.A.) of a prism in terms of the lateral area (L.A.) and the area (*B*) of a base of the prism.

Class Exercises

Find the lateral area and total area of each prism.

1.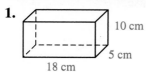
10 cm
5 cm
18 cm

2.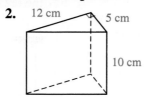
12 cm
5 cm
10 cm

3.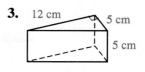
12 cm
5 cm
5 cm

Exercises

Use a calculator where appropriate.

Find the lateral area of each prism.

1.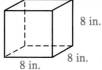
8 in.
8 in.
8 in.

2.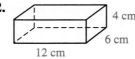
4 cm
6 cm
12 cm

3.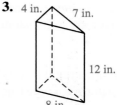
4 in. 7 in.
12 in.
8 in.

4.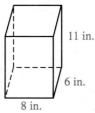
4 in. 7 in.
6 in.
8 in.

5.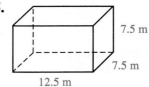
7.5 m
7.5 m
12.5 m

6.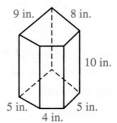
9 in. 8 in.
10 in.
5 in. 5 in.
4 in.

Find the lateral area and total area of each prism.

7.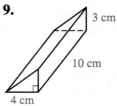
11 in.
6 in.
8 in.

8.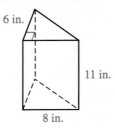
6 in.
11 in.
8 in.

9.
3 cm
10 cm
4 cm

10. Find the lateral area of a cube that is 5 in. on each edge.

11. Find the lateral area and total area of the figure.

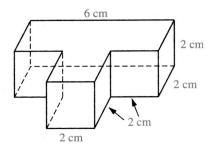

6 cm

2 cm

2 cm

2 cm

2 cm

12. a. Find the lateral area of the prism.
   b. The total area is 268 ft². What is the area of a base?

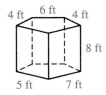

4 ft   6 ft   4 ft

8 ft

5 ft   7 ft

13. The top and all four sides of a toy box are to be painted.

   a. How many square inches of surface are to be painted?

   b. To the nearest square foot, how many square feet of surface are to be painted? (1 ft² = 144 in.²)

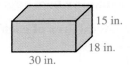

15 in.

18 in.

30 in.

14. Shown are a regular square pyramid and a pattern for constructing it. Find the lateral area and the total area of the pyramid.

9 in.

8 in.   8 in.

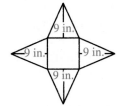

9 in.

9 in.   9 in.

9 in.

15. A drawing of a swimming pool and a side view of it are shown. Find each indicated area.

   a. one side wall

   b. the bottom

   c. the wall at the shallow end

   d. the wall at the deep end

   e. the combined area of the walls of the pool (Be sure to include both sides.)

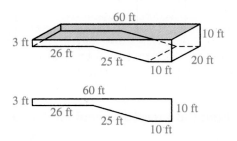

60 ft

3 ft   26 ft   25 ft   10 ft   20 ft   10 ft

60 ft

3 ft   26 ft   25 ft   10 ft   10 ft

16. Part of a barn is shown. (The roof is shaded.)

   a. Find the area of one end of the barn.

   b. Find the area of one side of the barn.

   c. Find the combined area of both ends and both sides of the barn.

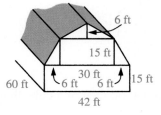

6 ft

15 ft

30 ft

60 ft   6 ft   6 ft   15 ft

42 ft

APPLICATIONS

17. Interior Decorating   The walls of a room that is 20 ft long, 13 ft wide, and 10 ft high are to be painted. There are four windows, each 5 ft by $2\frac{1}{2}$ ft, and one door that is 6 ft by $2\frac{3}{4}$ ft. Find the following.

   a. the combined area of the door and windows

   b. the combined area of the walls (including the door and windows)

   c. the wall area, excluding the door and windows

   d. the area of the ceiling

18. Painting   A gallon of paint covers about 400 ft^2 and costs $9.95. You are to paint the walls of a room that is 24 ft long, 18 ft wide, and 10 ft high. There are four windows, each $2\frac{1}{2}$ ft by 5 ft, and two doors, each $3\frac{3}{4}$ ft. by 8 ft. Find (a) the area to be painted, (b) the number of gallons of paint needed, and (c) the cost of the paint.

19. Wallpapering   A room is 13 ft long, 11 ft wide, and 9 ft high. There are two windows, one door, and a closet door.

   a. Find the combined area of the walls.

   b. A single roll of wallpaper will cover an area of about 30 ft^2 when you allow for waste. If one roll is deducted for every two openings (doors or windows), how many rolls must you purchase?

Test Yourself

Describe the size and shape of the faces of each prism. Then find the total area of the prism in Exercise 1 and the lateral area of the prism in Exercise 2.

1.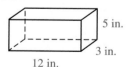
5 in.
3 in.
12 in.

2.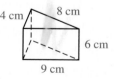
4 cm
8 cm
6 cm
9 cm

Find the lateral area and the total area of each prism.

3.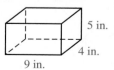
5 in.
4 in.
9 in.

4.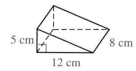
5 cm
8 cm
12 cm

5. Each edge of a cube is 3 in. long. Find the total area of the cube.

14.4 Volume of a Prism

Objective: To find the volume or a missing dimension (given the volume) of a right prism.

You can determine how much food a freezer will hold or the amount of storage space in a warehouse by finding its *volume*. The amount of space contained within, or occupied by, a three-dimensional figure is called its *volume (V)*. Using the cubic unit shown to measure volume, the volume of the rectangular prism at the right is 16 cubic units.

EXPLORING

1. The bottom of each rectangular prism shown has been covered by a layer of cubes.

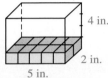 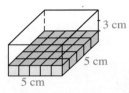

Prism A   **Prism B**   **Prism C**

For each figure find the following:
   a. the number of cubes in the bottom layer
   b. the number of layers needed to fill the prism
   c. the number of cubes needed to fill the prism

2. Without actually counting cubes, how can you find the following?
   a. the number of cubes in the bottom layer of a rectangular prism
   b. the number of cubes needed to fill a rectangular prism

3. Draw a picture of a base of the triangular prism. Then find the following information.
   a. the area of a base
   b. the height of the prism
   c. the volume of the prism

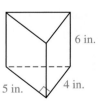

The **EXPLORING** activity demonstrates the following theorem.

THEOREM 14.2: **To find the volume (*V*) of any prism, multiply the area of a base (*B*) by the height (*h*). *V* = *Bh***

Example 1: **Find the volume of a room that is 8 m long, 5 m wide, and 3 m high.**

Solution: Area of base = 8 • 5 = 40 m²

The height of the room is 3 m.

$V = 40 \cdot 3 = 120$ m³

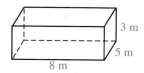

Example 2: **The volume of the triangular prism is 1,860 cm³. Find the height. A calculator may be helpful.**

Solution: $B = \frac{1}{2} \cdot 24 \cdot 10$

= 0.5 ☒ 24 ☒ 10 ▤ 120 cm²

$V = Bh$

1,860 = 120 • *h*

1,860 ➗ 120 ▤ 15.5 cm

The height is 15.5 cm.

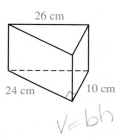

$V = bh$

Thinking Critically

1. Make a table that shows the volume and total area of each rectangular prism with dimensions shown at the right.

2. True or false?

   a. Rectangular prisms with equal volumes have equal total areas.

   b. Rectangular prisms with equal total areas have equal volumes.

   c. If one rectangular prism has a greater volume than a second rectangular prism, then it also has a greater total area.

   d. If one rectangular prism has a greater total area than a second rectangular prism, then it also has a greater total volume.

8 cm by 2 cm by 4 cm

7 cm by 3 cm by 3 cm

4 cm by 4 cm by 4 cm

4 cm by 4 cm by 3 cm

5 cm by 4 cm by 4 cm

12 cm by 2 cm by 2 cm

5 cm by 3 cm by 4 cm

10 cm by 2 cm by 3 cm

9 cm by 2 cm by 3 cm

Class Exercises

For each prism (a) draw a picture of a base, (b) find the area of the base, and (c) find the volume of the prism.

1.

6 in.   3 in.
15 in.

2.

6 cm   4 cm
9 cm

3.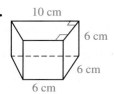

10 cm
6 cm
6 cm
6 cm

Find the indicated information for each prism, where:
V = volume of the prism, B = area of a base, and
h = height of the prism.

4. $B = 32$ cm^2
  $h = 4$ cm
  $V = \blacksquare$

5. $V = 80$ cm^3
  $h = 5$ cm
  $B = \blacksquare$

6. $B = 16$ cm^2
  $V = 96$ cm^3
  $h = \blacksquare$

Exercises

Use a calculator where appropriate.

Find each of the following.

1. a. *EH*
  b. *EA*
  c. *HG*
  d. *HD*
  e. lateral area
  f. area of a base
  g. total area
  h. volume

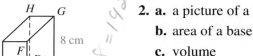

2. a. a picture of a base
  b. area of a base
  c. volume
  d. lateral area
  e. total area
  f. area of the largest face

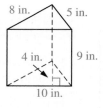

8 in.   5 in.
4 in.   9 in.
10 in.

For each prism (a) draw a picture of a base, (b) find the area of the base, and (c) find the volume of the prism.

3.

14 ft
7 ft
4 ft

4.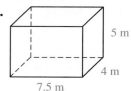

5 m
4 m
7.5 m

5.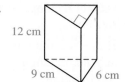

12 cm
9 cm   6 cm

6.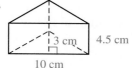
3 cm   4.5 cm
10 cm

7.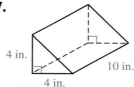
4 in.   4 in.   10 in.

8.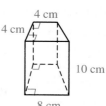
4 cm   4 cm   10 cm   8 cm

Find the indicated information for each prism, where V = volume of the prism, B = area of a base, and h = height of the prism.

9. $B = 15 \text{ cm}^2$
$h = 7 \text{ cm}$
$V = \blacksquare$

10. $h = 8 \text{ m}$
$B = 12.5 \text{ m}^2$
$V = \blacksquare$

11. $h = 3 \text{ ft}$
$V = 27 \text{ ft}^3$
$B = \blacksquare$

12. $h = 12 \text{ in.}$
$V = 144 \text{ in.}^3$
$B = \blacksquare$

13. $B = 45 \text{ yd}^2$
$V = 270 \text{ yd}^3$
$h = \blacksquare$

14. $V = 84 \text{ cm}^3$
$B = 24 \text{ cm}^2$
$h = \blacksquare$

15. Find the volume of a cube 5 in. on each edge.

Find the volume of a cube with each given total area.

16. 24 in.^2

17. 216 in.^2

18. 600 in.^2

19. Suppose a fan can move $3,375 \text{ ft}^3$ of air per minute. How many minutes will it take to move all the air in a room that is 27 ft long, 25 ft wide, and 10 ft high?

20. Under normal conditions air weighs 0.0807 lb/ ft^3. What is the weight of the air in a room that is 18 ft long, 12 ft wide, and 8 ft high?

21. A rectangular aquarium is 14 in. wide, 22 in. long, and 9 in. high.

  a. Find the volume of the aquarium in cubic inches.

  b. If the aquarium is filled to the top, how many gallons of water will it hold? (One gallon occupies 231 in.^3 of space.)

22. A prism has a base area of 18 cm^2 and a height of 14 cm. Find the volume of the prism.

23. A prism with a square base 4 ft on a side has a volume of 128 ft^3. Find the height of the prism.

24. A rectangular prism has a base 8 cm long and 4 cm wide. Its volume is 160 cm^3. Find the height of the prism.

Find the length of an edge of a cube with each given volume.

25. 8 in.^3

26. 27 in.^3

27. 125 in.^3

Find the indicated information for each figure.

28. Volume =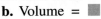

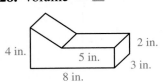

29. a. Total Area = ▦
   b. Volume = ▦

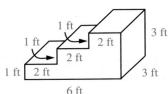

30. Volume = ▦

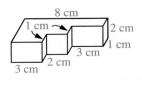

APPLICATIONS

31. Packaging   How many packages 2 in. by 3 in. by 1 in. will fit into a box that is 6 in. by 9 in. by 2 in.?

32. Packaging   Suppose box *A* has a volume of 30 cm³ and box *B* has a volume of 15 cm³. Will box *B* necessarily fit inside box *A*? Why or why not?

33. Firewood   A *cord* of wood is equal in volume to a stack of wood that is 8 ft long, 4 ft wide, and 4 ft high.

   a. A cord is equivalent to how many cubic feet?

   b. Find the volume in cubic feet of the stack of wood at the right.

   c. Find the number of cords of wood shown at the right.

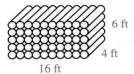

Seeing in Geometry

Each cube measures 1 cm on a side.
Find the volume.

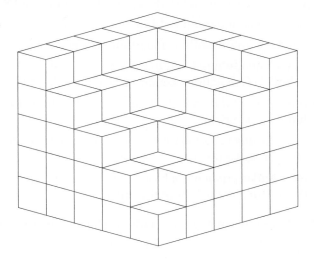

Paragraph Proofs

Proofs are sometimes written in paragraph form. When you write a *paragraph proof*, you need to include enough details so that someone reading your proof can follow your reasoning.

One possible paragraph proof for the theorem that was proved on page 392 is shown in the Example.

Example: Write a paragraph proof for the following. Begin with a diagram and statements of what is given and what is to be proven.

To be Proven: The two segments tangent to a circle from an outside point are congruent. (Theorem 13.5)

Solution: Given: $\overline{AB}$ and $\overline{AC}$ are tangent to $\odot O$.
Prove: $\overline{AB} \cong \overline{AC}$

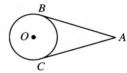

Proof: Draw $\overline{OA}$ and radii $\overline{OB}$ and $\overline{OC}$. Since $\overline{AB}$ and $\overline{AC}$ are tangent to $\odot O$, each is tangent to the radius drawn to the point of tangency. Thus $\angle OBA$ and $\angle OCA$ are right angles, and $\triangle OBA$ and $\triangle OCA$ are right triangles.

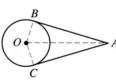

Since radii of a circle are congruent, $\overline{OB} \cong \overline{OC}$. Also, $\overline{OA} \cong \overline{OA}$. Thus $\triangle OBA \cong \triangle OCA$ by the HL Theorem. Since corresponding parts of congruent triangles are congruent, $\overline{AB} \cong \overline{AC}$.

Exercises

Write a paragraph proof for each of the following. Begin with a diagram and statements of what is given and what is to be proven.

1. Exercise 3, page 173
2. Exercise 5, page 174
3. Exercise 1, page 239
4. Exercise 2, page 239
5. Exercise 14, page 245
6. Corollary 2 of Theorem 7.5, page 202

Problem Solving Application: Choosing a Strategy

Objective: To solve problems by choosing an appropriate strategy.

In this lesson you will apply some of the problem solving strategies you have learned to problems involving the volume or lateral area of a prism. Recall the strategies of *Making a Diagram or Model*, *Working Backwards*, *Guess and Check*, *Finding a Pattern*, *Using Formulas*, and *Making a List*.

Example: A rectangular prism has a volume of 360 in.^{3}. The length is twice the width. The height is 1 in. less than the width. Find the length, width, and height of the prism.

Solution: *Guess and Check* is one strategy you can use to find the dimensions of a prism when certain information and the volume or lateral area are known. Try different values for the width and check the result. Record the information in a table.

Guess	Width w	Length $2w$	Height $w - 1$	Volume Bh	Result
1	10 in.	20 in.	9 in.	1,800 in.^{3}	too high
2	5 in.	10 in.	4 in.	200 in.^{3}	too low
3	6 in.	12 in.	5 in.	360 in.^{3}	correct

The length is 12 in., the width is 6 in., and the height is 5 in.

Class Exercise

1. A rectangular prism has a volume of 720 cm^3. The height is one-third the width. The length is 3 cm more than the width. Copy and complete the table to find the length, the height, and the width of the prism.

Guess	Width w	Length $w + 3$	Height $\frac{w}{3}$	Volume Bh	Result
1	▩	▩	▩	▩	▩

Exercises

Solve. Choose an appropriate strategy.

1. The length of a rectangular prism is 4 cm more than the width. The height is 3 cm less than the width. The volume of the prism is 4,160 cm³. Find the length, width, and height of the prism.

2. The length of a rectangular prism is three times the height. The height is 2 cm less than the width. The width is 8 cm. Find the volume of the prism.

3. The lateral area of a rectangular prism is 360 cm². The length is three times the height. The width is two times the height. Find the volume of the prism.

4. A designer produced a sample box shaped like a rectangular prism. The base area of the box was 12 in.². The manufacturer suggested that the height of the box be increased by 2 in. to give the box a volume of 108 in.³. What was the volume of the original box?

5. Three cubes have edges of length 2 in., 5 in., and 6 in., respectively.
   a. Find the volume of each cube.
   b. Find the volume of each cube after the length of an edge is doubled.
   c. Find the ratio of the volume of the new cube to the volume of the original cube. What do you discover?
   d. Use your results in part (c) to solve the following problem. The dimensions of a cube with edges of length 318.47 cm are doubled. What is the ratio of the volume of the new cube to the volume of the original cube?

6. A rectangular prism has a base area of 24 cm² and a volume of 72 cm³.
   a. What is the height of the prism?
   b. Using only whole-number lengths for the length and width, list all possible sets of dimensions for the base.

7. Suppose that twelve small boxes, each the same size, will fit exactly into the box shown.
   a. What is the volume of the box shown?
   b. What is the volume of each small box?
   c. Using only whole-number lengths, list some possible sets of dimensions for the small boxes.

6 in.

6 in.

8 in.

14.6 Equivalent Volumes

Objective: To change a measure of volume or capacity to another unit.

You can use diagrams to develop the relationships between units of volume.

 EXPLORING

1. What is the volume in cubic feet of the cube shown at the right?

2. Use a calculator to find the volume in cubic inches of the cube.

3. Complete:
    a. $1 \text{ ft}^3 = $ ▨ in.^3 b. $2 \text{ ft}^3 = $ ▨ in.^3

4. a. Draw a diagram to represent a cube with sides 1 yd long.
    b. What is the volume of the cube in cubic yards? in cubic feet?
    c. Complete: $1 \text{ yd}^3 = $ ▨ ft^3

5. Use a calculator to complete: $1 \text{ yd}^3 = $ ▨ in.^3

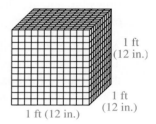

1 ft (12 in.)
1 ft (12 in.)
1 ft (12 in.)

A concept closely related to volume is that of *capacity*. **Capacity** is the measure of material, liquid, or gas that a container can hold. Some familiar units of capacity are the pint, quart, gallon, milliliter (mL), and liter (L).

A cube 10 cm on a side has a volume of 1,000 cm³ and a capacity of 1 L. Notice that the capacity of each small cube shown at the right is $\frac{1}{1000}$ of a liter. Therefore 1 mL of capacity is equivalent to 1 cm³ of volume.

$$1 \text{ L} = 1,000 \text{ cm}^3 \qquad 1 \text{ L} = 1,000 \text{ mL} \qquad 1 \text{ mL} = 1 \text{ cm}^3$$

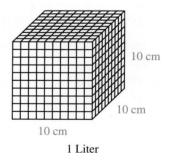

10 cm
10 cm
10 cm
1 Liter

Example 1: Complete:

   a. $3,500 \text{ cm}^3 = $ ▨ L b. $15 \text{ yd}^3 = $ ▨ ft^3

   Solution: a. $3,500 \div 1,000 = 3.5$ b. $15 \cdot 27 = 405$
   $$3,500 \text{ cm}^3 = 3.5 \text{ L} \qquad\qquad 15 \text{ yd}^3 = 405 \text{ ft}^3$$

Example 2: A sidewalk is to be 60 ft long, 3 ft wide, and 4 in. thick. How many cubic yards of concrete are needed?

Solution: The sidewalk is in the shape of a rectangular prism.

Change 4 in. to feet: 4 in. $= \frac{1}{3}$ ft

Find the volume of the prism: $V = Bh = (60 \cdot 3) \cdot \frac{1}{3} = 60$ ft^3

Change 60 ft^3 to cubic yards: $60 \div 27 = 2\frac{2}{9}$ yd^3

(This would be rounded up to the nearest half or full cubic yard. You would order $2\frac{1}{2}$ yd^3 or 3 yd^3 of concrete.)

Thinking Critically

1. What operation is used to change from one unit of measure to a larger unit? from one unit to a smaller unit?

2. State the dimensions (length, width, and height) of three different rectangular prisms having a capacity of 1 L.

3. Using only whole-number lengths, draw pictures of three different rectangular prisms having a volume of 1 yd^3.

4. Joe helped his parents find the amount of soil needed to cover a yard with 2 in. of topsoil. The yard measures 15 yd by 25 yd. Joe multiplied 15 • 25 • 2 and said they would need 750 yd^3. What mistake did Joe make? How many cubic yards of topsoil do they actually need?

Class Exercises

Change each measure to the indicated unit.

1. 4 yd^3 = ▨ ft^3 2. 7 yd^3 = ▨ ft^3 3. 54 ft^3 = ▨ yd^3

4. 135 ft^3 = ▨ yd^3 5. 5 L = ▨ mL   6. 3 L = ▨ cm^3

7. 4,000 mL = ▨ L   8. 100 mL = ▨ cm^3 9. 2,000 cm^3 = ▨ mL

10. 1,500 cm^3 = ▨ L   11. 10 ft^3 = ▨ in.^{3} 12. 8,640 in.^{3} = ▨ ft^3

Exercises

Use a calculator where appropriate.

Change each measure to cubic feet.

1. 2 yd^3 2. 5 yd^3 3. 10 yd^3 4. 3 yd^3

Change each measure to cubic yards.

5. 81 ft^3 **6.** 162 ft^3 **7.** 540 ft^3 **8.** 144 ft^3

Change each measure to the indicated unit.

9. 3 L = ▨ mL   **10.** 500 mL = ▨ L   **11.** 300 mL = ▨ cm^3

12. 1.5 L = ▨ mL   **13.** 2,000 mL = ▨ L   **14.** 750 cm^3 = ▨ mL

15. 2 L = ▨ cm^3 **16.** 1,500 cm^3 = ▨ L   **17.** 750 mL = ▨ L

18. 3 ft^3 = ▨ in.^{3} **19.** $\frac{1}{2}$ ft^3 = ▨ in.^{3} **20.** 13,824 in.^{3} = ▨ ft^3

21. A contractor is building a concrete driveway 54 ft long, 9 ft wide, and 4 in. thick.

    a. Find the number of cubic feet of concrete needed.

    b. Find the number of cubic yards of concrete needed.

22. The engine displacement of a motorcycle is usually given in cubic centimeters. What part of a liter is each of the following?

    a. 150 cm^3 **b.** 200 cm^3 **c.** 750 cm^3

23. Suppose you wish to cover the garden shown with 3 in. of topsoil.

    a. Find the number of cubic feet of topsoil needed.

    b. Find the number of cubic yards of topsoil needed.

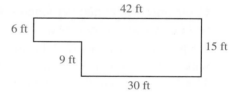

24. A prism-shaped container that is 11 in. high holds 1 gal of liquid. (1 gal = 231 in.^{3})

    a. Find the area of the base.

    b. Suppose the base is rectangular. State one possible set of values for the length and width of the rectangle.

    c. Suppose the base is triangular. State one possible set of values for the base length and height of the triangle.

25. Bricks are 8 in. long, $3\frac{1}{2}$ in. wide, and $2\frac{1}{4}$ in. high. To build a 16 ft by 21 ft rectangular patio, one layer of bricks is to be put on a 3 in. layer of sand.

    a. How many bricks are needed?

    b. How many cubic feet of sand are needed?

    c. The sand is moved in a wheelbarrow with a capacity of 4 ft^3. How many wheelbarrow loads will be needed?

26. Notice that the swimming pool shown is a right prism with one base *ABCDE*.

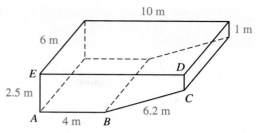

   a. Find the area of base *ABCDE*.

   b. Find the volume of the pool.

   c. What is the capacity of the pool in liters? (1 m³ = 1,000 L)

APPLICATIONS

27. Construction   A contractor is building a house. The hole for the foundation and basement will be 40 ft by 30 ft by 9 ft.

   a. Find the number of cubic feet of soil to be removed.

   b. Find the number of cubic yards of soil to be removed.

   c. Suppose a dump truck has a capacity of 6 yd³. How many truckloads of soil must be removed?

28. Estimation   One health standard requires that 200 ft³ of air space be available for each person in a room.

   a. Find the volume of your classroom in cubic feet.

   b. Find the maximum number of students this health standard would permit in your classroom.

   c. Would you like to have that many students in your classroom? Why or why not?

29. Hobbies   One type of tropical fish requires at least 250 in.³ of water for each inch of body length. Suppose that in the aquarium shown the water level is 2 in. from the top.

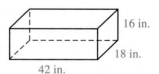

   a. What is the volume of the water in cubic inches?

   b. How much water is required by a fish of this type that is 2 in. long?

   c. How many fish 2 in. long could be kept in this aquarium?

Everyday Geometry

The weight of 1 cm³ of water is exactly 1 g. Because of this convenient relationship, it is easy to calculate the weight of water in a container.

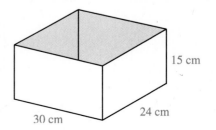

   1. What is the volume of the tank shown?

   2. What is the weight of the water in the tank?

Vocabulary and Symbols

You should be able to write a brief description, draw a picture, or give an example to illustrate the meaning of each of the following.

Vocabulary
capacity (p. 442)
prism
   lateral area (p. 429)
   total area (p. 426)
   volume (p. 434)

Symbols
L.A. (lateral area) (p.430)
T.A. (total area) (p. 430)
V (volume) (p. 434)
B (area of a base) (p.435)

Summary

The following list indicates the major skills, facts, and results you should have mastered in this chapter.

14.1 Describe the size and shape of the faces of a prism. (pp. 421–425)

14.2 Find the total area of a right prism. (pp. 426–428)

14.3 Find the lateral area of a right prism. (pp. 429–433)

14.4 Find the volume or a missing dimension (given the volume) of a right prism. (pp. 434–438)

14.5 Solve problems by choosing an appropriate strategy. (pp. 440–441)

14.6 Change a measure of volume or capacity to another unit. (pp. 442–445)

Exercises

Use a calculator where appropriate.

Describe the size and shape of the faces of each prism.

1.
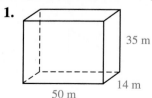

2.
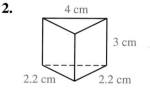

3.
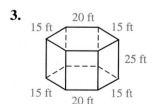

Find the total area of each prism.

4.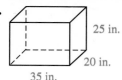
25 in.
20 in.
35 in.

5.
5 m   12 m
9 m
13 m

6.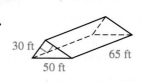
30 ft
50 ft
65 ft

7–9. Find the volume of each prism in Exercises 4–6.

Find the lateral area of each prism.

10.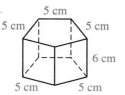
5 cm
5 cm   5 cm
6 cm
5 cm   5 cm

11.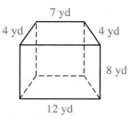
7 yd
4 yd   4 yd
8 yd
12 yd

12.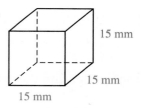
15 mm
15 mm
15 mm

Find the indicated information for each prism, where V = volume of the prism, B = area of a base, and h = height of the prism.

13. $B = 128$ mm^2
$h = 4$ mm
$V = $ ■

14. $V = 170$ m^3
$B = 34$ m^2
$h = $ ■

15. $V = 252$ ft^3
$h = 7$ ft
$B = $ ■

16. Find the volume of a cube 10 in. on each edge.

17. Find the volume of a cube with total area 54 mm^2.

Change each measure to the indicated unit.

18. 500 cm$^3 = $ ■ L

19. $2,500$ mL $ = $ ■ L

20. 8 yd$^3 = $ ■ ft^3

21. 135 ft$^3 = $ ■ yd^3

22. 450 mL $ = $ ■ cm^3

23. $6,912$ in.$^3 = $ ■ ft^3

PROBLEM SOLVING

24. Each dimension of a cube is increased by 1 cm. The resulting cube has volume 169 cm^3 greater than that of the original cube. Find the length of a side of the original cube.

Use a calculator where appropriate.

1. Describe the size and shape of the faces of the prism at the right.

2. Classify the prism at the right.

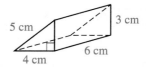

Find the lateral area and total area of each prism.

3.

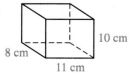

4.

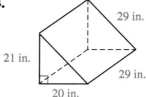

5. Base Area = 20 ft²

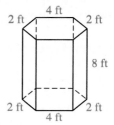

For each prism find the area of a base and the volume of the prism.

6.

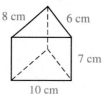

7.

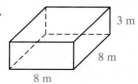

8.

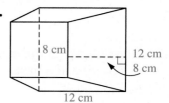

Find the indicated information for each prism, where V = volume of the prism, B = area of a base, and h = height of the prism.

9. $B = 14$ in.²
   $h = 6.5$ in.
   $V = \blacksquare$

10. $V = 816$ mm³
    $B = 102$ mm²
    $h = \blacksquare$

11. $V = 220$ in.³
    $h = 4$ in.
    $B = \blacksquare$

Change each unit to the indicated measure.

12. 14 yd³ = $\blacksquare$ ft³

13. $4,800$ cm³ = $\blacksquare$ L

14. 3.5 L = $\blacksquare$ mL

15. 216 ft³ = $\blacksquare$ yd³

16. 912 mL = $\blacksquare$ cm³

17. 1.5 ft³ = $\blacksquare$ in.³

PROBLEM SOLVING

18. The length of a rectangular prism is twice the width. The height is 5 cm less than twice the width. The volume of the prism is 504 cm³. Find the length, width, and height of the prism.

Chapter 15

Area and Volume of Other Space Figures

Containers designed for the storage of fluids — liquids and gases — or of powdery or granular solids, such as flour or sand, are often shaped like this silo, which has a curved wall.

Focus on Skills

Use a calculator where appropriate.

ARITHMETIC

Evaluate.

1. 2^2 2. 3^2 3. 8^2 4. 15^2 5. 7^2 6. 12^2

7. 3^3 8. 5^3 9. 1^3 10. 2^3 11. 4^3 12. 6^3

Find each square root.

13. $\sqrt{4}$ 14. $\sqrt{16}$ 15. $\sqrt{25}$ 16. $\sqrt{49}$ 17. $\sqrt{81}$ 18. $\sqrt{100}$

Multiply.

19. $\frac{1}{2} \cdot 28 \cdot 112$ 20. $3.14 \cdot 9^2 \cdot 5$ 21. $\frac{4}{3} \cdot 3.14 \cdot 3^3$

22. $\frac{1}{3} \cdot 252 \cdot 10$ 23. $\frac{22}{7} \cdot 14^2 \cdot 3$ 24. $\frac{1}{3} \cdot \frac{22}{7} \cdot 4^2 \cdot 21$

ALGEBRA

Solve.

25. $\frac{8}{15} = \frac{12}{x}$ 26. $\frac{9}{16} = \frac{z}{25}$ 27. $\frac{8}{27} = \frac{108}{y}$ 28. $\frac{16}{25} = \frac{t}{725}$

GEOMETRY

29. Find the exact diameter and circumference of a circle with radius 9 cm.

30. Find the area of a circle with diameter 42 mm. Use $\frac{22}{7}$ for π.

31. Find the circumference and area of a circle with radius 20 cm.

32. **a.** Find the length of $\overline{SW}$.

    b. Find the area of $\triangle RST$.

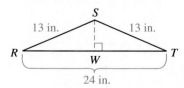

In the diagram, $\triangle ABC \sim \triangle EDF$.

33. Find the scale factor of $\triangle ABC$ to $\triangle DEF$.

34. Find the perimeter of each triangle.

35. Find the area of each triangle.

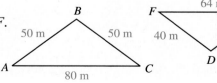

Area of a Cylinder

Objective: To find the lateral area or total area of a right circular cylinder.

A *right circular cylinder* is a three-dimensional figure with two congruent parallel circular bases. The bases are directly opposite each other and are connected by a curved lateral surface that, when flattened, is a rectangle. In this book we will use "cylinder" to mean "right circular cylinder."

EXPLORING

1. A pattern for a cylinder is shown below. What dimensions of the cylinder are equal to the dimensions (base length and height) of the rectangular region?

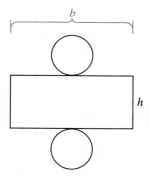

2. The rectangular region has the same area as the lateral surface of the cylinder. How would you find the lateral area of the cylinder?

3. How would you find the total area of the cylinder?

The **EXPLORING** activity demonstrates the following method for finding the lateral area of any cylinder.

THEOREM 15.1: The lateral area of a cylinder is the product of the circumference of the base and the height.

$$\text{Lateral Area } = \text{ Circumference of base} \cdot \text{Height}$$
$$\text{L.A. } = Ch = \pi dh = 2\pi rh$$

Example 1: $C = 12$ cm

$h = 4$ cm

L.A. = ■

Solution: L.A. = Ch

= $12 \cdot 4 = 48$ cm^2

Recall that in this book we will use 3.14 for π unless otherwise stated.

Example 2: For the cylinder shown find (a) the lateral area and (b) the total area.

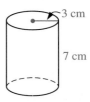

Solution: a. L.A. $= 2\pi rh$

$\approx 2 \boxed{\times} 3.14 \boxed{\times} 3 \boxed{\times} 7 \boxed{=} 131.88$ cm^2

b. First find the area B of each base.

$B = \pi r^2 \approx 3.14 \boxed{\times} 3 \boxed{\times} 3 \boxed{=} 28.26$ cm^2

T.A. $\approx 131.88 \boxed{+} 28.26 \boxed{+} 28.26 \boxed{=} 188.4$ cm^2

Thinking Critically

1. In Example 2 why is the symbol $\approx$ used?

2. The area of a base of a cylinder is often represented by B. Suggest a formula for the total area (T.A.) of a cylinder in terms of the lateral area (L.A.) and the area of a base (B).

Class Exercises

For each cylinder find (a) the circumference of a base, (b) the lateral area, (c) the area of a base, and (d) the total area.

1. Use $\frac{22}{7}$ for π.

2. Use 3.14 for π.

3. A cylinder has a lateral area of 120 cm^2. The circumference of each base is 15 cm. Find the height of the cylinder.

4. The height of a cylinder is 10 cm. The circumference of each base is 16 cm. Find the lateral area of the cylinder.

Exercises

Use a calculator where appropriate.

Find the indicated information for each cylinder.

1. $C = 6$ cm
   $h = 3$ cm
  L.A. = ▨

2. L.A. = 38 in.^{2}
    $h = 2$ in.
    $C =$ ▨

3. L.A. = 132 ft^2
   $C = 22$ ft
   $h =$ ▨

4. L.A. = ▨

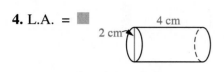

5. Use $\frac{22}{7}$ for π.
   L.A. = ▨

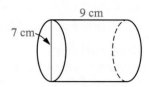

6. Use $\frac{22}{7}$ for π.
  a. L.A. = ▨
  b. T. A. = ▨

7. T. A. = ▨

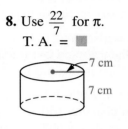

8. Use $\frac{22}{7}$ for π.
  T. A. = ▨

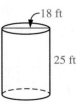

9. L.A. = 48π cm^2
   $h = 8$ cm
   $C =$ ▨

10. L.A. = 40π cm^2
   $C = 5π$ cm
   $h =$ ▨

11. L.A. = 48π cm^2
   $d = 4$ cm
   $C =$ ▨
   $h =$ ▨

12. A cylindrical tank has a height of 25 ft and a base diameter of 18 ft.
  a. What is the total area?
  b. If one gallon of paint covers 400 ft^2, how many gallons are needed to paint the outside surface of the tank? (Do *not* include the bottom of the tank.)

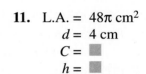

13. a. Find the exact lateral areas of cylinders *R*, *S*, and *T*.

  b. The bases of cylinders *R*, *S*, and *T* are congruent. Write statements comparing the lateral areas of the cylinders.

  c. The bases of cylinders *Y* and *Z* are congruent. The exact lateral area of *Y* is 12π cm^2. Find the exact lateral area of *Z*.

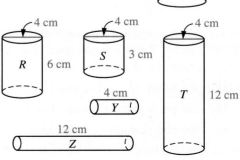

14. Two cylinders have the same lateral area. The height of the first cylinder is twice the height of the second cylinder. How do the radii of the bases of the cylinders compare?

APPLICATIONS

15. Packaging   The label on a can is wrapped around the can and glued with a $\frac{1}{2}$ in. overlap.

   a. Find the dimensions of the rectangular label.

   b. Find the area of the label.

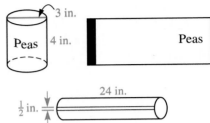

16. Metalworking   A stove pipe is to have a base circumference of 22 in. and a length of 24 in. Allowing $\frac{1}{2}$ in. for an overlapping seam, how many square inches of material are needed to make the pipe?

17. Landscaping   A lawn roller has a base 21 in. in diameter and is 3 ft long. How much surface is rolled in one full turn? (Use $\frac{22}{7}$ for π.)

18. Problem Solving   A lawn roller has a base 18 in. in diameter and is 2 ft long. It is to be used on a yard that is 60 ft by 100 ft.

   a. What information must you determine in order to answer the following questions?

   (1) How many revolutions are needed to roll one trip lengthwise (100 ft) across the yard?

   (2) How many lengthwise trips are needed to roll the yard?

   b. Find the answers to (1) and (2) of part (a).

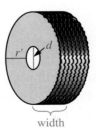

Everyday Geometry

A car tire resembles a cylindrical ring. One tire size is P205/75R14. The three numbers give the following information about the tire.

  205: the width of the tire in millimeters

   75: the percent of the width that gives r'

   14: the rim diameter (d) in inches

Find r' by finding 75% of 205 mm.

Thinking About Proof

Alternative Proofs

There are many different proofs of the Pythagorean Theorem. One proof was outlined in Exercise 22, page 373. That proof used similar right triangles and algebra.

The shortest "proof"—the figure shown below and the single word under it—was given by the twelfth century Indian mathematician Bhaskara.

Behold!

In the exercises, you will explore several ways to use areas of figures to prove the Pythagorean Theorem. Each exercise uses two or more congruent right triangles with legs of length a and b and hypotenuse of length c. You may need to use the following facts.

$$(a + b)^2 = a^2 + 2ab + b^2$$
$$(a - b)^2 = a^2 - 2ab + b^2$$

Exercises

1. **a.** What is the area of the large square at the right?

   b. What is the length of each side of the small square at its center?

   c. Find the sum of the areas of the triangles and the small square.

   d. Why is $a^2 + b^2 = c^2$?

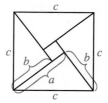

2. Several years before he was elected President of the United States, James Garfield used a diagram like the one at the right to prove the Pythagorean Theorem. Find the area of the trapezoid in two different ways. Explain why $a^2 + b^2 = c^2$.

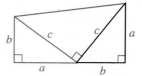

15.2 Volume of a Cylinder

Objective: To find the volume or a missing dimension (given the volume) of a right circular cylinder.

A cylinder is like a prism except that its bases are congruent circular regions rather than congruent polygonal regions. You find the volume of a cylinder in the same way you find the volume of a prism—by multiplying the area of the base by the height.

THEOREM 15.2: **The volume of a cylinder is the product of the area of a base and the height.**

$$\textbf{Volume} = \textbf{Area of base} \cdot \textbf{Height}$$
$$V = Bh = \pi r^2 h$$

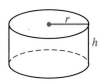

Example 1: **Find the volume of the cylinder shown. A calculator may be helpful.**

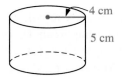

Solution: $V = \pi r^2 h$

$\approx 3.14 \; \boxed{\times} \; 4 \; \boxed{\times} \; 4 \; \boxed{\times} \; 5 \; \boxed{=} \; 251.2$

The volume is about 251.2 cm³.

Example 2: **Find the height of a cylinder with volume 120 in.³ and base area 30 in.².**

Solution: $V = Bh$

$120 = 30 \cdot h$

$h = 120 \div 30$

$= 4$ in.

EXPLORING

1. **a.** The bases of cylinders *A*, *B*, and *C* are congruent. Find the exact volume of each cylinder.

   b. Write statements comparing the volumes.

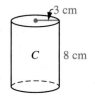

2. The bases of cylinders D and E are congruent. The exact volume of D is 48π cm^3. Find the exact volume of E.

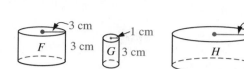

3. a. The heights of cylinders F, G, and H are equal. Find the exact volume of each cylinder.

b. Write statements comparing the volumes.

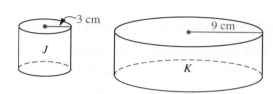

4. The heights of cylinders J and K are equal. The exact volume of J is 45π cm^3. Find the exact volume of K.

Thinking Critically

Answer the following questions for two cylinders that have the same volume.

1. If the height of the first cylinder is twice the height of the second cylinder, how do the areas of the bases compare?

2. If the area of a base of the first cylinder is twice the area of a base of the second cylinder, how do the heights compare?

3. If the radius of a base of the first cylinder is twice the radius of a base of the second cylinder, how do the areas of the bases compare? How do the heights compare?

4. If the height of the first cylinder is twice the height of the second cylinder, how do the radii of the bases compare?

Class Exercises

For each of the following find (a) the area of a base and (b) the volume.

1. Use $\frac{22}{7}$ for π.

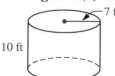

2. Use 3.14 for π.

Find the indicated information for each cylinder.

3. $B = 32$ in.^{2}
   $h = 4$ in.
   $V = $ ▨

4. $r = 4$ m
   $h = 5$ m
   $B \approx $ ▨
   $V \approx $ ▨

5. $V = 75.36$ cm^3
   $B = 12.56$ cm^2
   $h = $ ▨

Exercises

Use a calculator where appropriate.

Find the volume of each cylinder described.

1. The height is 4 ft and the radius of a base is 5 ft.

2. The height is 2 m and the radius of a base is 3 m.

3. The height is 10 cm and the diameter of a base is 8 cm.

4. The height is 9 in. and the area of a base is 45 in.^{2}.

Find the volume of each cylinder.

5. $B = 30$ cm^2

6 cm

6. Use $\frac{22}{7}$ for π.

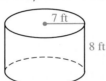

7 ft

8 ft

7.

10 cm

25 cm

8.

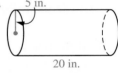

5 in.

20 in.

9. Can R has radius 2 in. and height 4 in.
   Can S has radius 4 in. and height 4 in.
   Can T has radius 2 in. and height 8 in.

   a. Find the exact volume of each can.

   b. Write statements comparing the volumes of the cans.

Find the indicated information for each cylinder.

10. $B = 30$ in.^{2}
   $h = 8$ in.
   $V = $ ▨

11. $V = 150$ cm^3
   $B = 25$ cm^2
   $h = $ ▨

12. $r = 6$ m
   $h = 10$ m
   $B \approx $ ▨
   $V \approx $ ▨

13. $V = 282.6$ cm^3
   $B = 28.26$ cm^2
   $h = $ ▨
   $r \approx $ ▨

Find the indicated information.

14. a. the volume of the top cylinder
    b. the volume of the bottom cylinder
    c. the combined volume of both cylinders

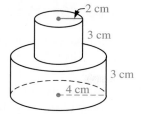

15. a. the volume of each cylinder
    b. Which holds more—the largest cylinder or the two smaller ones combined?

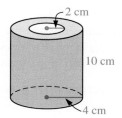

16. a. the volume of the outer cylinder
    b. the volume of the inner cylinder
    c. the volume of the shaded cylindrical ring

17. Find the total area of the cylindrical ring, including the inner surface.

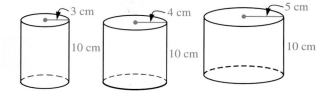

APPLICATION

18. Plumbing If all the pipes are the same length, which will hold more water, one 3-in. diameter pipe or two 2-in. diameter pipes? Explain.

Everyday Geometry

The roll of paper towels shown is wrapped around a cardboard cylinder with a radius of $\frac{3}{4}$ in. The radius of the large cylinder consisting of the paper towels *and* the inner cylinder is $2\frac{1}{2}$ in.

Round each answer to the nearest tenth.

  1. Find the volume of the inner cylinder.
  2. Find the volume of the large cylinder.
  3. Find the volume of paper in the roll.
  4. There are 120 sheets in the roll. Find the volume of each sheet.

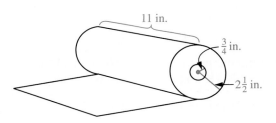

15.3 Area of Similar Figures

Objective: To use the relationship between the areas (and perimeters) of similar figures.

You can determine a relationship between areas (and perimeters) of similar figures. In the next lesson you will investigate the relationships between areas and volumes of three-dimensional figures that have the same shape.

======================= EXPLORING =======================

Part A

1. Make a table like the one shown at the right. Find the perimeter *P* and area *A* of each rectangle with base length *b* and height *h*.

b	*h*	*P*	*A*
3 cm	2 cm	▪	▪
6 cm	4 cm	▪	▪
9 cm	6 cm	▪	▪
12 cm	8 cm	▪	▪

2. What happens to the perimeter of a rectangle when both the base length and height are multiplied by 2? by 3? by 4?

3. What happens to the area when both dimensions are multiplied by 2? by 3? by 4?

4. Suppose both dimensions of a rectangle were multiplied by 5. What would happen to the perimeter? to the area?

5. **a.** If two rectangles are similar with scale factor 1 : 2, what is the ratio of their perimeters? of their areas?

   b. If two rectangles are similar with scale factor 1 : 3, what is the ratio of their perimeters? of their areas?

Part B

1. Why are the triangles similar? What is the scale factor of △*ABC* to △*WXY* ?

2. Find the perimeter of △*ABC*, the perimeter of △*WXY*, and the ratio of the perimeters.

3. Find the lengths of $\overline{BD}$ and $\overline{XZ}$.

4. Find the area of △*ABC*, the area of △*WXY*, and the ratio of the areas.

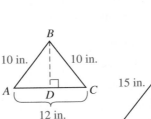

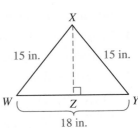

5. Compare the scale factor with the ratio of the perimeters and with the ratio of the areas.

The **EXPLORING** activity demonstrates the following theorem.

THEOREM 15.3: If the scale factor of two similar figures is $a : b$, then the ratio of corresponding perimeters is $a : b$ and the ratio of corresponding areas is $a^2 : b^2$.

Example: Two figures are similar, with scale factor 3 : 4.

a. If the perimeter of the larger figure is 36 cm, what is the perimeter of the smaller figure?

b. If the area of the smaller figure is 18 cm², what is the area of the larger figure?

Solution:

a. $\dfrac{\text{Perimeter of smaller figure}}{\text{Perimeter of larger figure}} = \dfrac{3}{4}$

$\dfrac{P}{36} = \dfrac{3}{4}$

$4P = 36 \cdot 3$

$P = 36 \boxed{\times} 3 \boxed{\div} 4 \boxed{=} 27 \text{ cm}$

b. The ratio of the areas is $3^2 : 4^2$, or 9 : 16.

$\dfrac{\text{Area of smaller figure}}{\text{Area of larger figure}} = \dfrac{9}{16}$

$\dfrac{18}{A} = \dfrac{9}{16}$

$18 \cdot 16 = 9A$

$A = 18 \boxed{\times} 16 \boxed{\div} 9 \boxed{=} 32 \text{ cm}^2$

Thinking Critically

1. a. Must two circles be similar?

b. If the radii of two circles are 2 cm and 4 cm, what are the exact circumferences and areas of the circles?

c. *Scale factor* has been defined for similar *polygons* only. How would you define *scale factor* for two circles? Would Theorem 15.3 apply to two circles?

2. a. The two figures shown are similar. Measure the length of each to determine the scale factor.

b. If the perimeter of the small figure is 7.5 cm, what is the perimeter of the large figure?

c. If the area of the large figure is 5 cm², what is the area of the small figure?

Class Exercises

1. Two polygons are similar, with scale factor 1 : 6.
   a. If the perimeter of the smaller polygon is 12 in., what is the perimeter of the larger polygon?
   b. If the area of the larger polygon is 180 in.^{2}, what is the area of the smaller polygon?

2. Two figures are similar, with scale factor 2 : 3. The perimeter and area of the smaller figure are 20 cm and 12 cm^2 respectively. What are the perimeter and area of the larger figure?

3. By what number is the area of a circle multiplied if its radius is multiplied by each of the following?
   a. 2   b. 4   c. 5

4. By what number is the circumference of a circle multiplied if its radius is multiplied by each of the following?
   a. 2   b. 4   c. 5

Exercises

Use a calculator where appropriate.

Complete for two similar figures.

	Scale Factor	Ratio of Perimeters	Ratio of Areas
1.	1 : 3	▩	▩
2.	4 : 5	▩	▩
3.	▩	4 : 9	▩
4.	▩	▩	4 : 25

5. The scale factor of two similar polygons is 1 : 2. The perimeter and area of the smaller polygon are 14 in. and 12 in.^{2} respectively. Find the perimeter and area of the larger polygon.

6. The scale factor of two similar polygons is 1 : 4. The perimeter and area of the larger polygon are 36 cm and 72 cm^2 respectively. Find the perimeter and area of the smaller polygon.

7. The scale factor of two similar triangles is 2 : 5. The perimeter and area of the larger triangle are 60 in. and 100 in.^{2} respectively. Find the perimeter and area of the smaller triangle.

8. The scale factor of two circles is 3 : 5. The area of the smaller circle is 45 cm^2. Find the area of the larger circle.

9. *ABCD ~ EFGH*

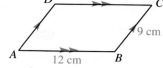

   a. Find the scale factor.

   b. Find *FG*.

   c. Find each perimeter.

   d. If the area of *ABCD* is 72 cm², what is the area of *EFGH*?

10. $\triangle RST \sim \triangle XYZ$, with *RS* = 12 cm, *ST* = 16 cm, *RT* = 20 cm, and *YZ* = 4 cm

   a. Find the scale factor of $\triangle RST$ to $\triangle XYZ$.

   b. Find the ratio of the perimeters.

   c. Find the ratio of the areas.

   d. Find the perimeter of $\triangle XYZ$.

11. a. What is the scale factor of $\triangle BEF$ to $\triangle BCD$?

   b. Find *EF*.

   c. If the area of $\triangle BCD$ is 27 cm², find the area of $\triangle BEF$.

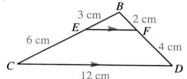

12. The areas of two similar regions are 144 in.² and 169 in.². Find the scale factor of the smaller region to the larger region.

APPLICATION

13. Advertising Comment on the advertisement. Do you think that the drawing represents the headline accurately? Why or why not?

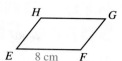

**PREFERRED BY THREE TIMES
AS MANY CUSTOMERS**

Computer

The LOGO procedure shown will compute the perimeter and the area of each of two similar rectangles when you input the ratio, *r*, of the larger rectangle to the smaller rectangle. Add steps to the program so that it will also calculate the ratio of the perimeters and areas.

```
To sim.rect :r :b :w
pr (se [rect.1 P = ] 2 * :b + 2 * :w)
pr (se [A = ] :b * :w)
pr (se [rect.2 P = ] 2 * :b * :r + 2 * :w * :r )
pr (se [A = ] :b * :r * :w * :r )
end
```

15.4 Area and Volume of Similar Space Figures

Objective: To use the relationships between the areas and volumes of similar space figures.

Just as for figures in a plane, three-dimensional figures are *similar* when they have the same shape. In particular, two prisms or two cylinders are similar when their bases are similar and their corresponding heights are in the same ratio as the scale factor of the bases.

Example 1: Are the figures similar? Explain.

a.

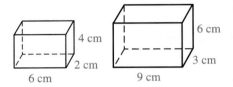

b.

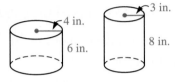

Solution: **a.** $\frac{6}{9} = \frac{2}{3}$, so the bases are similar, with scale factor 2:3. Since $\frac{4}{6} = \frac{2}{3}$, the prisms are similar, with scale factor 2:3.

b. The bases are similar, with scale factor 4:3. However, since the ratio of the first height to the second is 6:8, or 3:4, the cylinders are not similar.

EXPLORING

Part A

1. Make a table like the one shown at the right. Find the total area and volume of each cube with edge length *e*.

2. What happens to the total area of a cube when the length of each edge is multiplied by 2? by 3? by 4? by 5?

3. What happens to the volume of a cube when the length of each edge is multiplied by 2? by 3? by 4? by 5?

e	T.A.	*V*
1 cm	■	■
2 cm	■	■
3 cm	■	■
4 cm	■	■
5 cm	■	■

4. Start with a rectangular prism that is not a cube, like the one shown.

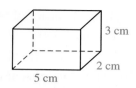

3 cm

2 cm

5 cm

  a. Find the total area and volume.

  b. What happens to the total area and volume when the length of each edge is multiplied by 2? by 3? by 4? by 5?

5. a. If two cubes or rectangular prisms are similar with scale factor 1 : 2, what is the ratio of their total areas? of their volumes?

  b. If two cubes or rectangular prisms are similar with scale factor 1:3, what is the ratio of their total areas? of their volumes?

  c. If two cubes or rectangular prisms are similar with scale factor 2 : 3, what is the ratio of their total areas? of their volumes?

Part B

1. a. Are the triangular prisms similar? What is the scale factor?

  b. Are the cylinders similar? What is the scale factor?

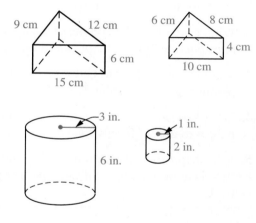

9 cm 12 cm   6 cm 8 cm

6 cm   4 cm

15 cm   10 cm

2. a. Find the lateral area and total area of each triangular prism.

  b. Find the exact lateral area and total area of each cylinder.

3 in.   1 in.

2 in.

6 in.

3. a. Find the volume of each triangular prism.

  b. Find the exact volume of each cylinder.

4. For each pair of figures compare the scale factor with the ratio of the lateral areas, the ratio of the total areas, and the ratio of the volumes.

The **EXPLORING** activity demonstrates the following theorem.

THEOREM 15.4: If the scale factor of two similar space figures is $a:b$, then the ratios of the lateral areas and of the total areas are $a^2 : b^2$ and the ratio of the volumes is $a^3 : b^3$.

Example 2: The scale factor of two similar space figures is 4 : 5.

        a. Find the ratio of the total areas.

        b. Find the ratio of the volumes.

        c. If the volume of the larger figure is 500 cm^3, what is the volume of the smaller figure?

Solution: **a.** The ratio of the total areas is $4^2 : 5^2$, or 16 : 25.

        b. The ratio of the volumes is $4^3 : 5^3$, or 64 : 125.

        c. $\dfrac{V}{500} = \dfrac{64}{125}$

            $125V = 500 \cdot 64$

               $V = 500 \; \boxed{\times} \; 64 \; \boxed{\div} \; 125 \; \boxed{=} \; 256 \text{ cm}^3$

Thinking Critically

1. What happens to the volume of a cube if only one dimension (length, width, or height) is doubled? if exactly two dimensions are doubled?

2. What happens to the volume of a cube if only one dimension is multiplied by 3? if exactly two dimensions are multiplied by 3?

3. Would your answers to Questions 1 and 2 be the same for rectangular prisms that are not cubes?

4. What happens to the volume of a cylinder if only the radius of the base is doubled? if only the height of the cylinder is doubled?

5. What happens to the volume of a cylinder if only the radius of the base is multiplied by 3? if only the height of the cylinder is multiplied by 3?

Class Exercises

1. Two prisms are similar, with scale factor 2 : 3.

    a. What is the ratio of the lateral areas?

    b. If the lateral area of the smaller prism is 20 in.^{2}, what is the lateral area of the larger prism?

    c. What is the ratio of the volumes?

    d. If the volume of the smaller prism is 24 in.^{3}, what is the volume of the larger prism?

2. Two prisms are similar, with scale factor 1 : 4. The total area and volume of the larger prism are 160 cm^2 and 128 cm^3, respectively. What are the total area and volume of the smaller prism?

3. By what number is the total area of a cylinder multiplied if both the radius of the base and the height of the cylinder are multiplied by each of the following?

    a. 2   **b.** 3   **c.** 5

4. By what number is the volume of a cylinder multiplied if both the radius of the base and the height of the cylinder are multiplied by each of the following?

    a. 2   **b.** 3   **c.** 5

Exercises

Use a calculator where appropriate.

Complete for two similar prisms.

	Scale Factor	Ratio of Lateral Areas	Ratio of Total Areas	Ratio of Volumes
1.	1 : 10	▦	▦	▦
2.	3 : 4	▦	▦	▦
3.	▦	36 : 1	▦	▦
4.	▦	▦	25 : 9	▦
5.	▦	▦	▦	8 : 125

Are the figures described similar? Explain.

6. two cubes with edges 2 cm and 6 cm long, respectively

7. a rectangular prism 6 in. long, 4 in. wide, and 8 in. high and a rectangular prism 9 in. long, 6 in. wide, and 12 in. long

8. a triangular prism with base area 5 cm^2 and height 4 cm and a rectangular prism with base area 10 cm^2 and height 8 cm

9. two cylinders, one with base area 10 cm^2 and height 3 cm, and the other with base area 40 cm^2 and height 6 cm

10. a. Find the total area and volume of the prism shown.

    b. If the dimensions of a second prism are twice those of the prism shown, find the total area and volume of the second prism.

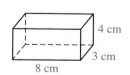

11. The volume of a rectangular prism is 10 in.^{3} and its total area is 34 in.^{2}. Suppose all dimensions are multiplied by 4.

    a. Find the new total area.   **b.** Find the new volume.

12. The total area of a cylinder is 20 in.^{2}. Find the total area of a cylinder with base radius and height that are four times as great.

13. The volume of a cylinder is 10 in.^{3}. Find the volume of a cylinder with base radius and height that are four times as great.

14. Two prisms are similar, with scale factor 1 : 3. The total area and volume of the smaller prism are 72 cm^2 and 30 cm^3, respectively. What are the total area and volume of the larger prism?

15. Two prisms are similar, with scale factor 3 : 4. The total area and volume of the smaller prism are 180 cm^2 and 54 cm^3, respectively. What are the total area and volume of the larger prism?

16. Two prisms are similar, with scale factor 2 : 3. The total area and volume of the larger prism are 198 in.^{2} and 162 in.^{3}, respectively. What are the total area and volume of the smaller prism?

17. The volumes of two similar space figures are 64 cm^3 and 1,000 cm^3. What is the scale factor of the larger figure to the smaller figure?

APPLICATION

18. **Biology**   Shown are the leg bones of two animals that have similar shapes. The dimensions of the larger bone are two times the corresponding dimensions of the smaller bone.

   a. Weight is proportional to volume. How do the weights of the animals compare?

   b. The strength of any structural material, such as bone, is proportional to its cross-sectional area. How much stronger is the larger bone?

Test Yourself

1. a. L.A. ≈ ▨

   b. T.A. ≈ ▨

   c. V ≈ ▨

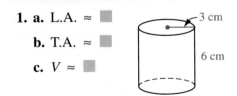

2. Complete for a cylinder:

   $V = 251.2$ cm^3

   $B = 50.24$ cm^2

   $h = $ ▨

3. If the scale factor of two similar figures is 2 : 3, find (a) the ratio of the perimeters and (b) the ratio of the areas

4. If the scale factor of two similar space figures is 1 : 2, find (a) the ratio of the total areas and (b) the ratio of the volumes.

5. Two prisms are similar, with scale factor 4 : 5. The total area and volume of the smaller prism are 240 cm^2 and 192 cm^3, respectively. Find (a) the total area and (b) the volume of the larger prism.

15.5 Area and Volume of Regular Pyramids and Cones

Objective: To find the lateral area, total area, and volume of a regular pyramid or right circular cone.

A **regular pyramid** is a polyhedron such that:

1. the *base* is a regular polygon,
2. exactly one vertex, called *the vertex* of the pyramid, is not a vertex of the base, and
3. the *lateral edges* are congruent.

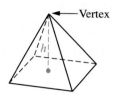

Vertex

The vertex of the pyramid is directly over the center of the base. The **height** (h) of the pyramid is the length of the segment joining the vertex and the center of the base.

Pyramids, like prisms, are named by the shape of the base.

EXPLORING

1. Describe the *lateral faces* of the regular pyramids shown. For each pyramid shown, are the lateral faces congruent? Why or why not?

2. What information is needed to find the area of one lateral face?

3. Suppose the area of one lateral face of each pyramid shown is 30 cm². Find the lateral area for each pyramid.

4. Describe how to find the lateral area and total area for any regular pyramid.

5. If s is the length of each side of a base and l is the height of each lateral face, what is the area of each lateral face?

6. If P is the perimeter of the base of a regular pyramid and l is the height of each lateral face, suggest a formula for the lateral area of the pyramid.

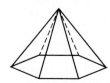

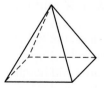

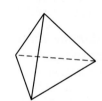

Regular triangular pyramid

Regular square pyramid

Regular hexagonal pyramid

The *slant height* (*l*) of a regular pyramid is the length of the altitude of each lateral face.

THEOREM 15.5: **The lateral area of a regular pyramid is half the product of the slant height and the perimeter of the base.** **L.A. $= \frac{1}{2} lP$**

Example 1: Find the lateral area and the total area of the regular square pyramid shown.

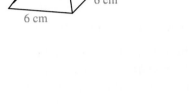

5 cm

6 cm

6 cm

Solution: L.A. $= \frac{1}{2} lP$

$\qquad = \frac{1}{2} \cdot 5(4 \cdot 6)$

$\qquad = 60 \text{ cm}^2$

$B = 6 \cdot 6 = 36 \text{ cm}^2$

T.A. $= 60 + 36 = 96 \text{ cm}^2$

A *right circular cone* is somewhat like a regular pyramid except that the base of a cone is a circular region. The *vertex* of a cone is directly over the center of the base. The *height* of the cone is the length of the segment joining the vertex and the center of the base. In this book we will use the word "cone" to mean "right circular cone."

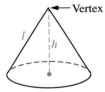

← Vertex

l | *h*

Right circular cone

The *slant height* (*l*) of a cone is the length of any segment joining the vertex and a point on the circle that determines the base. The formula for the lateral area of a cone is similar to that for a regular pyramid.

THEOREM 15.6: **The lateral area of a cone is half the product of the slant height and the circumference of the base.** **L.A. $= \frac{1}{2} lC = \pi lr$**

Shown on the left below are a square pyramid and a rectangular prism with the same base area. On the right are a cone and a cylinder with the same base area and height.

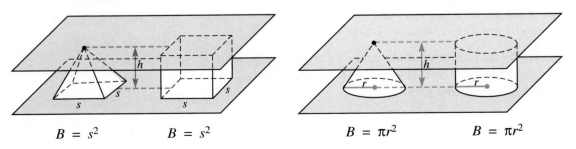

$B = s^2$ $\qquad$ $B = s^2$ $\qquad\qquad$ $B = \pi r^2$ $\qquad$ $B = \pi r^2$

Suppose models for the figures shown at the bottom of page 470 were filled with either a liquid or some material such as sand or sawdust. You would find that the prism holds three times as much as the pyramid and that the cylinder holds three times as much as the cone. This suggests that the volume of the pyramid is one-third the volume of the prism and that the volume of the cone is one-third the volume of the cylinder.

THEOREM 15.7: **The volume of a regular pyramid or cone is one-third the product of the area of the base and the height.** $V = \frac{1}{3}Bh$

Example 2: **Find each volume.**

a.

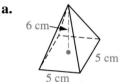

6 cm

5 cm

5 cm

b.

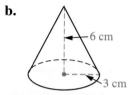

6 cm

3 cm

Solution: **a.** $B = 5 \cdot 5 = 25 \text{ cm}^2$

$V = \frac{1}{3}Bh$

$V = \frac{1}{3} \cdot 25 \cdot 6 = 50 \text{ cm}^3$

b. $B \approx 3.14 \boxed{\times} 3 \boxed{\times} 3 \boxed{=} 28.26 \text{ cm}^2$

$V \approx 1 \boxed{\div} 3 \boxed{\times} 28.26 \boxed{\times} 6 \boxed{=} 56.52 \text{ cm}^3$

The segment that joins the vertex of a regular pyramid or cone to the center of the base is perpendicular to each line in the base that goes through the center. That fact can help you find a missing dimension needed to find an area or volume.

Thinking Critically

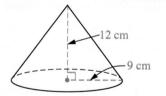

12 cm

9 cm

1. How could you find the slant height of the cone?

2. Find the exact lateral area and total area of the cone.

Class Exercises

Name each of the following for the regular square pyramid.

1. the vertex of the pyramid

2. the height of the pyramid

3. the slant height

4. a lateral edge

5. all segments congruent to $\overline{AE}$

6. four isosceles triangles

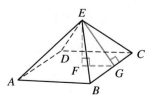

E

D   C

F   G

A

B

Find each of the following if AB = 12 cm, EF = 8 cm, **and** EG = 10 cm.

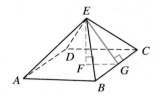

7. Lateral area   8. Area of base

9. Total area   10. Volume

Find each of the following for the cone.

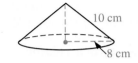

11. Lateral area   12. Area of base

13. Total area   14. Height

15. Volume

Exercises

Use a calculator where appropriate.

Classify each statement as true or false for the regular square pyramid.

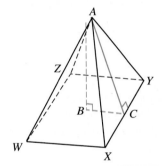

1. AC is the height of the pyramid.
2. If WX = 10 cm, then BC = 5 cm.
3. A is the vertex of the pyramid.
4. $\triangle AWX$ is equilateral.
5. $\triangle AWX \cong \triangle AXY$
6. $\overline{AW} \cong \overline{AX}$
7. The pyramid has four lateral faces.
8. The base of the pyramid is a square.

Find the volume of each figure described.

9. Cylinder: r = 4 in., h = 9 in.
11. Prism: B = 18 cm^2, h = 12 cm
13. Cone: B = 24 cm^2, h = 8 cm
15. Cube: each edge is 6 in. long

10. Cone: r = 4 in., h = 9 in.
12. Regular pyramid: B = 18 cm^2, h = 12 cm
14. Regular pyramid: B = 24 cm^2, h = 8 cm
16. Regular square pyramid: h = 6 in., base is a 6-in. by 6-in. square

Find the volume of each figure.

17.

18. B = 60 in.^{2}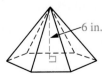

19.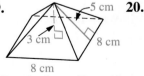
Regular Square Pyramid

20.

21. Find the lateral area and total area of the regular square pyramid in Exercise 19.

22. Find the lateral area and total area of the cone in Exercise 20.

23. A cone has a volume of 225 cm³. What is the volume of a cylinder with the same base area and height?

24. A prism holds 432 in.³. How much would a regular pyramid with the same base area and height hold?

25. A regular pyramid and a cone have the same height. Their bases have the same area. How do their volumes compare?

26. A cone has a base area of 25 in.² and a volume of 100 in.³. Find its height.

Draw a figure to fit each description.

27. regular triangular pyramid   **28.** regular square pyramid   **29.** right circular cone

30. Two pyramids are similar, with scale factor 2 : 5. The lateral area and volume of the smaller pyramid are 36 cm² and 48 cm³, respectively. What are the lateral area and volume of the larger pyramid?

APPLICATION

31. History The Great Pyramid at Giza, Egypt, was built about 2600 B.C. It was considered to be one of the wonders of the ancient world. The base is approximately 230 m square. The height was originally about 147 m. Find the original volume.

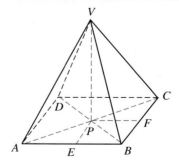

Seeing in Geometry

A *line is perpendicular to a plane* if and only if the line is perpendicular to all lines in the plane that pass through the point of intersection. In the figure at the right, $\overline{VP}$ is perpendicular to the base of the regular square pyramid. Name the lines shown that are perpendicular to $\overline{VP}$.

15.6 Area and Volume of a Sphere

Objective: To find the area or volume of a sphere.

Recall that a circle is the set of all points in a plane that are the same distance from some point called the center. Similarly, a *sphere* is the set of all points in space that are the same distance from some point called the **center**. Most terms used to describe circles are also used for spheres. For the sphere shown at the right, $\overline{OA}$, $\overline{OB}$, and $\overline{OC}$ are radii.

The following formulas can be used to find the area and volume of any sphere.

THEOREM 15.8:   **The area and volume of a sphere with radius r are given by the following formulas.**

$$A = 4\pi r^2 \qquad V = \tfrac{4}{3}\pi r^3$$

Example 1: **A sphere has a radius of 5 cm. Find (a) the exact area and volume and (b) the approximate area and volume.**

Solution: **a.** $A = 4\pi r^2$ $V = \dfrac{4}{3}\pi r^3$ **b.** $A = 100\pi$ $V = \dfrac{500\,\pi}{3}$

$\qquad\qquad = 4\pi(5)^2 \qquad\quad = \dfrac{4}{3}\pi(5)^3 \qquad\qquad \approx 100(3.14) \qquad \approx \dfrac{500\,(3.14)}{3}$

$\qquad\qquad = 4\pi(25) \qquad\qquad\qquad\qquad\qquad \approx 314 \text{ cm}^2 \qquad\qquad \approx 523.33 \text{ cm}^3$

$\qquad\qquad = 100\pi \text{ cm}^2 \qquad = \dfrac{4}{3}\pi(125)$

$\qquad\qquad\qquad\qquad\qquad\qquad = \dfrac{500\,\pi}{3} \text{ cm}^3$

Example 2: **The area of a sphere is 64π cm². Find its radius and diameter.**

Solution: $A = 4\pi r^2$

$\qquad\qquad 64\pi = 4\pi r^2$

$\qquad\qquad r^2 = \dfrac{64\,\pi}{4\,\pi} = 16$

$\qquad\qquad r = 4$

The radius is 4 cm and the diameter is 8 cm.

---

1. Find the exact area and exact volume of each sphere described. Record your answers in a table like the one shown.

Sphere	Radius	Area	Volume
A	1 cm	4π cm^2	$\frac{4\pi}{3}$ cm^3
B	2 cm	▪	▪
C	3 cm	▪	▪
D	4 cm	▪	▪

2. What happens to the area of a sphere when its radius is multiplied by 2? by 3? by 4?

3. What happens to the volume of a sphere when its radius is multiplied by 2? by 3? by 4?

4. **a.** Are all spheres similar? Why or why not?

   b. Does Theorem 15.4 on page 465 apply to two spheres?

   c. If the ratio of the radii of two spheres is 2 : 3, what is the ratio of their areas? of their volumes?

Thinking Critically

1. The intersection of a sphere and a plane that contains the center of the sphere is called a *great circle* of the sphere. If the radius of a sphere is 1 cm, what is the exact area of a great circle of the sphere?

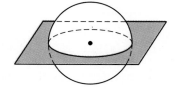

2. How does the area of a sphere compare to the area of a great circle of the sphere?

3. How many great circles does a sphere have?

Class Exercises

1. Name all radii shown for the sphere.
2. Name all diameters shown.
3. Find *XY*.
4. Find *XZ*.
5. Find the area.
6. Find the volume.
7. The volume of a sphere is 36π cm³. Find its radius and diameter.

$XW = 4$ cm

Exercises

Use a calculator in the exercises where appropriate.

1. If a radius of a sphere is 6 cm long, how long is each diameter?
2. If the diameter of a sphere is 10 cm, what is the radius?

For a sphere with each given radius or diameter, find (a) the exact area and volume and (b) the approximate area and volume.

3. $r = 8$ cm
4. $d = 30$ m

Copy and complete the following table about spheres.

	r	*d*	Exact Area	Exact Volume
5.	7 cm	▓	▓	▓
6.	▓	12 cm	▓	▓
7.	▓	▓	324π cm²	▓
8.	▓	▓	▓	$\dfrac{256\pi}{3}$ cm³

9. The approximate area of a sphere is 1,256 cm². Find the radius and the diameter of the sphere.
10. The approximate area of a sphere is 314 m². Find the radius and the approximate volume of the sphere.
11. The exact area of a sphere is π cm². Find the exact volume of the sphere.

12. In the diagram shown, the cylinder is just large enough to hold the sphere. Find each of the following.

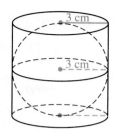

   a. the exact area of the sphere

   b. the height of the cylinder

   c. the exact lateral area of the cylinder

   d. the exact total area of the cylinder

   e. the exact volume of the sphere

   f. the exact volume of the cylinder

13. Repeat Exercise 12 for a sphere with radius 10 cm.

14. Suppose the radius of the sphere in Exercise 12 is r cm. Find each of the following ratios.

   a. the total area of the sphere to the lateral area of the cylinder

   b. the total area of the sphere to the total area of the cylinder

   c. the volume of the sphere to the volume of the cylinder

APPLICATIONS

Geography   Earth is approximately a sphere with a radius of 4,000 mi.

15. Find the approximate volume of Earth.

16. The equator is a great circle of Earth. Find the approximate distance around the equator.

17. A *hemisphere* is half a sphere. Find the approximate area of the northern hemisphere of Earth.

Everyday Geometry

At rest, all warm-blooded animals lose the same amount of heat per unit area of skin. Therefore, the amount of food they require is proportional to their total area, *not* their volume or weight, although the total surface area in relation to the volume or weight is a factor. Because a small animal has a relatively large total area compared to its volume or weight, the animal requires a relatively large amount of food each day.

Problem Solving Application: Combining Space Figures

Objective: To solve problems by combining areas and volumes.

Sometimes an apparently complex figure is actually a combination of several simpler figures. You can find its area and volume by combining those of the simpler figures.

Example: **Find the approximate total area and volume of the lunch box. Use 3.14 for π.**

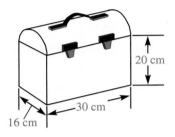

20 cm

30 cm

16 cm

Solution: The lunch box consists of half of a cylinder placed on top of a rectangular prism.

a. Area of top

$$= \frac{1}{2} (2\pi rh + 2\pi r^2)$$

$$\approx \frac{1}{2} (2 \cdot 3.14 \cdot 8 \cdot 30 + 2 \cdot 3.14 \cdot 8^2)$$

$$= 954.56 \text{ cm}^2$$

b. Area of bottom

$$= 2lh + 2wh + lw$$

$$= (2 \cdot 30 \cdot 20) + (2 \cdot 16 \cdot 20) + (30 \cdot 16)$$

$$= 2,320 \text{ cm}^2$$

c. Total area of lunch box $\approx 954.56 \text{ cm}^2 + 2,320 \text{ cm}^2$

$$= 3,274.56 \text{ cm}^2$$

d. Volume of top $= \frac{1}{2} \cdot \pi r^2 h$

$$\approx \frac{1}{2} \cdot 3.14 \cdot 8^2 \cdot 30$$

$$= 3,014.4 \text{ cm}^3$$

e. Volume of bottom $= Bh$

$$= 30 \cdot 16 \cdot 20$$

$$= 9,600 \text{ cm}^3$$

f. Volume of lunch box $\approx 3,014.4 \text{ cm}^3 + 9,600 \text{ cm}^3$

$$= 12,614.4 \text{ cm}^3$$

Class Exercises

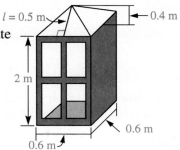

1. What two space figures have been combined to create the telephone booth?

2. Describe how you can calculate the total area and volume of the telephone booth.

3. Find the approximate total area and volume.

Exercises

Solve. Use 3.14 for π.

1. Find the total area and volume of the figure.

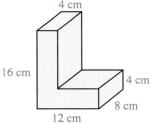

2. What is the approximate volume of the machine nut pictured?

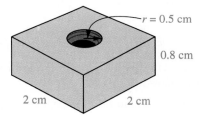

3. Find the amount of concrete needed to make this section of hollow pipe.

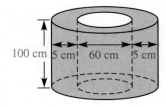

4. A feeding bowl for a pet is a solid plastic cylinder with a hemisphere scooped out. What is the approximate volume of plastic in the dish?

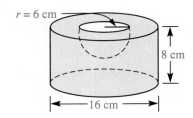

Vocabulary and Symbols

You should be able to write a brief statement, draw a picture, or give an example to illustrate the meaning of each term or symbol.

Vocabulary

regular pyramid (p. 469)
   height (p. 469)
   slant height (p. 470)
   vertex (p. 469)

right circular cone (p. 470)
   height (p. 470)
   slant height (p. 470)
   vertex (p. 470)
right circular cylinder (p. 451)
similar space figures (p. 464)

sphere (p. 474)
   center (p. 474)
   great circle (p. 475)
   hemisphere (p. 477)

Symbol

l (slant height) (p. 470)

Summary

The following list indicates the major skills, facts, and results you should have mastered in this chapter.

15.1 Find the lateral area or total area of a right circular cylinder. (pp. 451–454)

15.2 Find the volume or a missing dimension (given the volume) of a right circular cylinder. (pp. 456–459)

15.3 Use the relationship between the areas (and perimeters) of two similar figures. (pp. 460–463)

15.4 Use the relationships between the areas and volumes of similar space figures. (pp. 464–468)

15.5 Find the lateral area, total area, and volume of a regular pyramid or right circular cone. (pp. 469–473)

15.6 Find the area or volume of a sphere. (pp. 474–477)

15.7 Solve problems by combining areas and volumes. (pp. 478–479)

Exercises

Use a calculator where appropriate.

Find the indicated information.

  1. the height of a cylinder with volume 720 cm^3 and base area 45 cm^2

  2. the height of a regular pyramid with volume 80 in.^{3} and base area 40 in.^{2}

Find the indicated information for each figure.

3. L.A. = ▨; T.A. = ▨

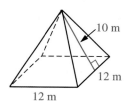

8 m

9 m

4. Use $\frac{22}{7}$ for π; V = ▨

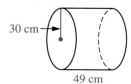

30 cm

49 cm

5. V = ▨

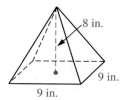

8 in.

9 in.

9 in.

6. L.A. = ▨; T.A. = ▨

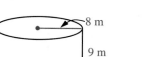

10 m

12 m

12 m

7. L.A. = ▨

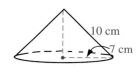

10 cm

7 cm

8. T.A. = ▨; V = ▨

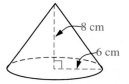

8 cm

6 cm

9. Find the exact area and volume of a sphere with radius 8 ft.

10. The scale factor of two similar polygons is 3 : 5. The perimeter and area of the larger polygon are 110 ft and 750 ft^2, respectively. Find the perimeter and area of the smaller polygon.

11. The scale factor of two similar prisms is 1 : 3. The total area and volume of the smaller prism are 56 in.^{2} and 40 in.^{3}, respectively. Find the total area and volume of the larger prism.

PROBLEM SOLVING

12. A hole is drilled in a wooden block. Find the approximate volume of wood remaining. Use 3.14 for π.

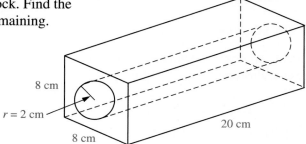

8 cm

$r = 2$ cm

8 cm

20 cm

Use a calculator where appropriate.

Find the indicated information.

1. the circumference of the base of a cylinder with lateral area 80π cm^2 and height 5 cm

2. the height of a cylinder with volume 180 cm^3 and base area 15 cm^2

Find the lateral area, total area, and volume of each figure. The figure in Exercise 4 is a regular square pyramid.

3.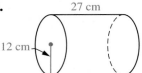
27 cm

12 cm

4.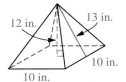
12 in.   13 in.
10 in.
10 in.

5.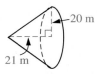
20 m
21 m

6. A sphere has diameter 20 m. Find (a) the exact area and volume of the sphere and (b) the approximate area and volume.

7. $\triangle JKL \sim \triangle XYZ$ and the scale factor of $\triangle JKL$ to $\triangle XYZ$ is 2:5. If $\triangle JKL$ has area 72 mm^2, find the area of $\triangle XYZ$.

8. The scale factor of two similar pyramids is 3 : 4. If the smaller pyramid has total area 216 cm^2 and volume 135 cm^3, find the total area and volume of the larger pyramid.

9. The volume of a cylinder is 448π cm^3. If the area of the base of the cylinder is 64π cm^2, find the height of the cylinder.

Are the figures described similar? Explain.

10. a rectangular prism 12 in. long, 6 in. wide, and 9 in. high and a rectangular prism 18 in. long, 9 in. wide, and 13.5 in. high

PROBLEM SOLVING

11. Find the approximate volume of the cold capsule illustrated. Use 3.14 for π.

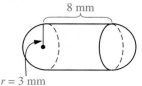
8 mm

$r = 3$ mm

Chapter 16

Trigonometry

The Gateway Arch towers over the St. Louis skyline. To estimate its height, you could use shadows to set up a proportion from similar triangles. Or you could use *trigonometry* (literally, "triangle measurement") based on similar right triangles.

Focus on Skills

ARITHMETIC

Express each fraction in simplest form.

1. $\frac{10}{26}$ **2.** $\frac{18}{24}$ **3.** $\frac{24}{26}$ **4.** $\frac{9}{15}$ **5.** $\frac{15}{12}$

6. $\frac{24}{51}$ **7.** $\frac{45}{51}$ **8.** $\frac{40}{58}$ **9.** $\frac{42}{40}$ **10.** $\frac{45}{24}$

Write each ratio as a fraction in simplest form.

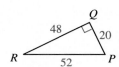

11. $JK : KL$ **12.** $JK : JL$ **13.** $KL : JL$

14. $KL : JK$ **15.** $PQ : QR$ **16.** $PQ : PR$

17. $RQ : PR$ **18.** $RQ : PQ$ **19.** $PR : PQ$

Write the decimal equivalent of each fraction. If necessary, round to the nearest thousandth. A calculator may be helpful.

20. $\frac{3}{4}$ **21.** $\frac{5}{8}$ **22.** $\frac{7}{12}$ **23.** $\frac{9}{15}$ **24.** $\frac{20}{29}$

25. $\frac{20}{30}$ **26.** $\frac{15}{45}$ **27.** $\frac{12}{20}$ **28.** $\frac{21}{29}$ **29.** $\frac{5}{13}$

GEOMETRY

Exercises 30–35 refer to the figures.
Name the leg opposite the given angle.

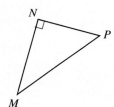

30. $\angle M$ **31.** $\angle Y$ **32.** $\angle Z$

Name the leg adjacent to the given angle.

33. $\angle Y$ **34.** $\angle M$ **35.** $\angle P$

The lengths of two sides of a right triangle are given. Find the third length if c is the length of the hypotenuse. Simplify each answer.

36. $a = 8, b = 15, c = $ ▇

37. $a = 42, c = 58, b = $ ▇

38. $b = 9, c = 18, a = $ ▇

39. $a = 10, b = 10, c = $ ▇

40. If $\angle R$ and $\angle T$ are complementary angles and $m\angle R = 42°$, find $m\angle T$.

41. $\triangle DEF$ has height 6 cm and base length 10 cm. Find the area of $\triangle DEF$.

16.1 Trigonometric Ratios

Objective: To find the sine, cosine, and tangent ratios for any acute angle of a right triangle, given the lengths of the sides of the triangle.

Trigonometry is based on properties of similar right triangles. The word "trigonometry" comes from two Greek words meaning "triangle measurement."

EXPLORING

1. Draw three noncongruent right triangles that contain a 37° angle.

2. Measure the sides of each triangle to the nearest millimeter.

3. For each triangle find each of the following ratios and express it in simplest form:

   a. $\dfrac{\text{length of leg opposite } 37° \text{ angle}}{\text{length of hypotenuse}}$

   b. $\dfrac{\text{length of leg adjacent to } 37° \text{ angle}}{\text{length of hypotenuse}}$

   c. $\dfrac{\text{length of leg opposite } 37° \text{ angle}}{\text{length of leg adjacent to } 37° \text{ angle}}$

4. Use a calculator to find a decimal approximation to the nearest thousandth for each ratio in Step 3.

5. Look at your results in Step 4. Which results are the same?

6. Draw three noncongruent right triangles that contain a 28° angle. Repeat Steps 2–5 using 28°.

7. If two right triangles have congruent acute angles, must the triangles be similar? Why or why not?

The **EXPLORING** activity demonstrates that each of the ratios in Step 3 depends only on the measure of the acute angle and not on the lengths of the sides. Because of the AA Postulate, this is true for all acute angles.

We can define the *sine*, *cosine*, and *tangent* ratios as follows for acute $\angle A$ in a right triangle. Note that the abbreviations sin, cos, and tan are used with the name of the angle. The abbreviations are read as sine, cosine, and tangent.

$$\sin A = \frac{\text{length of leg opposite } \angle A}{\text{length of hypotenuse}} \qquad \cos A = \frac{\text{length of leg adjacent to } \angle A}{\text{length of hypotenuse}}$$

$$\tan A = \frac{\text{length of leg opposite } \angle A}{\text{length of leg adjacent to } \angle A}$$

Example: **Express each ratio as a decimal to the nearest thousandth.**

    a. sin R, cos R, tan R **b.** sin T, cos T, tan T

Solution: **a.** $\sin R = \dfrac{ST}{RT} = \dfrac{5}{13} \approx 0.385$ **b.** $\sin T = \dfrac{RS}{RT} = \dfrac{12}{13} \approx 0.923$

$\cos R = \dfrac{RS}{RT} = \dfrac{12}{13} \approx 0.923$ $\cos T = \dfrac{ST}{RT} = \dfrac{5}{13} \approx 0.385$

$\tan R = \dfrac{ST}{RS} = \dfrac{5}{12} \approx 0.417$ $\tan T = \dfrac{RS}{ST} = \dfrac{12}{5} = 2.400$

Thinking Critically

1. Use your results from the Exploring activity to give a decimal approximation for each of the following.

    a. sin 37°   **b.** cos 37°   **c.** tan 37°

    d. sin 28°   **e.** cos 28°   **f.** tan 28°

    g. sin 53°   **h.** cos 53°   **i.** tan 53°

2. Explain how you could find x.

    a.

    b.

    c.

3. In the Example why is sin R = cos T?

4. What does "sin A" represent—a number, an angle, or an angle multiplied by a number?

Class Exercises

**Express each trigonometric ratio (a) in simplest form
and (b) as a decimal to the nearest thousandth.**

1. sin F **2.** cos F **3.** tan F

4. sin D **5.** cos D **6.** tan D

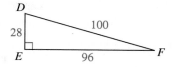

Exercises

Use a calculator where appropriate.

**Express each trigonometric ratio (a) in simplest form and (b) as a
decimal to the nearest thousandth.**

1. sin A

2. cos A

3. tan A

4. sin C

5. cos C

6. tan C

7. sin R

8. cos R

9. tan R

10. sin S

11. cos S

12. tan S

**Find the following information for each triangle shown. Express each
trigonometric ratio in simplest radical form and as a decimal to the
nearest thousandth.**

13. a. length of the hypotenuse

    b. tan 45°

    c. sin 45°

    d. cos 45°

14. a. length of the hypotenuse

    b. sin 30°   **c.** cos 30°

    d. tan 30°   **e.** sin 60°

    f. cos 60°   **g.** tan 60°

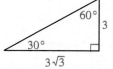

15. Look at your results for Exercise 13. Would your answers to parts
(b)–(d) be the same for any 45°-45°-90° triangle? Why or why not?

16. Look at your results for Exercise 14. Would your answers to parts
(b)–(g) be the same for any 30°-60°-90° triangle? Why or why not?

17. Classify each statement as true or false.

    a. sin A = cos B

    b. cos A = sin B

    c. tan A = tan B

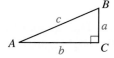

Explain why each statement is true.

18. It is impossible for the sine of an angle to be greater than 1.

19. It is impossible for the cosine of an angle to be greater than 1.

20. It is possible for the tangent of an angle to be greater than 1.

21. It is possible for the tangent of an angle to be less than 1.

22. $\cos 20° = \sin 70°$.

23. The sine of an acute angle is equal to the cosine of the complement of the angle.

24. If $\tan A = 1$, then $m\angle A = 45°$.

25. If $\tan A < 1$, then $m\angle A < 45°$.

26. If $\tan A > 1$, then $m\angle A > 45°$.

APPLICATION

27. Forestry To estimate the length of a usable log from a tree, a forester can stand 100 ft from the tree and sight to the highest point of the tree that is usable for lumber. Find BC to the nearest foot for each of the following angle measures.

a. $m\angle BAC = 37°$

b. $m\angle BAC = 28°$

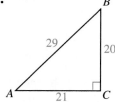

Calculator

1. Use your calculator to find $(\sin A)^2$, $(\cos A)^2$, and $(\sin A)^2 + (\cos A)^2$ for each triangle. Round each answer to the nearest thousandth.

a.

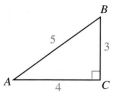

b.

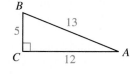

c.

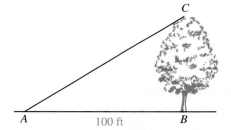

2. Use your results to complete this statement: The square of the sine of an angle plus the square of the cosine of the angle equals ▮.

16.2 Using a Table or Calculator

Objective: To use a table or calculator to find a trigonometric ratio for a given acute angle or to find the measure of a given acute angle of a right triangle, given the lengths of two sides of the triangle.

In the previous lesson you found the following trigonometric ratios.

$\sin 45° \approx 0.707$ $\qquad$ $\cos 45° \approx 0.707$ $\qquad$ $\tan 45° = 1$

$\sin 30° = 0.500$ $\qquad$ $\cos 30° \approx 0.866$ $\qquad$ $\tan 30° \approx 0.577$

$\sin 60° \approx 0.866$ $\qquad$ $\cos 60° = 0.500$ $\qquad$ $\tan 60° \approx 1.732$

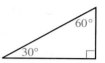

To find values of the trigonometric ratios for other acute angles, you could draw triangles, measure lengths of sides, and calculate ratios the way you did for 37° and 28° in the Exploring activity on page 485. However, that is very time consuming. Even small errors in drawing and measuring will lead to inaccurate results. Fortunately, accurate trigonometric tables such as the one on page 567 have been developed. Part of the table is shown below.

Angle	sin	cos	tan
71°	0.946	0.326	2.904
72°	0.951	0.309	3.078
73°	0.956	0.292	3.271
74°	0.961	0.276	3.487
75°	0.966	0.259	3.732

Example 1: **Find each value.**

    a. cos 74° $\qquad\qquad\qquad$ **b.** tan 72°

Solution: **a.** cos 74° = 0.276 $\qquad$ **b.** tan 72° = 3.078

Example 2: **Find $m\angle A$.**

    a. sin A = 0.946 $\qquad\qquad$ **b.** cos A = 0.292

Solution: **a.** $m\angle A$ = 71° $\qquad\qquad$ **b.** $m\angle A$ = 73°

The trigonometric ratio used to find the measure of an acute angle of a triangle depends on which side lengths are known.

Example 3: **a.** Find $m\angle R$.

b. Find $m\angle D$.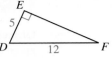

Solution: **a.** The tangent of $\angle R$ involves the two lengths known.

$$\tan R = \frac{7}{5} = 1.400$$

Using the tangent column, we find:

$\tan 54° = 1.376$ ⎫
⎬ difference: 0.024
$\tan R = 1.400$ ⎱
⎫ difference: 0.028
$\tan 55° = 1.428$ ⎭

Since 1.400 is closer to 1.376, $m\angle R \approx 54°$ to the nearest degree.

b. The cosine of $\angle D$ involves the two lengths known.

$$\cos D = \frac{5}{12} \approx 0.416666$$

Using the cosine column, we find that 0.416666 is closest to 0.423.

Thus $m\angle D \approx 65°$.

Scientific calculators have sin, cos, and tan keys. To find angle measures, as in Examples 2 and 3, you would use the second function key. If you have a scientific calculator, find out how to use it to do the Examples.

EXPLORING

You can use LOGO to find the sine or cosine of an angle. If you know the tangent of an angle, ARCTAN is the LOGO command that will output the measure of the angle.

1. Run each of the following procedures. Sketch each result, labeling each angle measure or side length of the triangle that is used in the procedure.

a. To triangle.1
rt 54 fd 50
lt 90 fd 70
home ht
end

b. To triangle.2
lt 90 fd 65
lt arctan 2.4 bk 25
lt 90 fd 60 ht
end

c. To triangle.3
fd 58
rt arctan 42 / 40 bk 40
rt 90 fd 42 ht
end

2. Write procedures to draw each triangle shown.

a.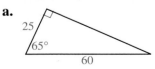

b.

c.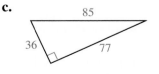

Thinking Critically

1. Which decimal values of trigonometric ratios for 30°, 45°, and 60° angles are exact? Which are approximations?

2. **a.** Find $m\angle A$.

   b. Name two ways to find $m\angle B$. Which do you prefer? Why?

3. **a.** Which numbered angle has the greatest sine?

   b. Which has the greatest cosine?

   c. Which has the greatest tangent?

Class Exercises

Use the table on page 567 to find each value.

1. sin 8°   2. sin 63°   3. cos 17°

4. cos 27°   5. tan 32°   6. tan 73°

Use the table on page 567 to find $m\angle A$ to the nearest degree.

7. sin A = 0.788   8. sin A = 0.450   9. cos A = 0.719

10. cos A = 0.670   11. tan A = 0.510   12. tan A = 2

Find $m\angle A$ to the nearest degree.

13.
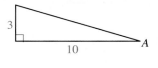

14.

15.
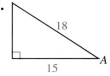

Exercises

Use the table on page 567 or a scientific calculator where appropriate.

State the trigonometric ratio—sine, cosine, or tangent—you would use to find $m\angle A$.

1.
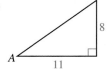

2.
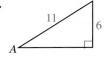

3.

State the trigonometric ratio—sine, cosine, or tangent—you would use to find $m\angle A$.

4.

5.

6.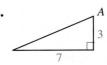

Find each value to the nearest thousandth.

7. sin 5°   **8.** sin 32°   **9.** sin 41°   **10.** sin 87°   **11.** cos 14°   **12.** cos 26°

13. cos 58°   **14.** cos 71°   **15.** tan 1°   **16.** tan 13°   **17.** tan 67°   **18.** tan 77°

Use the table on page 567 to locate the measure of each angle between two consecutive whole numbers.

19. If sin $A = 0.300$, then $m\angle A$ is between ▨ and ▨.

20. If cos $A = 0.750$, then $m\angle A$ is between ▨ and ▨.

21. If tan $A = 1.300$, then $m\angle A$ is between ▨ and ▨.

Find $m\angle A$ to the nearest degree.

22. sin $A = 0.866$ **23.** sin $A = 0.656$ **24.** sin $A = 0.750$

25. sin $A = \dfrac{9}{10}$ **26.** cos $A = 0.485$ **27.** cos $A = 0.974$

28. cos $A = 0.750$ **29.** cos $A = \dfrac{2}{5}$ **30.** tan $A = 0.700$

31. tan $A = 0.270$ **32.** tan $A = 2.5$ **33.** tan $A = 3$

34.

35.

36.

37.

38.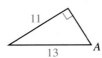

39.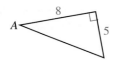

40. Does tan 60° = 2 • tan 30°?   **41.** Does cos 50° = cos 35° + cos 15°?

42. Complete: **a.** cos 22° = sin ▨   **43.** Which is greater, cos 70° or tan 20°?
                      b. sin 55° = cos ▨

Use the table on page 567 to answer each question.

44. For what angle is the tangent about twice the cosine?

45. For what angle is the cosine about three times the sine?

46. For what angle is the tangent about twice the sine?
Explain why this is true.

APPLICATIONS

47. Construction A ramp is to rise 7 ft over a horizontal distance of 100 ft. To the nearest degree, find the measure of the angle the ramp will make with the ground.

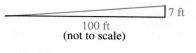

100 ft
(not to scale)

48. Computer Use LOGO to enter the procedure at the right. Then run the procedure for each of the following. Sketch each result.

a. triangle 30 40

b. triangle 40 30

c. triangle 30 80

d. triangle 80 30

```
To triangle :AB :BC
lt 90
fd sqrt ( :AB * :AB + :BC * :BC)
lt arctan :AB / :BC
bk :BC
lt 90
fd :AB
ht
end
```

49. Computer Describe how you would modify the procedure shown above so that it draws a triangle in each of the following positions, starting at point *A*.

a.

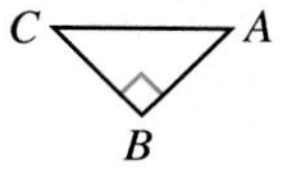

b.

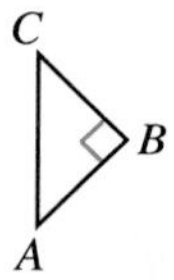

c.

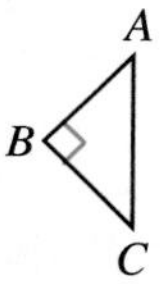

Test Yourself

Express each ratio as a fraction and as a decimal to the nearest thousandth.

1. sin *A*

2. cos *A*

3. tan *A*

4. cos *B*

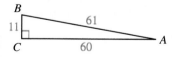

Use the table on page 567 or a scientific calculator.

Find each value to the nearest thousandth.

5. cos 59°

6. tan 19°

7. sin 31°

Find $m\angle A$ **to the nearest degree.**

8. sin *A* = 0.900

9. tan *A* = 1.5

10. cos *A* $= \dfrac{2}{9}$

11.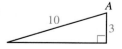

12.

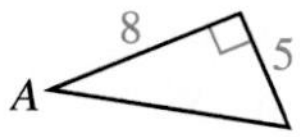

16.3 Applying Trigonometric Ratios

Objective: To find the length of each side of a right triangle, given the length of one side and the measure of an acute angle.

If you know the measure of one acute angle of a right triangle and the length of one side, you can find the length of either or both of the remaining sides. The trigonometric ratio you use depends on the information that is given and the length you want to find.

Example: **Find each length to the nearest whole number.**

    a. x **b.** y

Solution: **a.** For the 28° angle, the ratio that involves x and 9 is the tangent ratio.

$$\tan 28° = \frac{x}{9}$$

$$0.532 = \frac{x}{9}$$

$$x = 9 \boxed{\times} 0.532 \boxed{=} 4.788$$

$$x \approx 5$$

b. $\cos 28°$ involves y and 9.

$$\cos 28° = \frac{9}{y}$$

$$y(\cos 28°) = 9$$

$$y = \frac{9}{\cos 28°}$$

$$y = 9 \boxed{\div} 0.883$$

$$\boxed{=} 10.19$$

$$y \approx 10$$

If a person looks *from A to B*, then $\angle A$ is the ***angle of elevation***. If a person looks *from B to A*, then $\angle B$ is the ***angle of depression***.

The angle of elevation and the angle of depression are both determined by the *line of sight* and a horizontal ray. Notice that since $\angle A$ and $\angle B$ are alternate interior angles determined by parallel lines, they are congruent.

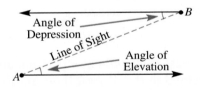

EXPLORING

1. A pilot flying a plane (P) saw two ships straight ahead and below at S_1 and S_2. Find the angle of depression of:

    a. S_1 **b.** S_2

2. Observers on the ships spotted the plane. Find the angle of elevation of the plane from:

    a. S_1 **b.** S_2

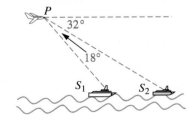

3. The plane is flying at an altitude of 10,000 ft. Find the distance between the ships. (*Hint:* Find the horizontal distance from directly below the plane to each of the ships.)

Thinking Critically

1. Does $5^2 + 9^2$ equal 10^2? Is the triangle in the Example on page 494 a right triangle? Explain why it *seems* not to satisfy the Pythagorean Theorem.

2. Describe three ways to find *x*.

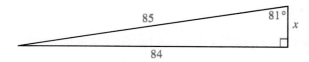

3. Describe the information about a right triangle that you must know in order to solve each problem described.

   a. Find the length of one side using the Pythagorean Theorem.

   b. Find the length of the hypotenuse using the sine ratio.

   c. Find the length of one leg using the tangent ratio.

   d. Find the measure of an acute angle using the cosine ratio.

   e. Find the measure of an acute angle using the Triangle Angle-Sum Theorem.

Class Exercises

State the trigonometric ratio — sine, cosine, or tangent — you would use to solve each problem.

Given:	Find:		Given:	Find:
1. *a, b*	*m∠A*		**4.** *b, m∠A*	*c*
2. *a, c*	*m∠B*		**5.** *a, m∠A*	*c*
3. *b, c*	*m∠B*		**6.** *a, m∠B*	*b*

Find the indicated measures to the nearest whole number.

7. a. *m∠D*   **8. a.** *r*

   b. *m∠F*   **b.** *t*

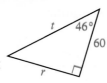

Exercises

Use a calculator where appropriate.

State the trigonometric ratio — sine, cosine, or tangent — you would use to solve each problem.

	Given:	Find:			Given:	Find:
1.	d, f	$m\angle F$		**6.**	$f, m\angle D$	e
2.	e, f	$m\angle D$		**7.**	$d, m\angle F$	f
3.	e, f	$m\angle F$		**8.**	$f, m\angle F$	d
4.	e, d	$m\angle D$		**9.**	$d, m\angle D$	e
5.	$f, m\angle F$	e		**10.**	$e, m\angle D$	f

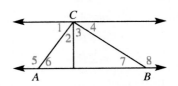

Select the equation that can be used to solve each problem.

11. $m\angle A = \blacksquare$

   a. $\sin A = \dfrac{8}{11}$

   b. $\sin A = \dfrac{11}{8}$

   c. $\cos A = \dfrac{8}{11}$

   d. $\cos A = \dfrac{11}{8}$

12. $m\angle A = \blacksquare$

   a. $\tan A = \dfrac{3}{7}$

   b. $\tan A = \dfrac{7}{3}$

   c. $\sin A = \dfrac{3}{7}$

   d. $\cos A = \dfrac{7}{3}$

13. $x = \blacksquare$

   a. $\sin 35° = \dfrac{x}{10}$

   b. $\cos 35° = \dfrac{x}{10}$

   c. $\sin 35° = \dfrac{10}{x}$

   d. $\cos 35° = \dfrac{10}{x}$

14. $a = \blacksquare$

   a. $\sin 40° = \dfrac{a}{12}$

   b. $\cos 40° = \dfrac{a}{12}$

   c. $\sin 40° = \dfrac{12}{a}$

   d. $\cos 40° = \dfrac{12}{a}$

15. Name the angle of elevation for each line of sight.

   a. from A to C **b.** from B to C

16. Name the angle of depression for each line of sight.

   a. from C to A **b.** from C to B

Find x to the nearest whole number.

17.

18.

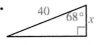

19.

20.

Find each measure to the nearest whole number or degree.

21. a. *MN*
   b. *NP*

22. a. *x*
   b. *y*

23. a. $m\angle B$
   b. $m\angle C$

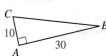

24. a. $m\angle R$
   b. $m\angle S$

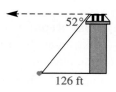

25. Find the measure of the angle formed by a diagonal and the shorter side of a rectangle that is 20 cm long and 8 cm wide.

26. From the top of a tower, the angle of depression to a rock on the ground is 52°. The rock is 126 ft from the foot of the tower. How tall is the tower? Find the answer to the nearest foot.

27. How far above the ground is the kite shown?

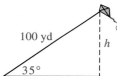

28. $\triangle ABC$ is isosceles and $m\angle ACB = 40°$. Find the height *h*.

29. *DEFG* is a rhombus and $m\angle GDE = 58°$. Find each of the following.
   a. $m\angle GDX$
   b. *GX*
   c. *GE*
   d. *DF*

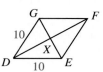

30. a. For $\triangle RST$, find *h*.
   b. Find the area of $\triangle RST$ to the nearest square centimeter.

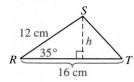

31. a. Write a formula for the area of $\triangle ABC$ in terms of *a* and *h*.
   b. Complete: $\sin C = \dfrac{\blacksquare}{\blacksquare}$, or $\blacksquare \cdot \sin C = \blacksquare$
   c. Substitute from part (b) into the formula in part (a) to get a formula for the area of $\triangle ABC$ that involves $\sin C$.
   d. Describe your formula in part (c) in words.

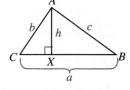

Use your formula from Exercise 31 to find the area of each triangle to the nearest square unit.

32.

33.

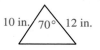

34.

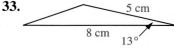

APPLICATIONS

35. **Surveying** The shadow of the tree shown is 15 m long. Find the height of the tree to the nearest meter.

50°
15 m

36. **Surveying** Find d, the distance across the river shown, to the nearest meter.

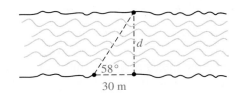

d
58°
30 m

37. **Home Repair** For safety, a ladder placed against a wall should make an angle of about 75° with the ground. Give your answers to these questions to the nearest foot.

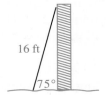

16 ft
75°

   a. How far from the bottom of a wall should a 16-foot ladder be?

   b. How far up the side of the wall will the ladder reach?

Everyday Geometry

A tree that shades a house blocks the sun's light and heat, so the house stays cooler in summer. But in winter, when people in cool climates want more heat, even a leafless tree blocks some of the sun's energy.

Trigonometry can be used to determine how close to a house a tree can be planted without casting a shadow on the house in winter. Because Earth is tilted with respect to the sun, the altitude (angle) of the sun at noon varies throughout the year; it is lowest on December 21. If $\angle A$ is the altitude of the sun at noon on December 21, then:

$$\tan A = \frac{\text{maximum height of tree}}{\text{minimum distance from house}}$$

Latitude (°N)	Cities Near This Latitude	$m\angle A$
40	Denver; Philadelphia	24°
45	Minneapolis; Toronto	19°

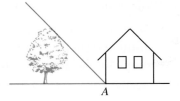

A

1. Suppose you live in Denver and want to plant a tree 50 ft from the house. What is the maximum height that the tree can attain?

2. Suppose you live in Minneapolis and want to plant a tree that will grow to be no more than 30 ft tall. How close to the house can you plant it?

Trigonometric Identities

A **trigonometric identity** is an equation involving trigonometric ratios that is true for *all* angles. You discovered the best-known trigonometric identity, $(\sin A)^2 + (\cos A)^2 = 1$, in the calculator feature on page 488.

Example: Use $\triangle ABC$ to show that $(\sin A)^2 + (\cos A)^2 = 1$ for any acute $\angle A$.

Solution: Since $\sin A = \frac{a}{c}$, $(\sin A)^2 = \left(\frac{a}{c}\right)^2 = \frac{a^2}{c^2}$.

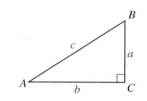

Since $\cos A = \frac{b}{c}$, $(\cos A)^2 = \left(\frac{b}{c}\right)^2 = \frac{b^2}{c^2}$.

So $(\sin A)^2 + (\cos A)^2 = \frac{a^2}{c^2} + \frac{b^2}{c^2} = \frac{a^2 + b^2}{c^2}$.

By the Pythagorean Theorem, $a^2 + b^2 = c^2$.

So $(\sin A)^2 + (\cos A)^2 = \frac{a^2 + b^2}{c^2} = \frac{c^2}{c^2} = 1$.

Exercises

1. a. Complete: $\cos A \tan A = \left(\dfrac{\blacksquare}{\blacksquare}\right)\left(\dfrac{\blacksquare}{\blacksquare}\right) = \dfrac{a}{c} = \blacksquare$

b. Write the identity that you showed to be true in part (a).

c. Use the identity in part (b) to complete this identity: $\dfrac{\sin A}{\cos A} = \blacksquare$

2. The Law of Sines says that if $\angle A$ and $\angle B$ are angles of *any* triangle, with a and b the lengths of the sides opposite $\angle A$ and $\angle B$, respectively, then:

$$\frac{\sin A}{a} = \frac{\sin B}{b}$$

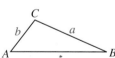

To show that the Law of Sines is true for any acute angles $\angle A$ and $\angle B$ of $\triangle ABC$, let h be the length of the altitude from vertex C. Complete:

a. $\sin A = \dfrac{\blacksquare}{\blacksquare}$ and $\sin B = \dfrac{\blacksquare}{\blacksquare}$

b. $\blacksquare \cdot \sin A = \blacksquare$ and $\blacksquare \cdot \sin B = \blacksquare$

c. Substitute in (b): $\blacksquare \cdot \sin A = \blacksquare \cdot \sin B$

d. When you divide both sides of the equation in part (c) by ab and simplify, what do you get?

16.4 Problem Solving Application: Trigonometry in Astronomy

Objective: To use trigonometry to estimate distances in the solar system.

You can use trigonometry to estimate sizes and distances for many of the objects in our solar system.

Example: Astronomers have found that angle E shown below has measure about $0.26°$. The distance d from Earth to the moon is approximately 240,000 mi. Find the approximate radius of the moon.

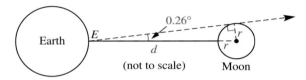

(not to scale)

Solution:

$$\sin 0.26° = \frac{r}{d + r}$$

$$\sin 0.26°(d + r) = r$$

$$(\sin 0.26°)\, d + (\sin 0.26°)\, r = r$$

$$(\sin 0.26°)\, d = r - (\sin 0.26°)\, r$$

$$= (1 - \sin 0.26°)\, r$$

$$\frac{(\sin 0.26°)d}{(1 - \sin 0.26°)} = r$$

Use a scientific calculator or the table shown below.

Substitute 240,000 for d and 0.004538 for $\sin 0.26°$.

$$1 - \sin 0.26° \approx 1 - 0.004538 = 0.995462$$

$$r \approx 0.004538 \;\boxed{\times}\; 240000 \;\boxed{\div}\; 0.995462 \;\boxed{=}\; 1094.085$$

$$\approx 1094$$

The radius is about 1,094 mi. (It is actually 1,080 mi.)

Angle	sin	cos	tan
0.17°	0.002967	0.999996	0.002967
0.26°	0.004538	0.999990	0.004538
0.37°	0.006458	0.999979	0.006458
0.68°	0.011868	0.999930	0.011869

Class Exercises

1. Find the approximate length of the tangent segment shown in the figure for the Example on page 500. Use 1,094 mi for the radius of the moon. Round your answer to the nearest thousand miles.

2. **a.** Suppose the measure of the angle in the Example were 0.25° instead of 0.26°. (The sine of 0.25° is about 0.004363.) What length would you find for the approximate radius of the moon now?

   b. Suppose the measure of the angle were actually between 0.25° and 0.26°, but closer to 0.26°. Describe the range of possible values for the radius of the moon.

Exercises

The distance d from Earth to the sun is approximately 95,000,000 mi. You have probably noticed that the sun appears to be the same size as the moon. That is because although it is much farther away than the moon, it is so much larger that its apparent size is the same. Therefore the measure of the angle formed by the solid line and the dashed tangent is the same as for the moon—about 0.26°.

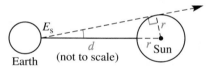

1. Use the correct trigonometric ratio and the table in the Example to find the approximate radius of the sun. Round your answer to the nearest thousand miles.

2. About how many moons could be lined up end-to-end across the diameter of the sun?

You can use your answer to Exercise 1 to find the distances from the planets Mercury, Venus, and Mars to the sun. The measure of the angle corresponding to $\angle E_s$ would be 0.68° from Mercury, 0.37° from Venus, and 0.17° from Mars.

3. Use a scientific calculator or the table in the Example to find the distance of each planet from the sun. Round to the nearest million miles.

   a. Mercury   **b.** Venus   **c.** Mars

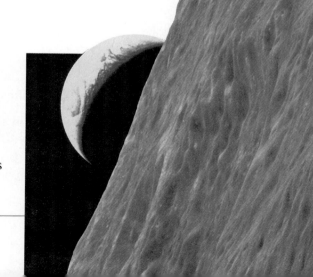

Vocabulary and Symbols

You should be able to write a brief statement, draw a picture, or give an example to illustrate the meaning of each term or symbol.

Vocabulary	Symbols		
angle of depression (p. 494)	cos A	(cosine of $\angle A$)	(p. 486)
angle of elevation (p. 494)	sin A	(sine of $\angle A$)	(p. 486)
cosine (p. 486)	tan A	(tangent of $\angle A$)	(p. 486)
sine (p. 486)			
tangent (p. 486)			
trigonometric identity (p. 499)			
trigonometry (p. 485)			

Summary

The following list indicates the major skills, facts, and results you should have mastered in this chapter.

16.1 Find the sine, cosine, and tangent ratios for any acute angle of a right triangle, given the lengths of the sides of the triangle. (pp. 485–488)

16.2 Use a table or calculator to find a trigonometric ratio for a given acute angle or to find the measure of a given acute angle of a right triangle, given the lengths of two sides of the triangle. (pp. 489–493)

16.3 Find the length of each side of a right triangle, given the length of one side and the measure of an acute angle. (pp. 494–498)

16.4 Use trigonometry to estimate distances in the solar system. (pp. 500–501)

Exercises

Use the table on page 567 or a scientific calculator where appropriate.

Express each trigonometric ratio (a) in simplest form and (b) as a decimal to the nearest thousandth.

1. sin S   **2.** cos S   **3.** tan S

4. sin T   **5.** cos T   **6.** tan T

Find the following information. Express each trigonometric ratio in simplest radical form and as a decimal to the nearest thousandth.

7. *XZ*

8. sin 45°

9. cos 45°

10. tan 45°

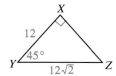

State the trigonometric ratio—sine, cosine, or tangent—you would use to find *m∠J*.

11.

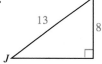

12.

13.

Find each value to the nearest thousandth.

14. sin 15°   **15.** sin 72°   **16.** cos 34°   **17.** cos 2°   **18.** tan 42°   **19.** tan 66°

Find *m∠Y* to the nearest degree.

20. sin *Y* = 0.809   **21.** sin *Y* = 0.282   **22.** cos *Y* = 0.684

23. cos *Y* = 0.948   **24.** tan *Y* = 5.671   **25.** tan *Y* = 0.750

Find each measure to the nearest degree or whole number.

26. a. *x*   **27. a.** *a*   **28. a.** *m∠H*   **29. a.** *m∠D*

    b. *y*   **b.** *b*   **b.** *m∠G*   **b.** *m∠E*

                              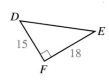

PROBLEM SOLVING

30. If the sun were viewed from the planet Jupiter, the measure of the angle formed by the distance and tangent lines would be about 0.0512°. Use the fact that sin 0.0512° = 0.000894 and the radius of the sun is approximately 433,000 mi to find the distance from Jupiter to the sun to the nearest million miles.

Chapter 16 Test

Use the table on page 567 or a scientific calculator where appropriate.

Express each ratio (a) in simplest form and (b) as a decimal to the nearest thousandth.

1. sin W
2. cos W
3. tan W
4. sin N
5. cos N
6. tan N

Find each value to the nearest thousandth.

7. cos 55° **8.** tan 19° **9.** sin 16° **10.** tan 80° **11.** sin 64° **12.** cos 48°

Find $m\angle D$ to the nearest degree.

13. cos D = 0.839
14. sin D = 0.993
15. sin D = 0.640
16. tan D = 4.329
17. cos D = 0.925
18. tan D = 0.708

Two people at points A and B sight a helicoptor hovering at point P directly over point Q. Name each angle for the given line of sight.

19. the angle of elevation from A to P
20. the angle of depression from P to B
21. the angle of elevation from B to P
22. the angle of depression from P to A
23. If AQ = 65 ft and $m\angle 6$ = 49°, find PQ.

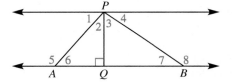

Select the equation that can be used to solve each problem.

24. $m\angle J = $ ▨

  a. cos $J = \dfrac{4}{9}$
  b. cos $J = \dfrac{9}{4}$
  c. sin $J = \dfrac{4}{9}$
  d. sin $J = \dfrac{9}{4}$

25. $m\angle P = $ ▨

  a. sin $P = \dfrac{8}{5}$
  b. sin $P = \dfrac{5}{8}$
  c. cos $P = \dfrac{5}{8}$
  d. cos $P = \dfrac{8}{5}$

PROBLEM SOLVING

26. If the sun were viewed from the planet Saturn, the measure of the angle formed by the distance and tangent lines would be about 0.0278°. Use the fact that sin 0.0278° = 0.000485 and the radius of the sun is approximately 433,000 mi to find the distance from Saturn to the sun to the nearest million miles.

Chapter 17

The Coordinate Plane

In many cities and towns the streets are arranged in a grid pattern formed by rows of parallel streets that are intersected by perpendicular cross-streets.

Focus on Skills

ARITHMETIC

Find each sum, difference, product, or quotient.

1. $-7 + 8$ **2.** $-5 + 2$ **3.** $6 - 9$ **4.** $-6 - 3$ **5.** $-2 - (-4)$

6. $-3 \cdot 6$ **7.** $2(-4)$ **8.** $-2(-4)$ **9.** $-2 \cdot 0$ **10.** $-5 - (-1)$

11. $-2 \div 2$ **12.** $-9 \div 3$ **13.** $-4 \div 8$ **14.** $\frac{1}{2} \cdot (-2)$ **15.** $-6 \div (-2)$

16. $-1 \div (-1)$ **17.** $-1 \div \frac{1}{2}$ **18.** $-1 \div \frac{2}{3}$ **19.** $-1 \div (-3)$ **20.** $-2 + (-5)$

21. $5 + (-10)$ **22.** $-\frac{1}{2}(-2)$ **23.** $-\frac{2}{3} \cdot 6$ **24.** $3 + (-6)$ **25.** $3 - (-6)$

Write each fraction in simplest form.

26. $\dfrac{12}{8}$ **27.** $\dfrac{6}{8}$ **28.** $\dfrac{-2}{8}$ **29.** $\dfrac{15}{20}$

30. $\dfrac{3}{-12}$ **31.** $\dfrac{12 - 4}{16 - 4}$ **32.** $\dfrac{-3 + 5}{6 - 8}$ **33.** $\dfrac{7 - 3}{5 - (-1)}$

Write each expression in simplest radical form.

34. $\sqrt{88}$ **35.** $\sqrt{121}$ **36.** $\sqrt{180}$ **37.** $\sqrt{64}$

38. $\sqrt{3^2 + 4^2}$ **39.** $\sqrt{5^2 + 12^2}$ **40.** $\sqrt{6^2 + 5^2}$ **41.** $\sqrt{7^2 + 3^2}$

ALGEBRA

Find the value of _y_ for each given value of _x_.

42. $y = 2x + 11; x = -1, 2, 5$ **43.** $y = 3x - 2; x = -2, 0, 2$

44. $y = \frac{1}{2}x + 1; x = -2, 0, 2$ **45.** $3x - 2y = 6; x = -2, 0, 2$

GEOMETRY

Refer to the simplified street map. In each exercise, assume a person starts at point _O_ and walks the path described. At what point does the person finish?

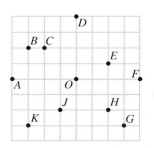

46. 2 blocks east, 1 block north

47. 3 blocks west, 3 blocks south

48. 4 blocks north

49. 2 blocks west, 2 blocks north

Describe the simplest path a walker could use to start at point _O_ and arrive at the given point.

50. _F_ **51.** _G_ **52.** _H_ **53.** _J_

17.1 Locating Points in a Plane

Objective: To graph points in the coordinate plane and name the coordinates of points in the coordinate plane.

We can use a pair of perpendicular number lines to develop a ***coordinate system*** for locating points in a plane. The horizontal number line is called the ***x-axis***. The vertical number line is called the ***y-axis***. The point where the two axes intersect is called the ***origin, O***.

The two axes divide the plane into four regions called ***quadrants***, which are numbered as shown. Every point is either in a quadrant or on an axis.

The coordinates of each point are an ***ordered pair*** of numbers. The first number is called the ***x-coordinate***. The second number is called the ***y-coordinate***.

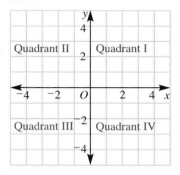

Example: **Graph $A(5, 2)$ and $B(-4, -3)$.**

Solution: To graph $A(5, 2)$, go right 5 units and up 2.

To graph $B(-4, -3)$, go left 4 units and down 3.

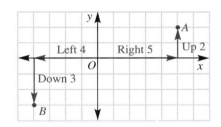

EXPLORING

1. What are the coordinates of the origin?

2. State the coordinates of each labeled point shown.

3. Which labeled points have the same x-coordinate?

4. Which labeled points have the same y-coordinate?

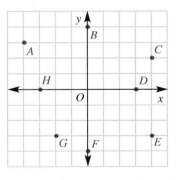

1. Give the coordinates of a point that is on both axes.

2. Are (5, 7) and (7, 5) coordinates of the same point? Why or why not?

3. Graph (2, 1), (5, 1), and (2, 4). If these are three vertices of a rectangle, what are the coordinates of the fourth vertex of the rectangle?

4. $\overline{AB}$, with endpoints $A(-2, 1)$ and $B(3, 4)$, is the hypotenuse of right $\triangle ABC$. Each leg of $\triangle ABC$ is parallel to one of the axes. Find all possible coordinates of point C.

Class Exercises

Give coordinates for three points to fit each description.

1. in Quadrant I
2. in Quadrant II
3. in Quadrant III
4. in Quadrant IV
5. on the x-axis
6. on the y-axis

State the quadrant or axis where each point is located.

7. (6, 1)
8. (−5, −2)
9. (0, 4)
10. (−6, 3)
11. (3, −5)
12. (3, 4)
13. (−6, 0)
14. (0, −2)

Draw a set of axes on graph paper. Graph each point.

15. $A(3, 1)$
16. $B(-3, 2)$
17. $C(0, 3)$
18. $D(0, -5)$
19. $E(4, 0)$
20. $F(3, -2)$
21. $G(-4, 0)$
22. $H(-3, -4)$

Exercises

Find the coordinates of each point.

1. A
2. B
3. C
4. D
5. E
6. F
7. G
8. H
9. I
10. J
11. K
12. L

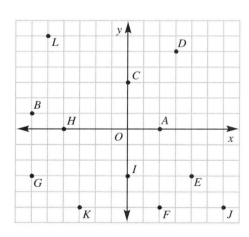

Draw a set of axes on graph paper. Graph each point.

13. $A(5, 2)$ **14.** $B(2, 5)$ **15.** $C(-3, 4)$ **16.** $D(4, -3)$

17. $E(0, 8)$ **18.** $F(8, 0)$ **19.** $G(9, -5)$ **20.** $H(-5, 9)$

21. $I(-10, 6)$ **22.** $J(6, -10)$ **23.** $K(-7, 0)$ **24.** $L(0, -7)$

25. $M(-5, -9)$ **26.** $N(-10, -10)$ **27.** $P(11, 6)$ **28.** $Q(11, -6)$

Find the distance between each pair of points. A sketch may be helpful.

29. (3, 0) and (12, 0)   **30.** (2, 0) and (−6, 0)   **31.** (0, 1) and (0, 7)

32. (1, 2) and (6, 2)   **33.** (−4, −3) and (6, −3)   **34.** (3, −2) and (3, 9)

Use a separate set of axes for each of Exercises 35–40.

35. a. Graph each of these points: (4, −4); (2, −2); (−1, 1); (3, 3); (−5, 5)

    b. Four of the points in part (a) are collinear. Which one is not on the line? Change one coordinate of that point so that it is on the line.

    c. If the x-coordinate of a point on the line is 15, what is its y-coordinate?

36. a. Graph each of these points: (−2, −4); (−4, −6); (0, −2); (6, 4); (3, 1)

    b. The points in part (a) are collinear. State the coordinates of three other points that are on that line.

37. a. Graph the following points and connect them in order:
(−1, 1); (1, 5); (7, 5); (5, 1); (−1, 1)

    b. What kind of figure is formed?

    c. Draw the diagonals of the figure. State the coordinates of the point of intersection of the diagonals.

38. a. Which points named below fit the following description?
"The sum of the x-coordinate and the y-coordinate is 6."
$A(5, 1)$; $B(-1, 7)$; $C(-4, -2)$; $D(0, 6)$; $E(-3, 3)$

    b. Graph the points you chose in part (a). (You should be able to draw one line through all these points.)

    c. State the coordinates of three other points that are on the line.

39. a. Draw a vertical line through the point (5, 2).

    b. State the coordinates of three other points on that line.

    c. Which coordinate, x or y, is the same for all points on a vertical line?

40. a. Draw a horizontal line through the point (5, 2).

    b. State the coordinates of three other points on that line.

    c. Which coordinate, x or y, is the same for all points on a horizontal line?

Graph the points. Then connect these points with segments in the order given. The direction "STOP" between two points means that you do not connect them.

41. **a.** (–9, 1) **b.** (–6, –2) **c.** (5, –2)
    d. (10, 1) **e.** (–9, 1) **f.** (1, 1)
    g. (1, 13) **h.** (–5, 3) **i.** (1, 3)

42. **a.** (–2, –1) **b.** (–2, 2) **c.** (2, 4)
    d. (6, 3) **e.** (–2, –1) **f.** (2, 4)
    STOP **g.** (–2, 2) **h.** (6, 3)

APPLICATIONS

43. **Design** Write a set of directions like those for Exercises 41– 42 that you could use to make a copy of the drawing of the triangular prism.

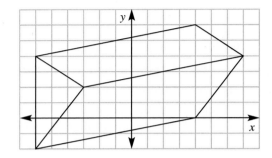

44. **Computer** Use LOGO to run the procedure at the right.

    a. Add commands to the procedure so that a smaller star is drawn inside of the larger one.

    b. Write a procedure that uses SETPOS commands to draw the x-axis and y-axis.

    c. Modify your procedure in part (b) to draw the axes and a triangle in Quadrant IV.

```
To star
pu setpos [0  60]
pd setpos [20  20]
setpos [60  0]
setpos [20 –20]
setpos [0 –60]
setpos [–20 –20]
setpos [–60  0]
setpos [–20  20]
setpos [0  60]
end
```

Computer

On graph paper, draw a three-dimensional figure, such as the pyramid shown. Add x- and y-axes that intersect the figure. Let each unit on your graph represent 10 units on the computer screen. Determine the coordinates of each vertex. Then write a procedure that uses SETPOS commands to draw the figure.

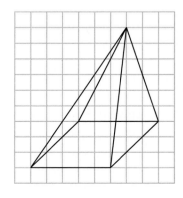

Distance and Midpoint Formulas

Objective: To find the distance between two points in the coordinate plane and to find the coordinates of the midpoint of a segment.

You can find the distance between any two points on either a horizontal or a vertical line by counting. For example, in the figure at the right, $AC = 2$ and $CB = 4$.

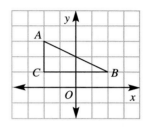

You cannot find AB simply by counting. However, since $\triangle ABC$ is a right triangle, you can use the Pythagorean Theorem.

$$AB^2 = 2^2 + 4^2 = 4 + 16 = 20$$

$$AB = \sqrt{20} = \sqrt{4} \cdot \sqrt{5} = 2\sqrt{5}$$

EXPLORING

1. Graph $A(-3, -2)$ and $B(3, 1)$. Draw $\overline{AB}$. Draw vertical and horizontal segments to form right $\triangle ABC$.

2. **a.** Find AC and BC.

   b. Find AB in simplest radical form.

3. Repeat Steps 1 and 2 for $A(4, 1)$ and $B(1, 5)$.

4. **a.** If two points lie on a horizontal line, how is the distance between them related to their coordinates?

   b. If two points lie on a vertical line, how is the distance between them related to their coordinates?

You can use the following formula to find the distance between any two points.

THEOREM 17.1 (The Distance Formula):

The distance d between any two points $A(x_1, y_1)$ and $B(x_2, y_2)$ is

$$d = \sqrt{(x_2 - x_1)^2 + (y_2 - y_1)^2}.$$

Example 1: **Find the distance between $C(-6, -4)$ and $D(-3, -7)$.**

Solution: $d = \sqrt{(-3-(-6))^2+(-7-(-4))^2}$

$= \sqrt{(-3+6)^2+(-7+4)^2}$

$= \sqrt{(3)^2+(-3)^2}$

$= \sqrt{9+9}$

$= \sqrt{18} = \sqrt{9} \cdot \sqrt{2} = 3\sqrt{2}$

EXPLORING

Part A

Graph each pair of points and draw the segment with those endpoints. Find the midpoint of the segment and determine its coordinates.

1. $A(-2, 3)$ and $B(4, 3)$ **2.** $C(6, 5)$ and $D(6, -5)$ **3.** $E(-3, -2)$ and $F(0, -2)$

4. The **average** of two numbers a and b is $\dfrac{a + b}{2}$.

   a. How are the coordinates of the midpoint of a horizontal segment related to the coordinates of the endpoints of the segment?

   b. How are the coordinates of the midpoint of a vertical segment related to the coordinates of the endpoints of the segment?

Part B

For $\triangle ABC$ shown at the right, $\overline{AC}$ is a horizontal segment, $\overline{BC}$ is a vertical segment, and M is the midpoint of $\overline{AB}$.

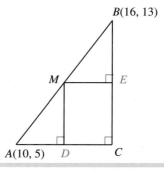

1. The vertical and horizontal lines through M intersect $\overline{AC}$ and $\overline{BC}$ at D and E respectively. $\overline{MD} \parallel \overline{BC}$ and $\overline{ME} \parallel \overline{AC}$. Why are D and E the midpoints of $\overline{AC}$ and $\overline{BC}$?

2. What are the coordinates of C? of D? of E?

3. What are the coordinates of M?

You can use the following formula to find the coordinates of the midpoint of any segment when you know the coordinates of the endpoints.

THEOREM 17.2 (The Midpoint Formula): If the coordinates of the endpoints of a segment are (x_1, y_1) and (x_2, y_2), then the midpoint of the segment has coordinates $\left(\dfrac{x_1 + x_2}{2} , \dfrac{y_1 + y_2}{2} \right)$.

Example 2: Find the coordinates of the midpoint of the segment with endpoints $R(-3, 4)$ and $S(5, -6)$.

Solution: $\left(\dfrac{-3 + 5}{2} , \dfrac{4 + (-6)}{2} \right) = \left(\dfrac{2}{2} , \dfrac{-2}{2} \right) = (1, -1)$

Thinking Critically

1. The endpoints of a segment are $P(-3, 0)$ and $Q(0, -4)$. Describe several ways to find the length of $\overline{PQ}$.

2. Suppose that you are given the coordinates of the vertices of quadrilateral $ABCD$.
   a. Describe how you could use the distance formula to determine whether $ABCD$ is a parallelogram.
   b. Describe how you could use the midpoint formula to determine whether $ABCD$ is a parallelogram.

Find the length of each side of each triangle. Describe each triangle. Give reasons for your answers.

3.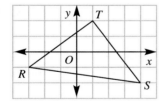

4.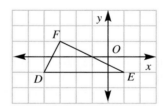

Class Exercises

Find the distance between the points with the given coordinates.

1. $(-3, -4)$ and $(-3, -7)$ 2. $(-1, 5)$ and $(3, 5)$

3. $(-5, 4)$ and $(3, -2)$ 4. $(-3, 1)$ and $(2, 4)$

Find the coordinates of the midpoint of $\overline{AB}$.

5. $A(-5, -2)$ and $B(3, 4)$ 6. $A(-3, 1)$ and $B(2, -2)$

Exercises

Write all answers in simplest radical form.

Find (a) *AB* and (b) the coordinates of the midpoint of $\overline{AB}$.

1. *A*(3, 2) and *B*(6, 6)

2. *A*(0, –2) and *B*(3, 3)

3. *A*(–4, –2) and *B*(1, 3)

4. *A*(–5, 2) and *B*(0, 4)

5. *A*(–3, –1) and *B*(5, –7)

6. *A*(–5, –3) and *B*(–3, –5)

7. *A*(–4, –5) and *B*(–1, 1)

8. *A*(2, 3) and *B*(4, –2)

9. *A*(–4, 2) and *B*(3, 0)

Graph $\triangle ABC$ with the given vertices. Then find (a) *AB*, (b) *BC*, (c) *AC*, (d) the perimeter of $\triangle ABC$, and (e) the area of $\triangle ABC$.

10. *A*(–3, 7)
    B(–3, 2)
    C(9, 2)

11. *A*(–2, –1)
    B(–2, 2)
    C(3, 2)

12. *A*(–4, 6)
    B(2, –2)
    C(6, 1)

13. *A*(–3, 3)
    B(–1, –1)
    C(3, 1)

Find the lengths of the diagonals of quadrilateral *RSTW*.

14. *R*(–2, –2), *S*(1, 2), *T*(3, 3), *W*(0, –1)

15. *R*(3, –3), *S*(5, 0), *T*(–1, 4), *W*(–3, 1)

Tell whether the triangle with the given vertices is isosceles.

16. *E*(–2, –1), *F*(–1, 3), *G*(1, 1)

17. *J*(–4, –3), *K*(–4, 7), *L*(2, 5)

Determine whether quadrilateral *ABCD* is a parallelogram by determining whether the diagonals have the same midpoint.

18. *A*(–4, –2), *B*(–3, 3), *C*(2, 1), *D*(1, –4)

19. *A*(2, 2), *B*(–7, 4), *C*(–9, –6), *D*(–1, –8)

20. Find the lengths of the medians $\overline{RM}$, $\overline{SN}$, and $\overline{TW}$ of the triangle with vertices *R*(–3, –4), *S*(1, 6), and *T*(5, 2).

APPLICATION

21. Trigonometry Right $\triangle ABC$ has vertices *A*(–3, –2), *B*(9, 3), and *C*(9, –2). Find sin *A*, cos *A*, and tan *A*. Use a calculator or the table on page 567 to find *m∠A* to the nearest degree.

Seeing in Geometry

1. *M*(3, 2) is the midpoint of $\overline{OB}$. What are the coordinates of *B*?

2. *O* is the midpoint of $\overline{NM}$. What are the coordinates of *N*?

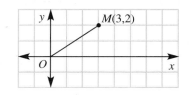

17.3 Slope of a Line

Objective: To find the slope of a given line.

The *slope* of a line is the measure of its steepness. The slope is given by the ratio of *rise* to *run*.

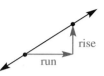

Example 1:

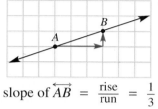

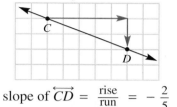

$$\text{slope of } \overleftrightarrow{AB} = \frac{\text{rise}}{\text{run}} = \frac{1}{3}$$

$$\text{slope of } \overleftrightarrow{CD} = \frac{\text{rise}}{\text{run}} = -\frac{2}{5}$$

You can use any two points on a line to find the slope of the line. For $\overleftrightarrow{AB}$ in Example 1, a rise of 1 unit will always correspond to a run of 3 units, a rise of 2 units will correspond to a run of 6 units, and so on. For $\overleftrightarrow{CD}$, if you start at D and go to C, you get a rise of 2 units and a run of -5 units. The slope would still be $-\frac{2}{5}$.

The rise for two points on a horizontal line is always 0. Thus the slope of every horizontal line is 0.

The run for two points on a vertical line is always 0. Since you cannot divide by 0, the slope of every vertical line is undefined.

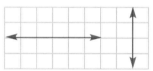

EXPLORING

1. For each line, pick two points. Then find the rise, run, and slope.

   a. 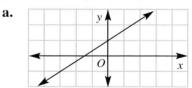 **b.**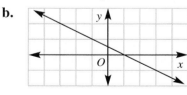

2. How is the rise related to the coordinates of the points you choose in Step 1? How is the run related to the coordinates of the points?

3. Suggest a formula you could use to find the slope of a line if you know that $A(x_1, y_1)$ and $B(x_2, y_2)$ lie on the line. Are there any restrictions on the formula?

The **EXPLORING** activity suggests that if $A(x_1, y_1)$ and $B(x_2, y_2)$ lie on a line and $x_1 \neq x_2$, then you can use the following formula to find the slope:

$$\text{slope} = \frac{y_2 - y_1}{x_2 - x_1}$$

Example 2: **Find the slope of the line through $C(-6, 5)$ and $D(-2, -1)$.**

Solution: $\text{slope} = \dfrac{-1 - 5}{-2 - (-6)} = \dfrac{-6}{-2 + 6} = \dfrac{-6}{4} = -\dfrac{3}{2}$

Thinking Critically

1. a. When a line rises from left to right, like line l shown below, what must be true of its slope?

b. When a line falls from left to right, like line n shown below, what must be true of its slope?

2. Find the slope of each line shown.

a.

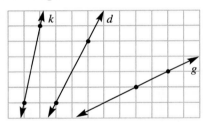

b.

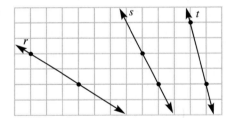

c. Which line in part (a) is steepest? Which line in part (b) is steepest? As lines become steeper, what happens to the slope?

Class Exercises

1. Tell whether each slope is positive, negative, zero, or undefined.

Find the slope of each line.

2. 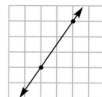   **3.**   **4.**   **5.**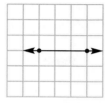

Find the slope of the line through each pair of points.

6. $(-4, 0)$ and $(-2, 5)$ **7.** $(3, 4)$ and $(5, 1)$ **8.** $(-2, 4)$ and $(3, 4)$

Exercises

1. Tell whether each slope is positive, negative, zero, or undefined.

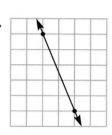

Find the slope of each line.

2.   **3.**   **4.** 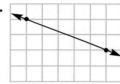   **5.** 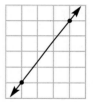   **6.**

Find the slope of the line through each pair of points.

7. $(1, 3)$ and $(5, 6)$ **8.** $(0, 0)$ and $(-3, -2)$ **9.** $(-4, -3)$ and $(1, -3)$

10. $(-3, 0)$ and $(-1, 6)$ **11.** $(0, 2)$ and $(3, -1)$ **12.** $(3, -2)$ and $(-2, 2)$

13. $(-4, -2)$ and $(-6, -3)$ **14.** $(-3, 4)$ and $(4, 2)$ **15.** $(-2, 5)$ and $(-2, 3)$

On graph paper, draw a line with the given slope.

16. 3   **17.** $-\frac{1}{2}$ **18.** $\frac{4}{5}$

19. Draw the line through $(3,1)$ that has slope $\frac{2}{5}$.

Find the missing coordinate. (A sketch may be helpful.)

20. The line through $(2, -1)$ with slope $\frac{3}{2}$ also passes through the point $(4, \blacksquare)$.

21. The line through $(3, 2)$ with slope $-\frac{3}{4}$ also passes through the point $(-1, \blacksquare)$.

22. a. Graph the following points: $A(1, 2)$; $B(4, 4)$; $C(1, -1)$; $D(4, 1)$
    b. Draw $\overleftrightarrow{AB}$ and $\overleftrightarrow{CD}$.
    c. Find the slopes of $\overleftrightarrow{AB}$ and $\overleftrightarrow{CD}$. How are the slopes related?

23. **a.** Graph the following points and connect them to form a triangle:
$A(1, 1)$; $B(6, 2)$; $C(3, 4)$

b. Use the distance formula to find the lengths of the sides of $\triangle ABC$. Classify $\triangle ABC$.

c. Find the slopes of $\overline{AC}$ and $\overline{BC}$. How are the slopes related?

APPLICATIONS

Home Improvement   Parts of a house exposed to the weather should be sloped to allow rainwater to run off. A driveway should be sloped, but not so steeply that cars slip in icy weather.

Some recommended slopes are shown in the diagram. In each case, the slope is the ratio $\frac{\text{rise}}{\text{run}}$.

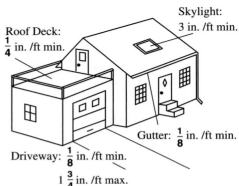

Skylight: 3 in. /ft min.

Roof Deck: $\frac{1}{4}$ in. /ft min.

Gutter: $\frac{1}{8}$ in. /ft min.

Driveway: $\frac{1}{8}$ in. /ft min.

$1\frac{3}{4}$ in. /ft max.

24. A roof deck 12 ft wide is to be built. What is the minimum rise?

25. A gutter 20 ft long is to be installed. What is the minimum rise?

26. A roof has a run of 15 ft and a rise of 4 ft. Can a properly sloped skylight be installed?

27. A driveway 30 ft long has a rise of 5 ft. Is its slope within the recommended range?

28. **Computer**   Write a LOGO procedure to draw a diagonal line from the upper right to the lower left of the computer screen. What is the slope of the line?

Everyday Geometry

The slope of a highway or railway is called the *grade*. A grade can be expressed in degrees or as a percent.

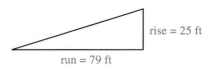

rise = 25 ft

run = 79 ft

The steepest street in the world is Filbert Street in San Francisco. Filbert Street rises 25 ft for every 79 ft of run.

1. Find the *percent* grade of Filbert Street by expressing $\frac{\text{rise}}{\text{run}}$ as a percent. Round to the nearest percent.

2. Find the *degree* grade of Filbert street by finding the angle that has $\frac{\text{rise}}{\text{run}}$ as its tangent. Round to the nearest degree.

Slope of Parallel and Perpendicular Lines

Objective: To find the slope of a line parallel to or perpendicular to a line with a given slope.

The slopes of parallel lines and perpendicular lines are related in special ways.

EXPLORING

Part A

In the figure at the right, *l* ∥ *n*.

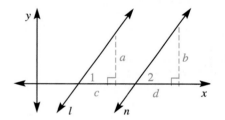

1. Why is ∠1 ≅ ∠2?

2. What is the tangent of ∠1? of ∠2?

3. What is the slope of line *l*? of line *n*?

4. Why is the slope of line *l* equal to the slope of line *n*?

5. Choose any number as a slope. Graph any two lines with that slope. How are the lines related?

Part B

1. How are each pair of lines shown below related? Use a protractor to verify your answer.

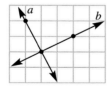
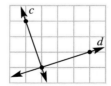
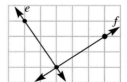

2. Find the slope of each line.

3. How are the slopes of each pair of lines related?

The **EXPLORING** activity demonstrates the following theorems.

THEOREM 17.3: Two nonvertical lines are parallel if and only if they have the same slope.

THEOREM 17.4: Two nonvertical lines are perpendicular if and only if the product of their slopes is –1.

Example: Line l has slope $\frac{4}{5}$. Find the slope of a line perpendicular to l.

Solution: Let m be the slope of the perpendicular line.

$$\frac{4}{5} \cdot m = -1$$

$$m = -1 \div \frac{4}{5} = -1 \cdot \frac{5}{4} = -\frac{5}{4}$$

Thinking Critically

1. What kind of line is perpendicular to a vertical line? What kind is parallel to a vertical line?

2. Will the product of the slopes of any two perpendicular lines always be equal to –1? Why or why not?

3. Explain why the word "nonvertical" is used in the statement of Theorems 17.3 and 17.4.

4. **a.** A, B, and C have coordinates $(-3, 4)$, $(-1, 1)$, and $(1, -2)$, respectively. Is $\overleftrightarrow{AB} \parallel \overleftrightarrow{BC}$? Why or why not?

   b. Consider points $R(-5, -1)$, $S(-2, -1)$, $T(1, -1)$, and $W(4, -1)$. Is $\overleftrightarrow{RS} \parallel \overleftrightarrow{TW}$? Why or why not?

Class Exercises

1. Line l has slope $-\frac{3}{5}$. Find the slope of each indicated line.

   a. a line parallel to l **b.** a line perpendicular to l

The slopes of two lines are given. Tell whether the lines are parallel, perpendicular, or neither.

2. $\frac{3}{4}$ and $\frac{4}{3}$ 3. $\frac{3}{7}$ and $\frac{3}{7}$ 4. $\frac{3}{5}$ and $-\frac{5}{3}$ 5. $\frac{2}{5}$ and $\frac{2}{5}$ 6. 3 and –3

Draw lines to fit each description.

7. two different lines, each with slope $-\frac{2}{5}$

8. a line with slope $\frac{2}{5}$ and a second line with slope $\frac{5}{2}$

9. a line with slope $\frac{2}{5}$ and a second line with slope $-\frac{5}{2}$

Exercises

1. Line l has slope $\frac{3}{4}$. Find the slope of each indicated line.

    a. a line parallel to l **b.** a line perpendicular to l

2. Line k has slope -4. Find the slope of each indicated line.

    a. a line parallel to k **b.** a line perpendicular to k

3. Line x is vertical.

    a. What kind of line is perpendicular to x?   **b.** What is the slope of x?

The slopes of two lines are given. Tell whether the lines are parallel, perpendicular, or neither.

4. $\frac{2}{3}$ and $\frac{6}{9}$ **5.** 1 and -1 **6.** 2 and -2 **7.** $\frac{3}{5}$ and $\frac{5}{3}$

8. $\frac{4}{7}$ and $\frac{7}{4}$ **9.** 1 and 1   **10.** $\frac{3}{5}$ and $-\frac{3}{5}$ **11.** 0 and 0

12. *WXYZ* shown below is a rhombus. The slope of $\overline{WZ}$ is $\frac{4}{3}$ and the slope of $\overline{ZX}$ is -2. Find the slope of each segment.

    a. $\overline{XY}$ **b.** $\overline{WY}$

13. *LMNO* shown below is a square. The slope of $\overline{LN}$ is $\frac{2}{3}$ and the slope of $\overline{ON}$ is $-\frac{1}{5}$. Find the slope of each segment.

    a. $\overline{OM}$ **b.** $\overline{NM}$ **c.** $\overline{LM}$ **d.** $\overline{LO}$

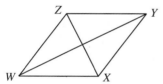

14. *ABCD* is a rectangle. The slope of $\overline{AD}$ is $\frac{1}{3}$. Find the slope of each segment.

    a. $\overline{BC}$ **b.** $\overline{DC}$ **c.** $\overline{AB}$

Draw lines to fit each description.

15. three different lines, each with slope $-\frac{3}{5}$

16. three different lines, each with slope $\frac{2}{7}$

17. a line with slope $\frac{3}{4}$ and a second line with slope $-\frac{4}{3}$

18. a line with slope $\frac{2}{3}$ and a second line with slope $\frac{3}{2}$

19. two different lines, each with slope 0

20. two different lines, each with undefined slope

21. the line through (3, 2) that is perpendicular to a line with slope $\frac{4}{3}$

22. The slopes of four lines are given:

a: $\frac{3}{5}$ b: $-\frac{10}{6}$ c: $-\frac{5}{3}$ d: $\frac{9}{15}$

Do these lines determine a rectangle? Why or why not?

23. The slopes of three lines that determine a triangle are given:

j: $\frac{5}{2}$ k: $\frac{3}{7}$ l: $-\frac{2}{5}$

What kind of triangle is it? Explain

24. Quadrilateral *ABCD* has vertices *A*(4, 5), *B*(8, –1), *C*(–1, –4), and *D*(–5, 2). Use slopes to determine whether *ABCD* is a parallelogram.

25. □*EFGH* has vertices *E*(– 4, 5), *F*(3, 2), *G*(6, –3), and *H*(0, –1). Determine whether *EFGH* is a rhombus by determining whether the diagonals are perpendicular.

26. △*ABC* has vertices *A*(7, 7), *B*(–3, 3), and *C*(3, 1). Find the slopes of altitudes $\overline{AD}$, $\overline{BE}$, and $\overline{CF}$ of △*ABC*.

27. Graph quadrilateral *RSTW* with vertices *R*(4, 7), *S*(9, –5), *T*(6, –8), and *W*(–6, –3). Give the best name for *RSTW*. Use slope and the distance formula to verify your answer.

APPLICATIONS

28. Distance to a Line   Find the distance from each point to the line shown. Express your answers in part (c) in simplest radical form.

a.
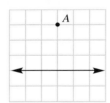

b.
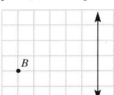

c.
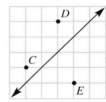

29. Reflections   On graph paper, draw a figure like the one shown at the right. Then find the reflection image of each point over line *l*. (*Hint*: For each point draw the line perpendicular to *l* from the point. Then count "diagonal units" to find the point on the other side of *l* that is the same distance from *l* as the given point.)

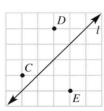

Computer

Write a LOGO procedure that draws two segments, each with slope $\frac{5}{3}$.

17.5 Equations of Lines

Objective: To find ordered pairs of numbers that satisfy a given linear equation and graph the equation.

Points with coordinates that fit certain patterns will lie on the same line. The line shown at the right contains $A(3, 0)$, $B(1, 2)$, $C(0, 3)$, and $D(-2, 5)$. Notice that for each of these four points, the sum of the x-coordinate and the y-coordinate is 3. To describe the relationship between the x-coordinate and the y-coordinate of *every* point on the line, you could use the equation $x + y = 3$.

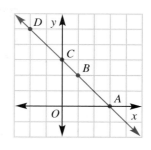

Equations such as $x + y = 3$ are called ***linear equations*** because their graphs are lines. A linear equation can always be written in the following form:

$Ax + By = C$, where A, B, C are constants and A and B are not both 0

If you subtract x from both sides of $x + y = 3$, you get $y = -x + 3$. The two equations are equations of the same line. Any equation of the following form is the equation of a line:

$$y = mx + b, \quad \text{where } m \text{ and } b \text{ are constants}$$

Coordinates that make an equation true are said to *satisfy* the equation.

Example 1: **Find four ordered pairs of numbers that satisfy the equation $y = -2x + 6$. Then graph the equation.**

Solution: Substitute values of x and find the corresponding values of y.

x	$y = -2x + 6$	y	(x, y)
0	$y = -2\,(0) + 6$	6	$(0, 6)$
1	$y = -2\,(1) + 6$	4	$(1, 4)$
2	$y = -2\,(2) + 6$	2	$(2, 2)$
3	$y = -2\,(3) + 6$	0	$(3, 0)$

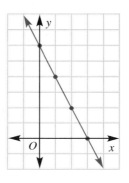

EXPLORING

1. Tell whether each equation is the equation of a line.

   a. $y = 4x$

   b. $4x = y$

   c. $x^2 + y^2 = 4$

   d. $2x - 4y = 5$

   e. $y = -5$

   f. $x = 6$

Find five ordered pairs of numbers that satisfy each equation. Then graph the line.

2. $y = 3x$

3. $y = 2x - 2$

4. $3x + y = 5$

5. Does $(2, -5)$ satisfy the equation $y = -5$? Name two other ordered pairs that satisfy the equation.

6. Name three ordered pairs that satisfy the equation $x = 6$.

To graph an equation such as $3x + y = 5$ in Step 4 above, many people find it easiest to solve first for y. To graph an equation such as $2x - 3y = 6$, you could solve first for y and get $y = \frac{2}{3}x - 2$. Then you could substitute multiples of 3 for x and find the corresponding values of y.

Another way to find ordered pairs of numbers that satisfy $2x - 3y = 6$ is to substitute values for *both* x and y as shown in Example 2.

Example 2: **Graph $2x - 3y = 6$.**

Solution: Choose convenient values for x and y.
Make a table of values.

Let $y = 0$.	Let $x = 0$.	Let $y = 2$.
$2x - 3(0) = 6$	$2(0) - 3y = 6$	$2x - 3(2) = 6$
$2x = 6$	$-3y = 6$	$2x = 12$
$x = 3$	$y = -2$	$x = 6$

x	y
3	0
0	-2
6	2

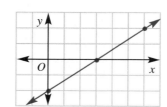

---

1. Only two points are needed to determine a line. When you are graphing a linear equation, why is it a good idea to find the coordinates of more than two points that lie on the line?

2. **a.** Is $x = 3$ the equation of a line? What kind of line?

   b. Is $3y = 2$ the equation of a line? What kind of line?

3. **a.** $4x + 3y = 12$ and $y = -\frac{4}{3}x + 4$ are equations of the same line. Graph the line.

   b. What is the slope of the line?

   c. What are the coordinates of the point where the line intersects the y-axis?

   d. How are the answers to parts (b) and (c) related to one of the equations for the line?

   e. Write any equation of the form $y = mx + b$ other than $y = -\frac{4}{3}x + 4$. Graph the line. Is your conclusion in part (d) true for this equation also?

4. What type of line does not have an equation of the form $y = mx + b$?

Class Exercises

Find four ordered pairs of numbers that satisfy each equation. Then graph the equation.

1. $y = 3x - 4$

2. $x = 5$

3. $y = 2$

4. $x + y = 5$

5. $2x + 5y = 15$

6. $x - 3y = 8$

Exercises

Select all coordinates that satisfy the given equation.

1. $y = 2x + 3$
   a. $(0, 3)$ **b.** $(1, 5)$ **c.** $(-2, 1)$

2. $y = x$
   a. $(-3, -3)$ **b.** $(0, 0)$ **c.** $(-2, -2)$

3. $y = -x$
   a. $(1, -1)$ **b.** $(0, 0)$ **c.** $(-1, 1)$

4. $x + 3y + 2 = 0$
   a. $(5, 1)$ **b.** $(1, -1)$ **c.** $(-5, 1)$

5. $2x - 3y = 1$
   a. $(2, 1)$ **b.** $(8, 5)$ **c.** $(14, 9)$

6. $2x + y = -5$
   a. $(-1, -3)$ **b.** $(2, 1)$ **c.** $(-4, 3)$

Select all coordinates that satisfy the given equation.

7. $y = 3$

   a. $(0, 3)$ **b.** $(3, 0)$ **c.** $(1, 2)$

8. $x = 2$

   a. $(0, 2)$ **b.** $(2, 0)$ **c.** $(1, 1)$

9. Select all equations of lines that contain the point with coordinates $(-2, 3)$.

   a. $y = x + 5$ **b.** $2x + y = -1$ **c.** $3x + 2y = 12$

   d. $x + 2y = 4$ **e.** $-3x = 2y$ **f.** $3y = 2x + 5$

Find three ordered pairs of numbers that satisfy each equation. Then graph the equation.

10. $x - y = 4$ **11.** $x - 2y = 8$

12. $y = 5x + 1$ **13.** $y = 2x$

14. $x = -2$ **15.** $y = 4$

16. a. For each equation below, find three ordered pairs that satisfy the equation. Graph all three equations on the same set of axes.

      $x + y = 4$ $x + y = 0$ $x + y = -4$

  b. Describe the relationship among the three lines.

17. a. For each equation below, find three ordered pairs that satisfy the equation. Graph all three equations on the same set of axes.

      $y = 2x$ $y = 2x + 4$ $y = 2x - 4$

  b. State the coordinates of the points where each line intersects the y-axis.

  c. Find the slope of each line.

  d. Describe the relationship among the three lines.

Graph each equation on a separate set of axes.

18. $y = -x$ **19.** $x = 3$ **20.** $y = -5$

21. $y = -3x + 1$ **22.** $y = 2x - 5$ **23.** $x = 3y$

24. $2x + y = 4$ **25.** $x - 2y = 6$ **26.** $3x + 4y = 8$

Write an equation that is satisfied by each set of four ordered pairs. (*Hint:* Look for a relationship between each x-coordinate and y-coordinate.)

27. $(1, 2), (3, 6), (-2, -4), (-5, -10)$ **28.** $(4, 7), (1, 4), (-1, 2), (-3, 0)$

29. $(5, 3), (0, -2), (4, 2), (-5, -7)$ **30.** $(0, 0), (-2, 1), (4, -2), (-6, 3)$

31. $(3, 7), (0, 1), (5, 11), (-2, -3)$ **32.** $(-5, 2), (0, -3), (7, -10), (1, -4)$

APPLICATIONS

33. Formulas The formula $d = \frac{1}{4}f$ gives the recommended slope for drainage pipes, where d = amount of drop in inches and f = length of pipe in feet.

If you label the horizontal axis as f and the vertical axis as d, you can graph the ordered pairs (4, 1), (8, 2), and (12, 3) that satisfy the formula. Then the red line shown is the graph of the formula.

f	d
4	1
8	2
12	3

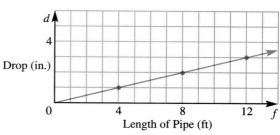

a. Could different scales be used to label the axes?

b. Why are only positive numbers graphed?

34. Formulas The formula $d = \frac{t}{5}$ gives the distance d in miles from a lightning flash in terms of the time t in seconds between the lightning flash and the thunder clap.

a. Find at least three pairs of values (t, d) that satisfy the formula.

b. Graph the formula.

Test Yourself

Draw a set of axes on graph paper. Graph each point.

1. $A(4, 3)$ **2.** $B(2, -3)$ **3.** $C(0, -4)$ **4.** $D(-5, 3)$ **5.** $E(-2, -2)$

Find (a) AB, (b) the coordinates of the midpoint of $\overline{AB}$, and (c) the slope of $\overleftrightarrow{AB}$. Write radicals in simplest radical form.

6. $A(-3, 4)$ and $B(3, 0)$ **7.** $A(-5, -7)$ and $B(3, -1)$

8. Draw the line through $(-2, 0)$ with slope $\frac{2}{5}$.

9. Draw the line through $(2, -1)$ with undefined slope.

10. Line l has slope $\frac{2}{3}$. Find (a) the slope of a line parallel to l and (b) the slope of a line perpendicular to l.

Find three ordered pairs of numbers that satisfy each equation. Then graph the equation.

11. $y = -4$ **12.** $y = -x + 2$ **13.** $2x - y = 3$

Thinking About Proof

Coordinate Proofs

You can use coordinates to prove theorems. The first step is to position a figure so that the computation is as easy as possible. Choose coordinates so that parallel segments have the same slope and congruent segments have the same length.

Example 1: **Find the missing coordinates.**

    a. *ABCD* is a rectangle.   **b.** *EFGH* is a parallelogram.

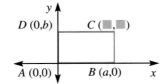 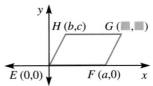

Solution: **a.** *C*(*a*, *b*)   **b.** *G*(*a* + *b*, *c*)

Example 2: **Use coordinates to show that the diagonals of a rectangle are congruent.**

Solution: Use the distance formula to find *AC* and *BD*.

$$AC = \sqrt{(a-0)^2 + (b-0)^2} = \sqrt{a^2+b^2}$$

$$BD = \sqrt{(a-0)^2 + (b-0)^2} = \sqrt{a^2+b^2}$$

Therefore *AC* = *BD*.

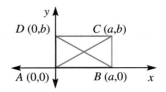

Exercises

Find the missing coordinates.

1. *ABCD* is a square.   **2.** *RSTW* is a square.   **3.** *M* and *N* are midpoints of *EF* and *GF*.

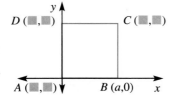 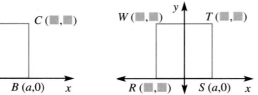 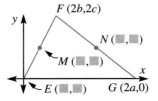

4. To be Proven: The diagonals of a parallelogram bisect each other.

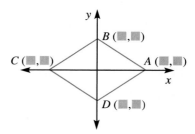

a. What are the coordinates of the midpoint of $\overline{EG}$? of $\overline{FH}$?

b. Why do $\overline{EG}$ and $\overline{FH}$ bisect each other?

5. To be Proven: The diagonals of a square are perpendicular to each other.

a. Draw a diagram like the one in Exercise 1. Include the coordinates of the vertices of *ABCD*. Draw $\overline{AC}$ and $\overline{BD}$.

b. What is the slope of $\overline{AC}$? of $\overline{BD}$?

c. Why is $\overline{AC} \perp \overline{BD}$?

d. Suppose you used the diagram for Exercise 2 instead. What would be the slope of $\overline{RT}$? of $\overline{SW}$?

6. To be Proven: A segment joining the midpoints of two sides of a triangle is parallel to the third side and half its length.

a. Draw a diagram like the one in Exercise 3. Include the coordinates of the five points shown. Draw $\overline{MN}$.

b. What is the slope of $\overline{EG}$? of $\overline{MN}$? Why is $\overline{MN} \parallel \overline{EG}$?

c. What are *EG* and *MN*? Is $MN = \frac{1}{2} EG$?

7. To be Proven: If the diagonals of a quadrilateral are perpendicular and bisect each other, then the quadrilateral is a rhombus.

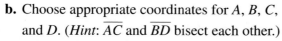

a. What part of the hypothesis allows you to position *ABCD* so that the diagonals lie along the *x*-axis and *y*-axis, as shown at the right?

b. Choose appropriate coordinates for *A*, *B*, *C*, and *D*. (*Hint*: $\overline{AC}$ and $\overline{BD}$ bisect each other.)

c. Show that *ABCD* is a rhombus.

8. To be Proven: The quadrilateral formed by joining the midpoints, in order, of the sides of any quadrilateral is a parallelogram.

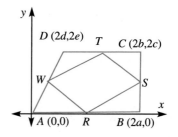

a. Draw a diagram like the one shown at the right, but larger. Include the coordinates of *R*, *S*, *T*, and *W*, the midpoints of the sides of *ABCD*.

b. Show that *RSTW* is a parallelogram.

Graphing Systems of Equations

Objective: To find the solution of a system of equations by graphing.

A *system of equations* is a set of two or more equations involving the same variables.

EXPLORING

1. For each of the equations at the right, find four pairs of numbers that satisfy the equation.

$$y = 2x$$
$$x + y = 3$$

2. Graph the equations on the same set of axes.

3. What is the point of intersection of the two lines? Do the coordinates of the point of intersection satisfy both equations?

4. Repeat Steps 1–3 for the pair of equations at the right.

$$x - 3y = 0$$
$$x - 2y = 3$$

When two lines intersect, the coordinates of the point of intersection satisfy the equations of the lines. The ordered pair is the *solution* of the system of equations.

Example: Solve the system of equations:

$$y = 2x + 1$$
$$y = 5$$

Solution: The graph of $y = 5$ is the horizontal line through $(0, 5)$.

To graph $y = 2x + 1$, find three ordered pairs:

x	$y = 2x + 1$	y	(x,y)
0	$y = 2(0) + 1$	1	$(0, 1)$
1	$y = 2(1) + 1$	3	$(1, 3)$
−1	$y = 2(−1) + 1$	−1	$(−1, −1)$

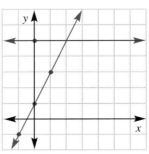

Graph the equations. The solution is $(2, 5)$.

1. **a.** Graph the equations $2x + y = 6$ and $y = -2x$ on the same set of axes. How are these lines related?

   b. What is the solution to this system of equations?

2. Graph the equations $x + 2y = 4$ and $2x + 4y = 8$ on the same set of axes. Describe the solution to this system of equations.

3. What is true of the coordinates of every point that lies on the line with equation $y = 2$? Describe a way to solve the system of equations at the right without graphing the equations. Then use your method to solve.

   $$y = 2$$
   $$3x + y = 5$$

4. What is the solution of the system at the right? How would you check that your solution satisfies each equation in the system?

   $$x = 3$$
   $$y = 2$$
   $$x + y = 5$$
   $$x - y = 1$$

Class Exercises

Find the solution, if any, of each system of equations.
For Exercises 1–3 use the graph at the right.

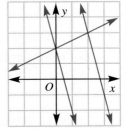

1. $4x + y = 2$
   $x - 2y = -4$

2. $x - 2y = -4$
   $4x + y = 11$

3. $4x + y = 2$
   $4x + y = 11$

4. $y = x + 2$
   $y = -2x - 1$

5. $3x - 2y = 6$
   $3x - 2y = -6$

6. $y = -3x + 3$
   $6x + 2y = 6$

Exercises

Graph to find the solution, if any, of each system of equations.

1. $x = 3$
   $y = -2$

2. $y = x - 2$
   $y = -x + 4$

3. $y = 2x - 5$
   $y = 5x - 2$

4. $x + 2y = 5$
   $3x - y = 1$

5. $3x + y = 4$
   $x + 3y = -4$

6. $y = 3x + 1$
   $y = 3x - 2$

7. $x + y = 2$
   $3x + 3y = 6$

8. $2x - 3y = 3$
   $x + 2y = 12$

9. $x + y = -1$
   $y = x + 5$
   $x + 3y = 3$

Find the solution of each system of equations by substituting.

10. $y = 4$
$y = 3x - 5$

11. $2x + 3y = 5$
$x = 10$

12. $x = y$
$3x + 2y = 10$

13. The graphs of the following equations determine a triangle.

$x = 3$ $y = x + 1$ $2x + 3y = 3$

a. Draw the graph of each equation on the same set of axes.

b. Find the coordinates of the vertices of the triangle.

c. Find the area of the triangle.

14. The graphs of the following equations determine a triangle.

$y = 2x - 1$ $y = -1$ $x + 2y = 8$

a. Draw the graph of each equation on the same set of axes.

b. Find the coordinates of the vertices of the triangle.

c. Find the length of each side of the triangle.

d. Show that the triangle is a right triangle.

15. The graphs of the following equations determine a parallelogram.

$y = 3x + 1$ $y = 4$ $y = 3x - 5$ $y = -2$

a. Draw the graph of each equation on the same set of axes.

b. Find the coordinates of the vertices of the parallelogram.

c. Find the slope of each side of the parallelogram.

d. Find the area of the parallelogram.

APPLICATION

16. Problem Solving   The length of a rectangular garden is two times its width ($l = 2w$). The perimeter of the garden is 30 m ($2l + 2w = 30$).

a. Draw the graph of each equation on the same set of axes. Use l for the horizontal axis and w for the vertical axis.

b. Find the coordinates of the intersection of the two lines.

c. Find the length and width of the garden.

Thinking in Geometry

Find the equation of the line through $(0, 0)$ that is parallel to the line with equation $y = x - 5$.

17.7 Equations of Circles

Objective: To find the center and radius of a circle from the equation of the circle and to write the equation of a circle, given the coordinates of the center and the radius.

Just as lines in the coordinate plane have equations, so do circles. You can use the distance formula, $\sqrt{(x_2 - x_1)^2 + (y_2 - y_1)^2}$, to develop an equation for a circle.

EXPLORING

1. What are the coordinates of the center C of the circle shown at the right? What is the radius?

2. If $P(x, y)$ is any point on the circle shown, what is the distance between C and P? Why?

3. Use Steps 1 and 2 to substitute in the distance formula to find the distance between C and $P(x, y)$:

    $$\sqrt{(\blacksquare - \blacksquare)^2 + (\blacksquare - \blacksquare)^2} = \blacksquare$$

4. Square both sides of your equation in Step 3 to write an equation without the radical sign.

5. Use the graph to name the coordinates of four points on the circle. Show that the coordinates of each of the four points satisfy your equation in Step 4.

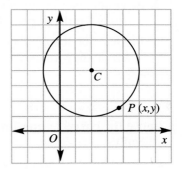

The **EXPLORING** activity demonstrates the following equation for any circle.

THEOREM 17.5: **The equation of a circle with center (h, k) and radius r is $(x - h)^2 + (y - k)^2 = r^2$.**

Example 1: **Find the equation of a circle with center $(4, -3)$ and radius 7.**

Solution: $(x - 4)^2 + (y - (-3))^2 = 7^2$

$(x - 4)^2 + (y + 3)^2 = 49$

Example 2: **Find the coordinates of the center and the radius of the circle with equation $(x + 5)^2 + y^2 = 36$.**

Solution:
$$(x + 5)^2 + y^2 = 36$$
$$(x - (-5))^2 + (y - 0)^2 = 6^2$$
center: $(-5, 0)$ radius: 6

Thinking Critically

1. What are the coordinates of the center and the radius of the circle with equation $x^2 + y^2 = \frac{1}{4}$?

2. In the equation for a circle, can h be negative? Can k? Can r?

3. **a.** Graph the system of equations at the right and find all ordered pairs that satisfy both equations.

                                                    $(x + 1)^2 + (y - 2)^2 = 25$
                                                             $2x - y = 1$

    b. Describe the number of points in which a line and a circle may intersect.

Class Exercises

Choose the correct equation of each circle with the given center and radius.

1. $(4, 2); 3$ **a.** $(x - 4)^2 + (y - 2)^2 = 3$ **b.** $(x + 4)^2 + (y + 2)^2 = 9$
                   c. $(x - 4)^2 + (y - 2)^2 = 9$ **d.** $(x + 4)^2 + (y + 2)^2 = 3$

2. $(3, -4); 4$ **a.** $(x - 3)^2 + (y - 4)^2 = 16$ **b.** $(x - 3)^2 + (y + 4)^2 = 16$
                   c. $(x + 3)^2 + (y - 4)^2 = 16$ **d.** $(x + 3)^2 + (y + 4)^2 = 16$

3. $(-2, -6); 5$ **a.** $(x + 2)^2 + (y + 6)^2 = 25$ **b.** $(x - 2)^2 + (y - 6)^2 = 25$
                   c. $(x + 2)^2 + (y + 6)^2 = 5$ **d.** $(x - 2)^2 + (y - 6)^2 = 5$

4. Find the coordinates of the center and the radius of the circle with equation $(x + 4)^2 + (y + 9)^2 = 36$.

Exercises

Find the coordinates of the center and the radius for each circle.

1. $(x - 5)^2 + (y - 8)^2 = 4$ 2. $(x - 3)^2 + (y - 6)^2 = 100$ 3. $(x + 5)^2 + (y - 4)^2 = 9$
4. $(x - 2)^2 + (y + 9)^2 = 16$ 5. $(x + 4)^2 + (y + 7)^2 = 64$ 6. $(x + 3)^2 + y^2 = 9$
7. $x^2 + (y + 6)^2 = 25$ 8. $x^2 + y^2 = 81$

Find the equation of each circle with the given center and radius.

9. $(2, 5)$; 3   **10.** $(3, -9)$; 6   **11.** $(-6, 4)$; 7

12. $(0, 0)$; 2   **13.** $(-1, -6)$; 5   **14.** $(0, 5)$; 4

Find the equation of each circle described.

15. The center is $(5, 3)$ and the diameter is 12.

16. The center is $(3, 3)$ and the circle is tangent to both the x-axis and y-axis.

17. The endpoints of a diameter are $(0, 0)$ and $(-8, 0)$.

Draw each circle. Use a separate set of axes for each exercise.

18. $x^2 + y^2 = 16$ **19.** $(x - 2)^2 + (y - 5)^2 = 9$

20. $x^2 + (y - 2)^2 = 25$ **21.** $(x + 1)^2 + (y + 4)^2 = 16$

The coordinates of a diameter of a circle are given. Find (a) the coordinates of the center, (b) the radius, and (c) the equation.

22. $(1, 1)$ and $(7, 9)$ **23.** $(1, 1)$ and $(7, 3)$

Describe the graph of each of the following.

24. $(x - 3)^2 + (y + 7)^2 > 16$ **25.** $(x - 3)^2 + (y + 7)^2 < 16$

26. $(x - 3)^2 + (y + 7)^2 \geq 16$ **27.** $(x - 3)^2 + (y + 7)^2 \leq 16$

28. Describe the graph of $x^2 + y^2 = 25$ if $y \geq 0$.

APPLICATION

29. Rotations   On one set of axes, graph the equations

$$(x - 2)^2 + (y - 3)^2 = 4 \quad \text{and} \quad (x - 2)^2 + (y - 3)^2 = 16.$$

Use your graph to give the coordinates of the rotation image of each given point under the given rotation about point $C(2, 3)$.

a. $(2, 5)$; 90° CW   **b.** $(2, 5)$; 180°   **c.** $(2, 5)$; 270° CW

d. $(2, -1)$; 180°   **e.** $(2, -1)$ 90° CCW   **f.** $(2, -1)$; 270° CCW

Seeing in Geometry

On one set of axes, draw the circles with equations

$$(x - 2)^2 + (y - 3)^2 = 9 \quad \text{and} \quad (x + 1)^2 + (y - 5)^2 = 9.$$

Lightly shade the region that is inside *both* circles. Estimate the area of the shaded region.

17.8 Problem Solving Strategy: Use a Graph

Objective: To solve problems by drawing graphs.

Sometimes you can solve a problem by drawing and interpreting a graph.

Example: A tree farmer measured the height of a tree at three different times. The results are shown in the table. The tree can be cut when it is 50 ft tall. If the tree grows at a constant rate, estimate when the tree will be tall enough to cut.

Age (y)	Height (ft)
4	14
6	20
9	29

Solution: Follow these steps.

1. Convert the data to ordered pairs and plot the points.

2. Draw a line connecting the points. Extend it as required.

3. Find the given coordinate 50 on the vertical axis. On the other axis, read the corresponding coordinate.

The graph contains (16, 50). The tree will be about 50 ft tall when it is 16 y old.

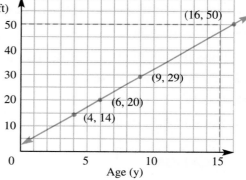

This method will work when the rate of growth or movement remains constant. Notice that you can use the completed graph to find the age or height of the tree at any given height or age. For example, the tree will be about 38 ft tall when it is 12 y old. The tree was about 26 ft high when it was 8 y old.

Class Exercises

The table below shows the lengths of time that were required to build three office buildings. The data is displayed on the graph at the right. Assume that all buildings are constructed at the same rate.

Construction Time (mo)	Height (stories)
2	11
5	23
6	27

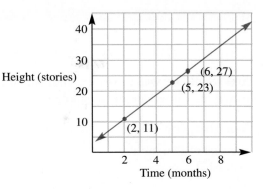

Height (stories)

Time (months)

1. How long will it take to build a 43–story building?

2. How tall a building can be constructed in 8 months? in 4 months?

Exercises

1. Temperature readings were taken at three different times in the morning. Draw a graph of the data. Assume that the temperature rises at a constant rate.

Time	Temperature (°C)
5 A.M.	16
7 A.M.	20
10 A.M.	26

2. Use your graph to find the temperatures at these times.
   a. 9 A.M.   **b.** Noon   **c.** 3 A.M.

3. Use your graph to find the time when each temperature reading occurred.
   a. 18°C   **b.** 29°C

4. The table displays the average price of one brand of calculator during three recent years. Draw a graph of the data. You may find it easier to label 1985 "Year 1" and relabel the remaining given years accordingly. Assume that the rate of change in the price of the calculator is constant.

Year	Price
1985	$28
1987	$25
1991	$19

5. Project the price of the calculator in 1997.

6. In what year will the projected price of the calculator be $14.50?

7. What is wrong with the assumption that the rate of change in the price of the calculator will remain constant?

Chapter *17 Review*

Vocabulary and Symbols

You should be able to write a brief statement, draw a picture, or give an example to illustrate the meaning of each term or symbol.

Vocabulary

average of two numbers (p. 512)

coordinate system (p. 507)

linear equation (p. 523)

ordered pair (p. 507)

origin (p. 507)

quadrant (p. 507)

rise (p. 515)

run (p. 515)

slope of a line (p. 515)

solution (p. 530)

system of equations (p. 530)

x-axis (p. 507)

x-coordinate (p. 507)

y-axis (p. 507)

y-coordinate (p. 507)

Symbol

O (the origin) (p. 507)

Summary

The following list indicates the major skills, facts, and results you should have mastered in this chapter.

17.1 Graph points in the coordinate plane and name the coordinates of points in the coordinate plane. (pp. 507–510)

17.2 Find the distance between two points in the coordinate plane and find the coordinates of the midpoint of a segment. (pp. 511–514)

17.3 Find the slope of a given line. (pp. 515–518)

17.4 Find the slope of a line parallel to or perpendicular to a line with a given slope. (pp. 519–522)

17.5 Find ordered pairs of numbers that satisfy a given linear equation and graph the equation. (pp. 523–527)

17.6 Find the solution of a system of equations by graphing. (pp. 530–532)

17.7 Find the center and radius of a circle from the equation of the circle and write the equation of a circle, given the coordinates of the center and the radius. (pp. 533–535)

17.8 Solve problems by drawing graphs. (pp. 536–537)

Exercises

Draw a set of axes on graph paper. Graph each point.

1. $A(-3, -2)$ **2.** $B(3, 4)$ **3.** $C(1, -3)$ **4.** $D(-5, -1)$

Find *JK*. Write all answers in simplest radical form.

   5. *J*(0, 6) and *K*(6, 14)   **6.** *J*(2, 4) and *K*(−3, 8)   **7.** *J*(−4, −7) and *K*(5, 5)

Find the coordinates of the midpoint of $\overline{PQ}$.

   8. *P*(9, 3) and *Q*(7, 7)   **9.** *P*(12, −4) and *Q*(2, 6)   **10.** *P*(−4, −3) and *Q*(5, −7)

Find the slope of the line through each pair of points.

   11. (3, 7) and (−3, 5)   **12.** (12, 4) and (8, 16)   **13.** (−4, −4) and (6, 2)

   14. Line *l* has slope $\frac{3}{4}$. Find (a) the slope of a line parallel to *l* and (b) the slope of a line perpendicular to *l*.

   15. Draw two different lines, each with slope 0.

   16. Draw a line with slope $\frac{2}{3}$ and a line with slope $-\frac{3}{2}$

Find three ordered pairs of numbers that satisfy each equation. Then graph the equation.

   17. *x* + *y* = 6   **18.** *y* = 3*x* − 2   **19.** 2*x* + *y* = 5

Graph to find the solution, if any, of each system of equations.

   20. *y* = *x* + 1   **21.** *x* + 2*y* = 5
        y = 2*x* − 1   *x* − *y* = 2

Find the coordinates of the center and the radius for each circle.

   22. $(x - 3)^2 + (y + 4)^2 = 100$

   23. $(x - 7)^2 + (y - 2)^2 = 36$

Find the equation of each circle described.

   24. The center is (1, 2) and the radius is 5.

   25. The endpoints of a diameter are (0, 4) and (0, −8).

PROBLEM SOLVING

   26. The table shows the charges for telephone calls between Acadia and Walker. Draw a graph of the data. Use your graph to find (a) the cost of a 2-minute call between the towns and (b) the length of a call that cost 95¢.

Length (min)	Cost (¢)
1	35
3	65
4	80

Draw a set of axes on graph paper. Graph each point.

1. A (4, 1)

2. B (0, 3)

3. C (−4, 2)

4. D (3, −1)

5. E (−2, −4)

6. F (−4, −2)

Find (a) AB in simplest radical form and (b) the coordinates of the midpoint of $\overline{AB}$.

7. $A(2, 1)$ and $B(10, 16)$

8. $A(1, 1)$ and $B(3, −1)$

9. $A(3, 5)$ and $B(7, 3)$

Find the slope of the line through each pair of points.

10. (4, 9) and (1, 3)

11. (−3, −2) and (3, 2)

12. (−2, 5) and (4, −1)

13. Line k has slope $-\dfrac{3}{2}$. Find (a) the slope of a line parallel to k and

(b) the slope of a line perpendicular to k.

14. Draw a line with slope $-\dfrac{1}{2}$.

Find three ordered pairs of numbers that satisfy each equation. Then graph the equation.

15. $x = -3$

16. $y = -2x + 3$

17. $2x - y = 1$

Graph to find the solution, if any, of each system of equations.

18. $x - y = 5$

    $x + y = -1$

19. $y = 2x - 4$

    $y = 2x + 3$

Find the coordinates of the center and the radius for each circle.

20. $x^2 + (y - 3)^2 = 49$

21. $(x - 8)^2 + (y + 2)^2 = 25$

22. Find the equation of the circle with center (−3, 4) and radius 10.

PROBLEM SOLVING

23. Jose Morales earns a base salary plus a commission for each computer he sells. The table shows the number of computers he sold and the amount he earned in three recent weeks. Draw a graph of the data. Use your graph to find (a) his basic weekly salary and (b) the number of computers he would need to sell in a week to earn $475.

Computers Sold	Amount Earned
6	$250
8	$300
12	$400

Chapter 18

Transformations and Coordinate Geometry

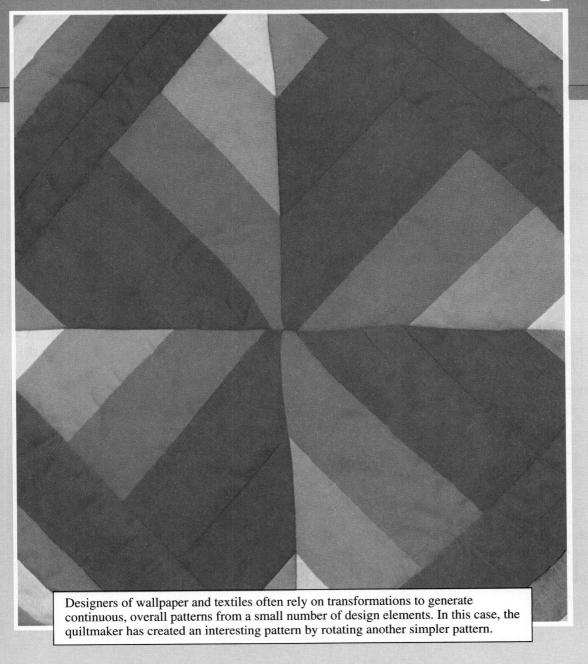

Designers of wallpaper and textiles often rely on transformations to generate continuous, overall patterns from a small number of design elements. In this case, the quiltmaker has created an interesting pattern by rotating another simpler pattern.

Focus on Skills

ARITHMETIC

Find each sum, difference, or product.

1. −4 + 2   **2.** −1 + 2   **3.** 2 − 4   **4.** −2 − 4   **5.** 3 − 5

6. −3 − 5   **7.** −3 + 5   **8.** 2 • 0   **9.** 2 (−4)   **10.** 3 (−2)

11. $\frac{1}{2}$ (−8)   **12.** $\frac{1}{3}$ (−6)   **13.** $\frac{2}{3}$ (−6)   **14.** $\frac{3}{4}$ (8)   **15.** $\frac{3}{5}$ (−10)

ALGEBRA

Find the value of each expression for $x = -2$.

16. $x - 6$ **17.** $2x + 4$ **18.** $-x + 1$ **19.** $-x + 4$ **20.** $-x - 4$

GEOMETRY

Graph the triangle with the given vertices.

21. $A(-3, 4)$, $B(-2, -1)$, $C(2, 5)$ **22.** $D(-4, -3)$, $E(5, 2)$, $F(0, -4)$

Copy each figure on graph paper. Use *right*, *left*, *up*, and *down* to describe the translation that takes B to B'. Then draw the image of $\triangle ABC$ under that translation.

23. 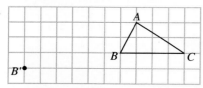   **24.**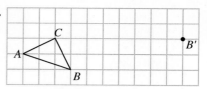

Copy each figure and find the indicated image.

25. reflection image over line l **26.** image under 90° CW rotation about O **27.** image under 90° CCW rotation about point C

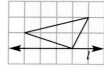

28. Must each of the following be congruent?

   a. a polygon and its reflection image   **b.** a polygon and its translation image

   c. a polygon and its rotation image   **d.** the diagonals of a rhombus

Graph each equation on a separate set of axes.

29. $y = -4$ **30.** $x = -3$ **31.** $y = x$ **32.** $y = -x$

18.1 Translations in the Coordinate Plane

Objective: To find translation images in the coordinate plane.

In Chapter 4 you learned that a *translation* slides each point of the plane the same distance in the same direction. It is especially easy to describe a translation using coordinates.

THEOREM 18.1: The image of (x, y) under a translation in the coordinate plane is $(x + a, y + b)$.

The rule for finding the image of (x, y) under translation T is sometimes written as $T: (x, y) \rightarrow (x + a, y + b)$.

Example: $\triangle A'B'C'$ is the image of $\triangle ABC$ under the translation $T: (x, y) \rightarrow (x + 3, y - 4)$.

$A(-4, 1) \rightarrow A'(-1, -3)$

$B(0, 1) \rightarrow B'(3, -3)$

$C(-1, 3) \rightarrow C'(2, -1)$

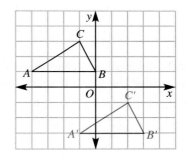

EXPLORING

For each translation shown, give the rule for finding the coordinates of the image of (x, y).

1.

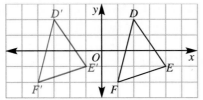

2.

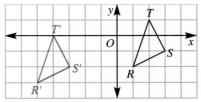

3. Use *right*, *left*, *up*, and *down* to describe each translation for which the given point is the image of (x, y).

    a. $(x, y + 4)$ **b.** $(x - 5, y)$ **c.** $(x + 7, y - 4)$

Thinking Critically

1. Points P' and A' in each part below are the images of $P(x, y)$ and A under a translation. Find the coordinates of A.

   a. $P'(x + 4, y); A'(3, 4)$ **b.** $P'(x, y - 3); A'(5, 2)$

   c. $P'(x + 4, y - 5); A'(6, -3)$ **d.** $P'(x + 6, y + 7); A'(4, 3)$

2. **a.** Suppose $P(x_1, y_1)$ and $Q(x_2, y_2)$ are any two points and P' and Q' are their images under the translation $T: (x, y) \rightarrow (x + a, y + b)$. What are the coordinates of P' and Q'?

   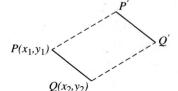

   b. Find the slopes of $\overline{PP'}, \overline{P'Q'}, \overline{QQ'}$, and $\overline{PQ}$.

   c. What kind of figure is $PP'Q'Q$? Why?

Class Exercises

Graph the triangle with the given vertices. Then graph the image of the triangle under the given translation.

1. $A(-4, -1); B(4, -1); C(-2, 4)$

   $T: (x, y) \rightarrow (x + 3, y + 2)$

2. $D(2, 1); E(5, -2); F(3, 2)$

   $T: (x, y) \rightarrow (x - 4, y - 1)$

Exercises

Graph the triangle with the given vertices. Then graph the image of the triangle under the given translation.

1. $A(-3, -1); B(1, 3); C(2, -2)$

   $T: (x, y) \rightarrow (x, y - 5)$

2. $D(-2, 1); E(4, 1); F(5, -2)$

   $T: (x, y) \rightarrow (x - 2, y - 3)$

3. $G(-5, -2); H(-4, 2); J(-2, 1)$

   $T: (x, y) \rightarrow (x + 3, y - 2)$

4. $K(-1, -2); L(1, 1); M(3, -1)$

   $T: (x, y) \rightarrow (x - 4, y + 5)$

Graph $\triangle ABC$ with the given vertices. Then graph $\triangle A'B'C'$, the image of $\triangle ABC$ under translation T_1, and $\triangle A''B''C''$, the image of $\triangle A'B'C'$ under translation T_2.

5. $A(-3, -2); B(-1, 1); C(0, -1)$

   $T_1: (x, y) \rightarrow (x + 4, y + 2)$

   $T_2: (x, y) \rightarrow (x - 2, y + 3)$

6. $A(2, 0); B(4, 1); C(3, -2)$

   $T_1: (x, y) \rightarrow (x - 2, y + 1)$

   $T_2: (x, y) \rightarrow (x - 3, y + 2)$

Give the rule for the translation that takes $\triangle ABC$ to $\triangle A''B''C''$ in the given exercise.

  7. Exercise 5   **8.** Exercise 6

  9. Translations T_1 and T_2 have rules T_1: $(x, y) \rightarrow (x + a_1, y + b_1)$ and T_2: $(x, y) \rightarrow (x + a_2, y + b_2)$. Describe the result of consecutive translations under T_1 and then T_2. Is the result the same for consecutive translations under T_2 and then T_1?

APPLICATIONS

10. Area   Let $(x + y, y)$ be the image of (x, y).

  a. Graph $\triangle BCD$ with vertices $B(1, 1)$, $C(4, 1)$, and $D(2, 5)$. Graph $\triangle B'C'D'$, the image of $\triangle BCD$. Find the area of each triangle.

  b. Graph $\triangle EFG$ with vertices $E(-1, 4)$, $F(-3, 4)$, and $G(-3, -2)$. Graph $\triangle E'F'G'$, the image of $\triangle EFG$. Find the area of each triangle.

  c. Must a triangle and its image under the given rule have the same area?

  d. Does this rule describe a translation? Why or why not?

11. Computer   Write a LOGO procedure that draws the triangle with vertices $(20, 10)$, $(60, 10)$, and $(30, 30)$. Have the procedure also draw the image of the triangle under the translation for which $(-10, 0)$ is the image of $(30, 30)$.

Computer

When congruent copies of a figure completely cover a plane without gaps or overlaps, the result is a *tessellation*.

  1. Use LOGO to run the procedures at the right. Then edit TESS.SQUARES so that the pattern covers more of the screen.

  2. Write a procedure that draws a tessellation based on the parallelogram shown below.

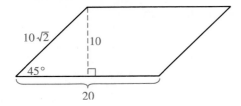

```
To square
repeat 4 [fd 30 rt 90]
end

To tess.squares
pu setpos [0  0] pd square
pu setpos [30  0] pd square
pu setpos [60  0] pd square
pu setpos [0 −30] pd square
pu setpos [30 −30] pd square
pu setpos [60 −30] pd square
end
```

18.2 Reflections in the Coordinate Plane

Objective: To find reflection images in the coordinate plane.

You can develop rules for the coordinates of images of points reflected over certain lines such as the *y*-axis, the *x*-axis, and the line with equation $y = x$.

The diagram at the right shows $\triangle A'B'C'$, the reflection image of $\triangle ABC$ over the line with equation $y = x$. Since the slope of the line is 1, the slope of each of $\overline{AA'}$, $\overline{BB'}$, and $\overline{CC'}$ is –1. Notice that each point and its reflection image are the same number of "diagonal units" from the line of reflection.

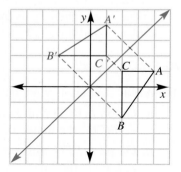

EXPLORING

For each of the following, draw a set of axes on graph paper. Graph the triangle with the given vertices. Then graph its reflection image over the indicated line.

1. $A(1, 3)$, $B(3, 6)$, $C(3, 3)$; the *y*-axis

2. $D(-3, -1)$, $E(-2, 3)$, $F(-1, 2)$; the *y*-axis

3. $A(1, 3)$, $B(3, 6)$, $C(3, 3)$; the *x*-axis

4. $R(-4, -3)$, $S(-1, -1)$, $T(1, -2)$; the *x*-axis

5. $A(1, 3)$, $B(3, 6)$, $C(3, 3)$; the line $y = x$

6. $N(-1, -3)$, $P(1, -3)$, $Q(1, 0)$; the line $y = x$

7. **a.** Make a table containing the coordinates of each point and its image in Steps 1 and 2. Compare the coordinates of each point with the coordinates of its image. Look for a pattern when a point is reflected over the *y*-axis.

   b. Look for a pattern when a point is reflected over the *x*-axis.

   c. Look for a pattern when a point is reflected over the line with equation $y = x$.

The **EXPLORING** activity suggests the following theorem.

THEOREM 18.2: If the coordinates of a point are (x, y), then the coordinates of its reflection image are:

a. $(-x, y)$ when the line of reflection is the y–axis

b. $(x, -y)$ when the line of reflection is the x–axis

c. (y, x) when the line of reflection is the line $y = x$

A translation is sometimes called a *glide*. The combination of a reflection and a translation parallel to the line of reflection is called a *glide reflection*. Footprints suggest a glide reflection.

Example: Graph $\triangle ABC$ with vertices $A(-1, -2)$, $B(1, -1)$, and $C(2, -2)$. Then graph its image under the glide reflection that combines reflection over the x-axis and the translation $T: (x, y) \rightarrow (x - 4, y)$.

Solution: $\triangle A'B'C'$ is the reflection image of $\triangle ABC$ over the x-axis.

$\triangle A''B''C''$ is the translation image of $\triangle A'B'C'$ and the result of the glide reflection.

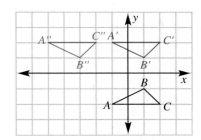

Thinking Critically

1. Does a glide reflection *always*, *sometimes*, or *never* reverse orientation? Explain.

2. Is a segment or angle congruent to its image under a glide reflection? Explain.

3. The diagram at the right shows $\triangle DEF$ and its reflection image over the line with equation $y = -x$. Compare the coordinates of each point with the coordinates of its image. Look for a pattern. What are the coordinates of the image when (x, y) is reflected over the line $y = -x$?

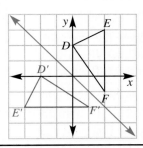

Class Exercises

Find the coordinates of the image of each point when it is reflected over (a) the y-axis, (b) the x-axis, and (c) the line y = x.

1. (5, 2)   **2.** (5, −3)

3. (−3, 4)   **4.** (−3, −2)

5. (0, 3)   **6.** (−2, 0)

Exercises

Find the coordinates of the image of each point when it is reflected over (a) the y-axis, (b) the x-axis, and (c) the line y = x.

1. (8, 5)   **2.** (4, −2)

3. (−6, 2)   **4.** (−5, −4)

5. (0, −6)   **6.** (−4, 0)

For each of the following, draw a set of axes on graph paper. Graph the triangle with the given vertices. Then graph its reflection image over the indicated line.

7. $A(-1, 2)$, $B(2, 3)$, $C(-2, 6)$;
the x-axis

8. $D(-3, 1)$, $E(1, 1)$, $F(-3, -3)$;
the line $y = x$

9. $D(-3, 1)$, $E(1, 1)$, $F(-3, -3)$;
the y-axis

10. $D(-3, 1)$, $E(1, 1)$, $F(-3, -3)$;
the x-axis

11. $A(-1, 2)$, $B(2, 3)$, $C(-2, 6)$;
the y-axis

12. $A(-1, 2)$, $B(2, 3)$, $C(-2, 6)$;
the line $y = x$

13. Graph $\triangle RSW$ with vertices $R(-4, -3)$, $S(-3, 1)$, and $W(0, -2)$. Then graph its image under the glide reflection that combines reflection over the y-axis and the translation $T: (x, y) \rightarrow (x, y + 2)$.

14. Graph $\triangle DEF$ with vertices $D(3, 2)$, $E(5, 4)$, and $F(4, 0)$. Then graph its image under the glide reflection that combines reflection over the line $y = x$ and the translation $T: (x, y) \rightarrow (x - 4, y - 4)$.

15. Draw the line with equation $y = 3$ on a set of coordinate axes. Graph $A(2, 0)$, $B(-1, 4)$, $C(4, 5)$, and $D(-3, 1)$ and the image of each point under reflection over the line $y = 3$.

16. Graph $\triangle RSW$ with vertices $R(-5, 1)$, $S(-2, 1)$, and $W(-2, 3)$. Graph $\triangle R''S''W''$ with vertices $R''(5, -5)$, $S''(2, -5)$, and $W''(2, -3)$. $\triangle R''S''W''$ is the image of $\triangle RSW$ under a glide reflection. Find the line of reflection and the rule for the translation.

A triangle and its reflection image are shown. Copy each diagram onto graph paper and find the line of reflection.

17.

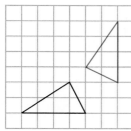

18.

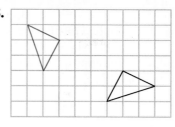

APPLICATIONS

19. Patterns   Look at your graph for Exercise 15. Which, if any, of the following are the coordinates of the image of (x, y) when it is reflected over the line with equation $y = 3$?

    a. $(x, y + 6)$ **b.** $(x, 6 - 2y)$ **c.** $(x, -y + 6)$ **d.** $(x, 2y + 6)$

20. Problem Solving   If (x, y) is reflected over the line with equation $x = -2$, what are the coordinates of the image?

21. Computer   Write a LOGO procedure that draws the isosceles triangle with vertices at $(10, 20)$, $(40, 20)$, and $(40, 50)$ and its reflection images over the x-axis and y-axis.

Test Yourself

Graph the triangle with the given vertices. Then graph the image of the triangle under the given translation.

  1. $A(-1, 1)$, $B(-2, -3)$, $C(1, -2)$
     $T: (x, y) \rightarrow (x + 4, y - 2)$

  2. $D(3, 1)$, $E(-1, -2)$, $F(2, -4)$
     $T: (x, y) \rightarrow (x - 3, y + 2)$

For each of the following, draw a set of axes on graph paper. Graph the triangle with the given vertices. Then graph its reflection image over the indicated line.

  3. $A(2, 1)$, $B(-1, 3)$, $C(-3, 2)$; the x-axis

  4. $D(0, 2)$, $E(-4, 1)$, $F(-3, -2)$; the y-axis

  5. $J(2, 3)$, $K(-1, 3)$, $L(-1, -1)$; the line $y = x$

18.3 Rotations in the Coordinate Plane

Objective: To find rotation images in the coordinate plane.

The rotations that are most appropriate for the coordinate plane are 90° CCW, 90° CW, and 180°.

EXPLORING

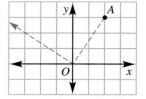

1. **a.** On a set of coordinate axes, graph $A(2, 3)$. Draw $\overline{OA}$.
   b. Find the slope of $\overline{OA}$. Draw the ray perpendicular to $\overline{OA}$ that is the image of $\overrightarrow{OA}$ under a 90° CCW rotation about O.
   c. On the new ray find the point A' such that $OA' = OA$.

2. Repeat Step 1 for $B(-2, 1)$.

3. Repeat Step 1 for $C(-4, -1)$.

4. Graph $A(2, 3)$ and its image under a 90° CW rotation about the origin, O.

5. Repeat Step 4 for $B(-2, 1)$.

6. Repeat Step 4 for $C(-4, -1)$.

7. Graph $A(2, 3)$, $B(-2, 1)$, and $C(-4, -1)$ and their images under a 180° rotation about O.

8. Use your graphs to make a table that shows the coordinates of A, B, and C and their images under the 90° CCW, 90° CW, and 180° rotations about O. For each rotation look for a pattern relating the coordinates of a point and its image.

The **EXPLORING** activity suggests the following theorem.

THEOREM 18.3: If the coordinates of a point are (x, y), then the coordinates of its rotation image about O are:
a. $(-y, x)$ for a 90° CCW rotation
b. $(y, -x)$ for a 90° CW rotation
c. $(-x, -y)$ for a 180° rotation

Example: Graph $\triangle ABC$ with vertices $A(-5, -6)$, $B(1, -6)$, and $C(-4, -3)$. Then graph $\triangle A'B'C'$, the image of $\triangle ABC$ under a 90° CCW rotation about O and graph $\triangle A''B''C''$, the image of $\triangle ABC$ under a 90° CW rotation about O.

Solution: $\triangle A'B'C'$ has vertices $A'(6, -5)$, $B'(6, 1)$, and $C'(3, -4)$.

$\triangle A''B''C''$ has vertices $A''(-6, 5)$, $B''(-6, -1)$, and $C''(-3, 4)$.

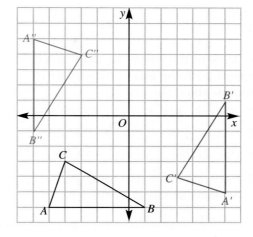

Thinking Critically

1. What are the coordinates of the point that is its own image under a 90° CCW rotation about the origin?

2. Describe the result of two consecutive 90° CCW rotations about the same point.

3. Which of the following preserve orientation? Which reverse orientation?
   reflection, translation, glide reflection, rotation

Class Exercises

Find the coordinates of the image of each point under each of the following rotations about O.

a. 90° CCW **b. 90° CW** **c. 180°**

  1. $(0, 4)$ **2.** $(-3, 0)$
  3. $(4, 5)$ **4.** $(-3, 5)$
  5. $(-2, -4)$ **6.** $(3, -2)$

Exercises

Find the coordinates of the image of each point under each of the following rotations about O.

a. 90° CCW   **b. 90° CW**   **c. 180°**

1. (6, 3)   **2.** (−5, 2)   **3.** (−4, 0)

4. (−3, −2)   **5.** (2, −6)   **6.** (0, −5)

For each of the following, draw a set of axes on graph paper. Graph the triangle with the given vertices. Then graph its image under the given rotation about the origin.

7. $A(1, -1)$, $B(4, 1)$, $C(4, -3)$; 90° CW   **8.** $D(3, 1)$, $E(4, 3)$, $F(-1, 4)$; 90° CCW

9. $G(1, -2)$, $H(-3, -1)$, $K(-3, -5)$; 180°   **10.** $L(-1, 2)$, $M(-2, -3)$, $N(-5, -3)$; 180°

11. $A(1, 2)$, $B(-5, 3)$, $C(-5, -4)$; 90° CCW   **12.** $O(0, 0)$, $D(3, 5)$, $E(-2, 4)$; 90° CW

13. Look at your graphs for Exercises 9 and 10. What seems to be true of a segment and its image under a 180° rotation about the origin?

Copy each diagram on graph paper. Find the image of $\overleftrightarrow{AB}$ under the given rotation about point C.

14. 180°   **15.** 90° CCW   **16.** 90° CW

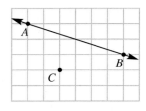   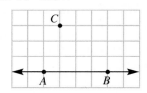   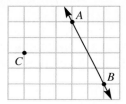

Copy each diagram on graph paper. Find the center of the rotation under which the red triangle is the image of the black triangle.

17.   **18.**   **19.**

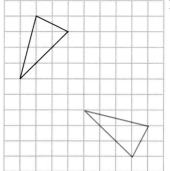   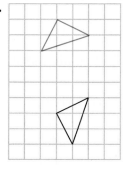   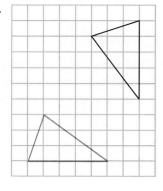

20. Perimeter and Area *R'S'T'W'* is the image of *RSTW* under a rotation about *O*.

   a. Find the perimeter of *RSTW*.

   b. Find the area of *RSTW*.

   c. Which of the vertices of *R'S'T'W'* have integers as coordinates?

   d. How could you find the lengths of $\overline{W'T'}$ and $\overline{S'T'}$?

   e. What are the perimeter and the area of *R'S'T'W'*? Why?

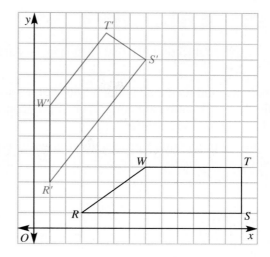

Seeing in Geometry

It can be shown in a more advanced course that if two coplanar triangles are congruent, then one is the image of the other under a reflection, translation, glide reflection, or rotation.

Each of the following shows two congruent triangles. Tell whether the red triangle is the image of the black triangle under a reflection, translation, glide reflection, or rotation.

1.

2.

3.

4.

5.

6.

7.

8.

9.

10.

11.

12.

18.4 Dilations

Objective: To find dilation images in the coordinate plane.

A polygon and its image under a translation, reflection, glide reflection, or rotation are congruent. A polygon and its image under a *dilation* are similar.

The image of a point P under a **dilation** with *center C* and positive *scale factor k* is the point P' on $\overrightarrow{CP}$ such that $CP' = k \cdot CP$.

$CP' = 4 \cdot CP$

EXPLORING

1. Use LOGO to run the procedure at the right. The line "setpc 5" causes the second quadrilateral to be drawn in blue. If you do not have a color monitor, you can delete the line.

2. The larger quadrilateral is the image of the smaller quadrilateral under a dilation with center $O(0, 0)$. What is the scale factor k of the dilation?

3. Change the last five lines of the procedure so that the scale factor is $\frac{1}{2}$. How does this change the image?

4. How are the coordinates of a point and its image under a dilation with center O related to the scale factor of the dilation?

5. Write a LOGO procedure that draws the triangle with vertices (30, 10), (–20, 30), and (–10, –20). The procedure should also draw the image of the triangle under a dilation with center $O(0, 0)$ and scale factor 3.

```
To dilation
pu setpos [40   20] pd
setpos [–20   40]
setpos [–20   0]
setpos [20  –40]
setpos [40   20]
setpc 5
pu setpos [80   40] pd
setpos [–40   80]
setpos [–40   0]
setpos [40  –80]
setpos [80   40]
end
```

The **EXPLORING** activity suggests the following theorem.

THEOREM 18.4: If the coordinates of a point are (x, y), then the coordinates of its image under a dilation with center O and scale factor k are (kx, ky).

Example: Graph $\triangle GHK$ with vertices $G(-2, 2)$, $H(4, 2)$, and $K(0, -2)$. Then graph its image under the dilation with center O and scale factor $\frac{3}{2}$.

Solution: $\frac{3}{2}(-2) = -3$ $\qquad$ $\frac{3}{2} \cdot 2 = 3$

$\frac{3}{2} \cdot 4 = 6$ $\qquad$ $\frac{3}{2} \cdot 0 = 0$

Thus $\triangle G'H'K'$ has the following vertices:
$G'(-3, 3), H'(6, 3), K'(0, -3)$

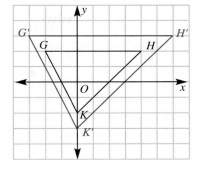

Thinking Critically

1. Does a dilation *always*, *sometimes*, or *never* reverse orientation?

2. a. Find the lengths of the sides of the triangles in the Example.
   b. Describe the relationship between a segment and its image under a dilation with scale factor k.

3. What is the relationship between an angle and its image under a dilation?

4. Is there any value of k for which a triangle and its image under a dilation with scale factor k will be congruent? Explain.

5. Is a point ever the same as its image under a dilation with scale factor $k \neq 1$? Explain.

Class Exercises

Find the coordinates of the image of each point under the dilation with center O and the indicated scale factor.

1. $(3, -4)$; $k = 2$ $\qquad\qquad$ **2.** $(-1, 3)$; $k = 4$

3. $(6, -9)$; $k = \dfrac{1}{3}$ $\qquad\qquad$ **4.** $(-6, -4)$; $k = \dfrac{1}{2}$

5. $(8, 12)$; $k = \dfrac{3}{4}$ $\qquad\qquad$ **6.** $(-2, 4)$; $k = \dfrac{5}{2}$

Exercises

Find the coordinates of the image of each point under the dilation with center O and the indicated scale factor.

1. $(5, 3)$; $k = 3$

2. $(-1, 3)$; $k = 5$

3. $(-8, -4)$; $k = \dfrac{1}{4}$

4. $(8, -10)$; $k = \dfrac{1}{2}$

5. $(-9, -3)$; $k = \dfrac{2}{3}$

6. $(-10, 15)$; $k = \dfrac{3}{5}$

On a set of coordinate axes, graph the figure with the given vertices. Then graph its image under the dilation with center O and the indicated scale factor.

7. $R(-2, 1)$, $S(-4, -3)$, $T(2, -3)$; $k = 2$

8. $E(-4, 2)$, $F(4, 2)$, $G(2, -2)$, $H(-2, -2)$; $k = \dfrac{5}{2}$

9. Find the areas of $\triangle RST$ and $\triangle R'S'T'$ in Exercise 7. What is the ratio of the area of $\triangle RST$ to the area of $\triangle R'S'T'$?

10. Find the areas of $EFGH$ and $E'F'G'H'$ in Exercise 8. What is the ratio of the area of $EFGH$ to the area of $E'F'G'H'$?

Copy each figure on graph paper. Then graph the image of the figure under the dilation with center C and the given scale factor.

11. $k = 4$

12. $k = \dfrac{1}{2}$

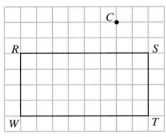

13. Point P' is the image of P under a dilation with center C. Find the scale factor of the dilation.

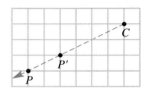

14. Points A' and B' are the images of A and B, respectively, under a dilation. Copy the diagram on graph paper and find the center of the dilation. Then determine the scale factor of the dilation.

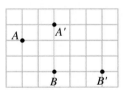

15. It is possible to define a dilation with a negative scale factor. Notice that if $k < 0$, then $-k > 0$, so $(-k) \cdot CP$ is a positive number. The image of point P under a dilation with center C and $k < 0$ is the point P' on the ray opposite $\overrightarrow{CP}$ such that $CP' = (-k) \cdot CP$.

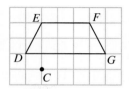

Copy the diagram on graph paper and graph the image of *DEFG* under the dilation with center C and scale factor -2.

APPLICATIONS

16. Computer Some LOGO commands and scale factors are given below. For each, write the LOGO command that would move the turtle to the point that is the image of the given point under the dilation with center O and the given scale factor.

a. setpos $[-6 \quad 9] \, k = 3$

b. setpos $[-12 \quad -10] \, k = 4$

c. setpos $[30 \quad -25] \, k = \dfrac{1}{5}$

d. setpos $[-20 \quad -60] \, k = \dfrac{1}{4}$

e. setpos $[22 \quad 55] \, k = \dfrac{2}{11}$

f. setpos $[-40 \quad -20] \, k = \dfrac{3}{5}$

17. Computer Write a LOGO procedure that draws the triangle with vertices $(-60, 0)$, $(30, 60)$, and $(120, -90)$ and its images under two dilations. The first dilation has center $O(0, 0)$ and scale factor $\frac{1}{2}$. The second dilation has center $(90, -60)$ and scale factor $\frac{1}{2}$. (*Hint*: First draw the triangle and its two images on graph paper. Let each unit on the graph paper represent 10 units on the computer monitor.)

Computer

Write a LOGO procedure using SETPOS commands that draws the black figure at the right. The procedure should also graph the image of the figure under the dilation with center $O(0, 0)$ and scale factor 5. For an unusual effect, edit the procedure so that it connects corresponding vertices of the figures.

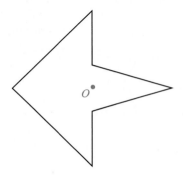

Objective: To tessellate the plane using polygons.

An arrangement of congruent figures that completely covers a plane without gaps or overlapping is called a *tessellation*. Some tessellations are based on translations. Others are based on rotations followed by translations.

Example: **Use each figure to show how to tessellate the plane.**

a.   b.

Solution: a. Rotate the triangle 180° about the midpoint of one side to form a parallelogram. Then use the parallelogram to tessellate the plane.

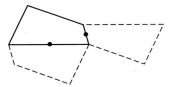 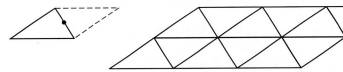

b. Rotate the quadrilateral 180° about the midpoints of two sides. Translate the original quadrilateral to form a concave polygon made up of four quadrilaterals. You can use this polygon to tessellate the plane.

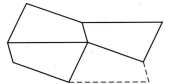

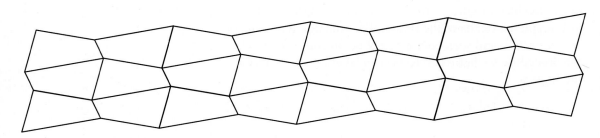

Class Exercises

Use at least eight of each figure to show how to tessellate the plane

1. a square
2. a right triangle

Exercises

Use at least eight of each figure to show how to tessellate the plane.

1. a rectangle
2. an equilateral triangle
3. an isosceles triangle
4. a parallelogram
5. an isosceles trapezoid
6. a regular hexagon

Copy each figure. Use it *at least* four times to show how to tessellate the plane.

7.
8.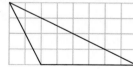
9.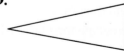

10.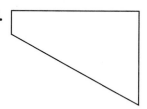
11.
12.

13. Part of a tessellation is shown. Explain how it demonstrates Theorem 8.10: A segment joining the midpoints of two sides of a triangle is parallel to the third side and half its length.

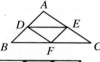

14. The figure shows a tessellation using right triangles. Explain how the shading of the figure illustrates the Pythagorean Theorem.

Vocabulary and Symbols

You should be able to write a brief statement, draw a picture, or give an example to illustrate the meaning of each term.

Vocabulary

glide reflection (p. 547)
dilation (p. 554)
center of a dilation (p. 554)
scale factor of a dilation (p. 554)
tessellation (pp. 545, 558)

Summary

The following list indicates the major skills, facts, and results you should have mastered in this chapter.

18.1 Find translation images in the coordinate plane. (pp. 543–545)

18.2 Find reflection images in the coordinate plane. (pp. 546–549)

18.3 Find rotation images in the coordinate plane. (pp. 550–553)

18.4 Find dilation images in the coordinate plane. (pp. 554–557)

18.5 Tessellate the plane using polygons. (pp. 558–559)

Exercises

Find the coordinates of the given image of each point.

1. $(3, -4)$; image under translation $T: (x, y) \rightarrow (x - 2, y + 2)$

2. $(-2, -1)$; reflection image over the y-axis

3. $(4, -4)$; reflection image over the x-axis

4. $(0, 5)$; reflection image over the line $y = x$

5. $(-5, -1)$; image under 90° CCW rotation about O

6. $(3, -2)$; image under 180° rotation about O

7. $(-6, 10)$; image under 90° CW rotation about O

8. $(-3, 1)$; image under the dilation with center O and scale factor 3

9. $(12, -8)$; image under the dilation with center O and scale factor $\frac{3}{4}$

Graph the triangle with the given vertices. Then graph the indicated image.

10. $B(-2, -3)$, $N(-4, 3)$, $K(0, 2)$
image under the translation
$T: (x, y) \rightarrow (x + 2, y + 1)$

11. $J(-2, -1)$, $G(3, 4)$, $L(1, -1)$
image under the translation
$T: (x, y) \rightarrow (x - 3, y + 2)$

12. $H(5, 1)$, $N(3, 3)$, $P(0, 1)$
reflection image over the x-axis

13. $K(-1, -3)$, $L(0, 3)$, $C(2, 2)$
reflection image over the y-axis

14. $E(-2, 2)$, $A(-2, -2)$, $R(0, 2)$
reflection image over the line $y = x$

15. $B(-2, 2)$, $C(1, 2)$, $D(1, 4)$
image under 90° CW rotation about O

16. $P(-2, 0)$, $Q(-4, 2)$, $R(-3, -2)$
image under 90° CCW rotation about O

17. $J(1, -2)$, $V(0, 3)$, $S(-3, 1)$
image under 180° rotation about O

18. $T(4, 0)$, $R(0, -2)$, $N(-2, 4)$
image under dilation with center O
and scale factor $\frac{1}{2}$

19. $O(0, 0)$, $N(3, 1)$, $B(-1, 2)$
image under dilation with center O
and scale factor 2

20. Graph $\triangle ABC$ with vertices $A(-4, -3)$, $B(-1, 0)$, and $C(1, -2)$. Then graph its image under the glide reflection that combines reflection over the x-axis and the translation $T: (x, y) \rightarrow (x + 4, y)$.

Each of the following shows two congruent triangles. Tell whether the red triangle is the image of the black triangle under a reflection, translation, glide reflection, or rotation.

21.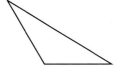

22.

23.

24.

PROBLEM SOLVING

Copy the figure. Then use the figure at least four times to show how to tessellate the plane.

25.

26.

27.

Graph the triangle with the given vertices. Then graph the image indicated.

1. $G(-2, -3)$, $H(3, -2)$, $J(3, 4)$
image under the translation
$T: (x, y) \rightarrow (x - 1, y + 1)$

2. $A(-3, -1)$, $B(-1, 2)$, $C(2, 0)$
image under the translation
$T: (x, y) \rightarrow (x + 2, y - 3)$

3. $M(-3, 3)$, $N(-3, 6)$, $O(0, 0)$
reflection image over the x-axis

4. $P(3, -2)$, $Q(-2, 1)$, $R(1, 4)$
reflection image over the y-axis

5. $S(-2, -1)$, $T(-1, 3)$, $U(1, 2)$
reflection image over the line $y = x$

6. $S(-2, -1)$, $B(-3, 2)$, $N(-1, 4)$
image under 90° CW rotation about O

7. $A(1, 2)$, $B(-3, -1)$, $C(-2, 4)$
image under 90° CCW rotation about O

8. $D(1, 3)$, $E(-2, 0)$, $F(-4, 1)$
image under 180° rotation about O

9. $R(-3, -1)$, $T(2, 2)$, $S(4, -1)$
image under dilation with center O and
scale factor 2

10. $M(-4, -4)$, $P(-6, 2)$, $G(2, 0)$
image under dilation with center O and
scale factor $\frac{1}{2}$

11. Graph $\triangle ABC$ with vertices $A(-1, 1)$, $B(-3, 2)$, and $C(-2, -2)$. Then graph its image under the glide reflection that combines reflection over the y-axis and the translation $T: (x, y) \rightarrow (x, y - 4)$.

PROBLEM SOLVING

Copy the figure. Then use the figure at least four times to show how to tessellate the plane.

12.

13.

14.

Select the best response.

1. $m\angle A = 54°$. $\angle B$ is a complement of $\angle A$. $\angle C$ is a supplement of $\angle A$. Choose the true statement.

   a. $m\angle B = 126°$ **b.** $m\angle B = 54°$

   c. $m\angle C = 126°$ **d.** $m\angle C = 36°$

2. In $\triangle GHJ$, $GH = 12$, $HJ = 9$, $JG = 15$, and $m\angle H = 90°$. Which phrase best describes $\triangle GHJ$?

   a. right isosceles

   b. right scalene

   c. acute scalene

   d. obtuse isosceles

3. What is a regular quadrilateral usually called?

   a. parallelogram **b.** rectangle

   c. square **d.** rhombus

4. In $\triangle XYZ$, $XY = 4$, $YZ = 5$, and $XZ = 8$. Name the largest angle.

   a. $\angle X$ **b.** $\angle Y$

   c. $\angle Z$ **d.** cannot be determined

5. Find $m\angle 1$.

   a. 35°

   b. 107°

   c. 72°

   d. 37°

   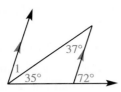

6. In $\triangle RST$, $m\angle R = 65°$ and $m\angle S = 50°$. Which of the following is *not* true?

   a. $\angle R \cong \angle T$ **b.** $m\angle T > m\angle S$

   c. $\overline{RS} \cong \overline{ST}$ **d.** $\overline{ST} \cong \overline{RT}$

7. $ABCD$ is a rhombus. Which statement is not necessarily true?

   a. $AC = BD$ **b.** $AB = BC$

   c. $\overline{AC} \perp \overline{BD}$ **d.** $\overline{AC}$ bisects $\overline{BD}$.

8. A parallelogram has base length 6 cm and height 4 cm. Find the area.

   a. 12 cm² **b.** 24 cm²

   c. 20 cm² **d.** 10 cm²

9. Choose the true statement.

   a. $x = 12.5$

   b. $y = 6$

   c. $x = 7.5$

   d. $x = 7$

10. Which lengths can be the lengths of the sides of a right triangle?

    a. $2^2, 3^2, 5^2$ **b.** $5, 5, 5\sqrt{3}$

    c. $5, 12, 17$ **d.** $3, 9, 3\sqrt{10}$

11. The lengths of the hypotenuse and one leg of a right triangle are 15 cm and 12 cm. Find the perimeter of the triangle.

    a. 9 cm **b.** 54 cm **c.** 36 cm **d.** 90 cm

12. Find the total area of the prism.

    a. 1,200 cm²

    b. 350 cm²

    c. 700 cm²

    d. 400 cm²

13. Find the volume of the prism shown in Exercise 12.

    a. 1,200 cm³ **b.** 350 cm³

    c. 700 cm³ **d.** 400 cm³

Cumulative Review

14. A cylinder has height 14 cm. The radius of a base is 8 cm. Which statement is *not* true?

   a. $V = 896\pi$ cm^3

   b. L.A. $= 224\pi$ cm^2

   c. T.A. $= 352\pi$ cm^2

   d. Base area $= 196\pi$ cm^2

15. The scale factor of pyramid A to similar pyramid B is 3 : 4. Pyramid B has volume 256 m^3. Find the volume of pyramid A.

   a. 108 m^3 **b.** 192 m^3

   c. 144 m^3 **d.** 455 cm^3

16. Find the exact volume of a sphere with radius 9 cm.

   a. 81π cm^3 **b.** 729π cm^3

   c. 324π cm^3 **d.** 972π cm^3

In Exercises 17 and 18, use the table on page 567 or a scientific calculator.

17. If $\cos R = 0.3123$, find m$\angle R$ to the nearest degree.

   a. 18°   **b.** 72°   **c.** 17°   **d.** 71°

18. Find sin 48° to the nearest thousandth.

   a. 0.743   **b.** 0.669

   c. 1.111   **d.** 1.111

19. Which equation could be used to find $m\angle N$?

   a. $\cos N = \dfrac{6}{9}$

   b. $\sin N = \dfrac{6}{9}$

   c. $\cos N = \dfrac{9}{6}$

   d. $\sin N = \dfrac{9}{6}$

20. Find CD for $C(-3, 5)$ and $D(2, 4)$.

   a. $\sqrt{26}$ **b.** $\sqrt{6}$ **c.** $\sqrt{2}$ **d.** $2\sqrt{13}$

21. Find the coordinates of the midpoint of $\overline{JK}$ for $J(4, 12)$ and $K(-6, -8)$.

   a. $(5, 10)$ **b.** $(-2, 4)$

   c. $(-1, 2)$ **d.** $(10, 20)$

22. Find the slope of the line through $(4, 6)$ and $(5, 4)$.

   a. -1 **b.** -2 **c.** $-\dfrac{1}{2}$ **d.** $\dfrac{1}{2}$

23. Find the equation of the circle with radius 4 and center $(2, 6)$.

   a. $(x + 2)^2 + (y + 6)^2 = 16$

   b. $(x - 2)^2 + (y - 6)^2 = 16$

   c. $(x + 2)^2 + (y + 6)^2 = 4$

   d. $(x - 2)^2 + (y - 6)^2 = 4$

24. Find the coordinates of the image of $(6, -9)$ under the dilation with center O and scale factor 3.

   a. $(2, -3)$ **b.** $(18, -27)$

   c. $(-2, 3)$ **d.** $(-18, 27)$

In Exercises 25–27, begin with $\odot O$ or $\triangle ABC$.

25. Draw P on $\odot O$ and construct the tangent to $\odot O$ at P.

26. Circumscribe a circle about $\triangle ABC$.

27. Sketch the locus of points whose distance from O is half the radius of $\odot O$.

28. Graph $\triangle RST$ with vertices $R(1, 3)$, $S(-2, 3)$, and $T(-2, -1)$. Then graph its reflection image over the line $y = x$.

29. Graph $\triangle ABC$ with vertices $A(-5, -4)$, $B(-2, -1)$, and $C(1, -4)$. Then graph its image under 90° CW rotation about the origin.

Table of Symbols

$\angle$ ($\measuredangle$) (p. 8)	angle (angles)
$\approx$ (p. 282)	approximately equal to
A (p. 259)	area
B (p. 435)	area of base
b (p. 263)	base length
cm (p. 16)	centimeter
$\odot$ (p. 39)	circle
C (p. 282)	circumference
CW (p. 113)	clockwise
$\cong$ (p. 20)	congruent to
$\leftrightarrow$ (p. 99)	corresponds to
cos (p. 486)	cosine
CCW (p. 113)	counterclockwise
$\degree$ (p. 43)	degree
d (p. 511)	distance
h (p. 263)	height
km (p. 16)	kilometer
L.A. (p. 430)	lateral area
AB (p. 16)	length of $\overline{AB}$
$\overleftrightarrow{AB}$ (p. 3)	line AB (A and B are points on the line.)
$\overset{\frown}{ABC}$ (p. 381)	major arc
$m\angle ABC$ (p. 43)	measure of angle ABC
$m\overset{\frown}{AB}$ (p. 381)	measure of arc AB
m (p. 16)	meter
mm (p. 16)	millimeter
$\overset{\frown}{AB}$ (p. 381)	minor arc
n^2 (p. 347)	number squared or square of a number
O (p. 507)	origin
$\parallel$ ($\nparallel$) (p. 12)	parallel to (not parallel to)
$\square$ ($\varolessthan$) (p. 221)	parallelogram (parallelograms)
P (p. 255)	perimeter
$\perp$ (p. 8)	perpendicular to
π (p. 282)	pi
(x,y) (p. 507)	point with coordinates x and y
$\sqrt{n}$ (p. 347)	positive square root of a number
r (p. 39)	radius
$a : b$ or $\dfrac{a}{b}$ (p. 295)	ratio of a to b
$\overrightarrow{AB}$ (p. 3)	ray AB (A is the endpoint.)
$\llcorner$ (p. 8)	right angle
k (p. 554)	scale factor
$\overline{AB}$ (p. 3)	segment AB (A and B are the endpoints.)
$\sim$ (p. 301)	similar to
sin (p. 486)	sine
l (p. 470)	slant height
tan (p. 486)	tangent
T.A. (p. 430)	total area
$\triangle$ ($\varowedge$) (p. 73)	triangle (triangles)
V (p. 434)	volume

Table of Squares and Square Roots

n	n^2	$\sqrt{n}$	n	n^2	$\sqrt{n}$	n	n^2	$\sqrt{n}$
1	1	1.000	51	2601	7.141	101	10,201	10.050
2	4	1.414	52	2704	7.211	102	10,404	10.100
3	9	1.732	53	2809	7.280	103	10,609	10.149
4	16	2.000	54	2916	7.348	104	10,816	10.198
5	25	2.236	55	3025	7.416	105	11,025	10.247
6	36	2.449	56	3136	7.483	106	11,236	10.296
7	49	2.646	57	3249	7.550	107	11,449	10.344
8	64	2.828	58	3364	7.616	108	11,664	10.392
9	81	3.000	59	3481	7.681	109	11,881	10.440
10	100	3.162	60	3600	7.746	110	12,100	10.488
11	121	3.317	61	3721	7.810	111	12,321	10.536
12	144	3.464	62	3844	7.874	112	12,544	10.583
13	169	3.606	63	3969	7.937	113	12,769	10.630
14	196	3.742	64	4096	8.000	114	12,996	10.677
15	225	3.873	65	4225	8.062	115	13,225	10.724
16	256	4.000	66	4356	8.124	116	13,456	10.770
17	289	4.123	67	4489	8.185	117	13,689	10.817
18	324	4.243	68	4624	8.246	118	13,924	10.863
19	361	4.359	69	4761	8.307	119	14,161	10.909
20	400	4.472	70	4900	8.367	120	14,400	10.954
21	441	4.583	71	5041	8.426	121	14,641	11.000
22	484	4.690	72	5184	8.485	122	14,884	11.045
23	529	4.796	73	5329	8.544	123	15,129	11.091
24	576	4.899	74	5476	8.602	124	15,376	11.136
25	625	5.000	75	5625	8.660	125	15,625	11.180
26	676	5.099	76	5776	8.718	126	15,876	11.225
27	729	5.196	77	5929	8.775	127	16,129	11.269
28	784	5.292	78	6084	8.832	128	16,384	11.314
29	841	5.385	79	6241	8.888	129	16,641	11.358
30	900	5.477	80	6400	8.944	130	16,900	11.402
31	961	5.568	81	6561	9.000	131	17,161	11.446
32	1024	5.657	82	6724	9.055	132	17,424	11.489
33	1089	5.745	83	6889	9.110	133	17,689	11.533
34	1156	5.831	84	7056	9.165	134	17,956	11.576
35	1225	5.916	85	7225	9.220	135	18,225	11.619
36	1296	6.000	86	7396	9.274	136	18,496	11.662
37	1369	6.083	87	7569	9.327	137	18,769	11.705
38	1444	6.164	88	7744	9.381	138	19,044	11.747
39	1521	6.245	89	7921	9.434	139	19,321	11.790
40	1600	6.325	90	8100	9.487	140	19,600	11.832
41	1681	6.403	91	8281	9.539	141	19,881	11.874
42	1764	6.481	92	8464	9.592	142	20,164	11.916
43	1849	6.557	93	8649	9.644	143	20,449	11.958
44	1936	6.633	94	8836	9.695	144	20,736	12.000
45	2025	6.708	95	9025	9.747	145	21,025	12.042
46	2116	6.782	96	9216	9.798	146	21,316	12.083
47	2209	6.856	97	9409	9.849	147	21,609	12.124
48	2304	6.928	98	9604	9.899	148	21,904	12.166
49	2401	7.000	99	9801	9.950	149	22,201	12.207
50	2500	7.071	100	10,000	10.000	150	22,500	12.247

Table of Trigonometric Ratios

Angle	sin	cos	tan	Angle	sin	cos	tan
1°	0.017	1.000	0.017	46°	0.719	0.695	1.036
2°	0.035	0.999	0.035	47°	0.731	0.682	1.072
3°	0.052	0.999	0.052	48°	0.743	0.669	1.111
4°	0.070	0.998	0.070	49°	0.755	0.656	1.150
5°	0.087	0.996	0.087	50°	0.766	0.643	1.192
6°	0.105	0.995	0.105	51°	0.777	0.629	1.235
7°	0.122	0.993	0.123	52°	0.788	0.616	1.280
8°	0.139	0.990	0.141	53°	0.799	0.602	1.327
9°	0.156	0.988	0.158	54°	0.809	0.588	1.376
10°	0.174	0.985	0.176	55°	0.819	0.574	1.428
11°	0.191	0.982	0.194	56°	0.829	0.559	1.483
12°	0.208	0.978	0.213	57°	0.839	0.545	1.540
13°	0.225	0.974	0.231	58°	0.848	0.530	1.600
14°	0.242	0.970	0.249	59°	0.857	0.515	1.664
15°	0.259	0.966	0.268	60°	0.866	0.500	1.732
16°	0.276	0.961	0.287	61°	0.875	0.485	1.804
17°	0.292	0.956	0.306	62°	0.883	0.470	1.881
18°	0.309	0.951	0.325	63°	0.891	0.454	1.963
19°	0.326	0.946	0.344	64°	0.899	0.438	2.050
20°	0.342	0.940	0.364	65°	0.906	0.423	2.145
21°	0.358	0.934	0.384	66°	0.914	0.407	2.246
22°	0.375	0.927	0.404	67°	0.921	0.391	2.356
23°	0.391	0.921	0.424	68°	0.927	0.375	2.475
24°	0.407	0.914	0.445	69°	0.934	0.358	2.605
25°	0.423	0.906	0.466	70°	0.940	0.342	2.747
26°	0.438	0.899	0.488	71°	0.946	0.326	2.904
27°	0.454	0.891	0.510	72°	0.951	0.309	3.078
28°	0.470	0.883	0.532	73°	0.956	0.292	3.271
29°	0.485	0.875	0.554	74°	0.961	0.276	3.487
30°	0.500	0.866	0.577	75°	0.966	0.259	3.732
31°	0.515	0.857	0.601	76°	0.970	0.242	4.011
32°	0.530	0.848	0.625	77°	0.974	0.225	4.331
33°	0.545	0.839	0.649	78°	0.978	0.208	4.705
34°	0.559	0.829	0.675	79°	0.982	0.191	5.145
35°	0.574	0.819	0.700	80°	0.985	0.174	5.671
36°	0.588	0.809	0.727	81°	0.988	0.156	6.314
37°	0.602	0.799	0.754	82°	0.990	0.139	7.115
38°	0.616	0.788	0.781	83°	0.993	0.122	8.144
39°	0.629	0.777	0.810	84°	0.995	0.105	9.514
40°	0.643	0.766	0.839	85°	0.996	0.087	11.430
41°	0.656	0.755	0.869	86°	0.998	0.070	14.301
42°	0.669	0.743	0.900	87°	0.999	0.052	19.081
43°	0.682	0.731	0.933	88°	0.999	0.035	28.636
44°	0.695	0.719	0.966	89°	1.000	0.017	57.290
45°	0.707	0.707	1.000				

Constructions

CONSTRUCTION 1   Construct a segment congruent to $\overline{AB}$. (p. 20)

CONSTRUCTION 2   Construct the perpendicular bisector of $\overline{AB}$. (p. 23)

CONSTRUCTION 3   Construct the perpendicular to a line at a point on the line. (p. 24)

CONSTRUCTION 4   Construct the perpendicular to a line from a point not on the line. (p. 24)

CONSTRUCTION 5   Construct an angle congruent to $\angle B$. (p. 52)

CONSTRUCTION 6   Construct the bisector of an angle. (p. 53)

CONSTRUCTION 7   Construct a triangle congruent to $\triangle ABC$ by constructing segments congruent to $\overline{AB}$, $\overline{AC}$, and $\overline{BC}$. (p. 155)

CONSTRUCTION 8   Construct a triangle congruent to $\triangle ABC$ by constructing an angle and segments congruent to $\angle A$, $\overline{AB}$, and $\overline{AC}$. (p. 159)

CONSTRUCTION 9   Construct a triangle congruent to $\triangle ABC$ by constructing a segment and angles congruent to $\overline{AB}$, $\angle A$, and $\angle B$. (p. 162)

CONSTRUCTION 10   Construct a triangle congruent to $\triangle ABC$ by constructing segments and an angle congruent to $\overline{AB}$, $\overline{AC}$, and right $\angle B$. (p.166)

CONSTRUCTION 11   Construct the line through P that is parallel to line l. (p. 197)

CONSTRUCTION 12   Divide $\overline{AB}$ into three congruent parts. (p. 338)

CONSTRUCTION 13   Construct a tangent to $\odot O$ at point A on the circle. (p. 389)

CONSTRUCTION 14   Construct the tangents to $\odot O$ from point A outside the circle. (p. 389)

CONSTRUCTION 15   Construct a circle circumscribed about $\triangle ABC$. (p. 397)

CONSTRUCTION 16   Construct a circle inscribed in $\triangle ABC$. (p. 398)

Postulates, Theorems, and Corollaries

POSTULATE 1	Two points determine exactly one line. (p. 5)
POSTULATE 2	Three noncollinear points determine exactly one plane. (p. 5)
POSTULATE 3	If two lines intersect, then their intersection is a point. (p. 12)
POSTULATE 4	If two planes intersect, then their intersection is a line. (p. 12)
POSTULATE 5	**Angle Addition Postulate:** If point W lies in the interior of $\angle XYZ$, then $m\angle XYW + m\angle WYZ = m\angle XYZ$. (p. 49)
THEOREM 2.1	**The Angle Bisector Theorem:** Each point on the bisector of an angle is equidistant from the sides of the angle. (p. 54)
THEOREM 2.2	**Vertical Angles Theorem:** Vertical angles are congruent. (p. 58)
THEOREM 4.1	Two consecutive reflections over parallel lines result in an image that is equivalent to a translation image. (p. 110)
THEOREM 4.2	Two consecutive reflections over intersecting lines result in an image equivalent to a rotation image. (p. 114)
THEOREM 5.1	**Isosceles Triangle Theorem:** The base angles of an isosceles triangle are congruent. (p. 129)
THEOREM 5.2	The line of symmetry for an isosceles triangle bisects the vertex angle and is the perpendicular bisector of the base. (p. 129)
THEOREM 5.3	If two angles of a triangle are congruent, then the sides opposite those angles are congruent. (p. 129)
THEOREM 5.4	An equilateral triangle is also equiangular. (p. 130)
THEOREM 5.5	Each of the three lines of symmetry for an equilateral triangle bisects an angle of the triangle and is the perpendicular bisector of the side opposite that angle. (p. 130)
THEOREM 5.6	An equiangular triangle is also equilateral. (p. 130)
THEOREM 5.7	The three altitudes of a triangle are concurrent. (p. 134)
THEOREM 5.8	The three medians of a triangle are concurrent. (p. 134)
THEOREM 5.9	If a point lies on the perpendicular bisector of a segment, then it is equidistant from the endpoints of the segment. (p. 138)
THEOREM 5.10	If a point is equidistant from the endpoints of a segment, then it lies on the perpendicular bisector of the segment. (p. 138)
THEOREM 5.11	The perpendicular bisectors of the three sides of a triangle intersect in a point that is equidistant from the vertices of the triangle. (p. 139)
THEOREM 5.12	The three angle bisectors of a triangle intersect in a point that is equidistant from the sides of the triangle. (p. 139)

THEOREM 5.13   If the two sides of a triangle are unequal, then the measures of the angles opposite those sides are unequal and the larger angle is opposite the longer side. (p. 144)

THEOREM 5.14   If the measures of two angles of a triangle are unequal, then the sides opposite those angles are unequal and the longer side is opposite the larger angle. (p. 144)

THEOREM 5.15   **The Triangle Inequality:** The sum of the lengths of any two sides of a triangle is greater than the length of the third side. (p. 145)

THEOREM 5.16   **The Hinge Theorem:** If two sides of one triangle are congruent to two sides of another triangle and the measures of the included angles are unequal, then the sides opposite those angles are unequal and the longer side is opposite the larger angle. (p. 147)

THEOREM 5.17   If two sides of one triangle are congruent to two sides of another triangle and the third sides are unequal, then the measures of the angles opposite the third sides are unequal and the larger angle is opposite the longer side. (p. 148)

POSTULATE 6   **SSS Postulate:** If three sides of one triangle are congruent to the corresponding sides of another triangle, then the triangles are congruent. (p. 155)

POSTULATE 7   **SAS Postulate:** If two sides and the included angle of one triangle are congruent to the corresponding sides and angle of another triangle, then the triangles are congruent. (p. 159)

POSTULATE 8   **ASA Postulate:** If two angles and the included side of one triangle are congruent to the corresponding angles and side of another triangle, then the triangles are congruent. (p. 162)

THEOREM 6.1   If two angles of one triangle are congruent to two angles of another triangle, then the third pair of angles are congruent. (p. 163)

THEOREM 6.2   **AAS Theorem:** If two angles and a non-included side of one triangle are congruent to the corresponding angles and side of another triangle, then the triangles are congruent. (p.163)

THEOREM 6.3   **HL Theorem:** If the hypotenuse and one leg of a right triangle are congruent to the hypotenuse and one leg of another right triangle, then the right triangles are congruent. (p. 167)

POSTULATE 9   **Alternate Interior Angles Postulate:** If two parallel lines are cut by a transversal, then each pair of alternate interior angles are congruent. (p. 188)

THEOREM 7.1   **Same-Side Interior Angles Theorem:** If two parallel lines are cut by a transversal, then each pair of same-side interior angles are supplementary. (p. 188)

POSTULATE 10	If two lines are cut by a transversal so that one pair of alternate interior angles is congruent, then the lines are parallel. (p. 189)
THEOREM 7.2	If two lines are cut by a transversal so that one pair of same-side interior angles is supplementary, then the lines are parallel. (p. 189)
THEOREM 7.3	**Corresponding Angles Theorem:** If two parallel lines are cut by a transversal, then each pair of corresponding angles are congruent. (p. 193)
THEOREM 7.4	If two lines are cut by a transversal so that one pair of corresponding angles are congruent, then the lines are parallel. (p. 194)
COROLLARY	If two coplanar lines are perpendicular to a third line, then the two lines are parallel to each other. (p. 194)
THEOREM 7.5	**Triangle Angle-Sum Theorem:** The sum of the measures of the interior angles of any triangle is 180°. (p. 202)
COROLLARY 1	The acute angles of any right triangle are complementary. (p. 202)
COROLLARY 2	The measure of each interior angle of an equilateral triangle is 60°. (p. 202)
THEOREM 7.6	In a triangle, the measure of each exterior angle is equal to the sum of the measures of its two remote interior angles. (p. 207)
COROLLARY	In a triangle, the measure of each exterior angle is greater than the measure of either of its remote interior angles. (p. 207)
THEOREM 7.7	The sum of the measures of the interior angles of any quadrilateral is 360°. (p. 210)
THEOREM 7.8	**Interior Angle Sum Theorem:** The sum of the measures of the interior angles of a polygon with n sides is $(n - 2)$ 180°. (p. 210)
THEOREM 7.9	**Exterior Angle Sum Theorem:** The sum of the measures of the exterior angles, one at each vertex, for any polygon is 360°. (p. 211)
THEOREM 8.1	The opposite sides of a parallelogram are congruent. (p. 221)
THEOREM 8.2	The opposite angles of a parallelogram are congruent. (p. 221)
THEOREM 8.3	The consecutive angles of a parallelogram are supplementary. (p. 221)
THEOREM 8.4	The diagonals of a parallelogram bisect each other. (p. 221)
THEOREM 8.5	Each diagonal of a rhombus bisects a pair of opposite angles. (p. 225)
THEOREM 8.6	Each diagonal of a rhombus is the perpendicular bisector of the other. (p. 225)
THEOREM 8.7	The diagonals of a rectangle are congruent. (p. 228)
THEOREM 8.8	The median of any trapezoid is parallel to the bases and has a length equal to half the sum of the base lengths. (p. 233)
THEOREM 8.9	Base angles of an isosceles trapezoid are congruent. (p. 233)
THEOREM 8.10	A segment joining the midpoints of two sides of a triangle is parallel to the third side and half its length. (p. 233)

THEOREM 8.11	Parallel lines are everywhere equidistant. (p. 237)
THEOREM 8.12	If the diagonals of a quadrilateral bisect each other, then the quadrilateral is a parallelogram. (p. 242)
THEOREM 8.13	If the diagonals of a quadrilateral bisect each other and (**a**) are congruent, then the quadrilateral is a rectangle; (**b**) are perpendicular, then the quadrilateral is a rhombus; (**c**) are perpendicular and congruent, then the quadrilateral is a square. (p. 242)
THEOREM 8.14	If both pairs of opposite sides of a quadrilateral are congruent, then the quadrilateral is a parallelogram. (p. 243)
THEOREM 8.15	If both pairs of opposite angles of a quadrilateral are congruent, then the quadrilateral is a parallelogram. (p. 243)
THEOREM 8.16	If two sides of a quadrilateral are both parallel and congruent, then the quadrilateral is a parallelogram. (p. 243)
THEOREM 9.1	The area (A) of a rectangle is the product of its base length (b) and its height (h). $A = bh$ (p. 264)
THEOREM 9.2	The area (A) of a parallelogram is the product of its base length (b) and its height (h). $A = bh$ (p. 267)
THEOREM 9.3	The area (A) of any triangle is half the product of its base length (b) and height (h). $A = \frac{1}{2}bh$ (p. 271)
THEOREM 9.4	The area of a trapezoid is half the product of the height (h) and the sum of the base lengths ($b_1 + b_2$). $A = \frac{1}{2}h(b_1 + b_2)$ (p. 276)
THEOREM 9.5	The area of a rhombus is half the product of the lengths of the diagonals. $A = \frac{1}{2}d_1 d_2$ (p. 276)
THEOREM 9.6	The area of a regular polygon is half the product of the apothem and the perimeter. $A = \frac{1}{2}aP$ (p. 276)
THEOREM 9.7	The circumference of any circle is the product of its diameter (d) and π. $C = \pi d = 2\pi r$ (p. 282)
THEOREM 9.8	The area (A) of any circle is the product of π and the square of the radius (r). $A = \pi r^2$ (p. 286)
POSTULATE 11	**AA Postulate:** If two angles of one triangle are congruent to two angles of another triangle, then the triangles are similar. (p. 321)
POSTULATE 12	**SSS Similarity Postulate:** If the lengths of the corresponding sides of two triangles are proportional, then the triangles are similar. (p. 325)
POSTULATE 13	**SAS Similarity Postulate:** If the lengths of two pairs of corresponding sides of two triangles are proportional and the corresponding included angles are congruent, then the triangles are similar. (p. 325)

THEOREM 11.1	**Triangle Proportionality Theorem:** If a line is parallel to one side of a triangle and intersects the other two sides, then the triangle formed is similar to the original triangle and the sides of the original triangle are divided proportionally. (p. 330)
THEOREM 11.2	If a line divides two sides of a triangle proportionally, then the line is parallel to the third side of the triangle. (p. 331)
THEOREM 11.3	If three or more parallel lines intersect two or more transversals, then the segments intercepted on the transversals are proportional. (p. 337)
COROLLARY	If three or more parallel lines intercept congruent segments on one transversal, then they intercept congruent segments on every transversal. (p. 337)
THEOREM 12.1	If the lengths a, b, and c of the three sides of a triangle are related by the formula $a^2 + b^2 = c^2$, then the triangle is a right triangle. (p. 353)
THEOREM 12.2	**Pythagorean Theorem:** For any right triangle, the square of the length of the hypotenuse is equal to the sum of the squares of the lengths of the legs. $c^2 = a^2 + b^2$ (p. 354)
THEOREM 12.3	In every 45°- 45°- 90° triangle, (**a**) both legs are congruent, and (**b**) the hypotenuse is $\sqrt{2}$ times as long as each leg. (p. 367)
THEOREM 12.4	In every 30°- 60°- 90° triangle, (**a**) the hypotenuse is twice as long as the side opposite the 30° angle (the shorter leg), (**b**) the side opposite the 30° angle (shorter leg) is half as long as the hypotenuse, and (**c**) the side opposite the 60° angle (longer leg) is $\sqrt{3}$ times as long as the side opposite the 30° angle (shorter leg). (p. 367)
THEOREM 12.5	The altitude to the hypotenuse of a right triangle forms two triangles that are similar to the original triangle and to each other. (p. 371)
COROLLARY 1	The length of the altitude drawn to the hypotenuse of a right triangle is the geometric mean between the lengths of the segments of the hypotenuse. (p. 371)
COROLLARY 2	When the altitude is drawn to the hypotenuse of a right triangle, the length of each leg is the geometric mean between the length of the hypotenuse and the length of the adjacent segment of the hypotenuse. (p. 371)
THEOREM 13.1	In the same circle (or congruent circles) two minor arcs are congruent when their central angles are congruent. (p. 382)
THEOREM 13.2	The measure of an inscribed angle is equal to half the measure of its intercepted arc. (p. 385)
COROLLARY	An inscribed angle that intercepts a semicircle is a right angle. (p. 385)
THEOREM 13.3	If a line is tangent to a circle, then the line is perpendicular to the radius drawn to the point of tangency. (p. 388)

THEOREM 13.4	If a line is in the same plane as a circle and is perpendicular to a radius at its endpoint on the circle, then the line is tangent to the circle. (p. 388)
THEOREM 13.5	The two segments tangent to a circle from a point outside the circle are congruent. (p. 389)
THEOREM 13.6	In a circle (or congruent circles), congruent chords determine congruent arcs and congruent arcs determine congruent chords. (p. 393)
THEOREM 13.7	In a circle, a diameter that is perpendicular to a chord bisects the chord and its arc. (p. 393)
THEOREM 13.8	In a circle, congruent chords are equidistant from the center and chords equidistant from the center are congruent. (p. 394)
THEOREM 13.9	The measure of an angle formed by a tangent and a chord is half the measure of the intercepted arc. (p. 403)
THEOREM 13.10	The measure of an angle formed by two chords that intersect inside a circle is half the sum of the measures of the intercepted arcs. (p. 403)
THEOREM 13.11	The measure of an angle formed by two secants, a tangent and a secant, or two tangents drawn from a point outside the circle is half the difference of the measures of the intercepted arcs. (p. 403)
THEOREM 13.12	If two chords of a circle intersect, then the product of the lengths of the segments of one chord is equal to the product of the lengths of the segments of the other chord. (p. 407)
THEOREM 13.13	If two secants intersect in the exterior of a circle, then the product of the lengths of one secant segment and its external segment is equal to the product of the lengths of the other secant segment and its external segment. (p. 408)
THEOREM 13.14	If a secant and a tangent intersect in the exterior of a circle, then the product of the lengths of the secant segment and its external segment is equal to the square of the length of the tangent segment. (p. 408)
THEOREM 14.1	The lateral area (L.A.) of a prism is the product of the perimeter (P) of a base and the height (h). L.A. $= Ph$ (p. 430)
THEOREM 14.2	To find the volume (V) of any prism, multiply the area of a base (B) by the height (h). $V = Bh$ (p. 435)
THEOREM 15.1	The lateral area of a cylinder is the product of the circumference of the base and the height. L.A. $= Ch = \pi dh = 2\pi rh$ (p. 451)
THEOREM 15.2	The volume of a cylinder is the product of the area of a base and the height. $V = Bh = \pi r^2 h$ (p. 456)
THEOREM 15.3	If the scale factor of two similar figures is $a : b$, then the ratio of corresponding perimeters is $a : b$ and the ratio of corresponding areas is $a^2 : b^2$. (p. 461)

THEOREM 15.4 If the scale factor of two similar space figures is $a : b$, then the ratios of the lateral areas and of the total areas are $a^2 : b^2$ and the ratios of the volumes is $a^3 : b^3$. (p. 465)

THEOREM 15.5 The lateral area of a regular pyramid is half the product of the slant height and the perimeter of the base. L.A. $= \frac{1}{2} lP$ (p. 470)

THEOREM 15.6 The lateral area of a cone is half the product of the slant height and the circumference of the base. L.A. $= \frac{1}{2} lC = \pi lr$ (p. 470)

THEOREM 15.7 The volume of a regular pyramid or cone is one-third the product of the area of the base and the height. $V = \frac{1}{3} Bh$ (p. 471)

THEOREM 15.8 The area and volume of a sphere with radius r are given by the following formulas: $A = 4\pi r^2$, $V = \frac{4}{3} \pi r^3$ (p. 474)

THEOREM 17.1 **The Distance Formula** The distance (d) between any two points $A(x_1, y_1)$ and $B(x_2, y_2)$ is $d = \sqrt{(x_2 - x_1)^2 + (y_2 - y_1)^2}$. (p. 511)

THEOREM 17.2 **The Midpoint Formula** If the coordinates of the endpoints of a segment are (x_1, y_1) and (x_2, y_2), then the midpoint of the segment has coordinates $\left(\dfrac{x_1 + x_2}{2}, \dfrac{y_1 + y_2}{2} \right)$. (p. 513)

THEOREM 17.3 Two nonvertical lines are parallel if and only if they have the same slope. (p. 520)

THEOREM 17.4 Two nonvertical lines are perpendicular if and only if the product of their slopes is -1. (p. 520)

THEOREM 17.5 The equation of a circle with center (h, k) and radius r is $(x - h)^2 + (y - k)^2 = r^2$. (p. 533)

THEOREM 18.1 The image of (x, y) under a translation in the coordinate plane is $(x + a, y + b)$. (p. 543)

THEOREM 18.2 If the coordinates of a point are (x, y), then the coordinates of its reflection image are **(a)** $(-x, y)$ when the line of reflection is the y-axis, **(b)** $(x, -y)$ when the line of reflection is the x-axis, and **(c)** (y, x) when the line of reflection is the line $y = x$. (p. 547)

THEOREM 18.3 If the coordinates of a point are (x, y), then the coordinates of its rotation image about O are **(a)** $(-y, x)$ for a 90° CCW rotation, **(b)** $(y, -x)$ for a 90° CW rotation, and **(c)** $(-x, -y)$ for a 180° rotation. (p. 550)

THEOREM 18.4 If the coordinates of a point are (x, y), then the coordinates of its image under a dilation with center O and scale factor k are (kx, ky). (p. 554)

Glossary

acute angle *See* angle.

acute triangle (p. 73) A triangle in which all angles are acute angles.

adjacent angles (p. 49) Two angles in the same plane with a common vertex and a common side but no interior points in common.

alternate interior angles *See* transversal.

altitude of a parallelogram (p. 267) A segment from a point on one side perpendicular to the line containing the opposite side, called the *base*. The length of an altitude is called the *height* of the parallelogram.

altitude of a three-dimensional figure *See* prism, regular pyramid, right circular cone, and right circular cylinder.

altitude of a triangle (p. 133) A segment from one vertex perpendicular to the line containing the opposite side, which is called the *base*. The length of an altitude is called the *height* of the triangle.

amount of turn (p. 40) A unit of angle measure used when an angle is viewed as the figure formed by turning a ray about its endpoint.

angle ($\angle$) (p. 8) An angle may be viewed as either two rays with a common endpoint, or the figure formed by turning a ray about its endpoint. The common endpoint of the rays is the *vertex* (p. 8) of the angle. The rays forming the angle are the *sides* (p. 8) of the angle. An angle can be named by the vertex (if only one angle pictured has that point as vertex), by the vertex and one point on each side, or by a lowercase letter or number written inside the picture of the angle. An *acute angle* (p. 43) is an angle with a measure less than 90°, or less than $\frac{1}{4}$ turn. An *obtuse angle* (p. 43) is an angle with a measure between 90° and 180°, or between $\frac{1}{4}$ and $\frac{1}{2}$ turn. A *straight angle* (p. 8, 43) is an angle with a measure of 180°, or $\frac{1}{2}$ turn. A *right angle* (p. 8, 43) is an angle with a measure of 90°, or $\frac{1}{4}$ turn.

angle bisector (p. 53) A ray that divides an angle into two congruent angles.

angle of depression, angle of elevation (p. 494) One of two alternate interior angles determined by parallel horizontal rays transversed by the line of sight.

apothem (p. 276) The measure of the radius of a circle inscribed in a regular polygon.

arc ($\frown$)(p. 381) A part of a circle determined by two points. An arc is named by its endpoints. A third point is used to name arcs larger than a semicircle. *Congruent arcs* (p. 382) are arcs of the same circle (or congruent circles) that have the same measure. The *length* (p. 381) of an arc is a fractional part of the circumference of the circle. A *major arc* $\overset{\frown}{PQR}$ (p. 381) is longer than a semicircle. A *minor arc* $\overline{AB}$ (p. 381) is shorter than a semicircle. The *measure* (p. 381) of a minor arc is defined as the measure of its central angle. The measure (p. 381) of a major arc is 360° minus the measure of the associated minor arc. The measure of a semicircle is 180°.

area (*A*) (p. 259) The amount of surface of any two-dimensional region. Area is usually measured in square units.

auxiliary line (p. 247) A line or segment added to a diagram to help prove a theorem.

average of two numbers (p. 512) The sum of two numbers divided by two.

base of a parallelogram (*b*) (p. 267) Any side of a parallelogram can be considered the base. This is true for all parallelograms including rectangles, rhombuses, and squares.

base of a rectangle (*b*) (p. 263) Any side of a rectangle can be called the base.

base of a three-dimensional figure *See* prism, regular pyramid, right circular cone, and right circular cylinder.

base of a triangle (p. 74, 271) Any side of a triangle may be called the base. In an isosceles triangle, it is the side opposite the vertex angle.

bisector of a segment (p. 21) Any line, segment, or ray that passes through the midpoint of a segment. A segment has infinitely many bisectors. A *perpendicular bisector* (p. 23) is a line which is perpendicular to a segment at its midpoint.

capacity (p. 442) A concept closely related to volume. Capacity is the amount of material, liquid or gas, that a container will hold. Some commonly used units of capacity are the pint, quart, gallon, liter, and fluid ounce.

center of gravity (p. 136) The balance point of an object. The *centroid* is the center of gravity of a triangular region.

center of a regular polygon (p. 276) The point equidistant from all vertices.

centimeter *See* meter.

central angle (p. 40) An angle whose vertex is at the center of a circle.

centroid (p. 134) The point of intersection of the three medians of a triangle.

chord *See* circle.

circle (⊙) (p. 39) All points in a plane which are the same distance from some point called the *center*. A circle is named by its center. A *chord* (p. 39) is a segment whose endpoints are any two points of a circle. A diameter is a special type of chord. A *radius* (*r*) (p. 39) is a segment joining the center of a circle and any point on the circle. Also, the distance from the center of a circle to any point on the circle. A *diameter* (*d*) (p. 39) is a chord that passes through the center of a circle. Also, the length of a chord that passes through the center of a circle. A *semicircle* (p. 381) is half a circle. The diameter of a circle divides the circle into two semicircles.

circumcenter (p. 139) The intersection of extended perpendicular bisectors of the sides of a triangle.

circumference (*C*) (p. 282) The distance around (perimeter of) a circle.

circumscribed circle (p. 397) A circle is circumscribed about a polygon if every vertex of the polygon is on the circle.

circumscribed polygon (p. 397) A polygon is circumscribed about a circle if every side of the polygon is tangent to the circle.

clockwise (**CW**) (p. 105) Rotation of an object to the right of its original orientation.

collinear points (p. 4) A set of points all on the same line.

common tangent (p. 388) A line that is tangent to two or more circles. An *internal common tangent* (p. 388) is a line that is tangent to two coplanar circles and intersects the segment joining the centers of the circles. An *external common tangent* (p. 388) is a line that is tangent to two coplanar circles but does not intersect the segment joining the centers of the circles.

complementary angles (p. 57) Two angles whose measures add up to 90°.

concurrent lines (p. 134) Three or more lines that intersect in a point.

congruent (≅) (p. 20) Being the same or equal to.

congruent angles (p. 52) Angles with equal measure.

congruent arcs *See* arc.

congruent circles (p. 382) Circles that have the same radii.

congruent figures (p. 99) Figures that have the same size and shape.

congruent polygons (p. 100) Polygons whose corresponding parts are congruent.

congruent segments (p. 20) Two or more segments that are equal in length.

consecutive (adjacent) angles *See* polygon.

consecutive (adjacent) sides *See* polygon.

consecutive (adjacent) vertices *See* polygon.

construction (p. 20) A geometric drawing made with only straightedge and compass.

converse (p. 200) The result of the interchange of the hypothesis and conclusion of a conditional statement.

coordinate (p. 507) A number used to name (locate) a point on a number line. An *origin* (p. 507) is the zero point on a number line.

coordinate axes (p. 507) The pair of perpendicular number lines used to name points on a plane with ordered pairs of numbers. The horizontal axis is called the *x-axis* and the vertical axis is called the *y-axis*. The point where the axes intersect is called the *origin*. A *quadrant* (p. 507) is one of the four regions, labeled I–IV, into which the coordinate axes divide the plane.

coordinates (*x, y*) (p. 507) An ordered pair of numbers used to name (locate) a point in a plane. The *x-coordinate* (p. 507) is the distance of a point right or left of the origin. The *y-coordinate* (p. 507) is the distance of a point above or below the origin. In an *ordered pair* (p. 507), the *x*-coordinate is always listed first and the *y*-coordinate second.

coordinate system (p. 507) A pair of perpendicular lines used to locate points in a plane. The horizontal number line is called the *x-axis*. The vertical number line is called the *y-axis*. The point where the two axes intersect is called the origin (*O*).

coplanar points (p. 4) A set of points all in the same plane.

corollary (p. 194) A statement that follows directly from a theorem.

corresponding angles *See* transversal.

corresponding parts *See* similar figures.

cosine *See* trigonometric ratios.

counterclockwise (CCW) (p. 105) Rotation of an object to the left of its original orientation.

counterexample (p. 27) A statement or question used to disprove the conclusion reached using inductive reasoning.

cube (p. 82) A rectangular prism in which all faces are congruent squares.

deductive reasoning (p. 27) The process of reasoning from known or accepted facts to a new conclusion.

degree (°) (p. 43) A unit of angle measure equivalent to $\frac{1}{360}$ turn.

diagonal of a polygon (p. 70) A segment joining a pair of nonconsecutive vertices.

diameter *See* circle.

dilation (p. 554) A transformation that results in a size change. A dilation has a *center* and a non-zero *scale factor*.

distance from a point to a line (p. 134) The length of the perpendicular segment from the point to the line.

divided proportionally (p. 330) Segments are said to be divided proportionally when the ratios of corresponding lengths are equal.

edge (p. 81) A segment formed by the intersection of two faces of a three-dimensional figure.

equiangular triangle (p. 73) A triangle with three congruent angles. Every equiangular triangle is also equilateral.

equidistant (p. 54) The same distance. If *AB* = *AC*, then *A* is equidistant from *B* and *C*.

equilateral triangle (p. 73) A triangle with three congruent sides. Every equilateral triangle is also equiangular.

exterior angles *See* transversal.

exterior angle of a triangle *See* interior angles of a triangle.

extremes (p. 298) The first and fourth terms of a proportion.

face (p. 81) A part of a plane forming a side of a three-dimensional figure.

glide reflection (p. 547) The combination of a reflection and a translation parallel to the line of reflection.

graph of an equation (p. 523) Line determined by an equation that can be written in the form $y = mx + b$.

great circle (p. 475) The circle determined by the intersection of a sphere and a plane that contains the center of the sphere.

height of a rectangle (*h*) (p. 263) The length of each of the sides perpendicular to the base.

height of a parallelogram (*h*) (p. 267) The length of an altitude.

height of a triangle (*h*) (p. 271) The length of the corresponding altitude from the opposite vertex to the base.

hemisphere (p. 477) Half a sphere

hexagon *See* polygon.

horizon line (p. 91) The line where the sky seems to meet the earth.

hypotenuse (p. 159) The side opposite the right angle of a right triangle.

incenter (p. 139) The point of intersection of the angle bisectors of a triangle.

included angle of a triangle (p. 147) For two sides of a triangle, the angle formed by those sides.

included side of a triangle (p. 147) For two angles of a triangle, the side of the triangle that is a side of both angles.

inductive reasoning (p. 27) The process of reasoning from specific examples to a general conclusion.

inscribed angle (p. 385) An angle whose vertex is on a circle and whose sides contain chords of the circle.

inscribed circle (p. 397) A circle is inscribed in a polygon if every side of the polygon is tangent to the circle.

inscribed polygon (p. 397) A polygon is inscribed in a circle if every vertex of the polygon is on the circle.

intercepted arc (p. 385) An arc determined by the points of intersection of a circle and the sides of a central or inscribed angle.

interior angles *See* transversal.

interior angles of a triangle (p. 202) The angles formed by the sides of a triangle. An *exterior angle of a triangle* (p. 206) is an angle formed by one side of the triangle and the extension of the adjacent side. A triangle has six exterior angles, two at each vertex. For each exterior angle, the interior angle to which it is adjacent is called the *adjacent interior angle* (p. 206) and the two interior angles to which it is not adjacent are called the *remote interior angles* (p. 206).

intersecting lines (p. 12) Lines that meet in a point.

isosceles trapezoid (p. 232) A trapezoid whose legs are congruent.

isosceles triangle (p. 73) A triangle with at least two congruent sides. The two congruent sides are called the *legs* (p. 74). The *vertex angle* (p. 74) is the angle formed by the legs. The *base angles* and *base* (p. 74) are the other angles and side. An equilateral triangle is a special type of isosceles triangle.

kilometer *See* meter.

lateral area (p. 429) The area of all faces, except the bases, of a three-dimensional figure.

lateral face (p. 81) Any face of a three-dimensional figure that is not a base.

legs of a right triangle (p. 159) Sides that determine the right angle.

length (p. 16) The distance between two identified points.

length of an arc *See* arc.

line ($\leftrightarrow$) (p. 3) A line is straight, has no thickness, and extends indefinitely in two directions.

linear equation (p. 523) An equation whose graph is a line. Example: $x + y = 3$.

line of reflection (p. 105) The perpendicular bisector of the segments formed by connecting the vertex to its image.

line of symmetry (p. 108, 120) A line that divides a figure into two parts that match exactly.

locus (p. 411) The set of all points that satisfy a set of conditions.

major arc *See* arc.

means (p. 298) The second and third terms of a proportion.

measure of an angle (p. 43) Shown by the notation $m\angle ABC$.

measure of a minor arc *See* arc.

measure of a major arc *See* arc.

median of a triangle (p. 133) A segment joining a vertex and the midpoint of the opposite side.

meter (m) (p. 16) The basic unit of length in the metric system. A meter is about the height of a doorknob from the floor. A *centimeter* (cm) (p. 16) is equivalent to $\frac{1}{100}$ or 0.01 of a meter (1 m = 100 cm or 1 cm = 0.01 m). A *millimeter* (mm) (p. 16) is equivalent to $\frac{1}{1000}$ or 0.001 of a meter (1m = 1000 mm or 1 mm = 0.001 m). A *kilometer* (km) (p. 16) is equivalent to 1000 meters (1 km = 1000 m or 1 m = 0.001 km).

midpoint of a segment (p. 21) A point that divides a segment into two congruent segments.

millimeter *See* meter.

minor arc *See* arc.

noncollinear points (p. 4) A set of points through which a single straight line cannot be drawn.

noncoplanar points (p. 4) A set of points through which a single plane cannot be drawn.

oblique prism *See* prism.

obtuse angle *See angle.*

obtuse triangle (p. 73) A triangle that contains one obtuse angle.

octagon *See* polygon.

one-point perspective (p. 91) A method of perspective drawing where parallel lines that intersect the horizon line seem to meet in a point on the horizon line.

opposite rays (p. 8) Two rays that have a common endpoint and form a line.

orientation (p. 105) The position of a figure with relation to its reflective image or other figures.

origin *See* coordinate and coordinate axes.

orthocenter (p. 133) The point at which the lines containing the three altitudes (or their extensions) of a triangle meet.

parallel lines (∥)(p. 12) Coplanar lines that do not intersect.

parallel planes (p. 13) Two or more planes that do not intersect.

parallelogram (▱) (p. 77) A quadrilateral with two pairs of opposite sides parallel. Rectangles, rhombuses, and squares are special types of parallelograms.

pentagon *See* polygon.

pentomino (p. 87) A figure formed by five same-size squares arranged edge to edge.

perfect square *See* square root.

perimeter (*P*) (p. 255) The distance around a two-dimensional figure.

perpendicular bisector *See* bisector of a segment.

perpendicular (⊥) (p. 8) Lines, segments, or rays that meet to form right angles.

perspective drawing (p. 91) The use of a special technique involving a *horizon line* and *vanishing point(s)* to represent a three-dimensional object on a flat surface.

pi (π) (p. 282) The ratio of the circumference of a circle to its diameter. The most commonly used approximations for π are $\frac{22}{7}$ and 3.14.

plane (p. 3) A flat surface which has no thickness and extends indefinitely in all directions.

point (p. 3) A point has no size and indicates a definite location.

point symmetry (p. 117, 121) The characteristic of a figure which coincides with its image under a rotation of 180°.

polygon (p. 69) A simple closed two-dimensional figure formed only by line segments that meet at points called *vertices*. The segments are called *sides. Consecutive (adjacent) angles of a polygon* (p. 70) are a pair of angles that have one side containing the same side of the polygon. *Consecutive (adjacent) sides of a polygon* (p. 70) are a pair of sides that intersect. *Consecutive (adjacent) vertices of a polygon* (p. 70) are a pair of vertices that are endpoints of the same side of a polygon. *Nonconsecutive angles of a polygon* (p. 70) are a pair of angles whose sides contain different sides of the polygon. *Nonconsecutive vertices of a polygon* (p. 70) are vertices that are not endpoints of the same side. A *quadrilateral* (p. 70) is a polygon with four sides. A *pentagon* (p. 70) is a polygon with five sides. A *hexagon* (p. 70) is a polygon with six sides. An *octagon* (p. 70) is a polygon with eight sides. A *convex polygon* is a polygon for which no line containing a side intersects the interior. A *concave polygon* is a polygon for which at least one line containing a side intersects the interior. *Congruent polygons* are polygons whose corresponding parts are congruent.

polyhedron (p. 81) A three-dimensional figure in which each surface is shaped like a polygon. A *face* (p. 81) is part of a plane forming a side of the three-dimensional figure. An *edge* (p. 81) is a segment formed by the intersection of two faces of the three-dimensional figure. A *vertex* (p. 81) is a point at which three or more edges intersect.

postulate (p. 5) A statement that is accepted without proof.

prism (p. 81) A three-dimensional figure that has two parallel congruent polygonal regions as bases and lateral faces that are parallelogram regions. Prisms are named by the shape of their *bases* . An *edge* of a prism is a segment formed by the intersection of two faces and a *vertex* is a point formed by the intersection of two edges. Any segment perpendicular to both bases and with one endpoint in each base is an *altitude* of the prism. A *right prism* (p. 82) is a prism in which all lateral faces are rectangles. An *oblique prism* (p. 82) is a prism that is not a right prism. The *total area* (T.A) of a prism is the sum of all the faces. The *lateral area* (L.A.) of a prism is the sum of the areas of its faces not including the bases.

proof (p. 171) A convincing argument.

proportion (p. 298) A statement that two ratios are equal.

proportional (p. 305) When numbers can be arranged so that a true proportion can be written, the numbers are said to be proportional.

protractor (p. 46) An instrument used to measure angles.

pyramid (p. 82) A solid figure having a polygonal base, the sides of which form the bases of triangular surfaces meeting at a common vertex.

Pythagorean theorem (p. 354) For any right triangle, the square of the length of the hypotenuse is equal to the sum of the squares of the lengths of the legs.

Pythagorean triple (p. 357) Three positive whole number lengths such as 3, 4, and 5 that determine a right triangle.

quadrant *See* coordinate axes.

quadrilateral *See* polygon.

radius *See* circle.

ratio (p. 295) A comparison of two numbers by division. If $b \neq 0$, the ratio of a to b is denoted by $\frac{a}{b}$, $a : b$, or a to b.

ray ($\longrightarrow$) (p. 3) Part of a line with one endpoint. It extends indefinitely in one direction. A ray is named by its endpoint and any other point on the ray. The endpoint is named first.

rectangle (p. 77) A parallelogram, with four right angles. A square is a special type of rectangle.

reflection image (p. 104) The image of an object in a mirror or a pond.

regular polygon (p. 78) A polygon in which all sides are congruent and all angles are congruent.

regular pyramid (p. 469) A three-dimensional figure formed by joining the vertices of a regular polygon (*base*) to a point (*vertex*) that is not in the same plane as the base and that lies directly above the center of the base. A segment from the vertex perpendicular to the base is an *altitude* of the pyramid. The *slant height* (p. 470) in a regular pyramid is the length of an altitude of a lateral face.

rhombus (p. 77) A parallelogram with four congruent sides. A square is a special type of rhombus.

right angle *See* angle.

right circular cone (p. 470) A three-dimensional figure formed by joining a circle (*base*) to a point (*vertex*) that is not in the same plane as the base and that lies directly above the center of the base. The segment from the vertex perpendicular to the base is the *altitude* of the cone. The *slant height* (*l*) of a cone is the length of any segment joining the vertex and a point on the circle that determines the base.

right circular cylinder (p. 451) A three-dimensional figure having two parallel congruent circular bases and a curved lateral surface connecting them. The bases are directly above or opposite each other. Any segment perpendicular to both bases and with one endpoint in each base is an *altitude* of the cylinder.

right triangle (p. 73) A triangle that contains one right angle. The side opposite the right angle (longest side) is the *hypotenuse* of the right triangle. The sides forming the right angle are the *legs* of the right triangle.

right prism *See* prism.

rise *See* slope.

rotation (p. 113) The turning motion of an object about a point.

rotation image (p. 113) A repeated, congruent figure, rotated clockwise or counterclockwise about a point.

rotational symmetry (p. 120) This condition exists if there is some rotation less than 360° about a point *C* for which the figure and its image coincide.

run *See* slope.

same-side exterior angle (p. 186) Angles on the same side of the transversal line, but between the two lines intersected by the transversal.

same-side interior angle (p. 186) Angles on the same side of the transversal line, but outside the two lines intersected by the transversal.

scale (p. 311) Ratio of a dimension of a scale drawing to the actual dimension of the object being represented. A *scale drawing* (p. 311) is an enlargement or reduction.

scale drawing *See* scale.

scale factor (p. 305) The ratio of the lengths of proportional corresponding sides.

scalene triangle (p. 73) A triangle with no congruent sides.

secant (p. 402) For a circle, a line that contains a chord. *Secant segment* (p. 407). *External secant segment* (p. 407).

segment (—) (p. 3) Part of a line with two endpoints. A segment is named by its endpoints.

semicircle *See* circle.

side *See* angle.

similar figures (~) (p. 301) Figures that have the same shape but not necessarily the same size. Congruent figures are also similar. A *similarity statement* (p. 301) is a statement asserting that two figures are similar. *Corresponding parts* (p. 301) are any pair of sides or angles in two congruent or similar polygons having the same relative position.

similar polygons (p. 305) Polygons such that corresponding angles are congruent and corresponding side lengths are proportional.

similar space figures (p. 464 Three-dimensional figures are similar when they have the same shape.

simpler form (p. 350) A square root expression can be written in simpler form if the number under the radical sign has at least one perfect square factor other than 1.

simplest radical form (p. 361) A square root expression is said to be in simplest radical form if there are no perfect-square factors other than 1 under a radical sign, the denominator does not contain a radical, and there is no fraction under the radical sign.

sine *See* trigonometric ratios.

skew lines (p. 13) Noncoplanar lines that also do not intersect. ——

slant height *See* regular pyramid.

slope (p. 515) The measure of the steepness of a line. It is the ratio of *rise* to *run*. The *rise* is the vertical distance from a given point to a second given point. The *run* is the horizontal distance from a given point to a second given point.

solution (p. 530) The ordered pair that is the coordinates of the point of intersection of two lines.

sphere (p. 474) The set of all points in space that are the same distance from some point called the *center*.

square (p. 77) A parallelogram with four right angles and four congruent sides. A square may also be defined as a parallelogram that is both a rectangle and a rhombus.

square of a number (n^2) (p. 347) A number multiplied by itself. The exponent 2 is used to represent the multiplication. For example, "3 squared" can be written 3^2 and $3 \cdot 3 = 9$.

square root ($\sqrt{}$) (p. 347) The positive number which when squared gives the original number as a product. The square root of 25 is 5 because $5 \cdot 5 = 25$. A *radical sign* (p. 347) is a symbol used to indicate the positive square root of a number. A *perfect square* (p. 347) is a number that can be written as the product of two equal factors. For example, 16 is a perfect square because $4 \cdot 4 = 16$.

straight angle *See* angle.

supplementary angles (p. 57) Two angles whose measures add up to 180°.

symmetric figure (p. 120) Any figure which has at least one line of symmetry. If a symmetric figure is folded over a line of symmetry, the two halves will match exactly. All lines of symmetry divide a figure into two congruent parts.

system of equations (p. 530) A set of two or more equations involving the same variables.

tangent *See* trigonometric ratios.

tangent (p. 388) A line is tangent to a circle if it is coplanar with the circle and intersects the circle in exactly one point. The point is the *point of tangency*. Such a line is called a *tangent line* or *tangent*. *Tangent segment* (p. 407).

tessellation (p. 558) An arrangement of congruent figures that completely covers a plane without gaps or overlapping.

theorem (p. 28) A statement that can be proved.

three-dimensional figures (p. 81) Figures that enclose space.

total area (p. 426) The sum of the areas of all faces of a three-dimensional figure.

translation (p. 109) A sliding of a geometric figure, without turning, from one position to another.

translation image (p. 110) The image of a figure under a translation.

transversal (p. 185) A line intersecting two or more coplanar lines each at a different point. *Alternate interior angles* (p. 185) are a pair of angles formed by a transversal and the two lines it intersects. Alternate interior angles have different vertices and are located on opposite sides of the transversal and inside, or between, the two lines it intersects. *Corresponding angles* (p. 185) are a pair of angles formed by a transversal and the two lines it intersects. Corresponding angles have different vertices and are located in corresponding positions with respect to the two lines. *Exterior angles* (p. 185) are four angles that are located outside the two lines crossed by a transversal. *Interior angles* (p. 185) are four angles located inside, or between, the two lines crossed by a transversal.

trapezoid (p. 77) A quadrilateral with exactly one pair of opposite sides parallel. The *bases* (p. 232) of a trapezoid are the parallel sides. The *base angles of a trapezoid* (p. 232) are a pair of consecutive angles whose included side is a base. The *legs* (p. 232) of a trapezoid are the nonparallel sides. The *median* (p. 232) is the segment that joins the midpoints of the legs.

triangle ($\triangle$) (p. 70) A polygon with three sides. The point of intersection of two sides is a *vertex*.

trigonometry (p. 485) Mathematical principles based on properties of similar right triangles.

trigonometric identity (p. 499) An equation involving trigonometric ratios that is true for *all* angles.

trigonometric ratio (p. 486) Ratio of corresponding sides in similar right triangles. The *cosine* (cos) is the ratio of the length of the leg adjacent to an acute angle of a right triangle to the length of the hypotenuse. The *sine* (sin) is the ratio of the length of the leg opposite an acute angle of a right triangle to the length of the hypotenuse. The *tangent* (tan) is the ratio of the length of the leg opposite an acute angle of a right triangle to the length of the adjacent leg.

two-point perspective (p. 92) A method of perspective drawing that uses two vanishing points.

vanishing point (p. 91) The point at which parallel lines seem to meet on the horizon line.

vertex *See* angle.

vertex (p. 69) The intersection of two sides.

vertical angles (p. 57) Two angles whose sides form two pairs of opposite rays. Vertical angles are congruent.

vertices *See* polygon.

volume (*V*) (p. 434) The amount of space within, or occupied by, a three-dimensional figure. Volume is usually measured in cubic units.

x-axis *See* coordinate system.

x-coordinate *See* coordinates.

y-axis *See* coordinate system.

y-coordinate *See* coordinates.

1.5 Congruent Segments (p. 21)

EXPLORING

1. On graph (grid) paper, select any two intersection points that are not on the same grid line. Label the lower point A and the upper point B. Draw $\overline{AB}$.

2. Draw $\overrightarrow{BY}$ along the vertical grid line that contains B. Draw $\overrightarrow{AX}$ along the horizontal grid line that contains A. Label the intersection of $\overrightarrow{AY}$ and $\overrightarrow{AX}$ as C.

3. **a.** Count squares to locate the midpoints, D and E, of $\overline{AC}$ and $\overline{BC}$, respectively.

   b. Locate the intersection of the vertical line through D and the horizontal line through E. Label it M. Is M on $\overline{AB}$?

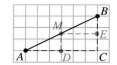

4. Measure $\overline{AB}$, $\overline{AM}$, and $\overline{MB}$ with a ruler. How are these measures related? What conclusion can you draw about M?

5. Test your conclusion. Repeat Steps 1–4 for other segments. Do you get the same results?

2.2 Degree Measure (p. 43)

1. Draw any segment $\overline{AB}$. Use Construction 2 to construct the perpendicular bisector of $\overline{AB}$.

2. Using the perpendicular lines as a guide, draw an acute angle, an obtuse angle, and a straight angle.

2.6 Pairs of Angles (p. 58)

EXPLORING

Look at the lines that intersect to form two pairs of vertical angles. Do not use a protractor for Steps 1–4.

1. Is ∠1 supplementary to ∠2?

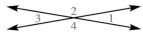

2. Is ∠3 supplementary to ∠2?

3. How do the measures of ∠1 and ∠3 compare? Why?

4. How do the measures of ∠2 and ∠4 compare? Why?

5. Now, use a protractor to measure ∠1, ∠2, ∠3, and ∠4.

6. Do your measurements agree with your answers in Steps 1–4?

7. How are the measures of vertical angles related?

Alternate Explorings

3.2 Classifying Triangles (p. 73)

EXPLORING

1. Judging by appearance only, classify △*ABC*.

2. Copy △*ABC*. Use Construction 2 to locate the midpoint, *D*, of side $\overline{BC}$. Draw $\overline{AD}$.

3. Judging by appearance only, what kind of triangle is *ABD*? *ADC*?

4. Repeat Steps 1–3 for △ *JKL*. Label the midpoint of $\overline{KL}$ as *M*. What kind of triangle is *JKM*? *JML*? How do triangles *JKM* and *JML* compare?

3.3 Quadrilaterals and Other Polygons (p. 78)

EXPLORING

1. Draw five different quadrilaterals. Include at least one rectangle, one square, one trapezoid, and one rhombus.

2. For each figure, mark the midpoint of each side.

3. Draw segments joining the midpoints of the adjacent sides in each figure.

4. What kind of figure is formed by connecting the midpoints?

5. Draw two or three other quadrilaterals. Make some of them strangely shaped. Follow Steps 1–3 for these new quadrilaterals. Do you get the same result?

5.2 Altitudes and Medians of Triangles (p. 133 top)

EXPLORING

1. Draw a large scalene triangle of each type indicated. Construct the three altitudes of each triangle.

   a. acute triangle   **b.** right triangle   **c.** obtuse triangle

2. For each triangle, do the altitudes meet in a point? If so, describe the location of the intersection point for each triangle. (*Hint*: Extend the altitudes if necessary.)

5.2 Altitudes and Medians of Triangles (p. 133 bottom)

EXPLORING

1. Draw a large scalene triangle of each type indicated. For each triangle, construct the midpoint of each side and draw the three medians.

   a. acute triangle   **b.** right triangle   **c.** obtuse triangle

2. For each triangle, do the medians meet in a point? If so, describe the location of the point of intersection of the three medians of each triangle.

5.3 Perpendicular Bisectors and Angle Bisectors (p. 138)

Part A

1. Draw any segment and label it as $\overline{AB}$. Construct the perpendicular bisector of $\overline{AB}$.

2. Mark four or five points on the perpendicular bisector. Measure the distance from each point to A and to B. Compare these distances.

3. Repeat Steps 1 and 2 for a segment longer than $\overline{AB}$ and for one shorter than $\overline{AB}$. Are the results the same?

6.1 SSS (p. 156)

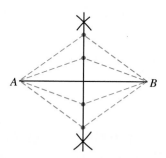

EXPLORING

1. Repeat Construction 7, but in Step 2 put your compass point on Q and in Step 3 put your compass point on P.

2. Again, repeat Construction 7, but this time draw the arcs in Steps 3 and 4 below $\overline{PQ}$ instead of above it.

3. Compare the two triangles just constructed in Steps 1 and 2 above with the one originally done in Construction 7. Are they congruent?

4. Did constructing the sides in a different order change the size and shape of the triangle?

6.3 ASA and AAS (p. 162)

========= EXPLORING

1. Using the same pair of angle measures but different side lengths, use Construction 9 to construct two triangles in which two angles of one triangle are congruent to two angles of the other triangle.

2. Measure the third angle in each triangle. Compare the measures.

3. Repeat Steps 1 and 2 using a different pair of angle measures and different side lengths. Are the results the same?

7.3 Parallel Lines and Corresponding Angles (p. 193)

========= EXPLORING

1. Draw any line a, transversal t, and point P on t as shown. Use Construction 5 to construct $\angle 1 \cong \angle 3$ at point P. Label $\angle 2$ and line b as shown.

2. Is $\angle 2 \cong \angle 3$? Why?   3. Is $\angle 1 \cong \angle 2$? Why?   4. Is $a \parallel b$? Why?

7.5 Interior Angle Measures of a Triangle (p. 201)

========= EXPLORING

(Use the second Exploring on p. 201.)

8.3 Properties of Rectangles and Squares (p. 228)

════════ **EXPLORING** ════════

Use a large rectangle that is not a square.

1. How many lines of symmetry does the rectangle have? What is the relationship of each line of symmetry to the sides of the rectangle? Verify your answers by folding.

2. Draw the diagonals of the rectangle. What is true of the diagonals of the rectangle because the rectangle is a parallelogram?

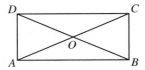

3. Label your rectangle as shown. Fold the rectangle so that $\overline{AD}$ and $\overline{BC}$ coincide. Compare the lengths of $\overline{OD}$ and $\overline{OC}$. How do the lengths of the diagonals compare?

8.6 Finding the Quadrilaterals that Are Parallelograms (p. 242)

════════ **EXPLORING** ════════

Draw segments to fit each description. Connect the endpoints of the segments. Name the special type of quadrilateral—parallelogram, rectangle, rhombus, or square—formed.

1. two noncongruent, nonperpendicular segments that bisect each other

2. two congruent, nonperpendicular segments that bisect each other

3. two noncongruent segments that are the perpendicular bisectors of each other

4. two congruent segments that are the perpendicular bisectors of each other

9.5 Area of a Triangle (p. 271)

1. Draw any acute scalene $\triangle RST$.

2. Cut out $\triangle RST$ and copy it on another piece of paper. Draw the image $\triangle R'S'T'$ of $\triangle RST$ under a 180° rotation about the midpoint of one of its sides.

3. What type of figure is formed by the combination of $\triangle RST$ and $\triangle R'S'T'$? How does the area of this figure compare to the area of $\triangle RST$?

4. Repeat Steps 1–3 with three different triangles. Do you get similar results?

5. Based on your observations, write a formula for the area of a triangle.

9.8 Circumference of a Circle
(p. 282)

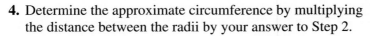

EXPLORING

1. Draw a large circle.

2. Draw two radii that form a small angle. Measure the angle and determine how many angles of this size are in the circle.

3. Measure the distance between the intersections of these two radii with the circle.

4. Determine the approximate circumference by multiplying the distance between the radii by your answer to Step 2.

5. Find the diameter. Divide the circumference by the diameter. What whole number is this close to?

6. Repeat Steps 1–5 on several different sized circles.

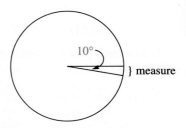

10° } measure

10.3 Similar Figures (p. 301)

EXPLORING

Part A

Which figures appear to be similar?

  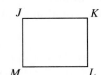

10.4 Similar Polygons (p. 305)

EXPLORING

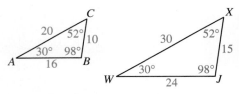 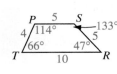

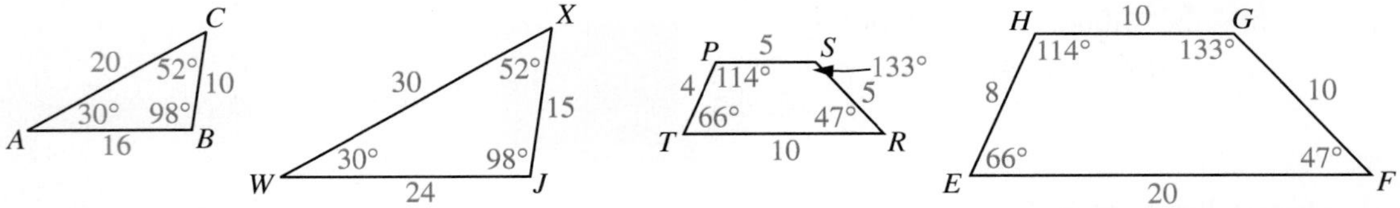

$\triangle ABC \sim \triangle WJX$ $TRSP \sim EFGH$

1. For each pair of similar polygons name the pairs of corresponding angles.

2. Compare the measures of corresponding angles.

3. Find each ratio in simplest form.

   a. $AB : WJ$ **b.** $BC : JX$ **c.** $CA : XW$ **d.** $PT : HE$

   e. $TR : EF$ **f.** $RS : FG$ **g.** $SP : GH$

4. For each pair of similar polygons compare the ratios of the lengths of corresponding sides.

11.2 SSS and SAS Similarity
Postulates (p. 325)

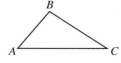

EXPLORING

**Draw a large scalene triangle. Label it as △ABC.
Use this triangle for Part A and Part B.**

Part A

1. Bisect $\overline{AB}$, $\overline{BC}$, and $\overline{AC}$, using
   Construction 2.

2. Draw a line. Construct $\overline{DE}$ such that
   $DE = \frac{1}{2} \cdot AC$.

3. With the compass point on D, draw an arc with radius
   $\frac{1}{2} \cdot AB$. With the compass point on E, draw an arc with
   radius $\frac{1}{2} \cdot BC$ that intersects the first arc in some point F.

4. Draw $\overline{DF}$ and $\overline{EF}$.

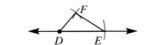

5. Measure the angles of △ACB and △DEF.
   Why is △DEF ~ △ACB?

Part B

1. Construct $\angle X \cong \angle A$.

2. On the sides of $\angle X$ construct $\overline{XY}$ and $\overline{XZ}$ so that
   $XY = 2 \cdot AC$ and $XZ = 2 \cdot AB$. Draw $\overline{YZ}$.

3. Measure $\angle XYZ$ and $\angle XZY$. Why is
   △XYZ ~ △ACB?

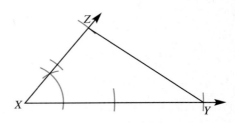

11.3 Triangles and Proportional Segments (p. 330)

EXPLORING

1. Draw any large triangle *ABC*.

2. At any point *D* on $\overline{AB}$, construct $\overline{DE}$ parallel to $\overline{BC}$ using Construction 11.

3. Measure the sides of triangles *ABC* and *ADE* to find the following ratios. How do the ratios compare?

   a. $\dfrac{AD}{AB}$ **b.** $\dfrac{AE}{AC}$ **c.** $\dfrac{DE}{BC}$

4. Why is $\triangle ADE \sim \triangle ABC$?

5. Measure to find the following ratios. How do these ratios compare?

   a. $\dfrac{AD}{DB}$ **b.** $\dfrac{AE}{EC}$

12.3 Square Roots and Right Triangles (p. 353)

EXPLORING

1. Using the lengths given in the table below, construct triangles having sides of length a, b, and c. (Note that c is the length of the longest side for each triangle.)

2. Copy and complete the table.

Triangle	a	b	c	a^2	b^2	c^2	$a^2 + b^2$
A	2 cm	3 cm	4 cm	▩	▩	▩	▩
B	3 cm	4 cm	5 cm	▩	▩	▩	▩
C	4 cm	4 cm	6 cm	▩	▩	▩	▩
D	6 cm	8 cm	10 cm	▩	▩	▩	▩
E	5 cm	12 cm	13 cm	▩	▩	▩	▩
F	6 cm	9 cm	12 cm	▩	▩	▩	▩
G	9 cm	12 cm	15 cm	▩	▩	▩	▩

3. Which are right triangles?

4. For each triangle compare the numbers in the last two columns. What seems to be true for the right triangles that is not true for the other triangles?

Alternate Explorings

12.3 Square Roots and Right Triangles (p. 354)

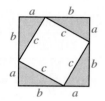

1. Copy the right triangle with sides of length a, b, and c. Use it as a pattern to draw eight congruent right triangles. Cut out the triangles.

2. Use four of the congruent right triangles to form the pattern shown on the left. With the other four, form the pattern shown on the right.

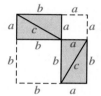

3. If the four shaded triangles are removed from each design, the regions shown at right are formed. Why are the areas of these regions equal? Why is $a^2 + b^2 = c^2$?

4. Suppose you used a different-size right triangle in Step 1 to form the designs in Step 2. Would you get the same results?

12.8 Similar Right Triangles
(p. 370)

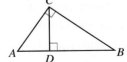

1. Draw any right triangle *ABC*. Construct the altitude to the hypotenuse.

2. Make a table that gives the measures of the angles of the three triangles formed.

3. Use your table to write a similarity statement for each pair of triangles.

4. One of your similarity statements in Step 3 should involve two triangles that have $\overline{CD}$ as a side. Use that statement to complete the following proportion:

$$\frac{\blacksquare}{CD} = \frac{CD}{\blacksquare}$$

5. Use two of your similarity statements in Step 3 to complete the following proportions:

$$\frac{\blacksquare}{AC} = \frac{AC}{\blacksquare} \qquad\qquad \frac{\blacksquare}{BC} = \frac{BC}{\blacksquare}$$

13.2 Inscribed Angles (p. 385)

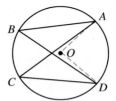

Draw a large ⊙ O.

1. Draw two inscribed angles, ∠*ABD* and ∠*ACD*, that intercept the same arc, $\overarc{AD}$. Use a protractor to measure ∠*ABD* and ∠*ACD*.

2. Draw central angle *AOD*. Measure ∠*AOD*. What is $m\overarc{AD}$? Why?

3. What seems to be the relationship between the measure of an inscribed angle and the measure of its intercepted arc?

4. Test your conclusion in Step 3 by repeating Steps 1 and 2 for two inscribed angles that intercept a different arc.

16.2 Using a Table or Calculator
(p. 490)

EXPLORING

Use the given information to find the measure of an acute angle of right triangle *ABC*, with right angle C.

Then use a protractor and a metric ruler to draw △*ABC*. Label the known lengths and measures.

1. $AC = 2.3$ cm
   $AB = 2.6$ cm

2. $BC = 2.3$ cm
   $AB = 3.1$ cm

18.4 Dilations (p. 554)

EXPLORING

1. Draw a set of axes on graph paper. Graph △*DEF* with vertices $D(-6, 0)$, $E(8, 6)$, and $F(-4, 6)$. Draw dashed segments $\overline{OD}$, $\overline{OE}$, and $\overline{OF}$. Let D' be the midpoint of $\overline{OD}$. Locate E' and F', the midpoints of $\overline{OE}$ and $\overline{OF}$, respectively. Draw △*D'E'F'*.

2. a. Describe the center and scale factor of the dilation in Step 1 for which △*D'E'F'* is the image of △*DEF*.

   b. How are the coordinates of *D*, *E*, and *F* related to the coordinates of *D'*, *E'*, and *F'* ?

   c. Describe the center and scale factor of the dilation for which △*DEF* is the image of △*D'E'F'*.

3. Draw a set of axes on graph paper. Graph △*RST* with vertices $R(3, 3)$, $S(-3, 9)$, and $T(-6, 3)$. Draw dashed rays $\overrightarrow{OR}$, $\overrightarrow{OS}$, and $\overrightarrow{OT}$. On $\overrightarrow{OR}$, locate R' so that $OR' = 3 \cdot OR$. On $\overrightarrow{OS}$, locate S' so that $OS' = 3 \cdot OS$. On $\overrightarrow{OT}$, locate T' so that $OT' = 3 \cdot OT$. Draw △*R'S'T'*.

4. Repeat Step 2 for △*RST* and △*R'S'T'*.

Index

Index

601

Selected Answers

Chapter 1 The Language of Plane Geometry

Focus on Skills: Arithmetic 1. 1.46 **3.** 0.8
5. 4.42 **7.** 24.5 **9.** 17 **11.** 136 **13.** 300 **15.** 630
17. 525,000 **19.** 7,800 **21.** 0.3 **23.** 0.63 **25.** 0.004
27. 1.6 **29.** 0.006 **31.** 8.2 **Algebra 1.** 36 **3.** 36 **5.** no
7. 450 ft **9.** $\frac{1}{2}m$ = 18; m = 36 in.

Lesson 1.1 1. yes **3.** $\overrightarrow{SR}$ **5.** $\overline{RS}$ ($\overline{SR}$), $\overline{RT}$ ($\overline{TR}$), $\overline{RU}$
($\overline{UR}$), $\overline{ST}$ ($\overline{TS}$), $\overline{SU}$ ($\overline{US}$), $\overline{TU}$ ($\overline{UT}$) **7.** yes **9.** yes **11.** yes
13. yes **15.** yes **17.** no **19.** true **21.** true
For 23–33, position of points may vary.

23. A B   C D   **25.**

27. H   I   **29.**   H   I

31. H   I   **33.**   A   B C

or

A C   B

35. Stars are represented by points, with segments
connecting the points to outline a figure.

Lesson 1.2 1. N; $\overrightarrow{NL}$, $\overrightarrow{NM}$ **3.** T; $\overrightarrow{TJ}$, $\overrightarrow{TK}$ **5.** Angle
names may vary. $\angle 1$, $\angle 2$, $\angle 3$, $\angle AOC$, $\angle BOD$, $\angle AOD$
7. $\overrightarrow{OA}$, $\overrightarrow{OD}$ **9.** $\angle R$, $\angle BRM$, $\angle MRB$ **11.** $\angle 1$, $\angle Q$, $\angle BQX$
13. $\angle TXR$, $\angle RXT$ **15.** $\angle 1$, $\angle RZN$, $\angle NZR$
For 17, 19, and 23, measure of angle may vary.

17.   R   B   T   **19.**   B   **21.**   S   R   T

23.   **25.**   D   S   R

27. Answers may vary. Emblems should not have
curved lines, circles, etc.

Lesson 1.3 1. $\overleftrightarrow{AC}$ and $\overleftrightarrow{DB}$ **3.** $\overleftrightarrow{AC}$ and $\overleftrightarrow{AB}$; $\overleftrightarrow{DB}$ and $\overleftrightarrow{AB}$
5. $\overleftrightarrow{GF}$ **7.** Answers may vary. For example: C, E, G
9. $\overline{TS}$, $\overline{XW}$, $\overline{UV}$ **11.** $\overline{TX}$, $\overline{SW}$, $\overline{VW}$, $\overline{UX}$ **13.** planes $QTXU$,

$STXW$ **15.** parallel **17.** intersecting For 19 and 21,
angle of intersection may vary.

19.   X   Z   Y   **21.**   E   B   A   D   C   G   F

23. line m ∥ line l **25.** $\overleftrightarrow{AB}$ and $\overleftrightarrow{CD}$ intersect at point X
27. 4 lines: 0, 1, 3, 4, 5, or 6; 5 lines: 0, 1, 4, 5, 6, 7, 8,
9, or 10 points **29.** Answers may vary. For example:
m, l; X, Y; p, q; X, Z; p, m

Test Yourself 1. Answers may vary. For
example: A, E, D, and C **2.** A, E, B **3.** $\overrightarrow{AB}$ **4.** A, E, B, C
5. true **6.** false **7.** true **8.** false **9.** false **10.** false

11.   X   Y   **12.**   l   m

13. Measure of angle may vary.   **14.**
B   X   R   C   A   B

Lesson 1.4 1. 5 cm **3.** 2 cm **5.** 5 cm; 54 mm;
5.4 cm **7.** 2 cm; 23 mm; 2.3 cm **9.** 17 **11.** 8.2 **13.** 136
15. 32 **17.** 40 **19.** Use a metric ruler to draw a segment 7 mm long. **21.** Use a metric ruler to draw a segment 6.7 cm long. **23.** 99 cm; 1 m; 2,000 mm; 0.3 km
25. m or cm **27.** m **29.** mm **31.** mm **33.** cm **35.** cm
or m **37.** no **39.** no **41.** no **43.** a **45.** b **47.** 50 h

Lesson 1.5 1. Use Construction 1. **3.** $\overline{MC}$, $\overline{MD}$; $\overline{MP}$,
$\overline{MQ}$ **5.** no **7.** true **9.** false **11.** true **13.** false **15.** Use
Construction 1 three times to draw a line with length
3d. **17.** $EF = \frac{1}{2} AB$

Lesson 1.6 1. infinitely many **3.** infinitely many
5. Use Construction 2. **7.** Use Construction 3.
9. Bisect $\overline{MN}$ at R. Bisect $\overline{RN}$ at U. Length of MU is
$\frac{3}{4} \cdot MN$. **11.** Use Constructions 1 and 3. **13.** Extend $\overline{BC}$.
Use Construction 4. **15.** A plumb line helps you
determine that the first piece of wallpaper you hang is
straight (vertical), thereby ensuring that the other
pieces will also be straight so any pattern will line up
correctly.

Lesson 1.7 1. The Chapter 7 Test will have 20–30 questions; inductive. **3.** The sum of two odd numbers is even; inductive. **5.** Answers may vary. For example: $\frac{4}{3} > 1$ **7.** yes **9.** *ABCD* is a parallelogram. **11.** △*DEF* has no sides the same length. **13.** Lines *a* and *b* do not intersect. **15. a.** Yes; the sum of two integers is an integer. **b.** Yes; the product of two integers is an integer. **c.** yes; yes **17. a.** 23, 27, 31 **b.** 6.5, 6.8, 7.1 **c.** 13, 21, 34 **d.** ◁, △, ▷

Lesson 1.8 1. only the figure on the right (6-sided) **3.** 155 ft **5.** Answers may vary. For example: Place 9 cm rod next to 2 cm rod to get 11 cm. Place 7 cm rod next to 5 cm rod to get 12 cm. The difference between 12 cm and 11 cm is 1 cm.

Chapter 1 Review 1. $\overrightarrow{CB}$, $\overrightarrow{CD}$ **2.** $\overline{AE}$ ($\overline{EA}$), $\overline{AC}$ ($\overline{CA}$), $\overline{BC}$ ($\overline{CB}$), $\overline{BD}$, ($\overline{DB}$), $\overline{CD}$ ($\overline{DC}$) **3.** A, E, C **4.** Answers may vary. For example: $\overleftrightarrow{AE}$ ∥ $\overleftrightarrow{BC}$ **5.** ∠*BCA* (∠*ACB*) **6.** ∠*BCD* For Exercises 7–10, answers may vary. For example: **7.** plane *X*, plane *Y* **8.** plane *X*, plane *Z* **9.** A, E, D, F **10.** $\overleftrightarrow{AB}$ **11.** 3 cm; 34 mm; 3.4 cm **12.** 6 cm; 58 mm; 5.8 cm **13.** 3 **14.** 320 **15.** 520 **16.** Use Construction 1. **17.** Use Construction 2. **18.** Use Construction 4. **19.** A square has four right angles; deductive. **20.** 40 posts

Chapter 2 Angles

Focus on Skills Arithmetic Answers for 1–9 may vary. For example: **1.** $\frac{6}{10}$, $\frac{9}{15}$, $\frac{12}{20}$ **3.** $\frac{6}{16}$, $\frac{9}{24}$, $\frac{12}{32}$ **5.** $\frac{1}{4}$, $\frac{6}{24}$, $\frac{9}{36}$ **7.** $\frac{4}{5}$, $\frac{40}{50}$, $\frac{80}{100}$ **9.** $\frac{6}{14}$, $\frac{9}{21}$, $\frac{12}{28}$ **11.** $\frac{2}{3}$ **13.** $\frac{3}{4}$ **15.** $\frac{8}{15}$ **17.** $\frac{1}{6}$ **19.** $\frac{5}{36}$ **21.** $\frac{1}{10}$ **23.** $\frac{9}{20}$ **25.** 5 **27.** 135 **Algebra 29.** 5 **31.** 34 **Geometry**
33. **35.**

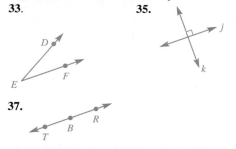

37.

Lesson 2.1 1. radii $\overline{AB}$, $\overline{AC}$, $\overline{AD}$; chords $\overline{CD}$, $\overline{EF}$; diameter $\overline{CD}$ **3.** radii $\overline{OA}$, $\overline{OH}$; chords $\overline{AM}$, $\overline{AH}$, $\overline{AT}$; diameter $\overline{AH}$ **5.** 5 cm **7.** $\frac{5}{16}$ turn **9.** $\frac{9}{36}$ or $\frac{1}{4}$ turn **11.** a

13. c **15.** d **17.** true **19.** true **21.** true
23. **25.** **27.**

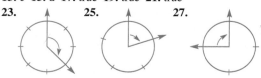

29. a. yes **b.** no **c.** There are no restrictions on its length. **31. a.** $\frac{1}{60}$ turn **b.** $\frac{10}{60}$ or $\frac{1}{6}$ turn **c.** $\frac{25}{60}$ or $\frac{5}{12}$ turn **d.** $\frac{30}{60}$ or $\frac{1}{2}$ turn
Lesson 2.2 1. 30° **3.** $\frac{1}{15}$ turn **5.** $\frac{1}{18}$ turn **7.** 60° **9.** $\frac{1}{8}$ turn **11.** right **13.** perpendicular **15.** acute **17.** right **19.** obtuse **21.** acute **23.** acute
25. (between 90° and 180°) **27.**

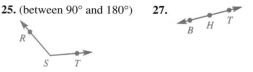

29. 60° **31.** 45° **33.** $\frac{1}{3}$ turn; 120°; obtuse **35.** $\frac{5}{16}$ turn; 112.5°; obtuse **37.** $\frac{1}{4}$ turn; 90° **39.** 174°
Lesson 2.3 Answers for 1–7 may vary. Actual measurement is given. **1.** 58° **3.** 15° **5.** 45° **7.** 37° **9.** 10° **11.** 80° **13.** 165° **15.** 40° **17.** 140° **19.** 100°
21. **23.**

25. **27.**

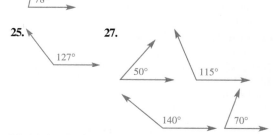

29. 2400 mils **31.** 3200 mils
Lesson 2.4 1. 120° **3.** 85° **5. a.** 180° **b.** 50° **c.** 180° **7.** ∠*AOD* **9.** ∠*AOB* **11.** *m*∠*ABE* = 55°; *m*∠*EBC* = 110°
Test Yourself 1. $\overline{CX}$, $\overline{CY}$, $\overline{CZ}$ **2.** $\overline{RS}$, $\overline{XY}$ **3.** $\overline{XY}$ **4.** 5 cm **5.** $\frac{1}{360}$ **6.** $\frac{1}{6}$ **7.** $\frac{5}{12}$ **8.** 360° **9.** 36° **10.** 135° **11.** 110° **12.** 60° **13.** Draw a 90° angle. **15.** Draw a 68° angle. **17.** Draw any angle between 0° and 90°.
Lesson 2.5 1. true **3.** false **5.** true **7, 9.** Use Construction 5 and then Construction 6. **11.** 4 **13.** Use Construction 5 twice. **15.** 35° **17.** 40°; 40° **19.** Draw a figure larger than the one shown. Then use

Construction 6 twice. **21.d.** The triangles all appear to have the same size and shape. **23.** 0.5 mi

Lesson 2.6 1. 55°, 145° **3.** 41°, 131° **5.** not possible; 61° **7.** 18°, 108° **9.** false **11.** true **13.** false **15.** false **17.** h **19.** a **21.** b **23.** d **25.** neither **27.** supplementary **29.** complementary **31.** neither **33. a.** ∠2, ∠3 (or ∠3, ∠4) **b.** ∠1, ∠3 **c.** ∠2, ∠4 **d.** ∠2, ∠4 **35.** 90° **37. a.** 90° **b.** 30° **c.** 90° **39.** 30° **41.** Answers for b, c, and d may vary. For example: **a.** ∠10, ∠22; ∠3, ∠11 **b.** ∠1, ∠21; ∠2, ∠4 **c.** ∠3, ∠22; ∠6, ∠7 **d.** ∠13, ∠16 **e.** No; they do not have a common vertex, and they share interior points.

Lesson 2.7 1. 130° **3.** 24 cm
5.

Chapter 2 Review 1. $\overline{CD}$, $\overline{CG}$, $\overline{CH}$, $\overline{CE}$ **2.** $\overline{DG}$, $\overline{HE}$ **3.** ∠HCD, ∠ECG **4.** ∠DCE, ∠HCG; ∠DCH, ∠ECG **5.** 10 cm **6.** $\frac{3}{8}$ turn; 135° **7.** $\frac{4}{9}$ turn; 160° **8.** $\frac{5}{12}$ turn; 150°
9. obtuse **10.** acute

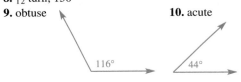

11. complement: 29°; supplement: 119°
12. no complement; supplement: 66° **13.** complement: 58°; supplement: 148° **14.** no complement; supplement: 90° **15.** 42° **16.** 73° **17.** Use Constructions 5 and 6. **18.** 8:35 A.M.

Chapter 3 Polygons and Polyhedrons

Focus on Skills Geometry 1. ∠ABC, ∠CBA **3.** Answers may vary. For example: ∠ABC, ∠ABD; ∠CBE, ∠DBE **5.** ∠ABC, ∠DBE **7.** obtuse **9.** acute **11.** obtuse **13.** $\overleftrightarrow{AB}$ ∥ $\overleftrightarrow{CD}$; $\overline{AC}$ ≅ $\overline{CD}$ **15.** $\overleftrightarrow{HI}$ ∥ $\overleftrightarrow{KJ}$; $\overline{HI}$ ≅ $\overline{KJ}$ **Algebra 17.** $x + y = 12$

Lesson 3.1 1. not a polygon **3.** not a polygon **5.** quadrilateral **7.** pentagon **9.** P, Q, R, S, T, U, V, W **11.** $\overline{WP}$, $\overline{QR}$ **13.** ∠W, ∠Q **15.** $\overline{QS}$, $\overline{QT}$, $\overline{QU}$, $\overline{QV}$, $\overline{QW}$ **17.** Draw any quadrilateral with vertices F, G, H, I. **19.** Draw any triangle. **21.** Draw any 5-sided polygon. **23.** Draw any 8-sided polygon. **25.** Draw any 4-sided

polygon that is not convex. **27.** Draw any polygon with 4 noncongruent sides. **29.** Draw any polygon with 3 or more sides. **31. a.** octagon; **b.** triangle; **c.** quadrilateral

Lesson 3.2 1. $\overline{CA}$, $\overline{CB}$; $\overline{AB}$; ∠C; ∠A, ∠B **3.** $\overline{FD}$, $\overline{FE}$; $\overline{DE}$; ∠F; ∠D, ∠E **5.** △DEF **7.** △DEF **9.** △ABC **11.** △MNO, △DEF **13.** scalene **15.** obtuse **17.** acute **19.** Draw any △RAS with $\overline{RA}$ ≅ $\overline{AS}$. **21.** Draw any △ASM with m∠S between 90° and 180°. **23.** Draw any triangle with one angle that measures between 90° and 180°. **25.** Draw any triangle with a 90° angle between two congruent sides. **27.** Draw any triangle with a 90° angle and 3 noncongruent sides. **29.** △ABD, △CBD, △ADC **31.** △ABE, △AEC, △BDE, △CDE **33.** △XWY and △XZY are both obtuse scalene triangles. **35.** Use *Your Own* and SSS options.

Lesson 3.3 1. no **3.** yes **5.** yes **7.** no **9.** hexagon, polygon **11.** pentagon, polygon, regular polygon **13.** true; definition **15.** false; a parallelogram may have no right angles **17.** false; a rhombus may have no right angles **19.** false; a rhombus may have no right angles **21.** true; definition **23.** Draw any quadrilateral with two pairs of parallel opposite sides and no right angles. **25.** Draw any 6-sided polygon with at least one pair of noncongruent sides. **27.** Draw any quadrilateral with exactly one pair of opposite sides parallel and one pair of congruent sides. **29.** Draw any rhombus with no right angles.

Test Yourself 1. $\overline{AB}$, $\overline{DE}$ **2.** A, D **3.** $\overline{BE}$, $\overline{BD}$ **4.** $\overline{SR}$, $\overline{ST}$ **5.** ∠R, ∠T **6.** Draw any quadrilateral. **7.** Draw any triangle with 3 noncongruent sides. **8.** Draw any equilateral, equiangular pentagon. **9.** Draw any 4-sided polygon and the 2 segments connecting opposite vertices. **10.** Draw any triangle with one angle beween 90° and 180°. **11.** Draw any triangle with a 90° angle between 2 congruent sides. **12.** Draw any triangle with all angles less than 90°. **13.** Draw any 6-sided polygon that is not convex. **14.** false **15.** true **16.** polygon, quadrilateral, parallelogram, rectangle

Lesson 3.4 1. no **3.** no **5.** no **7. a.** 6, 12, 8 **b.** square (or rectangular) **c.** cube (or rectangular prism) **9. a.** 7, 15, 10 **b.** pentagonal **c.** pentagonal prism **11. a.** 5, 9, 6 **b.** triangular **c.** triangular prism **13. a.** 6, 12, 8 **b.** trapezoidal **c.** trapezoidal prism **15.** false **17.** true **19.** If a prism is set on one of its bases, its lateral faces are the "sides" of the prism.

Lesson 3.5 1. Draw any polyhedron with parallel, same-size, triangle-shaped bases and 3 lateral faces. **3.** Copy the figure shown. **5.** Increase the length of the 4 vertical edges by the same amount. **7.** pentagonal prism **9.** hexagonal prism **11.** octagonal prism **13.** triangular prism **15.** triangular prism **17.** Vary the choice of dashed segments.

19. a. **b.** 1

Lesson 3.6 1. clockwise; clockwise; clockwise **3.** one **5.**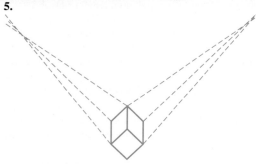

Lesson 3.7 1. two-point **3.** one-point **5.**

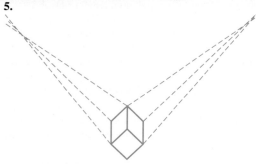

7. See the Exploring on p. 91. **9.** Answers may vary.

Chapter 3 Review 1. convex; triangle **2.** concave; pentagon **3.** convex; hexagon **4.** convex; quadrilateral **5.** isosceles right triangle **6.** equilateral triangle; equiangular triangle **7.** scalene right triangle **8.** obtuse scalene triangle **9.** rectangle **10.** trapezoid **11.** rectangle, rhombus **12. a.** faces: 7; edges: 15; vertices: 10 **b.** pentagonal **c.** pentagonal prism **13.** Change the choice of dashed edges. **14.** See the Explorings on pp. 91 and 92. **15.** Answers may vary. For example:

Chapter 4 Introduction to Transformations

Focus on Skills Geometry 1. Draw any segment with endpoints labeled C and D. **3.** Draw any line through points X and Y. **5.** Draw any segment with endpoints C and D, with $\overleftrightarrow{AB}$ intersecting $\overline{CD}$ at its midpoint. **7.** Draw any quadrilateral with vertices A, B, C, and D. **9.** Draw any angle that measures 60°. **11.** 62 mm **13.** 34 mm **15.** 37° **17.** 135° **19.** Use Construction 2. **21.** Use Construction 6.

Lesson 4.1 1. $A \cong H$; $B \cong I$; $C \cong E$; $F \cong G$ **3.** $\angle K \leftrightarrow \angle J$; $\angle M \leftrightarrow \angle L$; $\angle N \leftrightarrow \angle P$; $\overline{KM} \leftrightarrow \overline{JL}$; $\overline{MN} \leftrightarrow \overline{LP}$; $\overline{KN} \leftrightarrow \overline{JP}$ **5.** $\angle R \leftrightarrow \angle M$; $\angle S \leftrightarrow \angle N$; $\angle T \leftrightarrow \angle O$; $\angle U \leftrightarrow \angle P$; $\angle V \leftrightarrow \angle Q$; $\overline{RS} \leftrightarrow \overline{MN}$; $\overline{ST} \leftrightarrow \overline{NO}$; $\overline{TU} \leftrightarrow \overline{OP}$; $\overline{VU} \leftrightarrow \overline{PQ}$; $\overline{VR} \leftrightarrow \overline{QM}$ **7.** $\overline{PQ} \cong \overline{AB}$; $\overline{QR} \cong \overline{BC}$; $\overline{RS} \cong \overline{CD}$; $\overline{PS} \cong \overline{AD}$ **9. a.** $\angle CBD$ **b.** $\overline{CD}$ **c.** $\angle BCD$ **d.** $\overline{AC}$ **11.** Answers may vary. For example: $\triangle PRQ \cong \triangle PRS$ **13.** $\overline{MN} \leftrightarrow \overline{AB}$; $\overline{NO} \leftrightarrow \overline{BC}$, $\overline{OP} \leftrightarrow \overline{CD}$, $\overline{MP} \leftrightarrow \overline{AD}$ **15.** $\overline{AB} \leftrightarrow \overline{AB}$; $\overline{BC} \leftrightarrow \overline{BD}$; $\overline{AC} \leftrightarrow \overline{AD}$ **17.** $\triangle BGC$; $\triangle CHD$; $\triangle DIE$; $\triangle EJA$ **19.** $\triangle BHC$; $\triangle CID$; $\triangle EHD$; $\triangle AIE$; $\triangle CFB$; $\triangle BJA$; $\triangle EFA$; $\triangle DJE$; $\triangle DGC$ **21.** b

Lesson 4.2 1. reverse **3.** same **5.** G, E, F, none, B **7.** 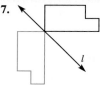 **9.**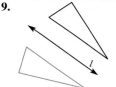

11. **13.**

15. Line of reflection is a line between the figures and the same distance from both. **17.** Line of reflection is a vertical line between the figures and the same distance from both.

19. **21.**

23.

25. yes **27.** no **29.** yes **31.** no **33.** SOS will not be the same when reflected, but the dot-dash code will be.

Lesson 4.3 1. b

3.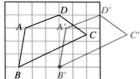

5.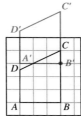

7.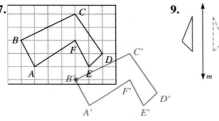

9.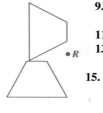

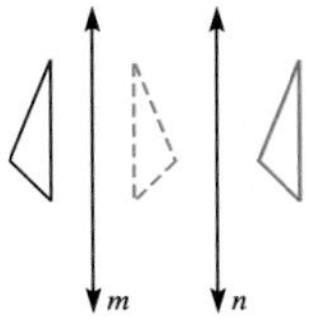

11.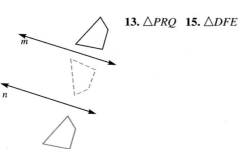

13. $\triangle PRQ$ **15.** $\triangle DFE$

Test Yourself **1.** $\angle A \cong \angle X$; $\angle B \cong \angle Y$; $\angle C \cong \angle Z$ **2.** $\overline{AB} \cong \overline{XY}$; $\overline{BC} \cong \overline{YZ}$; $\overline{AC} \cong \overline{XZ}$

3.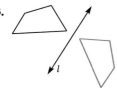

4. The line of reflection is a vertical line between the figures and the same distance from both.

5.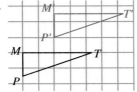

Lesson 4.4 **1.** a, b **3.** 135° CCW **5.** 45° CCW

7.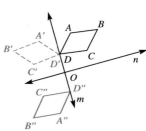

9. Image coincides with original figure.

11. 30° CW

13. 95° CW

15.

17., 19., 21. Image coincides with original figure.
23. 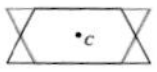 ⬡·c **25. a.** The first time it drew the same triangle twice. The second time it made a second triangle rotated 30° CW from the first triangle.
c. lt :angle measure **d.** It draws a square, then draws the rotation image of the square about the home point.

Lesson 4.5 **1.** Draw the reflection, H', of H over the right side of the playing area. Aim at the point where $\overline{BH'}$ intersects the right side. **3.** Draw the reflection, H', of H over the lower side of the playing area. Since there is still a corner in the way, draw the reflection, H'', of H' over the extension of the left side. Aim at the point where $\overline{BH''}$ intersects the left side.

Lesson 4.6 **1.** perpendiculars from three vertices to opposite sides **3.** all diameters **5.** line through vertical segment; perpendicular bisector of vertical segment **7.** line through center bar of "E" **9.** 1 (120°, 240°); 2 (90°, 180°, 270°); 3 (all $n°$, $0 < n < 360$); 5 (180°); 6(180°); 2, 3, 5, and 6 have point symmetry **11.** 180°; yes **13.** 120°, 240°; no **15. a.** the angle bisectors. **b.** The angle has measure half that of $\angle RPT$. Let $m\angle RPS = x$ and $m\angle SPT = y$. $\frac{1}{2}x + \frac{1}{2}y = \frac{1}{2}(x + y) = \frac{1}{2}m\angle RPT$. **17.** an equilateral triangle **19.** Answers will vary. The figure shows a hexagon with two lines of symmetry.

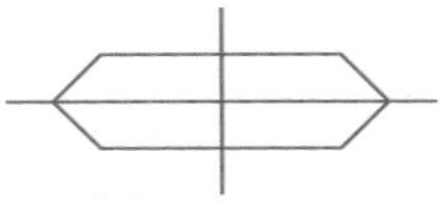

21. a trapezoid with nonparallel sides congruent **23. a.** vertical line of symmetry; no rotational symmetries **b.** no symmetry **c.** no lines of symmetry; 180° rotational symmetry; point symmetry

Chapter 4 Review 1. ∠R, ∠N; ∠T, ∠A; ∠M, ∠G 2. $\overline{RM}$, $\overline{NG}$; $\overline{RT}$, $\overline{NA}$; $\overline{TM}$, $\overline{AG}$ 3. △AGN; △NAG 4. $\overline{EF}$ 5. △HLK 6.

7.

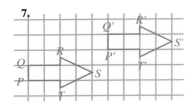

8.

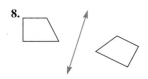

9.

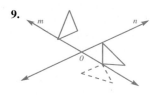

10. 4 lines of symmetry; yes; yes; 90°, 180°, 270°

11. Draw the reflection image, *H'*, of *H* over the lower side of the playing area. Aim at the point where $\overline{BH'}$ intersects the lower side.

Chapter 5 Triangles and Inequalities

Focus on Skills **Algebra** 1. > 3. < 5. 20 7. 17 9. 29 11. 12 **Geometry** 13. ∠D; $\overline{BN}$; ∠B, ∠N 15. Use Construction 2. 17. Use Construction 2. 19. *XY* = 11 cm, *YZ* = 16 cm, *XZ* = 11 cm

Lesson 5.1 1. false 3. false 5. false 7. ∠1, ∠3 9. ∠1 ≅ ∠3; ∠4 ≅ ∠6 11. a. 90° b. 60° c. 60°

d. 30° e. 30° f. 6 g. 6 h. 3 13. a. 50°; 100° b. 65° c. 6; 8 d. △GME; △EMO e. ∠1, ∠3; ∠4, ∠O; ∠1, ∠2; ∠2, ∠3 15. ∠J, ∠K 17. m∠B = m∠C = 50°; m∠A = 80° 19. *x* = 7

Lesson 5.2 1. $\overleftrightarrow{CA}$ 3. $\overleftrightarrow{YA}$ 5.

7, 9. In each triangle, use Construction 4 to construct the altitudes. The point where the three altitudes meet is the orthocenter. 11. Draw any triangle with 3 angles less than 90° and with a base length of 4 units and a height of 3 units. 13. Draw any triangle with one angle between 90° and 180° and with a base length of 5 units and a height of 4 units. 15. Answers may vary. For example, draw 1 right triangle, 1 acute triangle, 1 isosceles triangle. 17. equilateral triangle 19. isosceles non-equilateral triangle 21. obtuse isosceles triangle 23. The median; in an isosceles triangle, the altitude from the vertex angle and the corresponding median are the same segment. In an equilateral triangle, each altitude is also a median.

Lesson 5.3 1. 6 3. 7 5. isosceles 7. 3 9. 9 11. 6 13. 12 15. Draw a triangle with 3 non-congruent sides and 3 angles between 0° and 90°. Use Construction 2 twice. 17. Draw a triangle with a 90° angle between two congruent sides. Use Construction 2 twice. 19. Draw the same triangle as in exercise 15. Use Construction 6 twice. 21. Draw the same triangle as in Exercise 17. Use Construction 6 twice. 23. a. centroid, incenter b. The orthocenter and circumcenter of an obtuse triangle are in the exterior of the triangle. c. The orthocenter and circumcenter of a right triangle are on the triangle. 25. a. Construct the perpendicular bisector of any two sides, intersecting at *P*. b. Draw ⊙*P* with radius $\overline{PA}$ where *P* is the point equidistant from *A, B,* and *C*. 27. Use Construction 2 to construct the perpendicular bisector of the segment connecting Adams and Bakersville. Extend it to the shore of the lake. The picnic grounds will be at this point.

Test Yourself 1. 23° 2. 67° 3. 90° 4. 8 5. 3 6. Construct the three altitudes. 7. Construct the three medians.

Lesson 5.4 1. $\overleftrightarrow{KC}$ 3. *E* 5. 8 in.

Lesson 5.5 1. ∠P, ∠Q, ∠R 3. ∠I, ∠H, ∠G 5. $\overline{DE}$, $\overline{FD}$, $\overline{FE}$ 7. no 9. yes 11. ∠C, ∠A, ∠B 13. $\overline{RS}$, $\overline{ST}$, $\overline{RT}$ 15. 7 cm < *EF* <17 cm 17. a. 12 b. 8 c. 10

d. 30 **e.** 15 **19.** $\angle F$, $\angle E$, $\angle D$

Lesson 5.6 1. < **3.** $LN > KN$ **5.** $RT > XY$
7. $m\angle WST > m\angle WSR$ **9.** $m\angle B$; $x + (x + 4) +$
$(x + 8) = 39$, so $x = 9$. Then $2x - 3 = 15$ and
$x + 8 = 17$. Since $\overline{AB} \cong \overline{DE}$, $\overline{BC} \cong \overline{EF}$, and $AC >$
DF, $m\angle B > m\angle E$.

Chapter 5 Review 1. 24° **2.** 90° **3.** 66° **4.** 4
5. Use Construction 4 from point C to $\overline{AB}$. **6.** Use Construction 2 to locate the midpoint M of $\overline{DF}$. Draw $\overline{EM}$.
7. Use Construction 2 for all 3 sides. Extend the lines
if necessary, until they meet. **8.** Use Construction 6 for
all 3 angles. **9.** $\angle Q$; $\angle R$ **10.** $\overline{PQ}$; $\overline{RQ}$ **11.** 60°; 60°
12. 45 cm **13.** $\overrightarrow{DA}$ **14.** 7 cm **15.** $\angle R$, $\angle Q$, $\angle P$ **16.** $\overline{TU}$,
$\overline{ST}$, $\overline{SU}$ **17.** > **18.** < **19.** yes **20.** 45° and 135°

Chapter 6 Congruent Triangles

Focus on Skills Algebra 1. $x = 17$ **3.** $x = 13$
5. $a = 5$ **7.** $x = 6$; $y = 5$ **9.** yes **11.** no **13.** Draw
any triangle with 3 noncongruent sides and 3 angles
between 0° and 90°. Label vertices D, E, F. **15.** Draw
any triangle with exactly two congruent sides $\overline{RT}$ and
$\overline{RI}$. Use Construction 6 to draw $\overrightarrow{RA}$. **17.** Use
Construction 5. **19.** Use Construction 6.
21. $\overline{GH} \cong \overline{WX}$; $\overline{HK} \cong \overline{XY}$; $\overline{KG} \cong \overline{YW}$

Lesson 6.1 1. Use Construction 7. **3.** not possible
5. Use Construction 7. **7.** 4, 5 **9.** 2 **11.** 1 **13.** yes
15. yes **17.** $\overline{WL}$ **19.** SSS Postulate **21.** $\overline{MN}$; $\overline{NO}$; $\overline{MO}$
23. yes **25.** yes **27.** $\triangle BCA \cong \triangle ADB$ **29.** Measure
the sides of the first triangle. Use those measurements
to create a new triangle.

Lesson 6.2 1. Use Construction 8. **3.** Use
Construction 7. **5.** yes **7.** SAS **9.** SAS **11.** none
13. $\overline{XY}$; $\overline{YZ}$; $\overline{ZX}$ **15.** $\angle S$; $\overline{SR}$; $\overline{ST}$ **17.** $\overline{QP} \cong \overline{KJ}$; $\overline{RP} \cong \overline{LJ}$
19. $\overline{PQ} \cong \overline{JK}$, $\overline{QR} \cong \overline{KL}$ **21.** yes; $x = 3$, $y = 9$;
$\triangle RST \cong \triangle BXP$

Lesson 6.3 1. Use Construction 9. **3.** Use
Construction 8. **5.** AAS **7.** LA (or ASA) **9.** none
11. $\angle A \cong \angle D$, $\angle C \cong \angle F$ **13.** $\overline{BA} \cong \overline{ED}$, $\overline{CA} \cong \overline{FD}$
15. $\angle C \cong \angle F$ **17.** $\angle A \cong \angle D$, $\angle C \cong \angle F$; $\angle A \cong \angle D$,
$\angle B \cong \angle E$ **19.** $\overline{AB} \cong \overline{DE}$, $\angle A \cong \angle D$; $\overline{AB} \cong \overline{DE}$,
$\angle B \cong \angle E$ **21.** $\overline{AC} \cong \overline{DF}$, $\angle B \cong \angle E$; $\overline{BC} \cong \overline{EF}$,
$\angle A \cong \angle D$; $\overline{AC} \cong \overline{DF}$, $\angle A \cong \angle D$; $\overline{BC} \cong \overline{EF}$,
$\angle B \cong \angle E$ **23. a.** No; the given angles are not included
between the given sides. **b.** no **c.** No; answers may
vary. For example: Two sides and a non-included $\angle$ of

$\triangle ABC$ are $\cong$ the corr. sides and $\angle$ of $\triangle RST$, but the $\triangle$
are not congruent.

Test Yourself 1. Use Construction 7. **2.** Use
Construction 8. **3.** Use Construction 9. **4.** SAS
5. AAS **6.** none

Lesson 6.4 1. SSS **3.** HL **5.** AAS or LA **7.** HL
9. HL **11. a.** $\overline{EF} \cong \overline{EG}$ **b.** $\angle FDE \cong \angle GDE$
13. $\overline{GH} \cong \overline{CA}$, $\overline{KH} \cong \overline{BA}$ **15.** $\overline{KG} \cong \overline{BC}$, $\overline{KH} \cong \overline{BA}$;
$\overline{KG} \cong \overline{BC}$, $\overline{GH} \cong \overline{CA}$ **17.** $\angle G \cong \angle C$, $\overline{KH} \cong \overline{BA}$;
$\angle G \cong \angle C$, $\overline{KG} \cong \overline{BC}$; $\angle K \cong \angle B$, $\overline{GH} \cong \overline{CA}$;
$\angle K \cong \angle B$, $\overline{KG} \cong \overline{BC}$ **19.** $\triangle XWY$, $\triangle WXZ$
21. $\triangle ZWX$, $\triangle YXW$ **23.** $\triangle WXZ$, $\triangle XWY$
25. a. $\triangle ADC \cong \triangle CBA$ **b.** SSS **c.** $\angle 4 \cong \angle 1$,
$\angle 3 \cong \angle 2$, $\angle D \cong \angle B$ **27. a.** SSS **b.** CPCTC
c. Definition of angle bisector.

Lesson 6.5 1. a. Radii of congruent circles are
congruent. **b.** A segment is congruent to itself.
c. SSS Post. **d.** CPCTC **e.** Definition of angle
bisector **3. a.** Given **b.** Given **c.** Definition of angle
bisector **d.** A segment is congruent to itself. **e.** SAS
Post. **f.** CPCTC **5. a.** Given **b.** Definition of angle
bisector **c.** Given **d.** Definition of perpendicular
lines **e.** Definition of right triangle **f.** A segment is
congruent to itself. **g.** HA Thm. or AAS **h.** CPCTC
7. a. Given **b.** A segment is congruent to itself.
c. Given **d.** Definition of altitude **e.** Definition of
perpendicular lines **f.** Definition of right triangle.
g. HL Thm. **h.** CPCTC **i.** Definition of midpoint
j. Definition of median

Lesson 6.6 1. a. AAS Thm. or LA **b.** Measure
$\overline{TD}$, $\overline{TD} = \overline{CT}$ by CPCTC.

Chapter 6 Review 1. none **2.** HL **3.** SSS
4. AAS **5.** SAS **6.** ASA **7.** $\triangle WXR$, $\triangle ZYR$
8. $\triangle XZY$, $\triangle YWX$ **9.** $\triangle WXR$, $\triangle ZYR$ **10.** $\triangle YWX$,
$\triangle XZY$ **11.** Use Construction 7. **12.** Use Con-struction
9. **13.** Use Construction 8. **14.** $\overline{UV} \cong \overline{XY}$,
$\overline{VW} \cong \overline{YZ}$ **15.** $\angle U \cong \angle X$, $\angle V \cong \angle Y$ **16.** $\overline{VW} \cong \overline{YZ}$,
$\overline{UV} \cong \overline{XY}$ **17.** $\overline{WV} \cong \overline{ZY}$; $\overline{UV} \cong \overline{XY}$ **18.** $\overline{WV} \cong \overline{ZY}$,
$\overline{UV} \cong \overline{XY}$; $\overline{WV} \cong \overline{ZY}$, $\overline{UW} \cong \overline{XZ}$ **19. a.** Given **b.** A
segment is congruent to itself. **c.** SSS Post.
d. CPCTC **e.** A segment is congruent to itself.
f. SAS Post. **20.** $VW = 30$ yd; $\overline{WX} \cong \overline{YX}$,
$\angle WXV \cong \angle YXZ$ (vertical angles are congruent)
$m \angle VWX = 90° = m\angle ZYX$, so $\triangle VWX \cong \triangle ZYX$
(ASA) so $\overline{VW} \cong \overline{YZ}$ (CPCTC)

Chapter 7 Parallel Lines

Focus on Skills Algebra 1. a **3.** c, e **5.** $x = 21$
7. $z = 64$ **9.** $p = 23$ **Geometry 11.** Draw any nonsquare parallelogram with 4 congruent sides. **13.** Draw any quadrilateral that has two pairs of opposite parallel sides but no right angles. **15.** Use Construction 7, 8, or 9. **17.** Use Construction 4.

Lesson 7.1 1. $\angle 3$, $\angle 4$, $\angle 5$, $\angle 6$ **3.** $\angle 4$, $\angle 6$; $\angle 3$, $\angle 5$ **5.** $\angle 1$, $\angle 3$; $\angle 2$, $\angle 4$; $\angle 5$, $\angle 7$; $\angle 6$, $\angle 8$ **7.** $\angle 1$, $\angle 2$, $\angle 7$, $\angle 8$ **9.** $\angle 6$, $\angle 16$; $\angle 7$, $\angle 13$ **11.** $\angle 3$, $\angle 6$; $\angle 4$, $\angle 5$ **13.** c **15.** e **17.** d **19.** f **21. a.** $\overline{AD}$, $\overline{BC}$; $\overline{AC}$ **b.** $\overline{AB}$, $\overline{CE}$; $\overline{CB}$ **c.** $\overline{AB}$, $\overline{DE}$; $\overline{AC}$ **d.** $\overline{AC}$, $\overline{BE}$; $\overline{BC}$ **23.** correspond: to be equivalent or parallel/transverse: situated or lying across; crosswise

Lesson 7.2 1. $\angle 3$ **3.** $\angle 1$, $\angle 3$ **5.** $70°$ **7.** $110°$ **9.** $70°$ **11.** $70°$ **13. a.** $\angle 2$, $\angle 4$; $\angle 1$, $\angle 3$ **b.** $\angle 1$, $\angle 4$; $\angle 2$, $\angle 3$ **c.** $\angle 3$ **d.** $\angle 4$ **e.** $\angle 4$ **15. a.** $60°$ **b.** $45°$ **c.** $15°$ **17. a.** $48°$ **b.** $40°$ **c.** $92°$ **d.** $88°$ **19. a.** $55°$ **b.** $60°$ **c.** $65°$ **21.** $\overline{IF} \parallel \overline{HG}$ **23.** $r \parallel s$ **25.** $x \parallel y$ **27.** none **29.** $a \parallel b \parallel c$ **31.** $45°$, $45°$ **33.** $18°$, $18°$, $21°$ **35.** No; since $\overline{AD} \parallel \overline{BC}$, $\angle D$ and $\angle C$ are supplementary so $x = 11$. Then $m\angle B = 61°$ and $m\angle C = 120°$ so $\angle B$ and $\angle C$ are not supplementary.

Lesson 7.3 1. a. $\angle 1$, $\angle 5$; $\angle 2$, $\angle 6$; $\angle 3$, $\angle 7$; $\angle 4$, $\angle 8$ **b.** $\angle 3$, $\angle 5$, $\angle 7$ **c.** $\angle 2$, $\angle 4$, $\angle 6$, $\angle 8$ **3.** $105°$ **5.** $75°$ **7.** $75°$ **9.** $m\angle 1 = 40°$, $m\angle 2 = 70°$, $m\angle 3 = 110°$ **11.** $m\angle 1 = 90°$, $m\angle 2 = 120°$, $m\angle 3 = 120°$, $m\angle 4 = 60°$, **13. a.** $65°$ **b.** $115°$ **c.** $65°$ **15.** $n \parallel m$ **17.** $r \parallel s$ **19.** none **21.** none **23. a.** Use Construction 4. **b.** Use Construction 3. **c.** They are parallel. **25.** $x = 18$; $46°$ and $46°$

Test Yourself 1. $\angle 6$, $\angle 1$; $\angle 5$, $\angle 2$ **2.** $\angle 5$, $\angle 1$; $\angle 6$, $\angle 2$ **3.** $\angle 7$, $\angle 2$; $\angle 8$, $\angle 1$; $\angle 5$, $\angle 4$; $\angle 6$, $\angle 3$ **4.** $\angle 2$, $\angle 4$, $\angle 5$, $\angle 7$ **5.** $\angle 3$, $\angle 8$, $\angle 6$ **6.** $60°$, $120°$, $60°$ **7.** $115°$ **8.** $65°$ **9.** $115°$

Lesson 7.4 1. yes **3.** yes **5.** no **7.** Use Construction 11. **9.** Use Construction 11. **11.** Use Construction 1 to locate L on line a so that $KL = JK$. Draw $\overline{JK}$. Use Construction 11 to construct a line through J parallel to line a. Use Construction 11 to construct a line through L parallel to $\overline{JK}$. Label the point of intersection as M. **13.** Use Construction 4 to construct the perpendicular from E to line l. Label the point of intersection as F. Use Construction 1 to locate point G on line l so that $EF = FG$. Use Construction 3 to construct a perpendicular to line l at point G. Use Construction 3 to construct a perpendicular to $\overline{EF}$ at E.

Label the point of intersection H. **15.** Use Construction 3 to construct perpendiculars to $\overline{XY}$ at X and Y. Use Construction 1 to locate points W and Z, respectively, on the perpendiculars so that $WX = YZ = XY$. Draw $\overline{WZ}$. **17.** infinitely many

Lesson 7.5 Answers to Exercises 1–9 may vary. For example: **1.** True; The third angle would be 2°. **3.** True; measure of each angle equals 60° **5.** False; Corollary 2, Theorem 7.5 **7.** True; consider an equilateral triangle. **9.** False; sum of angle measures would be greater than 180°; contradicts Theorem 7.5 **11.** $m\angle 1 = 40°$ **13.** $m\angle 3 = 65°$; $m\angle 4 = 50°$ **15.** $m\angle 7 = 60°$ **17.** $m\angle 1 = 30°$; $m\angle 2 = 90°$; $m\angle 3 = 60°$ **19.** $m\angle 1 = 75°$; $m\angle 2 = 105°$; $m\angle 3 = 65°$ **21.** $m\angle 1 = 35°$; $m\angle 2 = 145°$; $m\angle 3 = 35°$; $m\angle 4 = 45°$; $m\angle 5 = 135°$ **23. a.** $27°$ **b.** $45°$ **c.** $45°$ **d.** $18°$ **25. a.** $35°$ **b.** $110°$ **27.** $62°$ **29. a.** $45°$ **b.** $55°$ **c.** scalene; Theorem 5.14 **31. a.** $m\angle D$; $m\angle E$ **b.** $m\angle D + m\angle E$ **c.** $m\angle C = 180° - (m\angle A + m\angle B)$; $m\angle F = 180° - (m\angle D + m\angle E)$ **d.** yes **e.** Definition of congruent angles. **33.** $15°$, $75°$.

Lesson 7.6 1. $\angle 4$; $\angle 5$, $\angle 6$ **3.** $\angle 6$; $\angle 5$, $\angle 4$ **5.** $\angle 2$, $\angle 4$, $\angle 6$, $\angle 8$, $\angle 10$, $\angle 12$ **7.** $\angle 3$; $\angle 5$, $\angle 9$ **9.** $\angle 2$, $\angle 4$; $\angle 1$, $\angle 3$; $\angle 6$, $\angle 8$; $\angle 5$, $\angle 7$; $\angle 10$, $\angle 12$; $\angle 9$, $\angle 11$ **11.** For example, extend $\overline{BA}$ upward. **13.** $m\angle 1 = 140°$; $m\angle 2 = 40°$ **15.** $m\angle 5 = 30°$; $m\angle 6 = 110°$ **17.** $m\angle 1 = 99°$; $m\angle 2 = 81°$ **19.** $m\angle 1 = 126°$; $m\angle 2 = 126°$; $m\angle 3 = 54°$; $m\angle 4 = 90°$; $m\angle 5 = 144°$ **21.** true **23.** false **25.** $m\angle 1 = m\angle 2 = 49°$ **27.** $m\angle 1 = 125°$; $m\angle 2 = 70°$ **29.** $m\angle C = 55°$; measure of exterior angle at C equals 125° **31.** Answers may vary. Two alternatives are given. **(I)** If $\overline{AB} \perp \overline{BC}$ and $\overline{AC} \perp \overline{BC}$, $m\angle ABD = 90°$ and $m\angle ACD = 90°$, which contradicts the corollary to Theorem 7.6. **(II)** $m\angle ABC = m\angle ACB = 90°$ so the sum of the measures of the interior angles is greater than 180°, which contradicts Theorem 7.5.

Lesson 7.7 1. $360°$ **3.** 9; $360°$ **5.** $22 \cdot 180° = 3{,}960°$; $360°$ **7.** $m\angle 3 = 80°$; $m\angle 2 = 80°$ **9.** $m\angle 1 = 108°$; $m\angle 2 = 72°$ **11.** $140°$; $40°$ **13.** 10; $36°$ **15.** 5 **17.** 10 **19.** 15 **21.** Draw any acute triangle. **23.** For example, draw any non-rectangular parallelogram.

25. Ex: **27.**

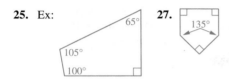

Lesson 7.8 **1.** 1,500 ft (or 1,400–1,600 ft);
2,800–3,000 ft **3.** more than 2,600 ft but less than
2,800 ft **5.** Draw a map with the contour lines very
close together.

Chapter 7 Review **1.** exterior **2.** interior
3. interior **4.** same-side exterior **5.** alternate interior
6. same-side interior **7.** corresponding
8. corresponding **9.** same-side exterior **10.** true
11. false **12.** true **13.** false **14.** true **15.** true **16.** true
17. true **18.** $a \parallel b$ **19.** $c \parallel d$ **20.** none **21.** $c \parallel d$
22. $a \parallel b$ **23.** $a \parallel b$ **24.** Use Construction 11 twice.
25. 60° **26.** $m\angle 2 = 70°$, $m\angle 3 = 80°$ **27.** 78°
28. 5,040° **29.** 14 **30.** between 100 ft and 150 ft

Chapter 8 Properties of Quadrilaterals

Focus on Skills **Algebra** **1.** $y = 5.25$ **3.** $b = 5$
5. $x = 2$ **7.** $x = 3$; $y = 5$ **9.** $x = 6$; $y = 6$
Geometry **11.** Draw the perpendicular bisectors of
the two pairs of opposite sides. **13.** Draw the
perpendicular bisectors of the two pairs of opposite
sides and the lines containing both diagonals. **15.** true
17. false **19.** true **21.** 99 **23.** perpendicular bisector
25. square

Lesson 8.1 **1.** false **3.** false **5.** false **7.** false
9. true **11.** false **13. a.** $\angle M$, $\angle K$; $\angle J$, $\angle L$ **b.** $\angle M$,
$\angle J$; $\angle M$, $\angle L$; $\angle K$, $\angle J$; $\angle K$, $\angle L$ **c.** $\overline{JK}$, $\overline{ML}$; $\overline{MJ}$, $\overline{LK}$
15. a. 120° **b.** 60° **c.** 40° **d.** 20° **17.** $m\angle P = 32°$;
$m\angle O = m\angle Q = 148°$ **19.** $m\angle N = 138°$; $m\angle O =$
$m\angle Q = 42°$ **21. a.** 4 **b.** 8 **c.** 12 **23.** no; $m\angle A \neq$
$m\angle C$ **25.** yes; by Postulate 10, $\overline{AB} \parallel \overline{CD}$ and $\overline{AD} \parallel$
$\overline{BC}$ **27. a.** SSS, SAS, ASA, AAS **b.** CPCTC
c. Isosceles Triangle Theorem
d. $m\angle 1 = m\angle 2 = m\angle 3 = 22°$

Lesson 8.2 **1.** true **3.** true **5.** true **7.** true **9.** true
11. false **13.** true **15.** true **17.** true **19.** true **21.** false
23. a. 8 **b.** 3 **c.** 4 **d.** 3 **e.** 37° **f.** 53° **g.** 53°
h. 90° **25. a.** $\angle 5$ **b.** $\angle 7$ **c.** $\angle BEA$ **d.** $\overline{BE}$ **e.** none
f. $\overline{AB}$ **27.** Draw any two rhombuses with congruent
corresponding angles but different length sides.

29. $x = 20$; $m\angle RSW = 40°$; $m\angle SRT = 50°$
Lesson 8.3 **1.** true **3.** false **5.** true **7.** true **9.** true
11. false **13.** 5 **15.** 6 **17.** 37° **19.** 37° **21.** $\overline{TS}$ **23.** $\overline{RS}$
25. $\overline{RT}$ **27.** $\angle 1$, $\angle 4$, $\angle 5$ **29.** $\triangle RPU$, $\triangle SPT$, $\triangle TPS$
31. $\triangle PUR$, $\triangle PRS$, $\triangle PST$, $\triangle PTU$ **33.** $\overline{UT}$, $\overline{TS}$, $\overline{RS}$
35. $\overline{US}$ **37.** $\triangle SPR$, $\triangle SPT$, $\triangle TPS$, $\triangle UPT$, $\triangle TPU$,
$\triangle UPR$, $\triangle RPU$ **39.** $\triangle TUR$, $\triangle URS$, $\triangle RST$, $\triangle STU$,
$\triangle RPS$, $\triangle SPT$, $\triangle UPT$, $\triangle UPR$ **41.** rectangle, square
43. rhombus, square **45.** rhombus, square
47. rhombus, square **49.** non-square, rectangle
51. square **53.** 18

Lesson 8.4 **1. a.** $\overline{AD}$, $\overline{BC}$ **b.** $\overline{AB}$, $\overline{DC}$ **c.** $\angle A$,
$\angle D$; $\angle B$, $\angle C$ **3. a.** $\overline{MQ}$, $\overline{NP}$ **b.** $\overline{MN}$, $\overline{QP}$ **c.** $\angle M$,
$\angle Q$; $\angle N$, $\angle P$ **5. a.** 10 **b.** 8 **c.** 7 **d.** 34° **e.** 102°
f. 44° **g.** 34° **h.** 44° **7. a.** 12 **b.** 42° **c.** 138°
d. 138° **9. a.** 6 **b.** 12 **c.** 12 **d.** 60° **e.** 95° **f.** 60°
11. a. 11.5 **b.** 20 **c.** 90° **d.** 110° **e.** 110° **f.** 90°
13. a. 72° **b.** 38° **c.** 34° **d.** 108° **e.** 38° **f.** 108°
15. $m\angle B = 70°$, $m\angle C = m\angle D = 110°$
17. $m\angle C = 120°$, $m\angle A = m\angle B = 60°$ **19.** 75° or
105° **21.** Base angles of an isosceles trapezoid are
congruent. **23.** definition of isosceles trapezoid
25. **27.** Another angle must be
supplementary to the right
angle, so if there is one rt. angle
there must be two.

29. Draw any trapezoid with two adjacent congruent
sides. **31.** 5; 10; 20

Test Yourself **1. a.** 120° **b.** 60° **c.** 35° **d.** 35°
2. a. 140° **b.** 70° **c.** 20° **d.** 90° **3. a.** 20° **b.** 70° **c.** 20°
d. 140° **4. a.** $\overline{YO}$ **b.** $\overline{XY}$ **c.** $\angle 5$ **5. a.** $\overline{YO}$ **b.** $\overline{WX}$, $\overline{XY}$, $\overline{ZY}$
c. $\angle 2$, $\angle 5$, $\angle 6$ **6. a.** $\overline{XO}$, $\overline{YO}$, $\overline{ZO}$ **b.** $\overline{XY}$ **c.** $\angle 8$, $\angle 4$, $\angle 5$
7. a. $\overline{XO}$, $\overline{YO}$, $\overline{ZO}$ **b.** $\overline{WX}$, $\overline{XY}$, $\overline{ZY}$ **c.** angles 2–8 **8. a.** 59°
b. 121° **c.** 59° **d.** 3 **e.** 6 **f.** 14 **9.** 135° and 70°

Lesson 8.5 **1. a.** given **b.** definition of rhombus
c. A segment is congruent to itself. **d.** SSS **e.** CPCTC
f. definition of angle bisector **3. a.** given **b.** definition of
isosceles trapezoid **c.** A segment is congruent to itself.
d. Base angles of an isosceles trapezoid are congruent.
e. SAS **f.** CPCTC **5. a.** given **b.** definition of rhombus
c. definition of rhombus **d.** The diagonals of a
parallelogram bisect each other. **e.** definition of bisector
f. A segment is congruent to itself. **g.** SSS **h.** CPCTC
i. Angles that are congruent and supplementary are right
angles. **j.** definition of a perpendicular bisector
7. Parallel lines are everywhere equidistant.

Lesson 8.6 **1.** yes; if both pairs of opposite sides
of a quadrilateral are congruent, then the quadrilateral

is a parallelogram. **3.** no **5.** no **7.** Draw any two
intersecting segments of length 6 cm and 4 cm that
bisect each other but do not form right angles.
Connect the four endpoints to form a parallelogram.
9. Draw two intersecting segments of length 6 cm and
4 cm that are perpendicular bisectors of each other.
Connect the four endpoints to form a rhombus.
11. Draw any two perpendicular segments that do not
bisect each other. Connect the four endpoints to form
a quadrilateral that is not a parallelogram.
13. a. given **b.** A segment is congruent to itself.
c. SSS **d.** CPCTC **e.** If 2 lines are cut by a transversal
so that one pair of alternate interior angles is
congruent, the lines are parallel. **f.** definition of a
parallelogram **15. a.** given **b.** Vertical angles are
congruent. **c.** SAS **d.** CPCTC **e.** If 2 lines are cut by
a transversal so that one pair of alternate interior
angles is congruent, the lines are parallel. **f.** definition
of parallelogram **17.** If the diagonals of a
quadrilateral bisect each other, the quadrilateral would
be a parallelogram. **19. b.** $\angle AOB \cong \angle COD$ (Vertical
angles are congruent.) **c.** $\triangle AOB \cong \triangle COD$ (SAS)
d. $\angle 1 \cong \angle 2$ and $\overline{AB} \cong \overline{CD}$ (CPCTC) **e.** $\overline{AB} \parallel \overline{DC}$
(If 2 lines are cut by a transversal so that one pair of
alternate interior angles is congruent, the lines are
parallel.) **f.** $ABCD$ is a parallelogram (Ex. 14).
21. The diagonals of $ABCD$ bisect each other so
$ABCD$ is a parallelogram and $\overline{DC} \parallel \overline{AB}$.

Lesson 8.7 1. The number of boxes in n rows is
n • n; 225 boxes **3. a.** 15 (Add a row of 5 dots.); 21
(Add a row of 6 dots.); **b.** 55

Chapter 8 Review 1. a. 55° **b.** 125° **c.** 7 **d.** 5
2. a. 25° **b.** 65° **c.** 130° **d.** 90° **3. a.** 10 **b.** 10
c. 124° **4.** rectangle, square **5.** rhombus, square
6. parallelogram, rhombus, rectangle, square
7. parallelogram, rhombus, rectangle, square
8. a. 108° **b.** 72° **c.** 11 **9. a.** 113° **b.** 67° **c.** 113°
10. a. given **b.** Definition of parallelogram
c. Alternate Interior Angles Postulate **d.** A segment is
congruent to itself. **e.** ASA **f.** CPCTC **11.** Yes; both
pairs of opposite sides are congruent. **12.** no **13.** Yes;
one pair of opposite sides are parallel and congruent.
14. no **15.** square **16.** 1 5 10 10 5 1;
1 6 15 20 15 6 1

Chapter 9 Perimeter and Area

Focus on Skills Arithmetic 1. 40.1 **3.** 33.12

5. $49\frac{2}{5}$ **7.** $25\frac{1}{8}$ **9.** 2,550 **11.** 302.12 **Algebra**
13. $x = 27$ **15.** 25 **17.** 100 **Geometry 19.** 164 mm
21. 12 in. **23.** 6 cm **25.** congruent, parallel
27. rectangle

Lesson 9.1 1. 8.5 cm **3.** 17.5 cm **5.** 32 in. **7.** $4\frac{3}{8}$
in. **9.** 20 in. **11.** 5 cm **13.** 24 cm **15.** 18 cm **17.** 12
19. 11 Answers to Exercises 21 and 23 may vary. For
example: **21.** length 2 units, width 4 units **23.**
25. a. $k = 7$ cm, $c = 3$ cm **b.** 60 cm
27. a. $e = 11$ in., $f = 7$ in. **b.** 80 in. **29.** $225 **31.** 35
Lesson 9.2 1. 8 sq. units **3.** $3\frac{1}{2}$ sq. units
5. 14 sq. units **7.** $8\frac{1}{2}$ sq. units **9.** 11 sq. units
11. a 10-by-2 rectangle **13.** a 5-by-5 rectangle
Answers to Exercises 15–21 may vary. For example:

17.Ex.

19.Ex.

21. Ex.

23. a.Ex.
b. the square **c.** yes **25.** 576
Lesson 9.3 1. 36 m² **3.** 189 cm² **5.** 56 cm²,
30 cm **7.** 2 cm, 20 cm **9.** 1.5 m² or 15,000 cm², 5 m
or 500 cm **11.** 2 ft, 16 ft **13. a.** 52 cm **b.** 88 cm²
15. 8 cm by 1 cm **17.** 6 units by 2 units **19.** Draw any
three rectangles for which the base lengths and heights
have the same sum but different products. For
example, rectangle A: $b = 1$, $h = 5$; rectangle B: $b = 2$,
$h = 4$; rectangle C: $b = 3$, $h = 3$. **21. a.** 48,000 ft²
b. $5,333\frac{1}{3}$ yd² **c.** greater
23. a. 360 in.²; 224 in.² **b.** the shelf **c.** No; both
dimensions of the turntable are longer than the shorter
dimension of the shelf. The turntable would hang over
the edge and be unstable.
Lesson 9.4 1. a. 8 sq. units **b.** 8 sq. units
c. 8 sq. units **d.** 8 sq. units **3.** $P = 32$ cm; $A = 50$ cm²

5. $P = 40$ cm; $A = 80$ cm^2 **7.** 19.22 cm^2 **9.** 6 yd^2 or 54 ft^2 **11.** 5 cm **13.** Draw a rectangle with a base 8 units long and a height 3 units long. **15.** altitude: $\overline{LM}$ (bases: $\overline{IF}$, $\overline{HG}$); altitude: $\overline{JK}$ (bases: $\overline{IH}$, $\overline{FG}$) **17.** altitude: $\overline{AB}$ (bases: $\overline{PO}$, $\overline{MN}$); altitude: $\overline{JY}$ (bases: $\overline{MP}$ and $\overline{ON}$) **19. a.** 50 ft^2 **b.** 25 ft^2 **21.** 60 ft^2 **23.** 60 in.^{2} **25. a.** 9,432 ft^2 **b.** 2 bags **c.** 1,048 yd^2

Test Yourself 1. $P = 32$ cm; $A = 64$ cm^2 **2.** $P = 10$ ft; $A = 6$ ft^2 **3.** 26.25 in.^{2} **4.** 12 cm **5.** 4 cm **6. a.** 5 cm **b.** 30 cm **7.** 6-unit square

Lesson 9.5 1. 38.5 cm^2 **3.** 9 ft^2 **5.** 30.375 ft^2 or 3.375 yd^2 **7.** 18 in.^{2} or 0.125 ft^2 **9.** 6 in.^{2} **11.** 4 ft **13.** altitude: $\overline{IZ}$ (base: $\overline{TR}$); altitude: $\overline{RX}$ (base: $\overline{TI}$); altitude: $\overline{TY}$ (base: $\overline{IR}$) **15.** altitude: $\overline{RZ}$ (base: $\overline{TI}$); altitude: $\overline{TY}$ (base: $\overline{RI}$); altitude: $\overline{IX}$ (base: $\overline{TR}$) **17.** Draw any triangle with one 90° angle between two legs of length 5 cm and 4 cm. **19.** Draw any triangle with 3 angles with measures between 0° and 90°. The triangle also must have one side that is 4 cm long, and the altitude to this side must be 3 cm long. **21.** Draw any triangle with 3 angles with measures between 0° and 90°. The triangle also must have one side that is 8 in. long, and the altitude to this side must be 3 in. long. **23.** Draw any triangle that has an area of 12 cm^2. For example: $b = 12$ cm and $h = 2$ cm. **25. a.** 36 cm^2 **b.** 18 cm^2 **c.** 18 cm^2 **27.** 34 cm^2 **29.** 25 in.^{2} **31.** 21 ft^2 **33. a.** area **b.** area **c.** area **d.** perimeter **e.** area **f.** area

Lesson 9.6 1. 111 cm^2 **3.** 36 cm^2 **5.** 61.5 cm^2 **7.** 45 cm^2 **9.** 32 cm^2 **11.** 75 in.^{2} **13.** 14 cm **15. a.** too little **b.** need length of ceiling **17.** Since $b_1 = b_2 = b$, $\frac{1}{2}h(b_1 + b_2) = \frac{1}{2}h(b + b) = \frac{1}{2}h(2b) = bh$ **19.** 1,224 ft^2

Lesson 9.7 1. $A = \frac{1}{2}aP$; $A = \frac{1}{2}h(b_1 + b_2)$; $A = bh$; about 166 in.^{2} **3.** 10.5 h **5.** 110; 137.5; 62; 71

Lesson 9.8 1. a. 16π cm **b.** about 50.24 cm **3. a.** 15π in. **b.** about 47.1 in. **5. a.** 56.4π m **b.** about 177.096 m **7. a.** 14π in. **b.** about 44 in. **9. a.** 70π cm **b.** about 220 cm **11.** $d = 14$ ft; $r = 7$ ft **13.** about 36.56 cm **15.** 16π mm (about 50.24 mm) **17.** about 25,160 mi

Lesson 9.9 1. a. 484π cm^2 **b.** about 1,519.76 cm^2 **3. a.** 0.64π cm^2 **b.** about 2.0096 cm^2 **5. a.** $1,369\pi$ ft^2 **b.** about 4,298.66 ft^2 **7. a.** 49π cm^2 **b.** about 154 cm^2 **9. a.** 100π in.^{2} **b.** about $314\frac{2}{7}$ in.^{2} **11. a.** $1\frac{24}{25}\pi$ ft^2 **b.** about $6\frac{4}{25}$ ft^2 **13.** 27.8 cm, 87.29 cm, 606.68 cm^2 **15.** 1 cm, 2 cm, 3.14 cm^2 **17.** about 254 mm^2

19. about 50 in.^{2} **21.** 3π in^2; about 9.42 in.^{2} **23.** 12π cm^2; about 37.68 cm^2 **25.** 8π in.^{2}; about 25.12 in.^{2} **27. a.** 1.5π in.^{2} **b.** The area of any circle is πr^2. The degree measure of a full turn is 360°. A central angle with measure $n°$ is $\frac{n}{360}$ of a full turn so it is reasonable to assume that the shaded region has area $\frac{n}{360}$ times that of the circle, or $(\frac{n}{360})\pi r^2$. **29. a.** about 4,415.625 ft^2 **b.** \$32.38 **31. a.** about 3,683.42 ft^2 **b.** \$27.01 **33. a.** 56 in, **b.** about 2,462 in.^{2}; about 1.9 yd^2 **c.** yes; yes

Chapter 9 Review 1. 14.5 in. **2.** 28 cm **3.** 28 cm **4.** 26 in. **5.** 12 m **6.** 8 cm **7.** 12 sq. units **8.** 10 sq. units **9.** $14\frac{1}{2}$ sq. units **10.** 576 ft^2 **11.** 90 cm^2 **12.** 12 in.^{2} **13.** 72 cm^2 **14.** 24 m^2 **15. a.** $C = 24\pi$ cm; $A = 144\pi$ cm^2 **b.** $C \approx 75.36$ cm; $A \approx 452.16$ cm^2 **16. a.** $C = 8\pi$ ft; $A = 16\pi$ ft^2 **b.** $C \approx 25.12$ ft; $A \approx 50.24$ ft^2 **17. a.** $C = 15\pi$ m; $A = 56.25\pi$ m^2 **b.** $C \approx 47.1$ m; $A \approx 176.625$ m^2 **18.** \$1,181.25

Chapter 10 Similarity

Focus on Skills Arithmetic 1. $\frac{2}{3}$ **3.** $\frac{4}{5}$ **5.** $\frac{3}{5}$ **7.** $\frac{2}{3}$ **9.** $\frac{5}{8}$ **11.** $\frac{4}{5}$ **13.** $\frac{1}{5}$ **15.** $\frac{7}{10}$ **17** $\frac{1}{6}$ **19.** 120 **21.** 90 **23.** 28 **Geometry 25. a.** $\triangle YXZ$ **b.** $\triangle YZX$ **27. a.** $\angle A \cong \angle S$; $\angle B \cong \angle T$; $\angle C \cong \angle R$ **b.** $\overline{AC} \cong \overline{SR}$; $\overline{BC} \cong \overline{TR}$; $\overline{AB} \cong \overline{ST}$ **c.** $\triangle ABC \cong \triangle STR$ **29.** $\angle L$; $\overline{IG}$

Lesson 10.1 1. 4:11 **3.** 4:15 **5.** Draw 2 rows (columns) of 5 squares. Shade any 3 squares. (Also, any multiple of the above.) **7.** 3:5 **9.** $\frac{1}{3}$ **11.** $\frac{3}{5}$ **13.** 1 to 3 **15.** 3 to 2 **17.** 2:5 **19.** 3:4 **21.** 2:3 **23.** 2:5 **25.** $\frac{3}{2}$ **27.** $\frac{5}{2}$ **29.** $\frac{2}{3}$ **31.** $\frac{3}{2}$ **33.** $\frac{2}{3}$ **35.** $\frac{3}{2}$ **37.** 2 to 5 **39.** 1 to 4 **41.** 3 to 10 **43.** 1 to 4 **45.** 1:3 **47.** 40°, 60°, 80°

Lesson 10.2 1. $x = 6$ **3.** $y = 10$ **5.** $z = 28$ **7.** $x = 7\frac{1}{2}$ **9.** $x = 6$ **11.** $n = 3\frac{1}{3}$ **13.** $b = 10$ **15.** $d = 6$ **17.** 9 **19.** $\frac{2}{8}$ **21.** $\frac{12}{9}$ **23.** $\frac{18}{16}$ **25.** not possible **27.** $x = 15$ **29.** 1,665 **31.** \$1,650

Lesson 10.3 1. false **3.** false **5.** false **7.** b **9.** b **11. a.** *CHGFD* **b.** *BTSNQ* **c.** *GHCDF* **d.** *NSTBQ* **13. a.** $\angle B \leftrightarrow \angle C$; $\angle Q \leftrightarrow \angle H$; $\angle N \leftrightarrow \angle G$; $\angle S \leftrightarrow \angle F$; $\angle T \leftrightarrow \angle D$ **b.** $\overline{BQ} \leftrightarrow \overline{CH}$; $\overline{QN} \leftrightarrow \overline{HG}$; $\overline{NS} \leftrightarrow \overline{GF}$; $\overline{ST} \leftrightarrow \overline{FD}$; $\overline{TB} \leftrightarrow \overline{DC}$ **15. a.** $\angle R \leftrightarrow \angle J$; $\angle S \leftrightarrow \angle K$; $\angle T \leftrightarrow \angle L$ **b.** $\overline{RS} \leftrightarrow \overline{JK}$; $\overline{ST} \leftrightarrow \overline{KL}$; $\overline{TR} \leftrightarrow \overline{LJ}$ **17. a.** $\angle A \leftrightarrow \angle A$; $\angle AMN \leftrightarrow \angle E$; $\angle ANM \leftrightarrow \angle F$ **b.** $\overline{AM} \leftrightarrow \overline{AE}$; $\overline{AN} \leftrightarrow \overline{AF}$; $\overline{MN} \leftrightarrow \overline{EF}$ **c.** Answers may vary. For example: $\triangle ANM \sim \triangle AFE$; $\triangle MAN \sim \triangle EAF$ **19.** Draw two triangles, one of which is an enlargement (reduction) of the other, so that $C \leftrightarrow P$, $D \leftrightarrow L$, and $E \leftrightarrow G$. **21.** Same as 19 except triangles have one common vertex (R), and $S \leftrightarrow W$, $T \leftrightarrow Y$. **23. a.** The area of the larger rectangle is 4 times that of the smaller. **b.** The area of the larger rectangle is 9 times that of the smaller. **c.** The area of the larger rectangle is 100 times that of the smaller.

Test Yourself 1. $\frac{3}{5}$ **2.** 2:3 **3.** $\frac{6}{5}$ **4.** 1 to 3
5. $y = 15$ **6.** $x = 20$ **7.** $c = 7$ **8.** $a = 64$
Lesson 10.4 1. a. 3 **b.** 30° **3.** 5 **5. a.** $\frac{1}{2}$ **b.** $\frac{1}{2}$ **c.** $\frac{1}{2}$
d. 7 **e.** $\frac{1}{2}$ **f.** $\frac{1}{2}$ **g.** 18 **h.** 48° **7. a.** $\frac{3}{1}$ **b.** 15 **c.** 35°
9. a. $\frac{3}{1}$ **b.** $\frac{1}{3}$ **c.** $\frac{1}{3}$ **d.** 18 cm **e.** 5 cm **f.** 3 **g.** $\frac{3}{1}$ **h.** 45°
i. $\frac{3}{1}$ **11.** $\frac{1}{2}$ **13.** $\frac{2}{5}$ **15.** 15 ft and 20 ft **17.** Draw any two rectangles such that one is an enlargement of the other. **19.** not possible **21.** Draw one triangle with two congruent angles. Draw a second triangle with two congruent angles whose measures differ from those of the first triangle. **23.** Draw two regular polygons of the same kind such that one is an enlargement of the other. **25.** 15 in. **27.** no; corresponding angles are not congruent; (step 1) To XYZ / (step 2) fd 75 rt 120 / (step 3) fd 75 rt 120 / (step 4) fd 75 / (step 5) end; The scale factor is 2 : 5.

Lesson 10.5 1. 16 ft **3.** 72 ft **5.** 8 in. **7.** 32 in.
9. 100 mi **11.** 20 mi **13. a.** 20 ft by 18 ft **b.** 360 ft² **15.** $2\frac{1}{4}$ in. by $2\frac{1}{4}$ in. **17.** $3\frac{3}{4}$ in. by 2 in. **19.** $\frac{1}{4}$ in. = 4 ft (or 1 in. = 16 ft) **21.** 1 cm = 15 m **23.** Answers may vary. For example: floorplans, blueprints.

Lesson 10.6 1. about 3.75 mi high **3. a.** $\dfrac{h}{1\frac{3}{4}} = \dfrac{72}{\frac{1}{2}}$
b. 252 ft

Chapter 10 Review 1. $\frac{2}{1}$ **2.** $\frac{1}{3}$ **3.** $\frac{1}{3}$ **4.** $\frac{2}{3}$
5. $\frac{3}{7}$ **6.** 5 : 3 **7.** 3 : 4 **8.** 1 to 5 **9.** 4 to 5 **10.** 1 to 6
11. $y = 18$ **12.** $x = 4$ **13.** $a = 4$ **14.** $z = 36$
15. $x = 1.5$ **16.** $n = 1\frac{2}{3}$ **17.** $c = 5$ **18.** $z = 15$ **19.** c
20. c **21.** $\overline{AE} \leftrightarrow \overline{BD}$; $\overline{AC} \leftrightarrow \overline{BC}$; $\overline{EC} \leftrightarrow \overline{DC}$
22. $\angle A \leftrightarrow \angle DBC$; $\angle E \leftrightarrow \angle BDC$; $\angle C \leftrightarrow \angle C$
23. $\triangle AEC \sim \triangle BDC$; $\triangle ECA \sim \triangle DCB$;
$\triangle CAE \sim \triangle CBD$ **24.** $\frac{2}{3}$ **25.** $\overline{LM}$ **26.** $\angle M$ **27.** $\frac{2}{3}$
28. 60° **29. a.** $\frac{2}{3}$ **b.** $\frac{2}{3}$ **c.** 8 **30.** 6 **31.** 7 m **32.** 12.25 m
33. 17.5 m **34.** 1.75 m **35.** 16.1 m **36.** 12 ft **37.** 6 ft
38. 9 ft **39.** 18 ft **40.** 36 ft **41.** about 1.7 mi

Chapter 11 Similar Triangles

Focus on Skills Arithmetic **1.** b **3.** e **5.** c
7. $y = 6$ **9.** $n = 6$ **Geometry 11. a.** $\angle RSW \cong \angle T$;
$\angle RWS \cong \angle Z$; $\angle R \cong \angle R$; **b.** $\frac{RW}{RZ}$; $\frac{SW}{TZ}$ **13.** Use Construction 5. **15. a.** Use Construction 2. **b.** Use Construction 1 three times. **c.** Find the midpoint, E, of $\overline{CD}$. Use Construction 1 to construct $\overline{RS} \cong \overline{CE}$ (or $\overline{ED}$).

Lesson 11.1 1. yes; $\triangle ABC \sim \triangle FED$ **3.** yes;
$\triangle RTS \sim \triangle WTX$ **5. a.** $\triangle RST$ **b.** WZ, RT **7. a.** 37°
b. $\triangle FDE$ **c.** 8 **d.** 6 **9.** yes **11.** Use Construction 2 to construct the $\perp$ bisector of $\overline{AB}$, intersecting $\overline{AB}$ at X. Use Construction 1 to construct $\overline{DE} \cong \overline{AX}$. Use Construction 5 to construct $\angle D \cong \angle A$ and $\angle E \cong \angle B$. Extend the sides of angles D and E to intersect at F.
13. a. $\triangle ABC \sim \triangle CBD$ **b.** CD; DB; CB
c. $\triangle ABC \sim \triangle ACD$ **d.** AD; DC; AC **15. a.** AA Postulate ($\angle TCS \cong \angle BCA$) **b.** 4:1 **c.** 160 ft

Lesson 11.2 1. SSS Similarity, SAS Similarity, AA **3.** AA **5.** none **7. a.** $\triangle CAB$; SSS Similarity **b.** 112° **c.** 28° **d.** 28° **9. a.** $\triangle XYT$; SAS Similarity Postulate **b.** 118° **c.** 36° **d.** 6 **e.** $\overline{RS} \parallel \overline{XY}$; if 2 lines are cut by a transversal so that one pair of alternate interior angles is congruent, the lines are parallel.
11. SAS Similarity **13.** SAS Similarity **15.** DE; GK, $\angle D$ **17.** $\dfrac{PQ}{XY} = \dfrac{QR}{YZ}$ **19. a.** 2 **b.** 2 **c.** 2 **d.** $\triangle MNO$; SSS Similarity Postulate **e.** 53° **f.** $\triangle YNO$; AA Postulate (or SAS Similarity Post.) **g.** 8 **h.** 2 **i.** 4 **j.** 2:1 **21. a.** AA **b.** DF **c.** The lengths of corresponding altitudes of similar triangles are proportional to the lengths of corr. sides of the $\triangle$.
23. a. $\overline{PY}$ is an altitude of $\triangle PAB$. **b.** $\frac{1}{4}$; $\frac{1}{4}$
c. 80 cm **d.** 2 cm

Lesson 11.3 1. BT **3.** RS **5.** $\frac{3}{8}$ **7.** $\frac{3}{11}$ **9.** $\frac{3}{11}$ **11. a.** 9
b. 10 **13.** 6 **15. a.** 24 **b.** 9 **17.** No; $\overline{DE}$ does not divide $\overline{AB}$ and $\overline{AC}$ proportionally. **19. a.** The median of a trapezoid is parallel to the bases. **b.** definition of the median of a trapezoid **c.** By Theorem 11.1, if a line is parallel to one side of a triangle and intersects the other two sides, the sides of the original triangle are divided proportionally. **d.** The median of a trapezoid bisects both diagonals. **21.** 25 ft
Test Yourself 1. SSS Similarity Postulate
2. SAS Similarity Postulate **3.** $\dfrac{DF}{GK} = \dfrac{EF}{HK}$
4. $\angle E \cong \angle H$ or $\angle F \cong \angle K$ **5.** 8 **6.** 10

Lesson 11.4 1. 6 m **3.** 12.5 ft **5.** about 481 ft

Lesson 11.5 1. $\frac{2}{5}$ **3.** $\frac{2}{7}$ **5.** $\frac{5}{7}$ **7.** 5 **9.** 7 **11. a.** 15
b. 35 **c.** 16 **d.** 28 **13.** In Step 1 of Construction 12, mark off *four* congruent segments on the ray and proceed. **15.** Divide $\overline{CD}$ into 3 congruent segments, $\overline{CX}$, $\overline{XY}$, $\overline{YD}$. $CY = \frac{2}{3} \cdot CD$ **17.** Corollary to Theorem 11.3 **19.** Use Construction 12 to divide each side of $\triangle ABC$ into 3 congruent segments. Using Construction 7, construct $\triangle DEF$ with sides $\frac{2}{3}$ as long as those of $\triangle ABC$. **21. a.** 13 **b.** Answers may vary. For example

use 15 lines or use 22 of the lines and mark every third line. **23.** 6; 6; 9

Chapter 11 Review
1. AA Postulate
2. SAS Similarity Postulate **3.** SSS Similarity Postulate **4.** SAS Similarity Postulate or AA Postulate
5. none **6.** SAS Similarity Postulate
7. AA Postulate **8.** SAS Similarity Postulate **9.** none
10. $\frac{PQ}{ST} = \frac{PR}{SU} = \frac{QR}{TU}$ **11.** $\frac{PQ}{ST} = \frac{PR}{SU}$ **12.** $\angle P \cong \angle S$ or $\angle Q \cong \angle T$ **13. a.** ED **b.** AD **14. a.** 9 **b.** 28 **15.** 12
16. 30 **17.** 15 **18.** 25 **19.** In Step 1, mark off 3 congruent segments on the ray and proceed. **20.** In Step 1, mark off 4 congruent segments on the ray and proceed. **21.** In Step 1, mark off 5 congruent segments on the ray and proceed. **22.** 57 ft

Chapter 12 Square Roots and Right Triangles

Focus on Skills **Arithmetic 1.** 5,832 **3.** 289
5. 4,225 **7.** $\frac{3}{10}$ **9.** $\frac{4}{9}$ **11.** 8.8 **13.** 19.3 **15.** 1.4 **17.** 0.1
19. 20 **Algebra 21.** $n = 15$ **23.** $y = \frac{2}{3}$ **Geometry**
25. yes **27.** no **29.** yes **31.** 1,024 m^2 **33.** 24 m^2
35. 60° **37.** 30° **39.** Legs: $\overline{RT}, \overline{RS}$; hypotenuse: $\overline{ST}$

Lesson 12.1 **1.** 3 cm **3.** $\sqrt{21}$ cm **5.** 32 **7. a.** no
b. yes **c.** no **d.** no **e.** yes **f.** no **g.** yes **h.** yes **9.** 81
11. 9 **13.** 3 **15.** 400 **17.** 8 **19.** 20 **21.** 4, 5 **23.** 5, 6
25. 11, 12 **27. a.** 3,136 **b.** 0.3136 **c.** 0.003136
d. 31,360,000 **e.** 15,129 **f.** 151.29 **g.** 1.5129
h. 0.015129 **i.** 313.29 **j.** 6.4516 **k.** 85.5625
l. 632,025

Lesson 12.2 **1.** no **3.** no **5.** yes **7.** yes **9.** no; $2\sqrt{6}$
11. no; $6\sqrt{3}$ **13.** $2\sqrt{5}$ **15.** $4\sqrt{3}$ **17.** $5\sqrt{3}$ **19.** 17
21. 9.274 **23.** 11 **25.** 25 **27.** 13,689 **29.** 625 **31.** 44
33. 100 **35.** 4,624 **37.** 3.5 **39.** 9.6 **41.** 0.5 **43.** 13.2
45. 27.4 **47.** 15.7 **49.** 43.6 **51.** 83.7 **53. a.** 3 **b.** 4 **c.** 7
d. 5 **55.** true **57.** false **59.** false **61.** 16 ft by 16 ft

Lesson 12.3 **1.** $x^2 = 3^2 + 5^2$ **3.** $x^2 = 6^2 - 3^2$
(or $x^2 + 3^2 = 6^2$) **5.** no **7.** yes **9.** no **11.** $c = 20$ **13.**
$y = 20$ **15.** $b = 8$ **17.** $2\sqrt{5}$ **19.** $6\sqrt{3}$ **21.** $3\sqrt{3}$ **23.**
a. 3 cm **b.** 33 cm^2 **25. a.** 8 cm, **b.** 6 cm, **c.** 24 cm^2,
d. 96 cm^2 **27.** 10.4 m **29.** 150 m^2 **31. a.** $8\sqrt{2}$ ft **b.**
11.3 ft **c.** 11 ft

Lesson 12.4 **1.** $\sqrt{42}$ **3.** $5\sqrt{2}$ **5.** $2\sqrt{6}$ **7.** 30 **9.** 5

11. 32 **13.** 18 **15.** 45 **17.** yes **19.** no **21.** $x = 13$
23. $z = 3\sqrt{5}$ **25.** $b = \sqrt{38}$ **27.** $5\sqrt{2}$ cm **29.** $x = 4$
31. a. 5 cm, **b.** $\frac{25\sqrt{3}}{2}$ cm^2 **33. a.** $2\sqrt{3}$ cm **b.** 2 cm
c. 4 cm **d.** $2\sqrt{3}$ cm^2 **35.** $x = \sqrt{a^2 + b^2}$;
$d = \sqrt{a^2 + b^2 + c^2}$

Lesson 12.5 **1.** yes **3.** no; $3\sqrt{3}$ **5.** no; $\frac{\sqrt{14}}{2}$ **7.** yes
9. yes **11.** $\frac{8\sqrt{3}}{3}$ **13.** $\sqrt{3}$ **15.** $\frac{\sqrt{35}}{5}$ **17.** $\sqrt{2}$ **19.** $\frac{\sqrt{30}}{6}$
21. a. 6 **b.** $2\sqrt{13}$ **23. a.** $x = 5$ **b.** 45°; By the
Isosceles Triangle Theorem, $m\angle R = m\angle S$.
By Corollary 1 to the Triangle Angle-Sum Theorem,
$m\angle R + m\angle S = 90°$; then $m\angle R = m\angle S = 45°$
25. a.–b. From a starting point, draw a line 5 units
straight up, then four units to the right, then four units
straight up, then 2 units to the right. Draw a segment
from the starting point to the finish point. **c.** $3\sqrt{13}$ mi;
10.8 mi

Test Yourself **1.** 7.810 **2.** 32 **3.** 6,889 **4.** 324
5. 4.243 **6.** $4\sqrt{2}$ **7.** $3\sqrt{7}$ **8.** $5\sqrt{10}$ **9.** $4\sqrt{3}$ **10.** 13
11. 21 **12.** $\frac{2\sqrt{3}}{3}$ **13.** $\frac{\sqrt{10}}{5}$ **14.** yes **15.** yes **16.** yes
17. $x = 5$ **18.** $y = 2\sqrt{7}$

Lesson 12.6 **1.** c **a.** 28 ft **b.** 27.3 ft **c.** 27.2 ft
3. b **a.** 26 ft **b.** 25.5 ft **c.** 25.6 ft

Lesson 12.7 **1. a.** 8 **b.** $4\sqrt{3}$ **3. a.** 6 **b.** 60°
c. 60° **d.** 30° **5. a.** $4\sqrt{2}$ **b.** $4\sqrt{2}$ **7. a.** $5\sqrt{2}$ cm
b. 50 cm^2 **9. a.** 60° **b.** 30° **c.** 10 cm **d.** $10\sqrt{3}$ cm
e. $100\sqrt{3}$ cm^2 **11. a.** 6 cm **b.** $2\sqrt{3}$ cm **c.** 60cm^2
d. $(20 + 8\sqrt{3})$ cm **13. a.** $6\sqrt{2}$ cm **b.** $6\sqrt{2}$ cm
c. $(6 + 6\sqrt{2})$ cm **d.** $(18 + 18\sqrt{2})$ cm^2 **15. a.** 1.4
b. 1.7 **c.** 7.1 **d.** 8.7 **e.** 70.7

Lesson 12.8 **1.** 8 **3.** $2\sqrt{10}$ **5.** $\triangle RWT$; $\triangle WST$
7. 6 **9.** $4\sqrt{5}$ **11.** 16 **13.** $h = 12, s = 15, r = 20$
15. $a = 2, h = \sqrt{14}, d = 3\sqrt{2}, e = 3\sqrt{7}$
17. $b = 2, h = 2, d = 2\sqrt{2}, e = 2\sqrt{2}$ **19.** $d = 4\sqrt{3}, b = 13, c = 16, e = 4\sqrt{13}$ **21.** 2.4 cm
23. Each time you run the procedure, the result is a
right triangle with an altitude drawn to the hypotenuse,
forming two triangles that are similar to the original
triangle and to each other. **25.** Answers may vary.
For example: (Step 1) To tri / (Step 2) fd 48
lt 90 fd 64 home / (Step 3) fd 48 lt 90 fd 36
home / (Step 4) end

Chapter 12 Review **1.** 144 **2.** 4 **3.** 169 **4.** 13
5. 64 **6.** 20 **7.** 625 **8.** 5 **9.** 25 **10.** Answers will vary.
For example: 4, 16, 25, 144, 225 **11.** $5\sqrt{3}$ **12.** $4\sqrt{6}$
13. $10\sqrt{2}$ **14.** $6\sqrt{3}$ **15.** 81 **16.** 5,329 **17.** 17,689

18. 2.646 **19.** yes **20.** yes **21.** yes **22.** $a = 20$
23. $x = 29$ **24.** $z = 2\sqrt{13}$ **25.** $x = 9; y = 9\sqrt{2}$
26. $a = 6\sqrt{3}$; $b = 6$ **27.** $c = d = \sqrt{2}$ **28.** $2\sqrt{3}$
29. 6 **30.** 17 **31.** 200 **32.** $\frac{5\sqrt{2}}{2}$ **33.** $\sqrt{6}$ **34.** $\frac{2\sqrt{5}}{5}$
35. 2 **36.** 10 **37.** $x = 3\sqrt{2}; y = 3\sqrt{3}$ **38.** $a = 12$;
$b = 16$; $c = 20$ **39.** about 12.08 ft

Chapter 13   Circles

Focus on Skills Arithmetic **1.** 91 **3.** 85
Geometry 5. Draw $\overline{AB}$. Use Construction 2. **7.** Draw
any angle less than 90°. Use Construction 6. **9.** 360
11. $\frac{1}{3}$; 120° **13.** 54 m² **15. a.** ∠P; an angle is
congruent to itself. **b.** △PSQ; AA Postulate **c.** PS;
PQ; Definition of similar polygons
Lesson 13.1 1. a. $\overparen{PL}, \overparen{LC}$ **b.** $\overparen{LCP}, \overparen{LPC}$ **c.** $\overparen{PLC}$
3. a. $\frac{1}{12}$ **b.** $\frac{1}{8}$ **c.** $\frac{1}{3}$ **5.** 120° **7. a.** 125° **b.** 235°
9. a. 45° **b.** 135° **c.** 120° **d.** 105° **e.** 180° **f.** 240°
11. a. 80° **b.** 50° **c.** 130° **d.** 130° **e.** 230° **f.** 310°
13. Draw two circles with the same radii.
15–19. Draw a circle and then draw central angles
with the indicated measures. **21.** 11 in. **23.** 245°
For 25 and 27, draw a radius, then the given central
angles. **25.** three 120° angles **27.** eight 45° angles
Lesson 13.2 1. Draw $\overline{PA}$ to form ∠BPA. **3.** Draw
$\overline{CB}, \overline{AC},$ and $\overline{DA}$ to form ∠BDA and ∠BCA. **5.** Draw $\overline{DA}$
to form ∠BDA (or $\overline{DC}$ to form ∠CDB). **7. a.** 220° **b.**
140° **9. a.** 50° **b.** 260° **11. a.** 25° **b.** 100° **13. a.** 60°
b. 180° **c.** 80° **d.** 20° **e.** 40° **f.** 240° **15.** $m\angle 1 = 30°$,
$m\angle 2 = 25°, m\angle 3 = 130°, m\angle 4 = 50°, m\angle 5 = 65°$,
$m\angle 6 = 45°, m\angle 7 = 25°, m\angle 8 = 65°, m\angle 9 = 45°$
17. a. Put one corner on the circle; mark the points
where the sides that form the angle intersect the circle.
The corner (a right angle) intercepts a semicircle, and
therefore determines a diameter. **b.** Locate a second
diameter. The intersection of the diameters is the
center of the circle. **19.** $x = 20$
Lesson 13.3 1. Use Construction 14. **3.** 90°; 90°
5. 20 **7.** 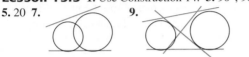 **9.**

11. There aren't any. **13. a.** 5 **b.** 5 **15. a.** 8
b. 24 square units **17. a.** 6 **b.** 12 **c.** 6 **d.** 12 **e.** 54
units **f.** 162 square units **19. a.** 35.8 km **b.** 50.6 km
c. 62 km
Lesson 13.4 1. 9; 18 **3.** 64°; 128° **5.** 65°; 65°

7. 17 **9.** $\overparen{TK}$ **11.** $\overparen{KN}$ **13.** $\overline{KL}$ **15.** Theorem 13.6
17. Theorem 13.7 **19.** yes; Theorem 13.6 **21.** no
23. a. 4 **b.** $4\sqrt{3}$ **25. a.** $4\sqrt{2}$ **b.** $2\sqrt{2}$ **c.** 45° **d.** 45°
e. $2\sqrt{2}$ **27.** $x = 15, y = 5$
Test Yourself 1. a. 140° **b.** 100° **c.** 50°
2. a. 130° **b.** 115° **c.** 5 **3. a.** 16 **b.** 20
4.
5. Draw any inscribed angle
that intercepts an arc greater
than a semicircle.

6. Use Construction 13. **7.** Use Construction 14.

Lesson 13.5 1. C **3.** F **5–8.** proceed as in Part
B(3), p. 398 to draw the central angles described. **5.** 6
60°-angles **7.** 4 right angles **9.** Use Construction 15.
11. Use Construction 16. Answers to exercises 13–17
may vary. For example:
13.  **15.** **17.**

19. Proceed as in Part A, p. 398, to locate 6 points .
Connect every other point. **21.** Construct ABCD as in
Ex 20. Construct the perpendicular bisectors of the
chords in order to locate the other 4 vertices. **23. a.** 10
b. 90° **c.** 8 **d.** 48 square units **e.** 25π square units **f.**
(25π − 48) square units **25.** Repeat Exercise 19. Use
Construction 13 three times. Extend the tangent
segments until they intersect. **27.** Choose 4 points
determining 2 chords. Construct the perpendicular
bisector of each chord. Their intersection is the center
of the original plate. **29. a.** Change line 4 to rt 135 **b.**
Change line 5 to: repeat 4[fd :r ∗ sqrt 2 rt 90]
Lesson 13.6 1. a. 55° **b.** 55° **3. a.** 120° **b.** 60°
5. a. 15° **b.** 110° **7. a.** 25° **b.** 115° **9. a.** 60° **b.** 30°
c. 75° **d.** 105° **11. a.** 30° **b.** 20° **c.** 20° **d.** 50° **e.** 70°
13. a. 150° **b.** 30° **15.** $x = 25$ **17.** $x = 22$
Lesson 13.7 1. $x = 10$ **3.** $x = 2$ **5.** $x = 4\frac{4}{5}$
7. $x = 6$ **9.** $x = 12$ **11.** $x = 13$ **13.** $x = 8$
Lesson 13.8 1. 2 lines parallel to l, on opposite
sides of l, both 3 cm from l **3.** a circle with center O
and radius 2 cm **5.** a capsule-shaped figure consisting
of 2 segments parallel to $\overline{AB}$, on opposite sides of $\overline{AB}$
and 3 cm from $\overline{AB}$, and 2 semicircles with centers A
and B and radius 3 cm **7.** line perpendicular to l
through P, not including point P **9.** The intersection of
2 circles, ⊙R with radius 3 cm and ⊙S with radius 4
cm. The locus consists of no points if RS > 7 cm, 1
point if RS = 7 cm, and 2 points if RS < 7 cm. **11.** a

line parallel to and 7.5 in. above the given surface

Lesson 13.9 **1.** car **3. a.** 126° **b.** 151.2° **c.** 28.8°
5.

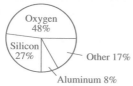

Chapter 13 Review **1.** 60° **2.** 180° **3.** 90°
4. 60° **5.** 30° **6.** 30° **7.** 12; 13 **8.** 85° **9.** $z = 6$
10. $x = 4$ **11.** $x = 38$ **12.** $t = 10$ **13.** 124° **14.** 30
15. $\overline{XY} \cong \overline{AB}$; $\overarc{XY} \cong \overarc{AB}$; (or $\overarc{AXB} \cong \overarc{XAY}$)
16. Use Construction 13. **17.** Use Construction 14.
18. Use Construction 15. **19.** Draw a diameter $\overline{AB}$.
Construct the perpendicular bisector of $\overline{AB}$, intersecting
the circle in C and D. Connect A, C, B, D. **20.** $\odot O$,
excluding points J and K
21.

Chapter 14 Area and Volume of Prisms

Focus on Skills **Arithmetic** **1.** 2,016 **3.** 3,360
5. 9,240 **7.** 1,582 **9.** 8.5 **11.** 4,200 **13.** 4
Geometry **15.** $c = 29$ cm **17.** $b = 15$ ft **19.** 420.25 m²
21. 384 cm² **23.** 13.5 m² **25.** 6; rectangular; rectangular
prism (cube) **27.** 6; trapezoidal; trapezoidal prism

Lesson 14.1 **1.** 6 6 cm by 6 cm squares
3. 4 15 m by 5 m rect.; 2 5 m by 5 m squares **5.** back:
8 in. by 9 in. rect.; left side: 6 in. by 9 in. rect.; right
side: 4 in. by 9 in. rect.; bases: triangles with legs
8 in., 6 in., and 4 in. long **7.** sides: 6 cm by 10 cm
rect., 8 cm by 10 cm rect., 4 cm by 10 cm rect., 7 cm
by 10 cm rect., 4 cm by 10 cm rect.; bases: pentagons
with sides 6 cm, 8 cm, 4 cm, 7 cm, and 4 cm long
9. 1. rectangular prism (cube), 2. rect. prism, 3. rect.
prism, 4. triang. prism, 5. triang. prism, 6. triang.
prism, 7. pent. prism, 8. hex. prism **11.** a, c, e **13. a.** 3
b. 4 **15.** $a = e = 8$ cm, $b = d = 3$ cm, $c = 4$ cm
17. $a = c = 7$ cm, $b = 8$ cm, $d = 9$ cm **19. a.** 176
in. **b.** 8 **21. a.** 408 cm **b.** 8
Lesson 14.2 **1.** 8 cm **3.** 10 cm **5.** 12 cm **7.** 9 in.

9. 3 in. **11.** 3 in. **13.** 12 m **15.** 9 m **17.** 12 cm by 10 cm
19. 9 in. by 3 in. **21.** 25 m by 15 m **23.** 25 m by 9 m
25. bases: rt. △ with legs 6 in. and 8 in. long; front: 10 in.
by 6 in. rect.; left rear: 8 in. by 6 in. rect; right rear: 6 in.
square; T.A. = 192 in.² **27.** bases: pentagons with sides
8 in., 6 in., 3 in., 6 in., and 3 in. long; left front: 5 in by 3
in. rect.; center front: 6 in.by 5 in. rect.; right front: 5 in.
by 3 in. rect; left rear: 8 in. by 5 in. rect.; right rear: 6
in.by 5 in. rect.; T.A. = 210 in.² **29.** 308.25 m² **31.** 248.5
in.² **33.** 150 in.² **35.** 5 ft **37.** 205 in.²

Lesson 14.3 **1.** 256 in.² **3.** 228 in.² **5.** 300 m²
7. 308 in.²; 404 in.² **9.** 120 cm²; 132 cm² **11.** 40 cm²;
72 cm² **13. a.** 1,980 in.² **b.** 14 ft² **15. a.** 334 ft²
b. 1,220 ft² **c.** 60 ft² **d.** 200 ft² **e.** 928 ft²
17. a. $66\frac{1}{2}$ ft² **b.** 660 ft² **c.** $593\frac{1}{2}$ ft² **d.** 260 ft²
19. a. 432 ft² **b.** 13
Test Yourself **1.** bases: 12 in. by 3 in. rect.;
front/back: 12 in. by 5 in. rect.; ends: 5 in. by 3 in.
rect.; T.A. = 222 in.² **2.** bases: △ with sides 4 cm, 8
cm, and 9 cm long; front: 9 cm by 6 cm rect.; left rear:
6 cm by 4 cm rect.; right rear: 8 cm by 6 cm rect.;
L.A. = 126 cm² **3.** 130 in.²; 202 in.² **4.** 240 cm²; 300
cm² **5.** 54 in.²

Lesson 14.4 **1. a.** 6 cm **b.** 8 cm **c.** 4 cm **d.** 8 cm
e. 160 cm² **f.** 24 cm² **g.** 208 cm² **h.** 192 cm³
3. a. Draw a 4 ft by 7 ft rectangle. **b.** 28 ft² **c.** 392 ft³
5. a. Draw a right triangle with 6 cm and 9 cm legs.
b. 27 cm² **c.** 324 cm³ **7. a.** Draw an isosceles right
triangle with 4 in. legs. **b.** 8 in.² **c.** 80 in.³ **9.** 105 cm³
11. 9 ft² **13.** 6 yd **15.** 125 in.³ **17.** 216 in.³ **19.** 2 min
21. a. 2,772 in.³ **b.** 12 gallons **23.** 8 ft **25.** 2 in. **27.** 5
in. **29. a.** 78 ft² **b.** 36 ft³ **31.** at most 18 **33. a.** 128
ft³ **b.** 384 ft³ **c.** 3 cords

Lesson 14.5 **1.** 20 cm, 16 cm, 13 cm
3. $V = 1,296$ cm³ **5. a.** 8 in.³; 125 in.³; 216 in.³
b. 64 in.³; 1,000 in.³; 1,728 in.³ **c.** If the length of an
edge is doubled, the volume is multiplied by 8. **d.** 8:1
7. a. 288 in.³ **b.** 24 in.³ **c.** Answers may vary.
For example: 4 in. by 2 in. by 3 in., 2 in. by 2 in. by
6 in., 4 in. by 1 in. by 6 in., 1 in. by 3 in. by 8 in.

Lesson 14.6 **1.** 54 ft³ **3.** 270 ft³ **5.** 3 yd³ **7.** 20 yd³
9. 3,000 **11.** 300 **13.** 2 **15.** 2,000 **17.** 0.75 **19.** 864
21. a. 162 ft³ **b.** 6 yd³ **23. a.** 130.5 ft³ **b.** $4\frac{5}{6}$ yd³
25. a. 1,728 **b.** 84 ft³ **c.** 21 **27. a.** 10,800 ft³ **b.** 400 yd³
c. 67 $(66\frac{2}{3})$ **29. a.** 10,584 in.³ **b.** 500 in.³ **c.** 21

Chapter 14 Review **1.** bases: 50 m by 14 m
rect.; front/back: 50 m by 35 m rect.; ends: 14 m by
35 m rect. **2.** bases: isos. △ with sides 4 cm, 2.2 cm,

and 2.2 cm long ; back: 3 cm by 4 cm rect.; front: 2
2.2 cm by 3 cm rect. **3.** bases: hexagons with sides
15 ft, 20 ft, 15 ft, 15 ft, 20 ft, and 15 ft long;
front/back: 20 ft by 25 ft rect.; sides: 4 15 ft by 25 ft
rect. **4.** T.A. = 4,150 in.^{2} **5.** T.A. = 330 m^2 **6.**
T.A. = 9,000 ft^2 **7.** V = 17,500 in.^{3} **8.** V = 270 m^3 **9.**
V = 39,000 ft^3 **10.** 150 cm^2 **11.** 216 yd^2 **12.** 900 mm^2
13. 512 mm^3 **14.** 5 m **15.** 36 ft^2 **16.** 1,000 in.^{3} **17.** 27
mm^3 **18.** 0.5 **19.** 2.5 **20.** 216 **21.** 5 **22.** 450 **23.** 4
24. 7 cm

Chapter 15 Area and Volume of Other Space Figures

Focus On Skills Arithmetic 1. 4 **3.** 64 **5.** 49
7. 27 **9.** 1 **11.** 64 **13.** 2 **15.** 5 **17.** 9 **19.** 1,568
21. 113.04 **23.** 1,848 **Algebra 25.** x = 22.5
27. y = 364.5 **Geometry 29.** d = 18 cm; C = 18π cm
31. $C \approx$ 125.6 cm; $A \approx$ 1,256 cm^2 **33.** 5 : 4
35. $\triangle ABC$: A = 1,200 m^2; $\triangle DEF$: A = 768 m^2

Lesson 15.1 1. 18 cm^2 **3.** 6 ft **5.** 198 cm^2
7. 113.04 cm^2 **9.** 6π cm **11.** 4π cm, 12 cm
13. a. R: 24π cm^2; S: 12π cm^2; T: 48π cm^2 **b.** L.A. of
cylinder T = 2(L.A. of cylinder R) = 4(L.A. of
cylinder S) **c.** 36π cm^2 **15. a.** $\approx$ 9.92 in. long by 4 in.
wide **b.** $\approx$ 39.68 in.^{2} **17.** 2,376 in.^{2} or 16.5 ft^2

Lesson 15.2 1. $\approx$ 314 ft^3 **3.** $\approx$ 502.4 cm^3
5. 180 cm^3 **7.** $\approx$ 7,850 cm^3 **9. a.** R: 16π in.^{3}; S: 64π
in.^{3}; T: 32 in.^{3} **b.** V of can S = 2 • V of can T = 4 • V
of can R **11.** 6 cm **13.** 10 cm; 3 cm **15. a.** $\approx$ 282.6
cm^3, $\approx$ 502.4 cm^3, $\approx$ 785 cm^3 **b.** The largest cylinder
holds the same amount as the two smaller cylinders
combined. **17.** $\approx$ 452.16 cm^2

Lesson 15.3 1. 1:3, 1:9 **3.** 4:9, 16:81 **5.** 28 in.;
48 in.^{2} **7.** 24 in.; 16 in.^{2} **9. a.** 3:2 **b.** 6 cm **c.** 42 cm;
28 cm **d.** 32 cm^2 **11. a.** 1:3 **b.** 4 cm **c.** 3 cm^2
13. No; each dimension of the car on the right is 3
times the corr. dimension of the car on the left, but
since the area is 9 times as great the appearance is
misleading.

Lesson 15.4 1. 1:100, 1:100, 1:1,000 **3.** 6:1,
36:1, 216:1 **5.** 2:5, 4:25, 4:25 **7.** Yes; the bases are
similar with scale factor 2:3; ratio of heights is 2:3.
9. Yes; scale factor is 1:2; ratio of radii is 1:2, which is
equal to the ratio of heights. **11. a.** 544 in.^{2}
b. 640 in.^{3} **13.** 640 in.^{3} **15.** 320 cm^2; 128 cm^3 **17.** 5:2
Test Yourself 1. a. 113.04 cm^2 **b.** 169.56 cm^2
c. 169.56 cm^3 **2.** 5 cm **3. a.** 2:3 **b.** 4:9 **4. a.** 1:4
b. 1:8 **5. a.** 375 cm^2 **b.** 375 cm^3

Lesson 15.5 1. false **3.** true **5.** true **7.** true
9. $\approx$ 452.16 in.^{3} **11.** 216 cm^3 **13.** $\approx$ 64 cm^3 **15.** 216
in.^{3} **17.** $\approx$ 527.52 cm^3 **19.** 64 cm^3 **21.** L.A. = 80
cm^2; T.A. = 144 cm^2 **23.** 675 cm^3 **25.** They are
equal.

27. Size and proportions
of figure may vary.

29. Size and proportions
of figure may vary.

31. about 2,592,100 m^3

Lesson 15.6 1. 12 cm **3. a.** 256π cm^2; $\frac{2,048\pi}{3}$ cm^3
b. 803.84 cm^2; 2,143.57 cm^3 **5.** 14 cm, 196π cm^2,
$\frac{1,372\pi}{3}$ cm^3 **7.** 9 cm, 18 cm, 972π cm^3 **9.** 10 cm, 20 cm
11. $\frac{\pi}{6}$ cm^3 **13. a.** 400π cm^2 **b.** 20 cm **c.** 400π cm^2
d. 600π cm^2 **e.** $\frac{4,000\pi}{3}$ cm^3 **f.** 2,000π cm^3 **15.** 2.68 •
10^{11} mi^3 or 268,000,000,000 mi^3 **17.** 100,480,000 mi^2

Lesson 15.7 1. 640 cm^2; 768 cm^3 **3.** about
102,050 cm^3

Chapter 15 Review 1. 16 cm **2.** 6 in.
3. 452.16 m^2; 854.08 m^2 **4.** 138,600 cm^3 **5.** 216 in.^{3}
6. 240 m^2; 384 m^2 **7.** 219.8 cm^2 **8.** 301.44 cm^2;
301.44 cm^3 **9.** 256π ft^2; $\frac{2,048\pi}{3}$ ft^3 **10.** 66 ft; 270 ft^2
11. 504 in.^{2}; 1,080 in.^{3} **12.** 1,028.8 cm^3

Chapter 16 Trigonometry

Focus on Skills Arithmetic 1. $\frac{5}{13}$ **3.** $\frac{12}{13}$ **5.** $\frac{5}{4}$
7. $\frac{15}{17}$ **9.** $\frac{21}{20}$ **11.** $\frac{4}{3}$ **13.** $\frac{3}{5}$ **15.** $\frac{5}{12}$ **17.** $\frac{12}{13}$ **19.** $\frac{13}{5}$ **21.** 0.625
23. 0.6 **25.** 0.667 **27.** 0.6 **29.** 0.385 **Geometry**
31. $\overline{XZ}$ **33.** $\overline{XY}$ **35.** $\overline{NP}$ **37.** 40 **39.** 10 $\sqrt{2}$ **41.** 30 cm^2
Lesson 16.1 1. a. $\frac{20}{29}$ **b.** 0.690 **3. a.** $\frac{20}{21}$ **b.** 0.952
5. a. $\frac{20}{29}$ **b.** 0.690 **7. a.** $\frac{35}{37}$ **b.** 0.946 **9. a.** $\frac{35}{12}$ **b.** 2.917
11. a. $\frac{35}{37}$ **b.** 0.946 **13. a.** 6 $\sqrt{2}$; 8.485 **b.** 1; 1.000
c. $\frac{\sqrt{2}}{2}$; 0.707 **d.** $\frac{\sqrt{2}}{2}$; 0.707 **15.** Yes; in any 45°-45°-
90° $\triangle$, each of these ratios is the same by Thm. 12.3.
17. a. true **b.** true **c.** false Use the following diagram
for Exercises 19, 21, 23, and 25:

19. The length of a leg of a right triangle cannot
exceed the length of the hypotenuse. Since $c > a$ and

$c > b$, $\frac{a}{c}$ and $\frac{b}{c}$ are both less than 1. **21.** For an angle less than 45°, the opposite leg is shorter than the adjacent leg. If $a < b$, $\tan A < 1$. **23.** $\angle A$ and $\angle B$ are complementary angles. $\cos \angle A = \frac{b}{c} = \sin \angle B$ **25.** If $\tan A = \frac{a}{b} < 1$, then $a < b$. Since $m\angle A + m\angle B = 90°$, $m\angle A < 45°$ by Thm. 5.13 **27. a.** 75 ft **b.** 53 ft

Lesson 16.2 **1.** tangent **3.** cosine **5.** sine **7.** 0.087 **9.** 0.656 **11.** 0.970 **13.** 0.530 **15.** 0.017 **17.** 2.356 **19.** 17°; 18° **21.** 52°; 53° **23.** 41° **25.** 64° **27.** 13° **29.** 66° **31.** 15° **33.** 72° **35.** 46° **37.** 66° **39.** 32° **41.** no **43.** $\tan 20°$ **45.** 18° **47.** 4° **49. a.** Change 4th line to: rt arctan :AB / :BC; change 6th line to: rt 90 **b.** Delete 2nd line. **c.** Change 2nd line to: lt 180 **Test Yourself** **1.** $\frac{11}{61}$; 0.180 **2.** $\frac{60}{61}$; 0.984 **3.** $\frac{11}{60}$; 0.183 **4.** $\frac{11}{61}$; 0.180 **5.** 0.515 **6.** 0.344 **7.** 0.515 **8.** 64° **9.** 56° **10.** 77° **11.** 73° **12.** 32°

Lesson 16.3 **1.** tangent **3.** sine **5.** sine **7.** tangent **9.** sine **11.** c **13.** b **15. a.** $\angle 6$ **b.** $\angle 7$ **17.** $x = 14$ **19.** $x = 14$ **21. a.** 10 **b.** 7 **23. a.** 18° **b.** 72° **25.** 68° **27.** 57 yd **29. a.** 29° **b.** 5 **c.** 10 **d.** 18 **31. a.** Area $= \frac{1}{2}ah$ **b.** $\frac{h}{b}$; b; h **c.** Area $= \frac{1}{2}ab \cdot \sin C$ **d.** The area of a triangle equals half the product of the lengths of two sides and the sine of the included angle. **33.** 4 cm² **35.** 18 m **37. a.** 4 ft **b.** 15 ft

Lesson 16.4 **1.** 433,000 mi **3. a.** 36,000,000 mi **b.** 67,000,000 mi **c.** 146,000,000 mi

Chapter 16 Review **1.** $\frac{24}{25}$; 0.96 **2.** $\frac{7}{25}$; 0.28 **3.** $\frac{24}{7}$; 3.429 **4.** $\frac{7}{25}$; 0.28 **5.** $\frac{24}{25}$; 0.96 **6.** $\frac{7}{24}$; 0.292 **7.** 12 **8.** $\frac{\sqrt{2}}{2}$; 0.707 **9.** $\frac{\sqrt{2}}{2}$; 0.707 **10.** 1 **11.** sine **12.** cosine **13.** tangent **14.** 0.259 **15.** 0.951 **16.** 0.829 **17.** 0.999 **18.** 0.900 **19.** 2.246 **20.** 54° **21.** 16° **22.** 47° **23.** 19° **24.** 80° **25.** 37° **26. a.** 13 **b.** 7 **27. a.** 29 **b.** 24 **28. a.** 46° **b.** 44° **29. a.** 50° **b.** 40° **30.** 484,000,000 miles

Chapter 17   The Coordinate Plane

Focus on Skills **Arithmetic** **1.** 1 **3.** −3 **5.** 2 **7.** −8 **9.** 0 **11.** −1 **13.** −$\frac{1}{3}$ **15.** 3 **17.** −2 **19.** $\frac{1}{3}$ **21.** −5 **23.** −4 **25.** 9 **27.** $\frac{3}{4}$ **29.** $\frac{3}{4}$ **31.** $\frac{2}{3}$ **33.** $\frac{2}{3}$ **35.** 11 **37.** 8 **39.** 13 **41.** $\sqrt{58}$ **Algebra** **43.** −8, −2, 4 **45.** −6, −3, 0 **Geometry** **47.** K **49.** C For Exercises 51 and 53, answers may vary. For example: **51.** 3 blocks east,

3 blocks south **53.** 1 block west, 2 blocks south

Lesson 17.1 **1.** (2, 0) **3.** (0, 3) **5.** (4, −3) **7.** (−6, −3) **9.** (0, −3) **11.** (−3, −5) For Exercises 13, 15, 19, 21, 23, 25, and 27 start at the origin. **13.** Go 5 units right and up 2 **15.** Go 3 units left and up 4 **17.** Go up 8 units **19.** Go 9 units right and down 5 **21.** Go 10 units left and up 6 **23.** Go 7 units left **25.** Go 5 units left and down 9 **27.** Go 11 units right and up 6 **29.** 9 **31.** 6 **33.** 10

35. a. 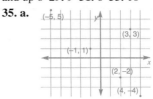 **b.** (3, 3); (−3, 3) or (3, −3) **c.** −15

37. a. 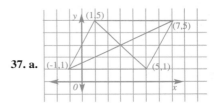 **b.** a parallelogram **c.** (3, 3)

39. a. 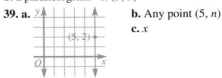 **b.** Any point (5, n) **c.** x

41.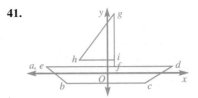

43. Answers may vary. For example: **a.** (−6, −2) **b.** (4, 0) **c.** (7, 4) **d.** (4, 6) **e.** (−6, 4) **f.** (−3, 2) **g.** (−6, −2) **h.** (−6, 4) STOP **i.** (−3, 2) **j.** (7, 4) **Lesson 17.2** **1. a.** 5 **b.** $(\frac{9}{2}, 4)$ **3. a.** $5\sqrt{2}$ **b.** $(\frac{-3}{2}, \frac{1}{2})$ **5. a.** 10 **b.** (1, −4) **7. a.** $3\sqrt{5}$ **b.** $(-\frac{5}{2}, -2)$ **9. a.** $\sqrt{53}$ **b.** $(-\frac{1}{2}, 1)$ **11. a.** 3 **b.** 5 **c.** $\sqrt{34}$ **d.** $8 + \sqrt{34}$

e. $7\frac{1}{2}$ square units **13. a.** $2\sqrt{5}$ **b.** $2\sqrt{5}$ **c.** $2\sqrt{10}$
d. $4\sqrt{5} + \sqrt{10}$ **e.** 10 square units
15. $SW = \sqrt{65}$; $RT = \sqrt{65}$ **17.** yes **19.** no
21. $\sin A = \frac{5}{13} \approx 0.385$; $\cos A = \frac{12}{13} \approx 0.923$; $\tan A = \frac{5}{12} \approx 0.417$; $m\angle A \approx 23°$

Lesson 17.3 1. a. positive **b.** negative **c.** 0
d. positive **e.** undefined **f.** negative **3.** $-\frac{2}{3}$ **5.** $\frac{4}{3}$ **7.** $\frac{3}{4}$
9. 0 **11.** -1 **13.** $\frac{1}{2}$ **15.** undefined **17.** Plot 2 or more
points on a piece of graph paper so that for a 1-unit
rise you have a run of -2 units. **19.** Plot (3,1). Plot 1
or more additional points so that for a 2-unit rise you
have a 5-unit run. **21.** 5
23. a.

b. $AB = \sqrt{26}$;
$AC = BC = \sqrt{13}$;
$\triangle ABC$ is an isosceles
right triangle
c. slope of $\overline{AC} = \frac{3}{2}$; slope
of $\overline{BC} = -\frac{2}{3}$; slope of $\overline{AC}$ •
slope $\overline{BC} = -1$
25. $2\frac{1}{2}$ in. **27.** no

Lesson 17.4 1. a. $\frac{3}{4}$ **b.** $-\frac{4}{3}$ **3. a.** a horizontal line
b. The slope is undefined. **5.** perpendicular **7.** neither
9. parallel **11.** parallel **13. a.** $-\frac{3}{2}$ **b.** 5 **c.** $-\frac{1}{5}$ **d.** 5

15.

17.

19.

21.

23. right triangle; j is perpendicular to l **25.** no, since
slope of $\overline{EG} = -\frac{4}{3}$ and slope of $\overline{FH} = 1$. **27.** $RSTW$ is an
isosceles trapezoid; slope of $\overline{ST} = 1 = $ slope of $\overline{RW}$;
$\overline{ST} \parallel \overline{RW}$; $TW = 13 = SR$ **29.**

Lesson 17.5 1. a, b **3.** a, b, c **5.** a, b, c **7.** a
9. a, b, d, e For Exercises 11,13, and 15, answers will
vary. For example: **11.** $(-2, -5)$, $(0, -4)$, $(4, -2)$
13. $(-1, -2)$, $(0, 0)$, $(1, 2)$ **15.** $(-2, 4)$, $(0, 4)$, $(2, 4)$

17. a.

b. $(0, 0)$; $(0, 4)$; $(0, -4)$
c. 2 **d.** The lines are
parallel.

19. The graph is a line parallel to the y-axis and 3 units
to the right of the y-axis.
For Exercises 21, 23, and 25, answers may vary.
Examples of ordered pairs of numbers are given.
21. $(-2, 7)$, $(1, -2)$, $(3, -8)$ **23.** $(-6, -2)$, $(0, 0)$, $(3, 1)$
25. $(-2, -4)$, $(0, -3)$, $(2, -2)$ **27.** $y = 2x$
29. $y = x - 2$ **31.** $y = 2x + 1$
33. a. Yes, for example 1 horizontal unit $= 2$ ft,
1 vertical unit $= 1$ in. Or both units could be in
inches. **b.** Length and distance are always positive.

Test Yourself 1–5.
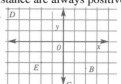

6. a. $2\sqrt{13}$ **b.** $(0, 2)$ **c.** $-\frac{2}{3}$ **7. a.** 10 **b.** $(-1, -4)$
c. $\frac{3}{4}$ **8., 9., 11–13.**
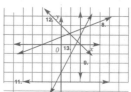

10. a. $\frac{2}{3}$ **b.** $-\frac{3}{2}$

Lesson 17.6
Solutions are given for Exercises 1, 3, 5, 7, and 9.
Your graph should show the solution.
1. $(3, -2)$ **3.** $(-1, -7)$ **5.** $(2, -2)$ **7.** all ordered pairs
with coordinates $(x, 2 - x)$ **9.** $(-3, 2)$ **11.** $(10, -5)$
13. a.

b. $(0, 1)$, $(3, -1)$, $(3, 4)$
c. $7\frac{1}{2}$ square units

15. a. **b.** $(-1, -2)$, $(1, 4)$, $(3, 4)$, $(1, -2)$
c. 2 sides have slope 0; 2 have slope 3.
d. 12 square units

15.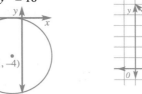

16.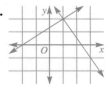

Lesson 17.7 **1.** $(5, 8)$; 2 **3.** $(-5, 4)$; 3 **5.** $(-4, -7)$; 8
7. $(0, -6)$; 5 **9.** $(x - 2)^2 + (y - 5)^2 = 9$
11. $(x + 6)^2 + (y - 4)^2 = 49$ **13.** $(x + 1)^2 + (y + 6)^2 = 25$
15. $(x - 5)^2 + (y - 3)^2 = 36$ **17.** $(x + 4)^2 + y^2 = 16$
19. **21.**

17–19. Answers may vary. Examples are given.
17. $(4, 2)$, $(3, 3)$, $(6, 0)$ **18.** $(1, 1)$, $(2, 4)$, $(0, -2)$

23. a. $(4, 2)$ **b.** $\sqrt{10}$ **c.** $(x - 4)^2 + (y - 2)^2 = 10$
25. All points in the interior of a circle with center $(3, -7)$ and radius 4. **27.** The circle with center $(3, -7)$ and radius 4 and its interior. **29. a.** $(4, 3)$ **b.** $(2, 1)$
c. $(0, 3)$ **d.** $(2, 7)$ **e.** $(6, 3)$ **f.** $(-2, 3)$

19. $(0, 5)$, $(1, 3)$, $(2, 1)$ **20.** $(2, 3)$

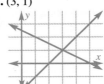 

Lesson 17.8

1.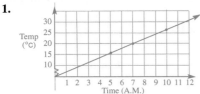

3. a. 6 A.M. **b.** 11:30 A.M. **5.** $10
7. First, prices are affected by many factors and so, fluctuate considerably. Second, that assumption leads to the conclusion that eventually the cost will hit 0.

21. $(3, 1)$

22. $(3, -4)$; 10
23. $(7, 2)$; 6
24. $(x - 1)^2 + (y - 2)^2 = 25$
25. $x^2 + (y + 2)^2 = 36$

Chapter 17 Review

1–4.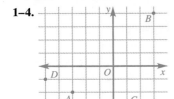

5. 10 **6.** $\sqrt{41}$
7. 15 **8.** $(8, 5)$
9. $(7, 1)$
10. $(0.5, -5)$
11. $\frac{1}{3}$ **12.** -3
13. $\frac{3}{5}$
14. a. $\frac{3}{4}$ **b.** $-\frac{4}{3}$

26. a. 50¢ **b.** 5 min

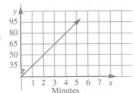

15 –16. Answers may vary. Examples are given.

Chapter 18 Transformations and Coordinate Geometry

Arithmetic **1.** –2 **3.** –2 **5.** –2
7. 2 **9.** –8 **11.** –4 **13.** –4 **15.** –6 **Algebra 17.** 0
19. 6 **Geometry 21.**

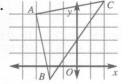

23. left 6, down 1

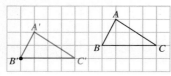

25.

27.

29. The graph is a line parallel to the *x*-axis through the point (0, –4). **31.** The graph is a line through the origin and with slope 1. Some coordinates are: (1, 1), (4, 4), (–2, –2)

Lesson 18.1 For Ex. 1 and 3 vertices of the image are given. **1.** *A′*(–3, –6) *B′*(1, –2) *C′*(2, –7)
3. *G′*(–2, –4) *H′*(–1, 0) *J′*(1, –1)
5.

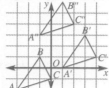

7. *T*: (*x*, *y*) → (*x* + 2, *y* + 5) **9.** The image of (*x*, *y*) is (*x* + a_1 + a_2, *y* + b_1 + b_2); yes **11.** Answers may vary. For example: (step 1) To triangle/ (step 2) pu setpos [20 10] pd/ (step 3) setpos [60 10] setpos [30 30] setpos [20 10]/ (step 4) end; (step 1) To image/ (step 2) pu setpos [–20 –20] pd/ (step 3) setpos [20 -20] setpos [–10 0] setpos [–20 –20]/ (step 4) end

Lesson 18.2 **1. a.** (–8, 5) **b.** (8, –5) **c.** (5, 8)
3. a. (6, 2) **b.** (–6, –2) **c.** (2, –6) **5. a.** (0, –6)

b. (0, 6) **c.** (–6, 0) For Exercises 7, 9, and 11, the coordinates of the vertices of reflected images are given. **7.** *A′*(–1, –2) *B′*(2, –3) *C′*(–2, –6) **9.** *D′*(3, 1) *E′*(–1, 1) *F′*(3, –3) **11.** *A′*(1, 2) *B′*(–2, 3) *C′*(2, 6)

13.

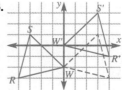

15.

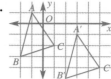

17.

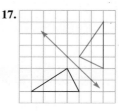

19. c **21.** Answers may vary. For example: (step 1) To isos.tri/ (step 2) rt 90 fd 30 lt 90 fd 30/ (step 3) lt 135 fd 30 ∗ sqrt 2 rt 135/ (step 4) end; (step 1) To refl.tri/ (step 2) pu setpos [10 20]/ (step 3) pd isos.tri/ (step 4) pu setpos [40 –50]/ (step 5) pd lt 90 isos.tri/ (step 6) pu setpos [–40 50]/ (step 7) pd lt 180 isos.tri/ (step 8) end

Test Yourself 1.

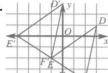

2.

3.

4.

5.

Lesson 18.3 **1. a.** (–3, 6) **b.** (3, –6) **c.** (–6, –3)
3. a. (0, –4) **b.** (0, 4) **c.** (4, 0) **5. a.** (6, 2)
b. (–6, –2) **c.** (–2, 6)

7.

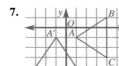

9.

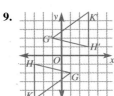

11.

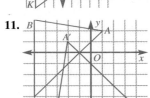

13. They are parallel.

15.

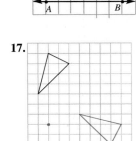

17.

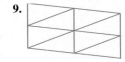

19. 5 units to the right of the bottom right vertex of the red triangle.

Lesson 18.4 1. (15, 9) **3.** (−2, −1) **5.** (−6, −2)

7.

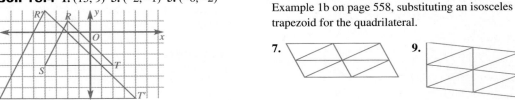

9. 12 square units; 48 square units; 1:4

11.

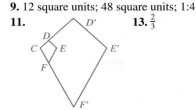

13. $\frac{2}{3}$

15. Each side of $D'E'F'G'$ is twice as long as the corresponding side of $DEFG$.

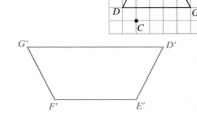

17. Answers may vary. For example: (step 1) To three.tri/ (step 2) pu setpos [−60 0]/ (step 3) pd setpos [30 60] setpos [120 −90] setpos [−60 0]/ (step 4) pu setpos [−30 0]/ (step 5) pd setpos [15 30] setpos [60 −45] setpos [−30 0]/ (step 6) pu setpos [15 −30]/ (step 7) pd setpos [60 0] setpos [105 −75] setpos [15 −30]/ (step 8) end

Lesson 18.5 1–5. Answers may vary. For example: **1.** Draw a figure in which congruent rectangles are placed next to/above each other without gaps or overlap. **3.** Draw a figure similar to that in the solution to Example 1a on page 558, substituting isosceles triangles for the scalene triangles.
5. Draw a figure similar to that in the solution to Example 1b on page 558, substituting an isosceles trapezoid for the quadrilateral.

7.   **9.**

11.

13. Because the triangles are all congruent. $\angle ADE \cong \angle DBF$, so $\overline{DE} \parallel \overline{BC}$. Similarly, $\overline{FE} \parallel \overline{AB}$ and $\overline{DF} \parallel \overline{AC}$. $DE = BF = FC$, so $DE = \frac{1}{2}BC$. Similarly, $FE = \frac{1}{2}AB$ and $DF = \frac{1}{2}AC$.

Chapter 18 Review **1.** (1, –2) **2.** (2, –1) **3.** (4, 4) **4.** (5, 0) **5.** (1, –5) **6.** (–3, 2) **7.** (10, 6) **8.** (–9, 3) **9.** (9, –6) For 10–19, coordinates of vertices of the image are given. **10.** $B'(0, -2)$ $N'(-2, 4)$ $K'(2, 3)$ **11.** $J'(-5, 1)$ $G'(0, 6)$ $L'(-2, 1)$ **12.** $H'(5, -1)$ $N'(3, -3)$ $P'(0, -1)$ **13.** $K'(1, -3)$ $L'(0, 3)$ $C'(-2, 2)$ **14.** $E'(2, -2)$ $A'(-2, -2)$ $R'(2, 0)$ **15.** $B'(2, 2)$ $C'(2, -1)$ $D'(4, -1)$ **16.** $P'(0, -2)$ $Q'(-2, -4)$ $R'(2, -3)$ **17.** $J'(-1, 2)$ $V'(0, -3)$ $S'(3, -1)$ **18.** $T'(2, 0)$ $R'(0, -1)$ $N'(-1, 2)$ **19.** $O'(0, 0)$ $N'(6, 2)$ $B'(-2, 4)$ **20.** The coordinates of the vertices of the image are: $A'(0, 3)$ $B'(3, 0)$ $C'(5, 2)$

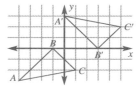

21. translation **22.** reflection **23.** rotation **24.** glide reflection

25.  **26.**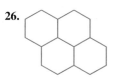

27.

Acknowledgments

Photo Credits

Front Cover: Thomas Leighton; Back Cover Left: Eugen Gebhardt/FPG; Back Cover Middle Left: David Young-Wolff/PhotoEdit; Back Cover Middle Right: David Sutherland/TSW; Back Cover Right: Fred Kong/PhotoEdit; 1, David Young-Wolff/PhotoEdit; 7 (both), Art Resource; 26, file photo; 32–33, Bob Daemmrich/The Image Works; 37, Jan Halaska/The Image Works; 53, Janice Sheldon/Photo 20–20; 63, Tony Freeman/PhotoEdit; 67, Fred Kong/PhotoEdit; 72, Nuridsany and Perennou/Photo Researchers, Inc.; 91, Superstock; 97, E.R. Degginger/H. Armstrong Roberts; 119, Tony Freeman/PhotoEdit; 127, Photo Researchers, Inc.; 143, Janice Sheldon/Photo 20–20; 153, Tony Freeman/PhotoEdit; 158, Joe Sohm/The Stock Market; 176–177, R. Thomas/FPG; 183, Superstock; 213, Otto Rogge/The Stock Market; 214, David Stoecklein/The Stock Market; 219, Peter Cole/Bruce Coleman, Inc.; 248, Eugene Mopsik/H. Armstrong Roberts; 253, Roy Morsch/The Stock Market; 255, Tony Freeman/PhotoEdit; 259, The Image Works; 280–281, Russ Lappa; 293, Felicia Martinez/PhotoEdit; 297, Myrleen Ferguson/PhotoEdit; 309, Robert Brenner/PhotoEdit; 314, NASA; 315, David Young-Wolff/PhotoEdit; 319, David Sutherland/TSW; 335, Robert Semeniuk/The Stock Market; 341, Superstock; 345, Shostal Associates/Superstock; 365, Superstock; 379, Eugen Gebhardt/FPG; 414, Superstock; 419, David Sutherland/TSW; 425 L, Charles M. Falco/Photo Researchers, Inc.; 425 TR, Janice M. Sheldon; 425 MR, M. Claye Jacana/Photo Researchers, Inc.; 425 BR, Mario Fantin/ Photo Researchers, Inc.; 441, Robert Huntzinger/ The Stock Market; 449, Jeff Gnass/The Stock Market; 454, Robert Neuman/The Stock Market; 473, George Holton/Photo Researchers, Inc.; 477, NASA; 478–479, Russ Lappa; 483, The Stock Shop; 500, NASA; 505, Henley and Savage/TSW; 536, Larry Kolvoord/The Image Works; 541, Coco McCoy/Rainbow; 558–559, Rochelle Newman.

Art Credit

215, Paul Foti/Boston Graphics, Inc.